Current Advanced Technologies in Catalysts/Catalyzed Reactions

Current Advanced Technologies in Catalysts/Catalyzed Reactions

Editors

Sagadevan Suresh
Is Fatimah

Basel • Beijing • Wuhan • Barcelona • Belgrade • Novi Sad • Cluj • Manchester

Editors
Sagadevan Suresh
Nanotechnology & Catalysis
Research Centre (NANOCAT)
University of Malaya
Kuala Lumpur
Malaysia

Is Fatimah
Chemistry Deparment
Universitas Islam Indonesia
Yogyakarta
Indonesia

Editorial Office
MDPI
St. Alban-Anlage 66
4052 Basel, Switzerland

This is a reprint of articles from the Special Issue published online in the open access journal *Catalysts* (ISSN 2073-4344) (available at: www.mdpi.com/journal/catalysts/special_issues/catalysts_catalyzed_reactions).

For citation purposes, cite each article independently as indicated on the article page online and as indicated below:

Lastname, A.A.; Lastname, B.B. Article Title. *Journal Name* **Year**, *Volume Number*, Page Range.

ISBN 978-3-7258-0870-0 (Hbk)
ISBN 978-3-7258-0869-4 (PDF)
doi.org/10.3390/books978-3-7258-0869-4

© 2024 by the authors. Articles in this book are Open Access and distributed under the Creative Commons Attribution (CC BY) license. The book as a whole is distributed by MDPI under the terms and conditions of the Creative Commons Attribution-NonCommercial-NoDerivs (CC BY-NC-ND) license.

Contents

About the Editors vii

Preface ix

Suresh Sagadevan and Is Fatimah
Current Advanced Technologies in Catalysts/Catalyzed Reactions
Reprinted from: *Catalysts* **2024**, *14*, 177, doi:10.3390/catal14030177 1

Mohd Nor Latif, Wan Nor Roslam Wan Isahak, Alinda Samsuri, Siti Zubaidah Hasan, Wan Nabilah Manan and Zahira Yaakob
Recent Advances in the Technologies and Catalytic Processes of Ethanol Production
Reprinted from: *Catalysts* **2023**, *13*, 1093, doi:10.3390/catal13071093 4

Wan Nor Roslam Wan Isahak, Lina Mohammed Shaker and Ahmed Al-Amiery
Oxygenated Hydrocarbons from Catalytic Hydrogenation of Carbon Dioxide
Reprinted from: *Catalysts* **2023**, *13*, 115, doi:10.3390/catal13010115 45

Asma Jabeen, Urooj Kamran, Saima Noreen, Soo-Jin Park and Haq Nawaz Bhatti
Mango Seed-Derived Hybrid Composites and Sodium Alginate Beads for the Efficient Uptake of 2,4,6-Trichlorophenol from Simulated Wastewater
Reprinted from: *Catalysts* **2022**, *12*, 972, doi:10.3390/catal12090972 85

Adama A. Bojang and Ho-Shing Wu
Production of 1,3-Butadiene from Ethanol Using Treated Zr-Based Catalyst
Reprinted from: *Catalysts* **2022**, *12*, 766, doi:10.3390/catal12070766 104

Chandhinipriya Sivaraman, Shankar Vijayalakshmi, Estelle Leonard, Suresh Sagadevan and Ranjitha Jambulingam
Current Developments in the Effective Removal of Environmental Pollutants through Photocatalytic Degradation Using Nanomaterials
Reprinted from: *Catalysts* **2022**, *12*, 544, doi:10.3390/catal12050544 123

Mehtab Parveen, Mohammad Azeem, Azmat Ali Khan, Afroz Aslam, Saba Fatima and Mansoor A. Siddiqui et al.
One-Pot Synthesis of Benzopyrano-Pyrimidine Derivatives Catalyzed by P-Toluene Sulphonic Acid and Their Nematicidal and Molecular Docking Study
Reprinted from: *Catalysts* **2022**, *12*, 531, doi:10.3390/catal12050531 151

Natthanan Rattanachueskul, Oraya Dokkathin, Decha Dechtrirat, Joongjai Panpranot, Waralee Watcharin and Sulawan Kaowphong et al.
Sugarcane Bagasse Ash as a Catalyst Support for Facile and Highly Scalable Preparation of Magnetic Fenton Catalysts for Ultra-Highly Efficient Removal of Tetracycline
Reprinted from: *Catalysts* **2022**, *12*, 446, doi:10.3390/catal12040446 168

Muhammad Nihal Naseer, Asad A. Zaidi, Hamdullah Khan, Sagar Kumar, Muhammad Taha bin Owais and Yasmin Abdul Wahab et al.
Statistical Modeling and Performance Optimization of a Two-Chamber Microbial Fuel Cell by Response Surface Methodology
Reprinted from: *Catalysts* **2021**, *11*, 1202, doi:10.3390/catal11101202 188

Omid Akbarzadeh, Solhe F. Alshahateet, Noor Asmawati Mohd Zabidi, Seyedehmaryam Moosavi, Amir Kordijazi and Arman Amani Babadi et al.
Effect of Temperature, Syngas Space Velocity and Catalyst Stability of Co-Mn/CNT Bimetallic Catalyst on Fischer Tropsch Synthesis Performance
Reprinted from: *Catalysts* **2021**, *11*, 846, doi:10.3390/catal11070846 . . . . **200**

Ganjar Fadillah, Is Fatimah, Imam Sahroni, Muhammad Miqdam Musawwa, Teuku Meurah Indra Mahlia and Oki Muraza
Recent Progress in Low-Cost Catalysts for Pyrolysis of Plastic Waste to Fuels
Reprinted from: *Catalysts* **2021**, *11*, 837, doi:10.3390/catal11070837 . . . . **215**

About the Editors

Sagadevan Suresh

Dr. Suresh Sagadevan is an Associate Professor at the Nanotechnology and Catalysis Research Centre, University of Malaya. He has published more than 450 research papers in the ISI's top-tier journals and Scopus. He has authored 15 international book series and 50 book chapters. He is an Editor, Guest Editor, and Editorial Board Member of many reputed ISI journals. He is a member of many professional bodies at the national/international level. He has been a recognized reviewer for many reputed journals. He is also working in various fields such as nanofabrication, functional materials, graphene, polymeric nanocomposite, glass materials, thin films, bio-inspired materials, drug delivery, tissue engineering, cell culture, supercapacitor, optoelectronics, photocatalytic, green chemistry, and biosensor applications.

Is Fatimah

Is Fatimah is a Professor and Head of the Chemistry Department at Universitas Islam, Indonesia. She leads the Laboratory of Materials for Energy and Environment Research Group. She has been actively involved in various research activities, particularly in the fields of nanomaterials and catalysis. She has published numerous papers in highly reputable international journals. She is a member of many professional bodies at national and international levels. She has presented many scientific papers as keynotes and invited speakers at various conferences/workshops, both nationally and internationally. She is an Editorial Board Member and reviewer for various high-impact-factor journals. Her main fields of research include the synthesis and characterization of nanocomposites for catalysis applications.

Preface

This Special Issue of "Current Advanced Technologies in Catalysts/Catalyzed Reactions" has shown remarkable progress and diversity within the field of catalysis. The present research highlights the crucial role that catalysts play in addressing some of the most pressing challenges of our time, from sustainable development and resource utilization to environmental remediation. The collection of articles within this Special Issue explores a wide range of advancements in catalyst design, reaction optimization, and material science. Several articles have addressed the development of efficient catalysts for clean energy production, biomass conversion, and CO_2 utilization. This reflects the growing recognition that catalysis is a key driver of the transition toward a more sustainable future. Articles in this Special Issue represent significant contributions to the field of catalysis. We would like to express our sincere gratitude to all authors for their valuable contributions to this Special Issue. We also thank the reviewers for their dedication and expertise in ensuring high quality in the published work.

Sagadevan Suresh and Is Fatimah
Editors

Editorial

Current Advanced Technologies in Catalysts/Catalyzed Reactions

Suresh Sagadevan [1,*] and Is Fatimah [2]

1. Nanotechnology & Catalysis Research Centre, University of Malaya, Kuala Lumpur 50603, Malaysia
2. Department of Chemistry, Faculty of Mathematics and Natural Sciences, Kampus Terpadu UII, Universitas Islam Indonesia, Jl. Kaliurang Km 14, Sleman, Yogyakarta 55784, Indonesia; isfatimah@uii.ac.id
* Correspondence: drsureshsagadevan@um.edu.my

Currently, catalysis represents an exciting research area. Catalysis underpins a diverse array of technologies that define our modern world from fuels that power our vehicles to medicines that improve our lives. This Special Issue delves into the cutting edge of this vibrant field, showcasing current advanced technologies that are revolutionizing the way we design, prepare, and utilize catalysts. The articles present a range of advancements in catalysis across diverse areas, including environmental remediation, energy production, and organic synthesis. This Special Issue provides an important snapshot of the current advancements in various areas of environmental and energy science.

Contribution 1. Wastewater Treatment:

The first contribution details the fabrication of $FeCl_3$-$NaBH_4$-modified mango seed shell (MS)-based hybrid composite ($FeCl_3$-$NaBH_4$/MS) and sodium alginate-modified mango seed shell (MS)-based composite (Na-Alginate/MS) beads for the adsorptive removal of 2,4,6-trichlorophenol from aqueous media.

Contribution 2. Biofuel Production:

The second contribution analyzes the catalytic activity of a commercial catalyst by treating it with oxalic acid, NaOH, and other essential acidic compounds. As the results show, the treated commercial catalyst yielded a better 1,3-butadiene production than the untreated catalyst. In addition, the presence of oxalic acid in combination with NaOH provides a good desilication process and increases the catalyst's acidic properties. This opens the door for new biofuel production routes.

Contribution 3. Pest Control and Drug Discovery:

The third contribution describes the novel one-pot synthesis of benzopyrano-pyrimidine derivatives with nematocidal properties, which was developed using P-toluene sulfonic acid as a catalyst. Molecular docking studies provided insights into their mode of action, potentially paving the way for new pest control strategies.

Contribution 4. Water Purification:

The fourth contribution discusses the preparation of magnetic sugarcane bagasse ash (MBGA) via a simple co-precipitation route. The results indicated that nearly 100% of the tetracycline (TC) concentration (or degradation of 40 mg of TC) could be achieved. This value is far higher than those reported in many studies. Furthermore, the catalyst exhibited a high degradation of TC, even after four cycles, with excellent magnetic properties being retained.

Contribution 5. Bioenergy:

The fifth contribution presents the RSM that was utilized to study the correlation between the power density output and the flow rate of the fuel feed to the anodic chamber,

Citation: Sagadevan, S.; Fatimah, I. Current Advanced Technologies in Catalysts/Catalyzed Reactions. *Catalysts* **2024**, *14*, 177. https://doi.org/10.3390/catal14030177

Received: 26 January 2024
Accepted: 17 February 2024
Published: 2 March 2024

Copyright: © 2024 by the authors. Licensee MDPI, Basel, Switzerland. This article is an open access article distributed under the terms and conditions of the Creative Commons Attribution (CC BY) license (https://creativecommons.org/licenses/by/4.0/).

the initial concentration of acetate in the anodic chamber, and the initial concentration of O_2 in the cathodic chamber. This study provides valuable insights into maximizing bioelectricity generation from organic waste.

Contribution 6. Fuel Synthesis:

The sixth contribution investigates cobalt–manganese bimetallic catalysts synthesized using acid and thermally treated CNT substrates through the SEA process. The efficiency of various percentage formulations of the Co-Mn catalyst supported on CNT was verified by the FTS reaction. The 95Co5Mn/CNT catalyst exhibited high stability for more than 45 h. It was concluded that the reaction variables had a high impact on the catalytic activities and product selectivities during the FTS process.

Contribution 7. Ethanol Production:

The seventh contribution explores recent advancements in ethanol production technologies and catalytic processes, highlighting promising avenues for sustainable and efficient biofuel production. The catalytic hydrogenation of CO_2 is a development direction for the production of ethanol that reduces environmental pollution problems. The limitations of CO_2, a fully oxidized, chemically inert, and thermodynamically stable molecule, should be considered when designing research because its conversion into chemicals requires large amounts of energy and H_2.

Contribution 8. CO_2 Utilization:

The eighth contribution highlights the progress made in the use of three-dimensional (3D) nanomaterials and their compounds and methods for their synthesis in the hydrogenation of CO_2. The development of 3D nanomaterials and metal catalysts supported on 3D nanomaterials is important for CO_2 conversion because of their stability and ability to continuously support catalytic processes, in addition to their ability to reduce CO_2 directly and hydrogenate it into oxygenated hydrocarbons.

Contribution 9. Environmental Remediation:

The ninth contribution focuses on metal oxides used as photocatalysts for the degradation of various types of pollutants. The progress of research on metal oxide nanoparticles and their application as photocatalysts in organic pollutant degradation is highlighted. The application of nano-based materials can be a new horizon for the use of photocatalysts for organic pollutant degradation.

Contribution 10. Plastic Waste Recycling:

The tenth contribution examines recent progress in developing low-cost catalysts for the pyrolysis of plastic waste into fuels. This research contributes to closing the loop on plastic waste management and promoting circular economy approaches. The development of low-cost catalysts is revisited to design better and more effective materials for plastic solid waste (PSW) conversion to oil/bio-oil products.

Overall, this Special Issue reflects a vibrant and diverse field of research that addresses crucial challenges in sustainable development, environmental remediation, and resource utilization. The development of new catalysts, reaction optimization, and material design shows strong potential for future advancements in various sectors. This highlights the potential of novel catalysts and catalytic processes to address pressing environmental and energy challenges while also paving the way for new applications in organic synthesis and other fields. Thus, we embark on this important exploration of current advanced technologies for catalysts/catalyzed reactions. Prepare to be inspired, informed, and empowered by the knowledge contained in these articles. The future of chemistry is catalyzed, and this Special Issue is a guide to its possibilities.

Conflicts of Interest: The authors declare no conflict of interest.

List of Contributions:

1. Jabeen, A.; Kamran, U.; Noreen, S.; Park, S.-J.; Bhatti, H.N. Mango Seed-Derived Hybrid Composites and Sodium Alginate Beads for the Efficient Uptake of 2,4,6-Trichlorophenol from Simulated Wastewater. *Catalysts* **2022**, *12*, 972. https://doi.org/10.3390/catal12090972.
2. Bojang, A.A.; Wu, H.-S. Production of 1,3-Butadiene from Ethanol Using Treated Zr-Based Catalyst. *Catalysts* **2022**, *12*, 766. https://doi.org/10.3390/catal12070766.
3. Parveen, M.; Azeem, M.; Khan, A.A.; Aslam, A.; Fatima, S.; Siddiqui, M.A.; Azim, Y.; Min, K.; Alam, M. One-Pot Synthesis of Benzopyrano-Pyrimidine Derivatives Catalyzed by P-Toluene Sulphonic Acid and Their Nematicidal and Molecular Docking Study. *Catalysts* **2022**, *12*, 531. https://doi.org/10.3390/catal12050531.
4. Rattanachueskul, N.; Dokkathin, O.; Dechtrirat, D.; Panpranot, J.; Watcharin, W.; Kaowphong, S.; Chuenchom, L. Sugarcane Bagasse Ash as a Catalyst Support for Facile and Highly Scalable Preparation of Magnetic Fenton Catalysts for Ultra-Highly Efficient Removal of Tetracycline. *Catalysts* **2022**, *12*, 446. https://doi.org/10.3390/catal12040446.
5. Naseer, M.N.; Zaidi, A.A.; Khan, H.; Kumar, S.; Owais, M.T.b.; Abdul Wahab, Y.; Dutta, K.; Jaafar, J.; Hamizi, N.A.; Islam, M.A.; et al. Statistical Modeling and Performance Optimization of a Two-Chamber Microbial Fuel Cell by Response Surface Methodology. *Catalysts* **2021**, *11*, 1202. https://doi.org/10.3390/catal11101202.
6. Akbarzadeh, O.; Alshahateet, S.F.; Mohd Zabidi, N.A.; Moosavi, S.; Kordijazi, A.; Babadi, A.A.; Hamizi, N.A.; Wahab, Y.A.; Chowdhury, Z.Z.; Sagadevan, S. Effect of Temperature, Syngas Space Velocity and Catalyst Stability of Co-Mn/CNT Bimetallic Catalyst on Fischer Tropsch Synthesis Performance. *Catalysts* **2021**, *11*, 846. https://doi.org/10.3390/catal11070846.
7. Latif, M.N.; Wan Isahak, W.N.R.; Samsuri, A.; Hasan, S.Z.; Manan, W.N.; Yaakob, Z. Recent Advances in the Technologies and Catalytic Processes of Ethanol Production. *Catalysts* **2023**, *13*, 1093. https://doi.org/10.3390/catal13071093.
8. Isahak, W.N.R.W.; Shaker, L.M.; Al-Amiery, A. Oxygenated Hydrocarbons from Catalytic Hydrogenation of Carbon Dioxide. *Catalysts* **2023**, *13*, 115. https://doi.org/10.3390/catal13010115.
9. Sivaraman, C.; Vijayalakshmi, S.; Leonard, E.; Sagadevan, S.; Jambulingam, R. Current Developments in the Effective Removal of Environmental Pollutants through Photocatalytic Degradation Using Nanomaterials. *Catalysts* **2022**, *12*, 544. https://doi.org/10.3390/catal12050544.
10. Fadillah, G.; Fatimah, I.; Sahroni, I.; Musawwa, M.M.; Mahlia, T.M.I.; Muraza, O. Recent Progress in Low-Cost Catalysts for Pyrolysis of Plastic Waste to Fuels. *Catalysts* **2021**, *11*, 837. https://doi.org/10.3390/catal11070837.

Disclaimer/Publisher's Note: The statements, opinions and data contained in all publications are solely those of the individual author(s) and contributor(s) and not of MDPI and/or the editor(s). MDPI and/or the editor(s) disclaim responsibility for any injury to people or property resulting from any ideas, methods, instructions or products referred to in the content.

Review

Recent Advances in the Technologies and Catalytic Processes of Ethanol Production

Mohd Nor Latif [1,2], Wan Nor Roslam Wan Isahak [1,3,*], Alinda Samsuri [4,5], Siti Zubaidah Hasan [1], Wan Nabilah Manan [1] and Zahira Yaakob [1]

1. Department of Chemical and Process Engineering, Faculty of Engineering & Built Environment, Universiti Kebangsaan Malaysia, Bangi 43600, Selangor, Malaysia
2. GENIUS@Pintar National Gifted Center, Universiti Kebangsaan Malaysia, Bangi 43600, Selangor, Malaysia
3. Research Centre for Sustainable Process Technology (CESPRO), Faculty of Engineering and Built Environment, Universiti Kebangsaan Malaysia, Bangi 43600, Selangor, Malaysia
4. Centre for Tropicalization, National Defence University of Malaysia, Kem Sungai Besi, Kuala Lumpur 57000, Malaysia
5. Department of Chemistry and Biology, Center for Defence Foundation Studies, National Defence University of Malaysia, Kem Sungai Besi, Kuala Lumpur 57000, Malaysia
* Correspondence: wannorroslam@ukm.edu.my; Tel.: +60-3-8911-8339

Citation: Latif, M.N.; Wan Isahak, W.N.R.; Samsuri, A.; Hasan, S.Z.; Manan, W.N.; Yaakob, Z. Recent Advances in the Technologies and Catalytic Processes of Ethanol Production. *Catalysts* 2023, 13, 1093. https://doi.org/10.3390/catal13071093

Academic Editors: Sagadevan Suresh and Is Fatimah

Received: 3 April 2023
Revised: 8 June 2023
Accepted: 10 June 2023
Published: 12 July 2023

Copyright: © 2023 by the authors. Licensee MDPI, Basel, Switzerland. This article is an open access article distributed under the terms and conditions of the Creative Commons Attribution (CC BY) license (https://creativecommons.org/licenses/by/4.0/).

Abstract: On the basis of its properties, ethanol has been identified as the most used biofuel because of its remarkable contribution in reducing emissions of carbon dioxide which are the source of greenhouse gas and prompt climate change or global warming worldwide. The use of ethanol as a new source of biofuel reduces the dependence on conventional gasoline, thus showing a decreasing pattern of production every year. This article contains an updated overview of recent developments in the new technologies and operations in ethanol production, such as the hydration of ethylene, biomass residue, lignocellulosic materials, fermentation, electrochemical reduction, dimethyl ether, reverse water gas shift, and catalytic hydrogenation reaction. An improvement in the catalytic hydrogenation of CO_2 into ethanol needs extensive research to address the properties that need modification, such as physical, catalytic, and chemical upgrading. Overall, this assessment provides basic suggestions for improving ethanol synthesis as a source of renewable energy in the future.

Keywords: carbon dioxide; catalytic hydrogenation; cascade reaction; ethanol production

1. Introduction

Nowadays, scientists, academics, policymakers, and environmental non-governmental organizations focused on global warming or climate change effect due to the impact of greenhouse gas emissions (GHG). Data show increment concentration patterns of methane (CH_4), carbon dioxide (CO_2), chlorofluorocarbons, and nitrous oxide in the environment every year, thus prohibiting the reradiation of solar heat and increasing the temperature of the surface of the earth. Global GHG emissions are from human activities, such as burning fuels for electricity generation systems [1–4], transportation [5,6], industry [7,8], and agriculture [9–11]. CO_2 is also known as the most important anthropogenic GHG for global warming or climate change that is associated with human activities [12–14]. The Global Carbon Budget 2021 reported that the global atmospheric CO_2 emissions growth is 5%. The global average amount of atmospheric CO_2 emissions in 2021 is 36 billion metric tons or equivalent to 415 ppm [15]. The increasing patterns of CO_2 concentration globally in the atmosphere are predicted to increase due to excessive industrialization, which leads to the development of heat retention since the introduction of the Fourth Industrial Revolution.

CO_2 is proven to be a recyclable, nonpoisonous, and inexpensive C_1 building block in the synthesis of high-value chemicals and fuels [16–19]. However, CO_2 is fully oxidized, chemically inert, and thermodynamically stable ($\Delta_f G_{298K} = -396$ kJ·mol^{-1}). Thus, its conversion into chemicals requires large energy and enormous H_2 resources [20–22]. Many CO_2 transformation approaches, e.g., hydration of ethylene, biomass residue, lignocellulosic materials, fermentation, electrochemical reduction, dimethyl ether (DME), reverse water gas shift (RWGS), catalytic hydrogenation, and other related processes, have been rapidly studied. Catalytic hydrogenation has been deemed a promising technology for generating a variety of products, such as hydrocarbons [23,24], alcohols [25,26], carboxylic acids [27,28], and aldehydes [29,30].

Ethanol is an ecological fuel that has an important advantage compared with conventional gasoline as a transportation fuel due to its properties of nontoxicity, accumulation of high oxygen content to promote improved combustion with reducing exhaust emissions, and high octane rating to give a high resistance to engine knock [31–34]. Hence, the establishment technology of fuel ethanol is needed in reducing environmental pollution problems [35,36]. Ethanol is produced from agricultural feedstock, such as corn (United States) and sugarcane (Brazil), and the European Union produces ethanol from wheat and sugar beet. The Renewable Fuels Association in 2021 reported that the main ethanol producer in the world is the United States, estimating ethanol production of more than 13,000 million gallons per year, which is more than half of the global ethanol production. Approximately 8000 million gallons per year of ethanol is produced in Brazil [37]. The use of agricultural feedstock especially corn in ethanol production is being criticized because of their importance as food. Food shortages and rising food prices will occur if agricultural feedstock is used as the raw material in ethanol production. Furthermore, the grain from corn will create environmental pollution problems, such as soil erosion, biodiversity loss, nitrogen oxide pollution, and emission of volatile organic compounds. The major constraint in commercial ethanol production is the disadvantage in energy balance and the area required for plantations [38,39]. Given these considerations, nations continue to look for new technologies and processes in reducing the cost of ethanol production without causing adverse effects on the environment.

2. Conventional Processes to Produce Ethanol

2.1. Ethanol Synthesis Based on the Hydration of Ethylene

In the petrochemical industry, the catalytic hydration of ethylene for ethanol production is a reversible exothermic reaction and is used commercially by Shell Oil Company in 1947 [40]. The process of reaction can be expressed by the following equation:

$$C_2H_4 + H_2O \rightleftharpoons C_2H_5OH. \tag{1}$$

The hydration of ethylene comprises three stages, i.e., reaction, recycling, and purification. Mohsenzadeh et al. [41] suggested that this process occurs in a fixed-bed catalytic reactor when ethylene is mixed with steam at a molar ratio of 0.6 at 250–300 °C, 70–80 bar, and the presence of a phosphoric acid catalyst (H_3PO_4/SiO_2) based on silica gel. The ethylene conversion is 4–25% with ethanol selectivity of 98.5 mol.%. A diagram of the hydration of ethylene is shown in Figure 1.

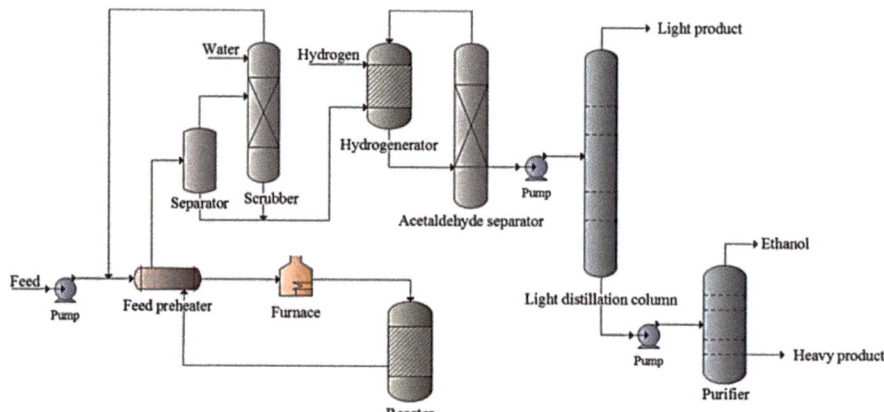

Figure 1. Hydration of ethylene [41].

A byproduct of this process, acetaldehyde, can be used directly in a cosmetic product or promoted to form ethanol through the hydrogenation process. The high-pressure separator is used to separate the unreacted reactants from the outlet stream mixture. Unreacted reactants are scrubbed with water to remove impurities before dissolution with ethanol. The molar ratio of ethylene to water is maintained at 1:0.6, and the mixture is combined with a recycle stream. When the bottom streams of the scrubber and the separator are fed to the hydrogenator, the nickel-packed catalyst promotes the formation of an ethanol mixture from acetaldehyde. The unreacted acetaldehyde in the separator column is removed and recycled in the hydrogenator. The ethanol concentration increases when the bottom stream is fed to light and heavy (purifier) columns [42].

The cost of a plant for the hydration of ethylene increases due to the formation of the ethanol-water mixture that will form an azeotrope mixture, which requires a special distillation process. The production of ethylene based on hydrocracking petroleum feedstocks, which are nonrenewable, is not economically feasible because of the market price of ethanol. The phosphoric acid catalyst that is used in this process is prone to leaching via vaporization, which causes the deactivation of catalysts and corrosion of the equipment. A solid acid catalyst, such as WO_3/ZrO_2 [43] and WO_3/TiO_2 [44], is introduced to overcome leaching issues; this route is still not favorable for the large-scale production of ethanol due to the price of ethylene and the rapid development of sugar fermentation during the hydration of ethylene.

The research on the gas-phase hydration of ethylene by using impregnated metal phosphates for catalytic activity at tin(IV) hydrogen phosphate ($Sn[HPO_4] \cdot 2H_2O$) showed that the weight-based rate is 0.94 $\mu mol \cdot min^{-1} \cdot gcat^{-1}$, which is approximately seven times higher than that at commercial H_3PO_4/SiO_2 catalyst [45]. The vapor-phase hydration of ethene has recorded a 93% selectivity for the tungsten trioxide (WO_3) monolayer loaded with titania. The co-presence of Brønsted and Lewis acid sites on the monolayer of WO_3 generated high selectivity for ethanol but the reaction process has issues from the environmental protection perspective [46]. The combination of vapor–liquid equilibria and the chemical stage equilibrium for the ethylene–water–ethanol ternary arrangement demonstrates that the crucial point of an azeotrope at 200 °C and 155 atm shows the active catalysis of H-pentasil zeolite for the maximum hydration of ethylene [47]. The ethanol production plant's simulated modeling was developed using the HYSYS software and the results of optimization over the catalytic hydration using zirconium-tungsten catalysts operating at 299 °C with column configuration for the extractive distillation which produced ethanol concentration of 99.7% and then linked to the Aspen Plus software [48].

The hydration process corresponds to petroleum-derived alkene over solid acid catalysts, which are limited by the low single-pass conversion (<5%), poor long-term stability, and strong dependence on crude oil. The chemical equilibrium conversion of the hydration of ethylene decreases when using high temperatures to increase the rate of reaction. As a result, the temperature setting of the reactor is needed to accommodate between thermodynamics and kinetics [46]. A high amount of energy to heat gases generates high pressure and uses crude oil, a nonrenewable resource. The purification method uses benzene to separate the azeotropes of ethanol and water that produce hazardous ethanol.

2.2. Ethanol Synthesis Based on Biomass Residues

Biofuel is produced from biomass residues and wastes for energy purposes, such as transportation fuels, renewable electricity, and thermal energy [49–51]. The three types of biomass residual resources are primary, secondary, and tertiary (Figure 2). Corn stalks, husks, stems, roots, leaves, cob, bagasse, and straw make up the main residue, which is described as outcomes of the cultivation of certain food crops and agroforestry in the agriculture sector. Then, the secondary residue is obtained by the processing of crops into the final form of a product or the production of other biomass-based materials. Examples of agricultural and food processing wastes are sawdust, wood chips, nutshells, palm kernel cake, fruit bunches, coffee husks, rice hulls, bark, and scrap wood. The tertiary residue consists of sewage sludge or wastewater derived after the consumption of biomass-based products, such as municipal solid waste. Human, animal, and industrial wastes have been identified as the main source of municipal solid waste [52–54].

Figure 2. Categories of biomass residues.

A study showed that wood processing residues, such as discarded logs, sawdust, and wood chips produced from sawmill and lumber processing, can be utilized as steamer fuels and feedstock for ethanol synthesis [53,55]. Tropical countries have been shown to apply the concept of economic utilization by using sugarcane residues, such as sugarcane bagasse and leaf residue, for ethanol production and value-added commercial product [56–58]. According to the research on delignified coconuts waste and cacti, an ethanol yield of 89.15% is recorded by utilizing the semisimultaneous saccharification and fermentation configuration, and this yield is higher compared with that obtained by simultaneous saccharification and fermentation (SSF) configuration [59]. The pretreatment mixture of 0.06 g·g^{-1} hydrogen peroxide to green liquor and furfural residues pretreated to cassava residue saccharification liquid with a ratio of 1:1 recorded an advantageous pretreatment method by producing a 93.6% yield of ethanol. This study shows that the increment of a high ethanol yield and lower byproduct concentrations occur when the proportion of lignocellulosic substrates was enhanced in the SSF of the substrate mixture of cassava residue and furfural residues [60]. The investigation of the carnauba straw residue by the SSF configuration process in a single reactor in the presence of *Kluyveromyces marxianus* ATCC-36907 and observed that cultivation at 45 °C results in the maximum ethanol concentration of 7.53 g·L^{-1} [61]. The mangosteen pericarp waste that has undergone popping pretreatment and enzymatic hydrolysis in the separate hydrolysis and fermentation (SHF) configuration method can achieve 75% ethanol [62]. The ethanol produced from the combination of *Salacca zalacca* and coconut sewage shows that the energy required for coordination obtained at 85 °C is 346.32 W, and the resulting ethanol is obtained at 40% and mass flow rate of 0.0655 kg·s^{-1} [63]. Rahman et al. [64] developed a green biorefinery concept to produce ethanol by integrating the pretreatment of fermentation and ethanol-assisted liquefaction in the presence of *Nannochloropsis* sp. The process increases the lipid content of fermented microalgae by 40%, whereas 10% of the required ethanol is produced through liquefaction. The utilization of wet algae increases the crude biodiesel yield threefold compared with the liquefaction of microalgae. The fermentation of over-ripened Indian blueberry at 33 °C, pH 5.2, and specific gravity of 0.875 obtain 6.5% ethanol [65].

However, biomass residue requires an energy-intensive process from large and specific machinery due to different types of biomass, thus increasing the cost of operation because of expensive machinery and the fuel needed for operation. The low density of biomass also influences the cost of operation by occupying increased volume and needing increased transportation for space. The other challenge is developing effective pretreatment technologies that cover physical, chemical, and biological pretreatments. The ideal pretreatment increases the rate of enzyme hydrolysis and decreases the amount of enzyme needed to convert the biomass into sugars in the presence of the microorganism. Issues regarding environmental pollution also need to be addressed because the conversion of biomass residue into ethanol produces a huge amount of CO_2.

2.3. Ethanol Synthesis Based on Lignocellulosic Materials

Lignocellulosic materials, which comprise nonedible feedstock from various agricultural and forestry residues, are abundantly available without geographical limitation and have a low cost. Extensive research showed the production of ethanol and value-added chemical by using different types of lignocellulosic sources, including waste paper [66,67], orange peel [68,69], sugarcane straw [70,71], corn stover [72,73], sugarcane bagasse [74,75], rice straw [76,77], wheat straw [78,79], sweet sorghum [80,81], oil palm empty fruit [82,83], and banana waste [84,85]. Figure 3 shows the composition of lignocellulosic materials that consist of three major fragments of cellulose (40–50%), hemicellulose (25–35%), and lignin (15–20%), which always exist beside other extracts and mineral traces [86–89]. However, the composition differs on the basis of the type of biomass, cultivation, and atmospheric conditions. The complex and rigid structure of lignocellulosic materials is made by noncovalent interactions with covalent cross-linkages [90].

Figure 3. Composition of lignocellulosic materials (an adaptation from [91]).

The conversion of lignocellulosic materials into ethanol involves pretreatment, enzymatic hydrolysis, fermentation, and distillation [92–95]. During the pretreatment process, various cutting-edge technologies are utilized to open the structure of lignocellulosic materials by physical, chemical, physicochemical, and biological techniques and separate complex interlinked structures among hemicellulose and lignin from the matrix [87,96–98]. Chemical pretreatment uses a variety of chemicals such as acid and alkaline chemicals to break down the structures present in the lignocellulosic biomass at a constant ambient temperature which subsequently enhances the biomass surface availability to enzymatic hydrolysis, permitting the cellulose and hemicellulose for further conversion of fermentable sugars into biofuels [99–101]. The research on the bioconversion of lignocellulosic byproduct corn stover into the value-added fermentative product L-lactic acid using the furfural tolerant *Enterococcus mundtii* WX1 and *Lactobacillus rhamnosus* SCJ9 showed that corn stover pretreated with 1% (v/v) sulfuric acid was selected for L-LA fermentation and shows the highest efficacy of fermentable sugar with the optimal conditions achieved for the release of glucose and xylose at 24.5 g/L and 11.2 g/L, respectively, from 100 g/L pretreated corn stover at 121 °C for 30 min [102]. A similar result was presented by other researchers reported in the study of tobacco stem waste [103], palm kernel shell [104], sugarcane bagasse [105], and oil palm frond bagasse [106] that the dilute acid for chemical pretreatment is effective to attain high reactivity and generates protons that have a quick diffusion which substantially enhances the hydrolysis of amorphous cellulose chains and the solubilization of hemicellulose.

Dilute acid pretreatment has received wide attention due to its cost-effective, non-toxicity, lower degradation products, corrosive, and hazardous processes that do not require as much corrosion-resistant equipment, making it easier to scale-up the operation process [107–109]. Alkaline pretreatment leads to the delignification of agricultural biomass by cleaving the intermolecular ester linkages between hemicelluloses and lignin fragments, increases the amorphous surface area of the cellulose as well as the porosity of the biomass, reduces the degree and crystallinity of the polymerization rate at low temperature and pressure, resulting in an enhanced hydrolysis and fermentation yield and a high amount of sugars [110–112]. In a comparison of various alkaline pretreatment techniques, alkaline hydrogen peroxide (AHP) pretreatment is the most effective as it increases the fermentation yield at mild conditions effectively by solubilizing lignin from the complex recalcitrant structure of the macromolecules because H_2O_2 could degrade to oxygen and H_2O without any residues left and increases the enzyme digestibility and fermentation efficiency of the feedstock required for subsequent processing [99,113–115]. The primary advantages of AHP pretreatment are environmentally friendly chemicals and reagent reusability, high effectiveness for various biomass concentrations providing high efficiency of enzymatic hydrolysis, high lignin, and hemicellulose solubilization values for the liquid fraction without a loss of carbohydrates retention, low energy consumption, less formation of toxic byproducts, no need for special reactors, compatibility with high solid loadings, and sterility conditions provided by alkaline H_2O_2 without a need to use antibiotics [116–118].

Organosolv pretreatment with aliphatic organic solvents is among the most promising pretreatments compared to acidic or alkaline pretreatment by producing a very distinctive separation of high-purity cellulose content from the remaining lignocellulosic constituents, such as lignin and hemicellulose at relatively low temperatures (below 180 °C), while preserving the integrity of the hemicellulose structure from thermal degradation kinetics [119–121]. Organosolv pretreatment using ethanol has also some advantages over other methods such as low toxicity and environmentally friendly nature, high delignification rates, high reaction stability, good solubility of lignin, miscibility with water, complete restoration of ethanol solvent due to its low boiling point and potentially provides substantial economic benefits [122–125].

The hydrolysis process breaks down the hemicellulose and cellulose components in the presence of cellulolytic enzymes or acids to form monosaccharides [126–128]. The conversion of sugars into ethanol by using a variety of potential microorganisms occurs during the fermentation process. *Saccharomyces cerevisiae*, a microorganism that is commonly used as baker's yeast on a large scale at the industrial level, has been identified to have a tolerance for ethanol production, robust ethanol dehydrogenase, and potential good resistance against inhibitors generated during the process. The use of this strain results in high ethanol productivity and efficient conversion of most of the sugars into ethanol rather than other byproducts [129,130]. The last step is distillation where the purification of the fermentation broth occurs. Distillation is an effective and favorable separation technique as the preferred choice for industrial application due to high alcohol recovery of 99.5% v/v purity, sufficient energy efficiency at moderate feed concentrations, and the ability to simulate the process using process simulation software which makes the integration of mass and energy in other processes easier to accomplish [131–133]. All formed byproducts and other impurities are removed during this process, and only imprints remain. Most energy-intensive units have remarkable effects on the gross energy demand that takes place during the distillation process. The cost of plant operation also increases due to the formation of an ethanol–water mixture that produces an azeotrope mixture where the simple distillation method cannot be used to change its composition [134,135].

The SHF process has been studied for ethanol production from waste paper. The pretreatment process of waste paper is applied using 0.5% (v/v) hydrogen peroxide at 121 °C for 30 min. The office paper that has been pretreated by hydrolytic enzymes produces 24.5 g·L^{-1} sugar equivalent to 91.8% hydrolysis efficiency. Then, the fermentation process that uses *S. cerevisiae* through hydrolysate obtains 11.15 g·L^{-1} ethanol with ethanol productivity of 0.32 g·L^{-1}·h^{-1} [67]. Oil palm trunk chips are introduced into a two-stage pretreatment, i.e., steam explosion and alkaline extraction. The steam explosion pretreatment shows that the reduction and isolation of hemicellulose occur in biomass recalcitrance. The alpha-cellulose content has been improved from 40.83% to 87.14% with alkaline extraction pretreatment at the conditions of 15% (w/v) NaOH at 90 °C for 60 min. By using *S. cerevisiae*, the ethanol concentration at SSF (44.25 g·L^{-1}) is prominent compared with that at prehydrolysis SSF (31.22 g·L^{-1}) [136].

The SHF process of the pretreatment of rice straw using *Saccharomyces tanninophilus* produces 9.45 g·L^{-1} ethanol with 83.5% yield. The saccharification of pretreated rice straw with *A. fumigatus* by using 1.0% NaOH in 200 FPU·mL^{-1} crude enzyme for 20 h of reaction obtains 22.15 g·L^{-1} limiting sugars, demonstrating high lignin-degrading manganese peroxidase activity and the activity of laccase enzymes [76]. The effectiveness of SHF and SSF techniques for the synthesis of ethanol originating from oil palm empty fruit bunch (OPEFB) with the conditions of 10% (w/v) loading of the substrate, pH 5, 1% (v/v) *K. marxianus* at 37.50 °C for 48 h of reaction is compared. SHF and SSF obtained 25.80 and 28.10 g·L^{-1} ethanol, respectively. The acid–alkali pretreatment of OPEFB is conducted by the loading of the substrate at 12.50% w/v with 0.2 M concentration of H_2SO_4 at 121 °C for 53 min followed by 5% (w/v) NaOH at 121 °C for 20 min. This result demonstrates that the acid–alkali pretreatment increases the cellulose yield to 72.10 wt.% and this process is a feasible method for eliminating hemicellulose and lignin from lignocellulosic biomass [82].

The cellulosic ethanol production by using *Issatchenkia orientalis* KJ27-7 in 90% wheat straw hydrolysate media for 24 h has obtained 10.3 g·L^{-1} ethanol corresponding to 0.50 g·g^{-1} glucose (97% of efficiency relative to the theoretical yield). The correlation of ethanol production with wheat straw hydrolysate concentrations is observed [78]. Studies on the effect of varying lignocellulosic feedstocks on technical performance for ethanol production that use the dilute acid pretreatment show that the switchgrass produces 46.2% energy efficiency of feedstock LHV, which is the highest carbohydrate content with the lowest forest residues compared to the *Eucalyptus globulus.*, *Birch* sp. residues, *Spruce* sp. residues, *Miscanthus*, corn stover, and wheat straw [137]. Cunha et al. [138] reported that the direct production of ethanol by using non-detoxified hemicellulose liquor by *S. cerevisiae* using hydrothermally pretreated corn cob without external hydrolytic catalysts results in 11.1 g·L^{-1} ethanol titer correlated with the ethanol yield of 0.328 g·g^{-1} potential sugar. The consolidated bioprocessing (CBP) of pretreated corn cob with the addition of commercial hemicellulases is more efficient than SSF in hemicellulosic ethanol production.

Diverse sources and seasonal lignocellulosic biomass affect the chemical characteristics of ethanol produced due to different harvesting times, resulting in the inconsistent composition of lignocellulosic biomass components. Pentose sugars are not fermented by the brewer's yeast, i.e., *S. cerevisiae*, during the hydrolysis of hemicellulose, thus compromising the ethanol production from total sugars in lignocellulosic materials. Hence, the energy consumption during the distillation process for ethanol recovery and treatment of a large amount of stillage increases due to the subsequent reduction of discharge. The pretreatment process produces lignin from inhibitors that act as limiting agents for high biomass loading and do not react productively with lignocellulosic materials. As a result, the ethanol production from cellulose is lower compared with that from grains. Other limitations of ethanol production from lignocellulosic materials, such as high capital, operational expenditure, dwindling price of gasoline, process uncertainty, low growth, and product yield, have also been identified.

2.4. Ethanol Synthesis Based on Fermentation

The major steps in ethanol production via the fermentation process are the treatment of a solution containing fermentable sugars, the formation of ethanol from sugars through fermentation, and distillation for the separation and purification of ethanol [134,139,140]. The main metabolic route involved in ethanol production by fermentation is glycolysis, which converts glucose into pyruvate that is further reduced to produce ethanol and CO_2 under anaerobic conditions [65,141]. Based on the stoichiometric equivalence, 1 mol glucose creates 2 mol CO_2, which is then expelled from the reactor as a weight loss and is proportional to ethanol yield.

$$C_6H_{12}O_6 \rightarrow 2C_2H_5OH + 2CO_2. \qquad (2)$$

The fermentation process can be produced in different systems, such as a batch, fed-batch, or continuous bioreactor. The batch bioreactor fermentation is a simple method with a closed culture system where both biomass and substrates are added to the fermenter in a single step of the procedure in which nothing is added or removed during the process and the products are only removed at the end of the process [142,143]. The system operation produces high cell densities, of which almost 99.5% is recycled in subsequent fermentation. The closed-loop design system that uses a high concentration of sugars generates a high concentration of ethanol [143]. In the conventional batch fermentation process at an ideal temperature and under anaerobic conditions, *S. cerevisiae* is used to convert glucose into ethanol. However, this process only occurs in hexose sugars but not in pentose sugars [97,129]. Although equipped with multiple vessels, the batch fermentation system is considered the simplest operation system due to its ability to complete the sterilization process, resulting in a low risk of contamination, low operation costs due to no labor required, easy control of feedstock processes, and flexibility for various product specifications. The disadvantage of this process is solvent inhibition, time consumption, difficulty in maintaining the sterilization of bioreactors, major downtime, long lag phase, and low productivity.

The fed-batch bioreactor process is a semicontinuous or partly open system that allows the addition of fermentation medium gradually or consistently during the process after the initial substrate has been used, overcoming the difficulties of substrate constraint in the batch bioreactor process. This process enables the overall proportion of substrate uptake to increase and sustain a low concentration of substrate within the fermentation vessel, thus decreasing the negative influence of osmotic pressure or rheology-related limitation linked with highly viscous substrates [143,144]. Knudsen and Rønnow [145] reported the highest ethanol production from wheat straw by using *S. cerevisiae* in the co-fermentation stage at the C5/C6 fermenting yeast, where glucose and xylose are fermented simultaneously. With the addition of urea and a primary yeast pitch of $0.2 \text{ g} \cdot \text{L}^{-1}$ completed broth in at least five fermenter volumes, the fed-batch fermentation process is stable, yielding an ethanol yield >90% during the experiment. The fed-batch fermentation process by using the mixture of sugarcane and molasses based on the Central Composite Design evaluates the effect of temperature at 27 °C, the concentration of sugar at 300 g·L^{-1}, and the concentration of cells at 15% (v/v) in the presence of *S. cerevisiae* for 30 h of reaction. This process has obtained ethanol concentration, productivity, and yield of 135 g·L^{-1}, 4.42 g·L^{-1}·h^{-1}, and 90%, respectively [146]. The fed-batch process is a cost-effective operation with an efficient cultivation strategy, short fermentation time, high dissolved oxygen concentration in the medium, and low toxic efficacy of medium constituents. However, the ethanol production in fed-batch fermentation is minimal; the concentration of cell mass and feed rate of the reaction thus provided a point of ingress for contamination and allowed the buildup of inhibitory agents and toxins. The outcome of high cell density numbers and product yields are difficult to deal with downstream, creating bottlenecks in the whole process.

The continuous fermentation process is carried out by continuously feeding substrates, new media, and nutrients into a bioreactor containing active microbes. This process concurrently harvests the used medium and cells, removes toxic metabolites, and replaces the consumed nutrients from the culture. As a result of the equivalency process of addition and removal, the culture volume in this process remains constant. Then, the maximum working volume of the vessel does not limit the amount of fresh medium or feed solution which can be added to the culture in the course of the process. However, the long cultivation period increases the risk of contamination and genetic changes in the cultures. This process also difficult to keep a constant population density over prolonged periods and the products of a continuous process cannot be neatly separated into batches for traceability. The production of high residual sugar and ethanol in this process is caused by the continuous exposure of yeast cells that may affect cell growth until biomass washout [147]. Margono et al. [148] developed the uncontrolled continuous fermentation process equipped with an integrated aerobic–anaerobic baffled reactor (IAABR) to study molasses in the presence of *S. cerevisiae* and generated 92.55 g·L^{-1} ethanol with a productivity of 4.63 g·L^{-1}·h^{-1} for a residence time of 19.2 h. The ethanol productivity with IAABR is 3.4% higher compared with that through the industrial batch process, and the maximum operation reaches 14 days of fermentation without contamination.

The cassava supernatant subjected to continuous ethanol production with a high cell density strategy at the dilution rate of 0.092 h^{-1} generated 104.65 g·L^{-1} ethanol and ethanol productivity of 9.57 g·L^{-1}·h^{-1}. The ethanol yield of this system is 96.96%, which is approximately 4.2% higher compared with that obtained by traditional fermentation with free cells. This research shows that cells sustain optimum condition activity by switching the flow direction in the in-series bioreactors and extend the long-term stability of continuous fermentation without any possibility of a contamination effect [149]. The continuous fermentation with a high cell density recycle operation demonstrates a better result compared with typical molasses-based batch fermentation by obtaining 0.44 g·g^{-1} ethanol from xylose and glucose and ethanol productivity of 3.4 g·L^{-1}·h^{-1} [150]. The continuous fermentation method yields an improved output in minor bioreactor volumes, has low operational costs by lowering production times, is cost-effective, allows for growth control via nutrient supply management, and is scale-up friendly. Some limitations of the continuous fermentation process, such as low product concentration, complicated downstream processes, difficulty in maintaining sterilization conditions, high risk of contamination with the extended culture time, limited yeasts' capacity to create ethanol and periodic handling, which may also increase the costs of operation, are observed.

Syngas fermentation is a biological carbon fixation process that uses a gaseous feedstock, primarily composed of a mixture of CO, CO_2, and H_2 which is obtained from biomass, coal, animal or municipal solid waste, and industrial CO-rich waste gases, that is a promising approach converted into valuable chemicals and fuels by microorganisms through a hybrid thermo/biochemical process [151,152]. Several Clostridium species are known to produce different bioproducts, but only a few of them use syngas as the sole carbon and energy source [153,154]. *Clostridium carboxidivorans* are acetogenic bacteria that are known to grow autotrophically with syngas and chemoorganotrophically with a wide range variety of sugars [155–157]. It is able to ferment these carbon sources to produce volatile fatty acids and alcohols that can be employed as platform chemicals or as feedstock for liquid fuel production qualifying it as an interesting microorganism for industrial production [152,158,159]. However, the issues that must be addressed in order to incorporate the syngas fermentation into an industrial-scale process include the gas-liquid mass transfer limitation brought to the low aqueous solubilities of the gaseous substrates that occur when cells have the capacity to process more gas than the bioreactor can supply. The resistance of gaseous substrate diffusion at the gas-liquid interface has been identified as the limiting step in syngas fermentation [160–163]. The other challenge identified as low carbon fixation yield, high production cost, and the effects of gaseous impurities such as NH_3, H_2S, and NO_x even at low concentrations by limiting microbial growth, enzyme activities or by changing physiochemical conditions led to the unintended accumulation of organic acids and decreased alcohol formation [152–154,159,164,165].

The SSF method combines enzymatic hydrolysis and fermentation in a single phase to produce value-added products. This method involves hydrolyzing cellulose and extracting sugars by using an enzymatic complex. These sugars are then used by microbes and transformed into value-added compounds [166]. The combination of the semicontinuous fermentation of sugarcane bagasse and SSF system produces 9.07% (v/v) ethanol with <1% residual glucose at the optimum conditions of 1% (w/v) NaOH, 160 °C, and 20 min of reaction. This study shows no remarkable variation throughout the whole process and that the system achieves a constant state [167]. Compared with SHF, SSF has several advantages. These advantages include the use of an individual vessel for fermentation and saccharification, which reduces the residence period and capital expenditure, and the reduction of the inhibitory composite from enzymatic hydrolysis, which enhances inclusive operational achievement. SSF has been intensively studied for the manufacture of ethanol from lignocellulosic and starchy raw materials because of these benefits. The optimal temperature for enzymatic hydrolysis is often higher than the fermentation temperature, and the SSF reaction is limited by the pH and temperature of the operation.

Meanwhile, in the pre-hydrolysis simultaneous saccharification and fermentation (PSSF) configuration, the pretreated material is pre-hydrolyzed at the optimum temperature of the cellulolytic enzyme, and the temperature is then lowered for further inoculation with no other additional step [168,169]. The main advantage of PSSF is significantly reduced overall fermentation time, environmentally friendly, increased initial velocity (V_0) of enzymes, and provided the optimum conditions for both the enzyme and yeast to utilize the substrate sufficiently that also reduces the production cost and favors the distillation process for high ethanol yields [170,171]. The banana peel with 25% (w/v) of high solid loading using commercial S. cerevisiae at 64 h of fermentation has been demonstrated as the promising feedstock for ethanol production by PSSF by achieving a maximum ethanol concentration of 32.6 g/L [172]. The bioconversion of barley straw to bioethanol was carried out by PSSF where the kinetic model was used as guidance in the choice of pre-hydrolysis time step. The highest ethanol concentration reached in the present study was 46.62 g/L at a high solid loading of 20% (w/v) of barley straw by applying 16 h of pre-hydrolysis. The mass balance of PSSF showed that the reduction in ethanol yield when solid loading increases could be attributed to the decrease in cellulose enzymatic conversion [173]. Under the PSSF strategy in the development of a process using *Sargassum* biomass at high pretreated solid loading 13% (w/v) was subjected to high-pressure technology for biomass fractionation recorded the maximum ethanol concentration of 18.14 g/L after 12 h of fermentation [174]. The ethanol production from potato peel waste subjected to the PSSF process allowed for reaching a maximum ethanol concentration of 104.1 g/L at high productivity with 54 h of fermentation [175].

The simultaneous saccharification and co-fermentation (SSCF) method breaks down cellulose into sugars called hexoses by using an enzyme complex. Specialized microorganisms with the capacity to ingest substrates consume these sweeteners generated in situ along with pentoses following a pretreatment to acquire a product of significance. The research on the corn stover subjected to temperature-profiled SSCF at 12% of glucan loading eliminated sugar accumulation and alleviated ethanol repression by process optimization, 59.8 g·L^{-1} ethanol. It suggested that the high-temperature resistant strain helped the xylose-utilizing strain maintain cell viability in SSCF at high temperatures (42 °C) which are higher compared with the threshold concentration for the economic distillation process [176]. Xylose utilization in the study of sugarcane bagasse by using SSCF with a thermotolerant *S. cerevisiae* at 40 °C demonstrated 99% xylose in the hydrolysate during the co-fermentation process, generating 36.0 g·L^{-1} ethanol [177]. The differences in pH, temperature, and other parameters required for the enzymatic hydrolysis and co-fermentation process have been recognized as limiting factors in the SSCF.

The enzymes that hydrolyze cellulose in the production of ethanol require high-temperature conditions, i.e., thermotolerant yeast and bacteria, to produce high enzyme contents. The thermotolerant microorganisms are beneficial in terms of efficiency improvement of processes by obtaining higher yields in saccharification, reducing costs associated with a cooling system while reducing the risk of bacterial contamination [178–181]. Many studies have examined various thermotolerant yeasts with their optimized temperature for ethanol production. The isolate *Pichia kudriavzevii* at 28 °C achieved the optimal ethanol concentration of 10.10 g/L and a productivity of 0.21 g/L/h in monoculture fermentation [182]. The optimal conditions of *Meyerozyma guilliermondii* at 45 °C using sugarcane bagasse as a substrate achieved the maximum ethanol concentration of 11.12 g/L and a productivity of 0.23 g/L/h [183]. The optimum fermentation conditions for ethanol production from sweet sorghum juice with the thermotolerant yeast *S. cerevisiae* at 37 °C revealed a maximum ethanol concentration of 99.75 g/L and a productivity of 2.77 g/L/h was achieved [184]. *K. marxianus* at 42 °C also effectively utilized biopretreated elephant grass hydrolysate and produced the maximum ethanol concentration of 14.65 g/L and a productivity of 0.62 g/L/h [185]. Fermentation or hydrolysis can be achieved under ideal specifications, and the microorganism must be specialized for both substrates and only applicable at high-temperature conditions. The process requirements also point to the necessity for the creation of specialized microorganisms. With the use of genetic engineering, the SSCF process offers several advantages, including the utilization of minimal equipment, short processing times, reduced contamination risk, and high ethanol production efficiency.

The development of consolidated bioprocess (CBP) of lignocellulosic biomass is the most integrated process for the bioconversion approach, where the process of hydrolysis, fermentation, and enzyme production occurs in a single reactor. The conversion of pretreated lignocellulose employs genetically modified single microorganisms or a microbial consortium capable of hydrolyzing biomass with enzymes produced on its own and fermenting monosugars into value-added products could provide the environmentally friendly, economically competitive by reducing costs for infrastructure, raw materials, and enzyme production [117,186,187]. The effective fermentation of monosugars obtained from lignocellulosic biomass is the next bottleneck in bioethanol production reaching an industrial scale. Several factors might affect its low conversion efficiency, including low enzyme concentrations at the start of the fermentation, temperature, time, pH, inoculum size, solid-to-liquid ratio, agitation rate, oxygen content, and rotation speed [188–190]. The isolated bacterium of *Hangateiclostridium thermocellum* in the study of pre-treated *Nannochloropsis gaditana* biomass converted into ethanol through CBP was investigated. In this study, the hemicellulose removal of dilute H_2SO_4 treatment was found to be best for the pretreatment of biomass at the concentration of 2.5% under 100 °C for 60 min, effectively disrupting a complex matrix of the holocellulose sample and removing the hemicellulose. The optimized conditions of the medium components and process parameters yielded a maximum ethanol concentration of 12.90 g/L [191]. The investigation on the potential of the fungus *Trichoderma asperellum* to produce ethanol and the physicochemical parameters required for paddy straw waste conversion via CBP using the numerical optimization was statistically validated by comparing the volume of ethanol produced, to the volume analyzed via Response Surface Methodology (RSM). The investigation proved that the fungus is a potential organism for on-site enzyme production with a maximum ethanol concentration of 0.94 g/L [192]. The maximal ethanol production capability of *Fusarium moniliforme* integrated biodelignification and CBP of Napier grass at solid-state conditions is 10.5 g/L by feeding the fungus with a surplus of glucose for fermentation. These results demonstrated the characteristics of a fungus for potential ethanol production from cellulose, mixed sugars, and lignocellulosic materials [193]. The study of the bioconversion of *Sargassum wightii* via CBP using the bacterial isolate, *Lachnoclostridium phytofermentans* shows excellent growing ability in the optimized production medium conditions with the maximum ethanol concentration of 13.75 g/L [194].

The pretreatment of feedstocks minimizes their size and makes following the procedures easy. Cellulose and hemicellulose are hydrolyzed into sugars that may be fermented. These carbohydrates are fermented into ethanol by using yeasts. *S. cerevisiae* is widely utilized as a yeast strain for the ethanol fermentation of lignocellulosic hydrolysates due to fast growth, high tolerance, efficient glucose anaerobic metabolism, high selectivity, cost-effective process for high ethanol yield, high rate of fermentation, low accumulation of byproducts, and use of a broad scale of disaccharides (e.g., sucrose and maltose) and hexoses (e.g., glucose, mannose, and galactose) [195,196]. Table 1 shows ethanol production by *S. cerevisiae* from a different type of feedstock at varying treatments.

Table 1. Ethanol production by *S. cerevisiae* from different types of feedstock at varying treatments.

Feedstock	Parameters			Ethanol Concentration (g/L)	Ref.
	Temperature (°C)	Agitation Speed (rpm)	Incubation Time (h)		
Galactose	30	200	28	96.90	[197]
Rice husk	43	150	96	15.63	[198]
Oil palm frond	30	152	15	4.79	[199]
Cellulose and sucrose/xylose	30	200	96	4.30	[200]
Papaya peels	30	200	48	0.51	[201]
Pineapple leaf	30	150	72	9.75	[202]
Pomegranate peel	30	100	24	5.58	[203]
Sweet sorghum	30	150	18	97.54	[204]
Sugarcane distillery waste	30	150	48	49.77	[205]
Corn starch	30	300	192	98.13	[206]
Rice straw	30	150	72	18.07	[207]
Sugarcane molasses	30	200	56	114.71	[208]
Corn stover	34	150	48	21.47	[209]
Oil palm trunk	30	150	18	44.25	[171]
Microalgae biomass	30	150	48	52.10	[210]
Sugar beet molasses	30	140	112	79.60	[211]
Cassava starch	30	200	72	81.86	[212]
Suweg starch	37	80	78	99.52	[213]
Frond Waste	50	150	96	33.15	[214]

Industrial ethanol production is efficiently produced from lignocellulosic hydrolysates by yeast strains with high hexose and pentose fermentation. This result is due to the high xylose and glucose contents in lignocellulosic biomass [215,216]. The modest acid stress caused by lignocellulosic materials also inhibits yeast fermentation. The presence of weak acids in deficient concentrations can boost ethanol synthesis through cellular division. In *S. cerevisiae*, weak acids are shown to increase glucose consumption, ethanol synthesis, and tolerance to 5-hydroxymethylfurfural and furfural [217,218]. Despite the fermentation route having been commercially realized, the cost is expensive due to the energy-intensive distillation steps and low yield to meet the market demand. Other remarkable obstacles to ethanol generation, such as excessive temperatures, prominent ethanol concentrations, and capacity to ferment pentose sugars remain in yeast fermentation. The main disadvantage is that yeasts grown in anaerobic states for an extended period lose the capacity to manufacture ethanol. Furthermore, at high dilution rates, which allow for high productivity, the substrate is not entirely utilized, resulting in low yield. The rate of development and metabolism of yeasts increases as the temperature rises until the optimal level is reached. The inhibition of microorganism expansion and viability can occur when ethanol concentrations rise during fermentation. The difficulty of S. cerevisiae growing on a medium with a high concentration of alcohol causes ethanol production to be inhibited. The limitation of S. cerevisiae is the inefficient fermentation of glucose and xylose. As a result, yeast strains that can ferment glucose and xylose or utilizing two separate yeast strains that can utilize these sugars individually should be found.

3. Future Directions of CO_2 Conversion into Ethanol

3.1. Ethanol Synthesis Based on Electrochemical Reduction

A carbon-neutral energy cycle is through the transformation of sunlight towards energy-dense fuels via electrolytic CO_2 reduction to fuels [17,219,220]. Based on the Nernst equation, electrochemical reduction potentials are translated into the reversible hydrogen electrode (RHE) range:

$$E_{RHE} = E_{Ag/AgCl} + 0.059 \times pH + E^0_{Ag/AgCl}, \tag{3}$$

where the potential, E_{RHE} vs. RHE, $E^0_{Ag/AgCl}$ = 0.198 V at 25 °C, and the potential measured, $E_{Ag/AgCl}$ vs. the reference electrode, Ag/AgCl.

The conversion of electrochemical CO_2 into ethanol involves a set voltage flow with steady or unnoticeable current conditions. This process determines the best voltage to convert CO_2 into ethanol via electrochemical synthesis easily. The conversion of CO_2 into ethanol is a nonspontaneous reaction (E^0 = negative) that requires an external voltage source from the power supply. A predetermined voltage flow with stable or imperceptible current states is used to convert electrochemical CO_2 into ethanol. This process makes identifying the appropriate voltage for the conversion of electrochemical CO_2 into ethanol easy. The conversion of CO_2 into ethanol is a nonspontaneous reaction (E^0 = negative) that needs a power supply voltage source. Splitting the process into two separate electrochemical stages is used to pursue ethanol synthesis. The intermediate product in the assembly cascade technique should be a stable species that can be easily isolated from the initial electrolyte. CO is chosen as the stable intermediate product from the start, as evidenced by CO_2 electroreduction at excessive faradaic efficiencies:

$$CO_2 + H_2O + 2e^- \rightarrow CO + 2OH^- \qquad E^0 = -0.10 \text{ V vs. RHE} \tag{4}$$

The poor solubility of CO causes the easy separation of intermediate products for transfer to the second-stage electrolyzer, leading to excessive current density due to the difficulty of the CO reduction process. Han et al. [221] concluded a feasible approach to solve the restricted CO coverage and deficiency in CO solubility in the catalytic position is to develop a cocatalyst for the formation of CO and reduction of CO_2 in the electrocatalytic reaction. Thus, the catalyst system can be prepared by coupling two sites, where one site efficiently reduces CO_2 to CO, which further distributes to the construction of C–C coupling in the formation of long carbon chain species that occur on the other sites of coupling. Yuan et al. [222], Kou et al. [223], and Ramírez-Valencia [224] reported that the CO-producing site's pyridinic N-doped carbon species components show excellent performance selectivity and high catalytic accomplishment for the reduction of CO_2 to CO. The development of N-doped porous carbon components influences the electronic order and size of the Cu catalyst and further improves the gas transport for enhanced availability to pyridinic N during the process of adsorption of CO_2 and reduction of CO. The electrocatalyst reaction for the direct transformation of CO_2 into ethanol demonstrates competitive faradaic efficiencies but prefers high current densities, low overpotentials, and poor selectivity with the long-term stability of the operation [225–227].

The overall cathode half-reaction for ethanol formation is as follows:

$$2CO_2 + 9H_2O + 12e^- \rightarrow C_2H_5OH + 12OH^- \qquad E^0 = 0.09 \text{ V vs. RHE} \qquad (5)$$

The advantage of the electrochemical synthesis of ethanol from CO_2 is the product selectivity generated on each electrode terminal. Then, the equipment and substance used are basic and have a low cost. The process is controllable and flexible with a safe and mild operating background and empowers the nonfossil energy from renewable energy sources with environmentally friendly coupling [224]. The kind of metal utilized on the electrode affects the electrochemical synthesis efficacy of converting CO_2 into ethanol. The electrocatalytic characteristics of metals employed as electrodes affect the transformation percentage of CO_2 and the distribution in overall compounds. The category of catalyst, reaction potential, properties of electrolyte solution, cell design, pH value, and reaction circumstances, such as temperature and pressure, influence the outcomes of electrochemical synthesis [228–230]. The variety of alkaline electrolytes ranging in pH from neutral to alkaline has shown the potential to improve C_2 products [231–233]. Some studies in the electrochemical synthesis operation for converting CO_2 into other chemicals revealed that the category of electrode used and the sensor preparation have a remarkable effect on the results produced [234,235].

In the electrochemical synthesis process, where the water oxidation reaction occurs, carbon is an inert electrode attribute that does not react when utilized as an anode. Carbon is not affected during electrochemical production because of its inert characteristics. Water is oxidized and becomes a source of protons and electrons because carbon is an inert compound, and the bicarbonate anion (HCO_3^-) does not oxidize in water. The mechanism of the reaction of ethanol synthesis at the cathode involves 12 protons and electron transfer, which aid the process of ethanol formation at the cathode [17,232,236–238]. In the electrolytic CO_2 reduction, normal metallic electrocatalysts only generate the C_1 building block but copper (Cu) elements have been identified to catalyze the manufacturing of low hydrocarbons at reasonable excessive faradaic efficiency (FE). The transformation of CO_2 into multicarbon alcohols via multiple electron transfer reactions facilitates C–C coupling reactions to produce C_2 products, resulting in decreased system's energetic competency and poor selectivity [221,239,240]. Various factors of the selectivity and activity, such as catalyst size, catalyst surface structure, catalyst oxidation state, structural morphology, crystallographic orientation, composition, type of electrolyte ions, pH, pressure, temperature, design of the electrochemical cell, and the existence and number of deficiencies (i.e., point fault, contamination, unorganized location, grain limits), enhance the catalytic performance of CO_2 electroreduction towards multicarbon products [230,239,241,242]. Nanocatalyst morphologies, supporting materials, nanograin boundaries, and catalyst surface changes can all have an impact on contrary reaction pathways.

Zhu et al. [243], Zhou and Yeo [244], and Chen et al. [245] believed that Cu-based catalyst arrays' structure elevates local pH, which favors CO generation and C–C coupling to generate C_2 products. Zhang et al. [246] agreed that on the Cu surface, the conversion of CO_2 to form C–H is difficult because it requires several electron reductions, protonation, and C–C coupling reactions. The restructuring of Cu facet coordination has stabilized facets on metal surfaces under electrolysis conditions, promoting the production of hydrogen. Compared with a traditional H-cell, a gas-fed flow cell improves FE toward CO_2 reduction products [239,247]. Jung et al. [248] also highlighted the importance to generate C_2 products selectively, which is critical to regulate and maintain the morphology of amorphous Cu nanoparticles. Density functional theory (DFT) calculation has shown that Cu is the preferable electrocatalyst for the formation of C_2 products [249–251]. CO_2 activation and CO dimerization to form C_2 products are remarkably improved by the linkage between the functional surfaces of Cu. The outcome also reveals that using the Cu complex as a precursor is critical for excellent performance because the Cu catalyst generated via direct electrodeposition has substantially low efficiency. The combination of copper with other metals produces higher catalytic activity for converting CO_2 into ethanol than pure copper metal. This finding is proven by the combination of Pd–Cu nanoparticles [252], copper-modified boron-doped diamond [253], copper–cuprous oxide [249], copper surface with a family of porphyrin-based metallic complexes [254], copper–silver composites [255], and cuprous oxide nanocubes with silver (Ag) nanoparticles [256]. Table 2 shows a summary of the experimental procedure applying copper-based catalysts in the transformation of CO_2 into ethanol.

Table 2. Summary of the experimental procedures applying copper-based catalysts in the transformation of CO_2 into ethanol.

Catalyst	Electrolyte	Cell Configuration	Current Density (h)	Overpotential	Faradaic Efficiency (%)	Total Current Density (mA cm^{-2})	Reference
Cu	1.0 M KOH 0.5 M KHCO$_3$	Electrochemical flow cell	4	−0.58 V vs. RHE	46	200	[257]
Cu	0.1 M KBr	H-type glass cell	3	−1.10 V vs. RHE	23	170	[258]
Cu	1.0 M KOH	Two compartment electrochemical H-cell	16	−0.95 V vs. RHE	32	126	[259]
Cu/Ag	1.0 M KOH	Electrochemical flow cell	not available	−0.70 V vs. RHE	25	300	[260]
Ce(OH)$_x$-doped-Cu	0.1 M KCl	Three-electrode electrochemical cell	6	−0.70 V vs. RHE	43	128	[261]
Cu	0.1 M KHCO$_3$	Flow cell reactor	6	−0.60 V vs. RHE	40	200	[262]
Cu/Ag	1.0 M KOH 1.0 M KHCO$_3$	Flow cell reactor	2	−0.67 V vs. RHE	41	250	[263]
Cu	1.0 M KHCO$_3$	MicroFlow® cell	20	−0.97 V vs. RHE	89	300	[264]
Cu	1.0 M KOH	Electrochemical flow cell	65	−0.71 V vs. RHE	90	520	[265]
FeTPP[Cl]/Cu	1.0 M KHCO$_3$	Electrochemical flow cell	12	−0.82 V vs. RHE	41	124	[254]
N-C/Cu	1.0 M KOH	Flow cell reactor	15	−0.68 V vs. RHE	52	156	[266]
zCu/Ni-N-C	1.0 M KOH	Electrochemical flow cell	103	−0.70 V vs. RHE	62	415	[267]
Cu$_2$O	1.0 M KHCO$_3$	Gas diffusion electrode flow cell	10	−0.85 V vs. RHE	76	300	[268]
np-Cu/VO$_2$	1.0 M KOH	Electrochemical flow cell	12	−0.80 V vs. RHE	38	102	[269]
ZnO/4Cu$_2$O	1.0 M KOH	Electrochemical flow cell	not available	−1.0 V vs. RHE	50	140	[270]
Cu$_{50}$/PTFE$_{15}$	1.0 M KOH	Gas diffusion electrode flow cell	2	−1.85 V vs. RHE	47	200	[271]
Cu$_2$O/Ag	1.0 M KOH	Gas diffusion electrode flow cell	not available	−1.18 V vs. RHE	73	243	[256]
Cu/C/PTFE	1.0 M KOH	Gas diffusion electrode flow cell	2	−1.0 V vs. RHE	76	250	[272]

Despite advancements in the electrochemical CO_2 reduction process, creating highly operative and selective nanocatalysts for the electrochemical CO_2 reduction reaction remains a major issue. The Cu-based catalyst of the nanostructure is chemically unstable, which demonstrates various catalytic performances via different procedures of operations due to the uncontrolled facet of oxidation and is related to the alternate in facet chemistry. The local reaction environment is further altered by electron reduction and protonation, making it challenging to stabilize the nanocatalyst. Facet coatings are a common way to improve the strength of nanocatalysts but affect the Cu's facet chemistry and its capability to convert CO_2 to form C–H hydrocarbon. MOFs are favorable support materials in stabilizing and improving the catalysts due to their electrical conduction. MOFs require pressured reactant supply and an outcome separation mechanism due to their porous nature. More experimental research should be conducted to stabilize Cu nanocatalysts for the discovery of their catalytically functional area and improved activity/selectivity. Other challenges include the effect of the poor transfer current densities causing the ineffective electron exchange rate of kinetics, the deactivation of electrodes, enormous overpotential (or low energy performance), restricting practical use, deficient selectivity of the product, which necessitates expensive separation processes, and technological commercialization.

3.2. Ethanol Synthesis from DME

The thermochemical approach to the production of ethanol from CO_2 via DME has two phases of reaction. The initial step is to make DME from CO_2 and H_2. Methanol synthesis and dehydration are the two phases in the traditional commercial DME synthesis from syngas, which includes CO and CO_2. However, a one-step experimental procedure on multifunctional catalysts is favorable due to its thermodynamic stability and operational cost-effectiveness [273,274]. DME is a cheap and bulk chemical with environmental acceptability, has high quality, and is an excellent replacement fuel for use in diesel engines [275–277]. DME is also used as a critical intermediary to bulk chemicals in the industrial sector by producing acetic acid, olefins, and hydrocarbons [278–280]. The commercialized process reaction of the DME synthesis reactor is based on the equation below.

CO Hydrogenation:

$$CO + 2H_2 \rightleftharpoons CH_3OH \qquad \Delta H^0{}_{298} = -90.8 \text{ kJ·mol}^{-1} \qquad (6)$$

CO_2 Hydrogenation:

$$CO_2 + 3H_2 \rightleftharpoons CH_3OH + H_2O \qquad \Delta H^0{}_{298} = -49.5 \text{ kJ·mol}^{-1} \qquad (7)$$

RWGS:

$$CO_2 + H_2 \rightleftharpoons CO + H_2O \qquad \Delta H^0{}_{298} = +41.2 \text{ kJ·mol}^{-1} \qquad (8)$$

Methanol dehydration:

$$2CH_3OH \rightleftharpoons CH_3OCH_3 + H_2O \qquad \Delta H^0{}_{298} = -23.4 \text{ kJ·mol}^{-1} \qquad (9)$$

Methanol catalysts, e.g., $Cu/ZnO/Al_2O_3$ (CZA), catalyze the reactions at Equations (6)–(8), whereas catalysts with acidic properties such as HZSM-5, zeolites or γ-alumina catalyze reactions at Equation (9) [281–284]. CZA is a dominant conventional catalyst in the DME reaction due to its capability to improve the catalytic performance and selectivity toward methanol production. Studies on the CZA catalyst for CO_2 hydrogenation to methanol for 720 h time-on-stream of the reaction demonstrated that the space–time output of methanol is reduced to 34.5% during long-term testing [285]. The addition of Zr in the CZA catalyst, which forms $CuO/ZnO/ZrO_2/Al_2O_3$ (CZZA) with HZSM-5, shows improvement stability with methanol production by reduction from 18.5% to 14.1% with more than 58.7% selectivity after 100 h of DME reaction [286]. The DME reaction by using $CuZn/Al_2O_3$ catalyst recorded optimum conditions at 250 °C and 40 bar, resulting in a methanol selectivity of 58% [287]. The optimum reaction condition for DME synthesis requires temperatures and pressures ranging from 200 °C to 300 °C and from 20 bar to 50 bar, respectively [288–290].

The direct production of DME in a one-step process involves the simultaneous completion of two stages of reactions, i.e., methanol generation (via CO_2 hydrogenation) and methanol dehydration to DME, in the same reactor by using hybrid/bifunctional catalysts in a closed system, avoiding the need for intermediate purification steps and transportation units to minimize the cost of operation [273,291,292]. The hybrid/bifunctional catalysts needed for direct DME production require the combination of metal sites with redox function properties for the selective CO_2 hydrogenation to methanol and acidic function for the transformation of methanol dehydration to produce DME. Based on the Le Chatelier principle, the high water content limits the production of methanol that occurs in the hydrogenation of CO_2. The dehydration reaction contributes to the production of water. If the reaction area is divided through the core-shell structure, the presence of water on metallic sites can be significantly limited [291,293,294]. Water molecules tend to be adsorbed on the surface of catalysts which deactivation of catalyst function by metal oxidation of the catalytic phase and constructs a faster metal sintering and destroys the structure of the acid catalysts blocking the production of methanol on the hydrogenation sites. Hence, an investigation is conducted to increase the stability of the catalyst implemented in the hydrogenation reaction. A remarkable improvement in catalytic stabilities has been recorded in In_2O_3/ZrO_2 [295], interlinkage of $CuO-ZnO-ZrO_2$ on the surface of the zeolite [296], zirconium-modified CZA [286], gallium nitride [297], Cu–Ho–Ga/γ-Al_2O_3 [298], $Cu/ZnO/ZrO_2$//H-FER 20 [299], and PdZn/TiO_2-ZSM-5 hybrid [300]. The introduction of membrane reactor technologies [292,301] and adsorbent material [288,289,302] has been proposed to limit the effectiveness of water in DME production. Additionally, stable acid sites are needed in DME production due to the presence of water because strong acidic sites catalyze secondary dehydration reactions that deposit carbon and form hydrocarbon [273,276].

The Cu-MOR@SiO_2 core–shell microcapsules catalyst in tandem with the ternary oxide CZA catalyst recorded the catalytic activity of DME conversion at 83.8% and ethanol selectivity at 48.7% over 50 h at 220 °C of reaction [303]. On the reaction of CZ@Cu-MOR microcapsule catalyst for 50 h at 400 °C, DME carbonylation converted to methyl acetate on the active sites of the zeolite subsequently hydrogenated the syngas to DME conversion, and ethanol selectivity of about 26.8% and 45.8%, respectively, was achievable [304]. In the optimal reaction condition of 24 h at 220 °C, the proximity effect in the two components of NMOR zeolite and CZA tandem catalysts exposed to syngas achieved a DME conversion of 66% along with ethanol selectivity of 43.4% [305]. The most challenging aspect of the direct production of DME from CO_2 by utilizing hybrid/bifunctional catalysts is ensuring the correct ratio with regulated metal and acid interaction that is required for methanol production and dehydration. The detriment of utilizing DME as an alternative fuel to diesel is due to its low viscosity, which causes a leak and component damage. Furthermore, DME has low heating point than diesel. Therefore, despite its higher energy performance, DME still requires fuel insertion every cycle of the reaction. DME also has low combustion enthalpy, low modulus of elasticity, and fuel tanks with low energy content. These disadvantages counteract the features of DME's low boiling point, and a pressured system must be used to keep the fuel in a liquid condition.

3.3. Ethanol Synthesis Based on RWGS

One of the most promising methods for CO_2 consumption as a renewable system delivering feedstock for nonfossil fuel synthesis and important chemical processes is the RWGS procedure [306–309]. The hydrogenation of CO_2 is converted into hydrocarbons via the RWGS reaction, which is catalyzed in tandem and subsequently modified by the Fischer–Tropsch synthesis (FTS) mechanism, where the intermediate product of the reaction is CO before hydrocarbons or alcohols are formed [310,311].

CO_2 hydrogenation:

$$nCO_2 + (3n+1)H_2 \rightarrow C_nH_{2n+2} + 2nH_2O \rightarrow \Delta_R H_{573K} = -128 \text{ kJ·mol}^{-1} \quad (10)$$

RWGS:

$$CO_2 + H_2 \rightarrow CO + H_2O \rightarrow \Delta_R H_{573K} = 38 \text{ kJ·mol}^{-1} \quad (11)$$

FTS:

$$nCO + (2n+1)H_2 \rightarrow C_nH_{2n+2} + nH_2O \rightarrow \Delta_R H_{573K} = -166 \text{ kJ·mol}^{-1} \quad (12)$$

This process is also known as the CO_2-FTS mechanism, where the CO produced from the RWGS reaction is inserted into the *CH_3 or *$CH_3(CH_2)_n$ generated from the CO-FTS mechanism to form methanol, ethanol, or other higher alcohol synthesis [312,313]. The RWGS reaction is an endothermic process favored at high temperatures, resulting in a high equilibrium conversion of CO_2 and performed at relatively low contact times [307,314].

The hydrogenation of the CO_2 reaction path is an inherent drawback because the FTS mechanism is an exothermic process that favors low temperatures [311,315]. Active sites for dissociating hydrogen and adsorbing CO_2 should be present in RWGS catalysts. The excellent performance of the precious metal-based catalyst in RWGS reactions has been recorded in a few studies due to their ability to dissociate hydrogen at low-temperature catalytic activities [316–318]. However, these catalysts are not suitable for industrial-scale promotion and application due to limitations of high prices and rare resources. In the RWGS reaction, Ni and Cu-based catalysts have demonstrated good activity and selectivity but are prone to sintering deactivation at high temperatures [318–321]. Transition-metal carbides are also favorable in RWGS reactions due to their dual functionality for dissociating hydrogen and C=O bond cleavages [322,323]. In ethanol production, Fe-based catalysts with the right combination of promoters, organized additives, or assistance form the active area of reaction. Alkali metals particularly Na and K elements have been identified as the most effective promoters of the catalytic performance of Fe-based by producing highly active for the FTS mechanism in improving the selectivity and CO_2 conversion during ethanol production [324,325]. Carbon support substances are natural support materials for Fe-based catalysts and show outstanding catalytic achievement for ethanol generation by enhancing selectivity and expanding the active dispersion phase in the FTS mechanism [326].

The catalytic performance of carbon support materials is based on the surface area, pore size, distribution, and pore structure [327,328]. Unfortunately, the use of Fe-based catalysts for CO_2 hydrogenation reactions produces highly toxic precursors and needs a long time of carbonization [326]. Cu-based catalysts in RWGS reactions have been extensively investigated due to their high stability, low cost, high-performance atmospheric pressure at very low temperatures, and excellent selectivity for CO [283,316,329,330]. When a considerable quantity of CO_2 in the feed is adsorbed on the surface, the oxidation and reduction processes of Cu-based catalysts demonstrate strong activity with a minimal number of undesirable products [283,331,332]. In the RWGS reaction, the hydrogenation of the CO_2 mechanism happens by using a Cu-based catalyst, with CO as an intermediate product before forming hydroxyl species on the surface, thus constraining the operative area for alcohol synthesis when CO_2 decomposes into water molecules. This process is the redox mechanism with the Cu-based catalyst involving CO_2 reduction, the rate-determining step, and active sites in RWGS reactions. CO_2 oxidizes into Cu^0 to produce Cu^+, which improves the CO selectivity by 10%, whereas H_2 reduces the Cu^+ to form Cu^0 to generate H_2O [306,333]. The investigation of the morphological effect shows that Cu/CeO_2 nanorods exhibit the highest CO_2 conversion compared with Cu/CeO_2 nanocubes and favor the strong link of metal–support interaction in generating a high density of oxygen vacancies under reducing conditions [334,335]. The rod-like morphology of CuO/CeO_2 demonstrates the highest catalytic activity and stability and achieves the thermodynamic equilibrium conversion at 350 °C [336]. In the RWGS reaction, the oxygen vacancies on the spinel oxide surfaces are vital in the adsorption and activation of CO_2 [337]. Based on the activation method, the adsorption of CO_2 on oxygen vacancies is the initial step of RWGS, which involves C=O bond cleavages under a high-temperature energy-driven process [338]. A study on the role of copper as a promoter has shown an indirect effect on catalyst activity. The study reported that the addition of Cu to the Mo_2C catalyst enhances the selectivity of CO yield [333]. The presence of Cu in MoO_3/FAU zeolite catalysts influences the reduction step of MoO_3 to MoO_2, thus improving the CO yield [332]. However, the major drawback of the Cu-based catalyst, which undergoes deactivation during the RWGS reaction because of poor thermal stability due to the fractional oxidation of the Cu metal, leads to the reduction in the surface area of the active sites and copper particle agglomeration at high temperatures. The effective metal stage and/or coke deposition are hampered by material sintering, which lowers the CO_2 transformation degree by restricting catalyst activity.

Some thermodynamic limits in RWGS reactions are present. The CO_2 reactant has the potential to damage the CO hydrogenation catalyst, and the water that is inevitably retained in the end product, generally 20–45% of the whole product, decreases product selectivity and catalytic activity. The endothermic nature of the RWGS reaction uses sophisticated catalysts that are frequently necessary to customize the cascade reactions, and a high temperature, typically above 300 °C, is required to drive these processes. Although methanol is used as an intermediary to make liquid hydrocarbon from CO_2 hydrogenation at high temperatures on some occasions, the end product has remarkable CO byproducts. The low activity and unstable C–C coupling formation in the FTS mechanism is another challenge in the CO_2 hydrogenation process that usually produces light hydrocarbons, particularly methane. Catalyst deactivation has been identified in the FTS mechanism by poising the catalyst in the presence of sulfur and nitrogen compounds, inactive metal support compound, hydrothermal sintering, and the formation of inactive catalytic phases as oxides.

3.4. Ethanol Synthesis Based on Catalytic Hydrogenation

Catalytic hydrogenation is one of the promising approaches to overcoming the obstacle in the chemical reduction activation of CO_2 [316,339,340]. The hydrogenation of CO_2 yields useful alcohols, such as methanol, ethanol, and higher alcohol, that have impressive energy density with broad applications to value-added chemicals, such as neat fuels, fuel additives, and raw chemicals [16,17,297,341]. However, due to a shortage of effective catalysts with excellent stability, the effective cleavage of the C–O bond, excessive strength barrier of C–C coupling, and generation of water as a byproduct in the process can simply inactivate several catalysts for CO_2 transformation, and the direct synthesis of ethanol via CO_2 hydrogenation is substantially more difficult than methanol synthesis [16,342,343]. The most efficient catalytic technique for producing ethanol directly from CO_2 should encourage partial CO_2 reduction, hydroxylation, and C–C bond formation at the same time [232]. According to theoretical investigations, minor catalyst effects improve CO_2 hydrogenation catalytic performance. The link between the structure and catalytic performance is established by regulating the catalyst structure of active sites, and constructing optimum catalysts is the most effective technique in managing carbon chain expansion with controlled alcohol arrangement [341,344]. The calcination process to synthesize the catalyst has also been identified to affect the performance of the catalyst. The maximum product yield of the CO_2 hydrogenation reaction is obtained from nickel(II) oxide supported on alumina and calcined at 700 °C with rod-like morphology and tiny crystallite size of nickel(II) oxide nanoparticles (12.7 nm) at facet (111) [345].

The fabrication of efficient heterogeneous Rh-based catalysts on TiO_2 nanorods should be beneficial in boosting ethanol selectivity due to a synergetic combination of surface hydroxyls and widely dispersed Rh nanoparticles. The use of promoters particularly Li and Fe is generally effective in enhancing the activity and ethanol selectivity by increasing the strength of adsorption of bridged-bond CO species and influencing the electronic condition of Rh [346]. Therefore, the $RhFeLi/TiO_2$ nanorod catalysts in CO_2 hydrogenation exhibit 30% ethanol selectivity, 15% CO_2 conversion, and stable performance for 20 h of operation. The effect on promotion is associated with the prominent density of the hydroxyl group on TiO_2 nanorods and the excessive distribution of Rh elements. Hydroxyl groups have been demonstrated to equalize the protonation of methanol and formate compounds, which are efficiently detached to form $*CH_x$. The production of CO from the RWGS reaction is then introduced to construct CH_3CO*, which is hydrogenated to further produce ethanol [347]. Rh-based catalysts may catalyze CO dissociation and CO insertion over their atomically neighboring Rh^0–Rh^{n+} species, increasing the possibility of coupling between *CO and $*CH_3$ in the synthesis of C_{2+} oxygenates from syngas, such as ethanol, acetic acid, and acetaldehyde [348]. The strong interaction in Rh/TiO_2 catalysts demonstrates excellent steady-state activity with 40% ethanol conversion at 120 min of reaction. The transformation of TiO_2 nanotubes into the anatase structure due to the acceleration of Rh nanowires has a remarkable impact on catalytic effects. The positive charge on Rh activates the CO_2 hydrogenation and promotes the further decomposition of formate intermediates [349]. Simulations through DFT reveal that the ionic liquid connects to the Rh species on TiO_2 with a binding energy from 0.69 eV to 1.19 eV. The turnover frequency (TOF) of the stabilized single atom Rh/TiO_2 is 800 h^{-1} in styrene hydroformylation and potentially recycled for five runs under harsh reaction conditions [350].

The quantity of vanadium oxide loaded and promoted on Rh-based catalysts enclosed in mesopore MCM-41 ($Rh-0.3VO_x$/MCM-41) has been demonstrated as extremely promising for ethanol synthesis, with CO_2 transformation and ethanol selectivity of 12% and 24%, respectively. This result is contributed by the equilibrium amount of CO dissociative adsorbed with nondissociative adsorbed affecting the yield and selectivity of ethanol synthesis. The electrical effect is thought to be responsible for the creation of Rh^+ species and the construction of interfacial VO_x–Rh active sites, which dissociate CO into $*CH_x$ and aid in the synthesis of ethanol following CO introduction [351]. The $2K20Fe5Rh–SiO_2$ catalyst in CO_2 hydrogenation has recorded 16% ethanol selectivity and 18% CO_2 conversion for 6 h of stability. The presence of K as a promoter in this experiment stabilizes the CO intermediate produced and C–H bond formation during CO_2 hydrogenation [346,352]. Rh10Se clusters supported on TiO_2 ($Rh10Se/TiO_2$) and treated in a fixed-bed reactor at 350 °C show optimized selectivity to ethanol synthesis at 83% and CO_2 conversion of 27%. The strong electrical interaction between Rh10 and Se is thought to hinder methane production and boost ethanol synthesis on Rh sites by encouraging C–C bond formation via CH_x and carbonyl coupling on the surface to produce acetate substances [353]. The scarcity and high cost of Rh-based catalysts restrict further development. Table 3 shows a summary of CO_2 hydrogenation over homogeneous and heterogeneous catalysts.

Pd-based catalysts have shown promise because they aid in C–C coupling, a crucial step in the formation of C_{2+} molecules, and precisely modify nanoparticle composition, nanoparticle structure, and support materials [354,355]. As a result, numerous Pd-based catalysts for direct ethanol synthesis from CO_2 hydrogenation with good selectivity in an autoclave reactor at a very high reaction temperature (>250 °C) and confined to standard disordered architectures have been created [313,347,356]. Ordered catalysts with considerable interaction with active sites have been proposed to increase charge transfer and regulate electronic effects. Other disordered catalysts may be outperformed by such a catalytic system [357,358]. Within 5 h of the experiment, the Pd_2/CeO_2 nanorod catalyst with a unique two-atom geometric feature arrangement enables the facile cleavage of the C–O bond and efficiently contributes to C–C coupling, yielding 99.2% ethanol selectivity, 9.2% CO_2 conversion, and TOF value of 211.7 h^{-1}. DFT results show that Pd dimers bind CO tightly, preventing CO desorption and forming the precursor of ethanol via connecting CO and CH_3 intermediates [343]. Similar to precious metal group materials, the Pd-based catalyst is scarce and expensive, thus hindering its large-scale application.

At 200 °C for 5 h, the $Ir_1–N_2O_3$ single-atom catalyst displays exceptional efficiency for CO_2 hydrogenation with 99% ethanol selectivity and a TOF value of 481 h^{-1}. The isolated monoatomic Ir atom interacts with the neighboring oxygen vacancy on In_2O_3 to create a Lewis acid-base pair. This phenomenon generates two independent catalytic centers that reduce CO_2 into active intermediate species of carbonyl (CO*) adsorbed on the Ir atom and subsequently contribute to the C–C coupling to ethanol production [344]. However, the scarcity and high cost of In-based catalysts render them unsuitable for practical application. Co-based catalysts may generate a high selectivity of alcohol products due to excellent CO insertion ability and catalyze C–C coupling. The synergistic impact of facets Co^0 and $Co^{\delta+}$ provides the remarkable achievement of the Co/La–Ga–O composite oxide catalyst in CO_2 hydrogenation, with 9.8% CO_2 conversion, 74.7% ethanol selectivity, and 88.1% ethanol [359]. The introduction of nickel into a Co-based catalyst exhibits high activity and selectivity in forming ethanol by CO_2 hydrogenation. The optimized $Co_{0.52}Ni_{0.48}AlO_x$ catalyst has recorded ethanol synthesis of 85.7% at 200 °C for 12 h on stream. This catalyst also shows high stability by remaining unchanged metal nanoparticle size and composition five times, whereas the selectivity of ethanol is maintained [360].

The study of the catalytic achievement of Na–Co/SiO$_2$ catalyst at 250 °C, 5 Mpa, gas hourly space velocity (GHSV) of 4000 h^{-1} for 300 h of reaction recorded 62.8% ethanol selectivity and 18% CO$_2$ conversion. The CO generated in the Co$_2$C active phase is injected into CH$_x$ intermediates, leading to the production of ethanol according to in situ DRIFTS data [313]. The interaction between Na and Co$_2$C produces ethanol efficiently with a CO$_2$ conversion of 53%. The electronic environment of Na–Co$_2$C active sites and the effects of CO activation are revealed using DFT calculations. The Bader charge analysis revealed that Na is a cation on the Co$_2$C surface with a Bader charge of 0.78 e, suggesting electron transfer from Na to Co$_2$C and the presence of contact between Na and Co$_2$C (Figure 4a). When CO$_2$ is adsorbed (Figure 4b), the most charge transferred and the stable adsorption structure on Na–Co$_2$C active sites reveal that the O–C–O bond of CO$_2$ is bent from linear in gaseous to 122.4, in which C and two O atoms contact with 2 Co atoms and 1 Na atom, showing that Na–Co$_2$C allows for easy adsorption and activation [361].

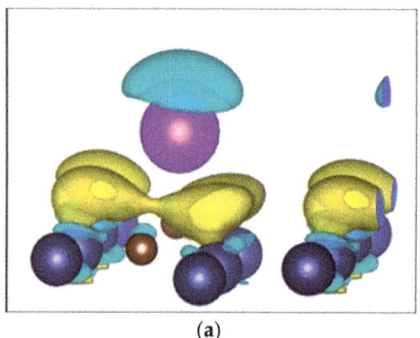
(a)

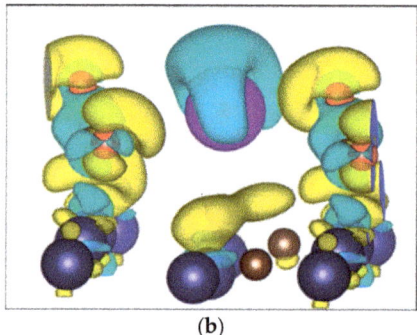
(b)

Figure 4. (a) Charge density calculations of Na adsorbed on the Co$_2$C surface. (b) Charge density calculations of CO$_2$ adsorbed on Na–Co$_2$C surface (blue: Co atom; grey: C atom; purple: Na atom) Reprinted with permission from Ref. [361] 2021 Copyright Elsevier.

The energy barriers of transition states on the Na–Co$_2$C surface are higher than that on the Co$_2$C surface regardless of CO-direct or H-assisted dissociation. A large number of alkyl species (CH$_x$) formed on the Co$_2$C surface (formate species or CO hydrogenation) and the adsorption energies of CH$_x$ species are constant on Na–Co$_2$C functional surface area. The increase in CO non-dissociative adsorption improves the CO/CH$_x$ ratio, which is conductive and allows for easier CO activation and coupling on Na–Co sites into the adjacent CH$_x$ on Co atoms to synthesize ethanol [361]. The use of CoMoC$_x$ catalyst prepared using the ionic liquid method as all-in-one precursors at 800 °C carbonized temperature results in the optimal catalytic performance of 97.4% ethanol selectivity for 6 h of the experiment. This catalyst also shows high stability performance in the seventh run without remarkable deactivation. This study also found that the most electron transfer and the largest shift towards the low binding energy occur at 180 °C and 2 Mpa and detected CO gas as a byproduct of the reaction. Water, a green solvent used in this reaction, is combined with CO$_2$ to form bicarbonate and accelerate CO$_2$ conversion [362]. However, the conventional Co$_2$C phase under H$_2$ above 220 °C usually suffers from the rate of deactivation in the long-term catalyst performance. The uncontrollable synergism between CO dissociative activation and CO insertion of metallic Co-based catalysts generates the high selectivity of methane in CO$_2$ hydrogenation.

Fe-based catalysts possess the ability of CO dissociation catalyzed by an operative composition, and the catalytic activity can be constructed in the formation of hydrocarbon products [363,364]. The improvement of the CO insertion procedure and match the alkyl species formed on Fe sites to make alcohol products, a Fe-based catalyst combined with a noble metal is required [347,356,365]. The catalytic activity of FeMnNa catalyst under experimental conditions of 340 °C, 2.0 Mpa, and $CO_2/H_2/Ar$ ratio of 24/72/4 has resulted in 35% CO_2 conversion and 31.7% ethanol selectivity. The temperature-programmed reduction analysis has also recorded excellent performance of the FeMnNa catalyst by promoting the formation of $MnCO_3$ from MnO in the presence of CO_2 and indicated that the Na component hinders the synergy between Fe and Mn in the reduction of FeMnNa catalyst for selective CO_2 hydrogenation to form ethanol [366]. The monometallic Fe-based catalyst co-modified with Na and S (FeNaS-0.6) achieves 16% ethanol selectivity, 32% CO_2 conversion, and high stability performance of over 100 h of the evaluation with no methanol formation [367]. In the CO_2 hydrogenation process, the sulfur present in the sulfate and its electron-withdrawing effects on Na-assisted Fe sites contribute to CO dissociation, nondissociative CO adsorption, boosting the hydrogenation barrier of $*CH_x$ compound and improving the production of ethanol [23,366].

The development of Mo-based catalysts has opened the door to a new strategy in the hydrogenation of CO_2, which promotes the C–C coupling in ethanol synthesis. The deposition reaction of atomic operative elements Rh and K successfully synthesizes the one-dimensional b-phase of Mo_2C nanowires with specified crystal facets (101). The modification of the $K_{0.2}Rh_{0.2}/b-Mo_2C$ complex catalyst results in a prominent production of 33.7 $\mu mol \cdot g^{-1} \cdot h^{-1}$ ethanol and ethanol selectivity of 72.1% at 150 °C [368]. Cu-based catalysts have been extensively studied in the hydrogenation of CO_2 to form ethanol products from syngas although their activity and selectivity are highly dependent on the support and promoter. Through local arrangement and fine-tuning of the catalytic centers by alkali promoters, the Zr12-bpdc-CuCs catalyst demonstrates ethanol synthesis with 99% selectivity in a 10 h assessment. With the help of alkali–metal promoters, the Cu-based catalyst facilitates H_2 activation and promotes direct C–C coupling and formyl species to offer an electron-rich environment for Cu-based catalysts and boost the stability and activity of a formyl intermediate [369]. Alkali promoters (K, Rb, and Cs) influence the achievement of CO_2 hydrogenation reactions made of precipitated iron-based catalysts; 1.5 Cs boosted catalyst has the best steady-state conversion stability of all the catalysts. Results show that these promoters have a synergistic impact that may result in improved CO_2 hydrogenation catalysts if balanced [370].

Table 3. Summary of CO_2 hydrogenation over homogeneous and heterogeneous catalysts.

Catalysts	Reactor	Reaction Temperature (°C)	Pressure (MPa)	Mixed Gas Ratio	Time Reaction (h)	CO_2 Conversion (%)	Ethanol Selectivity (%)	Ethanol STY (mmol g^{-1} h^{-1})	Brief Description	Ref
Na/Co$_2$C	Fixed-bed	250	5	$CO_2/H_2/N_2$ = 24.6/72.4/3	5	23.8	17.5	0.72	The active sites of Na/Co$_2$C improve the CO_2 and CO non-dissociative adsorption, then regulated the surface CO/CH$_x$ ratio to accelerate CO insertion in generating ethanol.	[361]
CoMoC	Fixed-bed	180	2	CO_2/H_2 = 1/3	6	n.d	97.4	0.53	The excellent stability of CoMoC$_x$ promotes the activation of H$_2$ and CO$_2$ and C-C coupling which is generated by the HCOO* and DMF species.	[362]
Pt/Co$_3$O$_4$	Fixed-bed	200	8	CO_2/H_2 = 1/3	15	n.d	82.5	0.42	Water protonates methanol followed by dissociation into CH$_3$*, OH*, and H* (or H$_2$O) species on the Pt/Co$_3$O$_4$ surface that promotes CH$_3$*–CO coupling to form ethanol.	[353]
Cs/CuFeZn	Fixed-bed	330	5	$CO_2/H_2/N_2$ = 24/72/4	3	36.6	20.7	1.47	The synergetic combination of Cu-Fe dual interfaces sites in the Cs/CuFeZn overrides methanol synthesis through a direct CO_2 hydrogenation route via HCOO* intermediates.	[356]
Ir/In$_2$O$_3$	Fixed-bed	200	6	CO_2/H_2 = 1/5	5	n.d	99.7	0.99	The Ir/In$_2$O$_3$ reduced a Lewis acid–base pair between Ir and adjacent oxygen vacancy to form a distinct catalytic center, which reduces CO_2 to active intermediates and facilitates the C–C coupling to form ethanol.	[344]

Table 3. Cont.

Catalysts	Reactor	Reaction Temperature (°C)	Pressure (MPa)	Mixed Gas Ratio	Time Reaction (h)	CO_2 Conversion (%)	Ethanol Selectivity (%)	Ethanol STY (mmol g^{-1} h^{-1})	Brief Description	Ref
Cu/Co$_3$O$_4$	Fixed-bed	200	30	$CO_2/H_2 = 1/3$	2	13.9	15.2	1.87	The adjacent oxygen vacancy on the surface of CoO promotes the CH$_3$O* intermediate dissociation is the rate-determining step for ethanol synthesis.	[371]
Co$_3$O$_4$	Fixed-bed	200	2	$CO_2/H_2/N_2 =$ 22/66/12	2	28.9	19.2	1.60	The metallic Co reduced from Co$_3$O$_4$ was the main activity site for CO$_2$ hydrogenation by promoting the growth of the C–C coupling for the production of ethanol.	[372]
CoAlO	Fixed-bed	200	4	$CO_2/H_2 = 1/3$	15	n.d	92.1	0.44	The CoAlO is attributed to the formation of acetate from formate with the insertion of *CH$_x$ which is an important intermediate to produce ethanol from CO$_2$ hydrogenation.	[373]
Pt/Co$_3$O$_4$	Fixed-bed	200	2	$CO_2/H_2/N_2 =$ 22.5/67.5/10	2	44.5	26.7	0.69	The synergic effect of Pt, Co nanoparticles, and oxygen vacancies of Co$_3$O$_4$ improved the adsorption of H$_2$ and CO$_2$ with stable CO$_2$ conversion in the synthesis of ethanol.	[374]

However, remarkable drawbacks to CO_2 hydrogenation for alcohol syntheses, such as CO_2 activation problems, a high energy barrier for C–O bond scission, and the creation of C1 by-products, remain. As a result, designing effective heterogeneous catalysts for ethanol generation is critical. The morphology of catalysts, such as the particle size and dispersion of deposited metal particles, influences the optimization of the metal/oxide interface to improve CO_2 conversion and product selectivity. The collision theory explained by increasing the surface area of a reactant created by high dispersion increases the frequency of collisions and increases the reaction rate. Reducing metal particle sizes leads to a high fraction of low-coordinate surface atoms at locations, such as corners and edges, especially when the particle size is smaller than 2–3 nm [353,375–377]. The small particle size content high in the surface area that generates by high dispersion is available for particles to collide, leading to improved catalytic performance of the reaction. Appropriate reducible metal oxide supports, such as TiO_2 and ZrO_2, have been used extensively to tailor the particle size. The Au/TiO_2 catalyst has good selectivity for ethanol from CO_2 reduction in DMF solvent due to the abundance of oxygen vacancies. With the addition of water, the bimetallic $Pd_2Cu/P25$ catalyst produces a high yield of ethanol. Other studies indicated that by manipulating the particle size, the electronic state of Rh for alcohol production from CO_2 hydrogenation can be adjusted. In CO hydrogenation, a promotion strategy based on hydroxyl groups is an effective way to improve alcohol selectivity at a high conversion rate and a wide range of operating temperatures.

4. Future Direction and Perspective

Ethanol is a fundamental chemical product, an important solvent, an industrial building block, and a promising renewable fuel. The chemical equilibrium conversion of the hydration of ethylene decreases at high temperatures to increase the rate of reaction. The excessive amount of energy to heat gases generates high pressure and utilizes crude oil, which is a nonrenewable resource. Ethanol synthesis from biomass residue produces a huge amount of CO_2 with a high cost of operation because of expensive machinery and fuel. The ethanol production from total sugars in lignocellulosic materials is inhibited by the action of pentose sugars, which are not fermentable by the brewer's yeast, i.e., *S. cerevisiae*, during the hydrolysis of hemicellulose. Although the fermentation route has been commercially realized, the cost of operation for this process is expensive due to the energy-intensive distillation steps at high dilution rates, which allow for high productivities, and the incomplete utilization of substrate, resulting in low yields to meet the market demand. Creating highly operative and selective nanocatalysts for the electrochemical CO_2 reduction reaction remains a major issue in the electrochemical CO_2 reduction process. DME is not suitable as an alternative fuel to diesel due to its lower viscosity, which causes leaking and component damage. The RWGS reaction uses sophisticated catalysts that are frequently necessary to customize the cascade reactions and require a high temperature (typically above 300 °C) to drive the processes still experiencing difficulty; low activity and unstable C–C coupling formation in the FTS mechanism is another challenge in the CO_2 hydrogenation process that are usually producing light hydrocarbons, methane in particular. The heterogeneous catalytic CO_2 hydrogenation approach is one of the initiatives to reduce CO_2 emissions as the source of GHG, which prompts the climate change or global warming that currently shows an effect worldwide, into the numerous high economic value-added chemicals and easily marketable fuel additives that have made substantial progress to explore as new concepts and opportunities for industrial manufacture. The hydrogenation of CO_2 into C_{2+} products occurs via a methanol-mediated route or modified FTS mechanism that involves two steps of the reaction, and these routes are distinguished by the intermediate's product. In the methanol-mediated route, CO_2 is hydrogenated into methanol and then converted into hydrocarbons, whereas CO_2 is reduced to CO via RWGS followed by chain propagation via a modified FTS mechanism. A recent development showed that the direct catalytic hydrogenation of CO_2 into ethanol in a single reactor is one of the promising strategies based on economic potential and energy efficiency. The formation of *CH_x is the rate-determining step in this process because the *HCOO generation and coupling steps are known to be accelerated in the CO_2-to-ethanol transformation. The reaction mechanism is also closely associated with multiple catalytic active sites controlling every elementary reaction over catalysts. Thus, catalysts for optimum active sites should be designed and prepared to improve CO_2 conversion and ethanol selectivity. The formation rate of ethanol increases with increasing reaction temperature because thermodynamic equilibrium is still not reached. However, many challenges, e.g., the control of the ratio of multiple active sites, regulation of interface sites, and device of catalyst structures need to be explored in future research to improve the conversion and selectivity in ethanol production.

5. Conclusions

In this exploration, we discuss current research discoveries in the improvement of technologies and operation procedures in ethanol production. The catalytic hydrogenation of CO_2 promises development direction in the production of ethanol and reduces environmental pollution problems. The limitations of CO_2, i.e., a fully oxidized, chemically inert and thermodynamically stable molecule, should be considered in designing the research because its conversion into chemicals requires large amounts of energy and H_2. The production of ethanol reduces CO_2 emissions globally, but reducing the dependence on conventional gasoline shows a decreasing pattern of production every year because ethanol has been identified as an ecological fuel due to its nontoxicity, accumulation of high oxygen content to promote better combustion with reduced exhaust emissions, and high-octane rating to giving high resistance to engine knock. The outcomes of this review will play a crucial and interesting role in further research development in providing the latest proposed approach for effective CO_2 hydrogenation to promote C–C coupling in ethanol production.

Author Contributions: M.N.L.: Literature collection and manuscript writing; A.S.: Literature collection and analysis; W.N.M.: Literature collection; S.Z.H.: Content design; W.N.R.W.I. and Z.Y.: Content design, revision, project administration, and funding acquisition. All authors have read and agreed to the published version of the manuscript.

Funding: This work was financially supported by Universiti Kebangsaan Malaysia under research code DIP-2022-010 and GENIUSpintar-2022-022.

Acknowledgments: Authors would like to thank the Ministry of Higher Education of Malaysia under the research code Fundamental Research Grant Scheme (FRGS/1/2020/TK0/UKM/02/31), Universiti Kebangsaan Malaysia under research code DIP-2022-010 and GENIUSpintar-2022-022.

Conflicts of Interest: The authors declare no conflict of interest.

References

1. Berrill, P.; Gillingham, K.T.; Hertwich, E.G. Drivers of Change in U.S. Residential Energy Consumption and Greenhouse Gas Emissions, 1990–2015. *Environ. Res. Lett.* **2021**, *16*, 034045. [CrossRef]
2. Li, J.; Tian, Y.; Deng, Y.; Zhang, Y.; Xie, K. Improving the Estimation of Greenhouse Gas Emissions from the Chinese Coal-to-Electricity Chain by a Bottom-up Approach. *Resour. Conserv. Recycl.* **2021**, *167*, 105237. [CrossRef]
3. Karmaker, A.K.; Rahman, M.M.; Hossain, A.M.; Ahmed, R.M. Exploration and Corrective Measures of Greenhouse Gas Emission from Fossil Fuel Power Stations for Bangladesh. *Clean. Prod.* **2020**, *244*, 118645. [CrossRef]
4. Babatundea, K.A.; Saida, F.F.; Nor, N.G.M. Reducing Carbon Dioxide Emissions from Malaysian Power Sector: Current Issues and Future Directions. *Eng. J.* **2018**, *1*, 59–69. [CrossRef] [PubMed]
5. Tchanche, B. Dynamics of Greenhouse Gas (GHG) Emissions in the Transportation Sector of Senegal. *Earth* **2021**, *2*, 1–15. [CrossRef]
6. Umar, M.; Ji, X.; Kirikkaleli, D.; Alola, A.A. The Imperativeness of Environmental Quality in the United States Transportation Sector amidst Biomass-Fossil Energy Consumption and Growth. *Clean. Prod.* **2021**, *285*, 124863. [CrossRef]
7. Olivier, J.G.J.; Peters, J.A.H.W. *Trends in Global CO_2 and Total Greenhouse Gas Emissions: Report 2019*; PBL Netherlands Environmental Assessment Agency: The Hague, The Netherlands, 2020; Volume 2020, p. 70.
8. Javadi, P.; Yeganeh, B.; Abbasi, M.; Alipourmohajer, S. Energy Assessment and Greenhouse Gas Predictions in the Automotive Manufacturing Industry in Iran. *Sustain. Prod. Consum.* **2021**, *26*, 316–330. [CrossRef]

9. Udmale, P.; Pal, I.; Szabo, S.; Pramanik, M. International Cereal Trade of Bangladesh: Implications for Virtual Land, Water, and GHG Emissions from Agriculture. *Int. Energy J.* **2021**, *21*, 107–118.
10. Ntinyari, W.; Gweyi-onyango, J. *Greenhouse Gases Emissions in Agricultural Systems and Climate Change Effects in Sub-Saharan Africa*; Springer Nature: Berlin, Germany, 2020; pp. 1–25.
11. Panchasara, H.; Samrat, N.H.; Islam, N. Greenhouse Gas Emissions Trends and Mitigation Measures in Australian Agriculture Sector—A Review. *Agriculture* **2021**, *11*, 85. [CrossRef]
12. Zhang, X.; Li, X.; Chen, D.; Cui, H.; Ge, Q. Overestimated Climate Warming and Climate Variability Due to Spatially Homogeneous CO_2 in Climate Modeling over the Northern Hemisphere since the Mid-19th Century. *Sci. Rep.* **2019**, *9*, 17426. [CrossRef]
13. Nunes, L.J.R.; Meireles, C.I.R.; Gomes, C.J.P.; Ribeiro, N.M.C.A. Forest Contribution to Climate Change Mitigation: Management Oriented to Carbon Capture and Storage. *Climate* **2020**, *8*, 21. [CrossRef]
14. Panda, R.; Maity, M. Global Warming and Climate Change on Earth: Duties and Challenges of Human Beings. *Res. Eng. Sci. Manag.* **2021**, *4*, 122–125.
15. Friedlingstein, P.; Jones, M.; Sullivan, M.O.; Hauck, J. Global Carbon Budget 2021. *Earth Syst. Sci. Data* **2021**, *14*, 1917–2005. [CrossRef]
16. Zhou, W.; Cheng, K.; Kang, J.; Zhou, C.; Subramanian, V.; Zhang, Q.; Wang, Y. New Horizon in C1 Chemistry: Breaking the Selectivity Limitation in Transformation of Syngas and Hydrogenation of CO_2 into Hydrocarbon Chemicals and Fuels. *Chem. Soc. Rev.* **2019**, *48*, 3193–3228. [CrossRef] [PubMed]
17. Kumaravel, V.; Bartlett, J.; Pillai, S.C. Photoelectrochemical Conversion of Carbon Dioxide (CO_2) into Fuels and Value-Added Products. *ACS Energy Lett.* **2020**, *5*, 486–519. [CrossRef]
18. Tan, X.; Sun, X.; Han, B. Ionic Liquid-Based Electrolytes for CO_2 Electroreduction and CO_2 Electroorganic Transformation. *Natl. Sci. Rev.* **2021**, *9*, nwab022. [CrossRef]
19. Hasan, S.Z.; Ahmad, K.N.; Isahak, W.N.R.W.; Pudukudy, M.; Masdar, M.S.; Jahim, J.M. Synthesis, Characterisation and Catalytic Activity of NiO Supported Al_2O_3 for CO_2 Hydrogenation to Carboxylic Acids: Influence of Catalyst Structure. *IOP Conf. Ser. Earth Environ. Sci.* **2019**, *268*, 012079. [CrossRef]
20. Xu, Y.; Shi, C.; Liu, B.; Wang, T.; Zheng, J.; Li, W.; Liu, D.; Liu, X. Selective Production of Aromatics from CO_2. *Catal. Sci. Technol.* **2019**, *9*, 593–610. [CrossRef]
21. Guzmán, H.; Russo, N.; Hernandez, S. CO_2 Valorisation towards Alcohols by Cu-Based Electrocatalysts: Challenges and Perspectives. *Green Chem.* **2021**, *23*, 1897–1920. [CrossRef]
22. Yao, B.; Xiao, T.; Makgae, O.A.; Jie, X.; Gonzalez-Cortes, S.; Guan, S.; Kirkland, A.I.; Dilworth, J.R.; Al-Megren, H.A.; Alshihri, S.M.; et al. Transforming Carbon Dioxide into Jet Fuel Using an Organic Combustion-Synthesized Fe-Mn-K Catalyst. *Nat. Commun.* **2020**, *11*, 6395. [CrossRef]
23. Yang, Q.; Skrypnik, A.; Matvienko, A.; Lund, H.; Holena, M.; Kondratenko, E.V. Revealing Property-Performance Relationships for Efficient CO_2 Hydrogenation to Higher Hydrocarbons over Fe-Based Catalysts: Statistical Analysis of Literature Data and Its Experimental Validation. *Appl. Catal. B Environ.* **2021**, *282*, 119554. [CrossRef]
24. Xing, Y.; Ma, Z.; Su, W.; Wang, Q.; Wang, X.; Zhang, H. Analysis of Research Status of CO_2 Conversion. *Catalysts* **2020**, *10*, 370. [CrossRef]
25. Duyar, M.S.; Gallo, A.; Regli, S.K.; Snider, J.L.; Singh, J.A.; Valle, E.; Mcenaney, J.; Bent, S.F.; Rønning, M.; Jaramillo, T.F. Understanding Selectivity in CO_2 Hydrogenation to Methanol for MoP Nanoparticle Catalysts Using in Situ Techniques. *Catalysts* **2021**, *11*, 143. [CrossRef]
26. Xu, D.; Wang, Y.; Ding, M.; Hong, X.; Liu, G.; Tsang, S.C.E. Advances in Higher Alcohol Synthesis from CO_2 Hydrogenation. *Chem* **2020**, *7*, 849–881. [CrossRef]
27. Weilhard, A.; Argent, S.P.; Sans, V. Efficient Carbon Dioxide Hydrogenation to Formic Acid with Buffering Ionic Liquids. *Nat. Commun.* **2021**, *12*, 231. [CrossRef]
28. Pandey, P.H.; Pawar, H.S. Cu Dispersed TiO_2 Catalyst for Direct Hydrogenation of Carbon Dioxide into Formic Acid. *J. CO_2 Util.* **2020**, *41*, 101267. [CrossRef]
29. Singh, D.; Gupta, S.K.; Seriani, N.; Lukacevic, I.; Sonvane, Y.; Gajjar, P.N.; Ahuja, R. Mechanism of Formaldehyde and Formic Acid Formation on (101)-TiO_2@Cu_4 Systems through CO_2 Hydrogenation. *Sustain. Energy Fuels* **2021**, *5*, 564–574. [CrossRef]
30. Yang, W.; Chernyshov, I.Y.; van Schendel, R.K.A.; Weber, M.; Müller, C.; Filonenko, G.A.; Pidko, E.A. Robust and Efficient Hydrogenation of Carbonyl Compounds Catalysed by Mixed Donor Mn(I) Pincer Complexes. *Nat. Commun.* **2021**, *12*, 12. [CrossRef]
31. Damyanov, A.; Hofmann, P. Operation of a Diesel Engine with Intake Manifold Alcohol Injection. *Automot. Engine Technol.* **2019**, *4*, 17–28. [CrossRef]
32. Zhao, L.; Wang, D. Combined Effects of a Biobutanol/Ethanol-Gasoline (E10) Blend and Exhaust Gas Recirculation on Performance and Pollutant Emissions. *ACS Omega* **2020**, *5*, 3250–3257. [CrossRef]
33. Elfasakhany, A. State of Art of Using Biofuels in Spark Ignition Engines. *Energies* **2021**, *14*, 779. [CrossRef]
34. Pang, J.; Zheng, M.; Zhang, T. Synthesis of Ethanol and Its Catalytic Conversion. *Adv. Catal.* **2019**, *64*, 89–191. [CrossRef]
35. Mizik, T.; Gyarmati, G. Economic and Sustainability of Biodiesel Production—A Systematic Literature Review. *Clean Technol.* **2021**, *3*, 19–36. [CrossRef]

36. Zhang, Z.; Lis, M. Modeling Green Energy Development Based on Sustainable Economic Growth in China. *Sustainability* **2020**, *12*, 1368. [CrossRef]
37. Cooper, G.; McCaherty, J.; Huschitt, E.; Schwarck, R.; Wilson, C. 2021 Ethanol Industry Outlook. *Renew. Fuels Assoc.* **2021**, 1–40.
38. Alalwan, H.A.; Alminshid, A.H.; Aljaafari, H.A.S. Promising Evolution of Biofuel Generations. Subject Review. *Renew. Energy Focus* **2019**, *28*, 127–139. [CrossRef]
39. Jeswani, H.K.; Chilvers, A.; Azapagic, A. Environmental Sustainability of Biofuels: A Review. *Proc. R. Soc. A* **2020**, *476*, 20200351. [CrossRef] [PubMed]
40. Momose, H.; Kusumoto, K.; Izumi, Y.; Mizutani, Y. Vapor-Phase Direct Hydration of Ethylene over Zirconium Tungstate Catalyst. I. Catalytic Behavior and Kinetics at Atmospheric Pressure. *J. Catal.* **1982**, *77*, 23–31. [CrossRef]
41. Mohsenzadeh, A.; Zamani, A.; Taherzadeh, M.J. Bioethylene Production from Ethanol: A Review and Techno-Economical Evaluation. *ChemBioEng Rev.* **2017**, *4*, 75–91. [CrossRef]
42. Zimmermann, H.; Walzl, R. Ethylene. In *Ullmann's Encyclopedia of Industrial Chemistry*; Wiley-VCH Verlag GmbH & Co. KGaA: Weinheim, Germany, 2000.
43. Chu, W.; Echizen, T.; Kamiya, Y.; Okuhara, T. Gas-Phase Hydration of Ethene over Tungstena-Zirconia. *Appl. Catal. A Gen.* **2004**, *259*, 199–205. [CrossRef]
44. Katada, N.; Iseki, Y.; Shichi, A.; Fujita, N.; Ishino, I.; Osaki, K.; Torikai, T.; Niwa, M. Production of Ethanol by Vapor Phase Hydration of Ethene over Tungsta Monolayer Catalyst Loaded on Titania. *Appl. Catal. A Gen.* **2008**, *349*, 55–61. [CrossRef]
45. Isobe, A.; Yabuuchi, Y.; Iwasa, N.; Takezawa, N. Gas-Phase Hydration of Ethene over Me(HPO$_4$)$_2$·NH$_2$O (Me = Ge, Zr, Ti, and Sn). *Appl. Catal. A Gen.* **2000**, *194*, 395–401. [CrossRef]
46. Gao, J.; Li, Z.; Dong, M.; Fan, W.; Wang, J. Thermodynamic Analysis of Ethanol Synthesis from Hydration of Ethylene Coupled with a Sequential Reaction. *Front. Chem. Sci. Eng.* **2020**, *14*, 847–856. [CrossRef]
47. Llano-Restrepo, M.; Muñoz-Muñoz, Y.M. Combined Chemical and Phase Equilibrium for the Hydration of Ethylene to Ethanol Calculated by Means of the Peng-Robinson-Stryjek-Vera Equation of State and the Wong-Sandler Mixing Rules. *Fluid Phase Equilib.* **2011**, *307*, 45–57. [CrossRef]
48. Roozbehani, B.; Mirdrikvand, M.; Imani, S.; Roshan, A.C.; Africa, S. Synthetic Ethanol Production in the Middle East: A Way to Make Environmentally Friendly Fuels. *Chem. Technol. Fuels Oils* **2013**, *49*, 115–124. [CrossRef]
49. Malico, I.; Nepomuceno, R.; Cristina, A.; Sousa, A.M.O. Current Status and Future Perspectives for Energy Production from Solid Biomass in the European Industry. *Renew. Sustain. Energy Rev.* **2019**, *112*, 960–977. [CrossRef]
50. Clauser, N.M.; González, G.; Mendieta, C.M.; Kruyeniski, J.; Area, M.C.; Vallejos, M.E. Biomass Waste as Sustainable Raw Material for Energy and Fuels. *Sustainability* **2021**, *13*, 794. [CrossRef]
51. Reid, W.V.; Ali, M.K.; Field, C.B. The Future of Bioenergy. *Glob. Chang. Biol.* **2020**, *26*, 274–286. [CrossRef]
52. Sudagar, S.; Adhisegar, C.; Bernadsha, P.; Balamurugan, R. A Comparative Study of Different Biomass Properties by Using Pyrolysis Process. *AIP Conf. Proc.* **2020**, *2225*, 040001. [CrossRef]
53. Lee, S.Y.; Sankaran, R.; Chew, K.W.; Tan, C.H.; Krishnamoorthy, R.; Chu, D.-T.; Show, P.-L. Waste to Bioenergy: A Review on the Recent Conversion Technologies. *BMC Energy* **2019**, *1*, 4. [CrossRef]
54. Debrah, J.K.; Vidal, D.G.; Dinis, M.A.P. Raising Awareness on Solid Waste Management through Formal Education for Sustainability: A Developing Countries. *Recycling* **2021**, *6*, 6. [CrossRef]
55. Haus, S.; Björnsson, L.; Börjesson, P. Lignocellulosic Ethanol in a Greenhouse Gas Emission Reduction Obligation System—A Case Study of Swedish Sawdust Based-Ethanol Production. *Energies* **2020**, *13*, 1048. [CrossRef]
56. Singh, K.; Kumar, R.; Chaudhary, V.; Sunil, V.; Arya, A.M.; Sharma, S. Sugarcane Bagasse: Foreseeable Biomass of Bio - Products and Biofuel: An Overview. *Pharmacogn. Phytochem.* **2019**, *8*, 2356–2360.
57. Formann, S.; Hahn, A.; Janke, L.; Stinner, W.; Sträuber, H.; Logroño, W.; Nikolausz, M. Beyond Sugar and Ethanol Production: Value Generation Opportunities through Sugarcane Residues. *Front. Energy Res.* **2020**, *8*, 579577. [CrossRef]
58. Ntimbani, R.N.; Farzad, S.; Görgens, J.F. Furfural Production from Sugarcane Bagasse along with Co-Production of Ethanol from Furfural Residues. *Biomass Convers. Biorefinery* **2021**, *12*, 5257–5267. [CrossRef]
59. Cotana, F.; Cavalaglio, G.; Gelosia, M.; Coccia, V.; Petrozzi, A.; Ingles, D.; Pompili, E. A Comparison between SHF and SSSF Processes from Cardoon for Ethanol Production. *Ind. Crop. Prod.* **2015**, *69*, 424–432. [CrossRef]
60. Ji, L.; Zheng, T.; Zhao, P.; Zhang, W.; Jiang, J. Ethanol Production from a Biomass Mixture of Furfural Residues with Green Liquor-Peroxide Saccarified Cassava Liquid. *BMC Biotechnol.* **2016**, *16*, 48. [CrossRef]
61. da Silva, F.L.; de Oliveira Campos, A.; dos Santos, D.A.; Batista Magalhães, E.R.; de Macedo, G.R.; dos Santos, E.S. Valorization of an Agroextractive Residue-Carnauba Straw-for the Production of Bioethanol by Simultaneous Saccharification and Fermentation (SSF). *Renew. Energy* **2018**, *127*, 661–669. [CrossRef]
62. Cho, E.J.; Park, C.S.; Bae, H.J. Transformation of Cheaper Mangosteen Pericarp Waste into Bioethanol and Chemicals. *J. Chem. Technol. Biotechnol.* **2020**, *95*, 348–355. [CrossRef]
63. Idris, M.; Novalia, U. Experimental Study of Bioethanol Production as Fuel from Salacca Zalacca Saste and Coconut Water Waste Combination. *IOP Conf. Ser. Mater. Sci. Eng.* **2019**, *506*, 012003. [CrossRef]
64. Rahman, Q.M.; Zhang, B.; Wang, L.; Joseph, G.; Shahbazi, A. A Combined Fermentation and Ethanol-Assisted Liquefaction Process to Produce Biofuel from *Nannochloropsis* sp. *Fuel* **2019**, *238*, 159–165. [CrossRef]

65. Chitranshi, R.; Kapoor, R. Utilization of over—Ripened Fruit (Waste Fruit) for the Eco - Friendly Production of Ethanol. *Vegetos* **2021**, *34*, 270–276. [CrossRef]
66. Al-Azkawi, A.; Elliston, A.; Al-bahry, S.; Sivakumar, N. Waste Paper to Bioethanol: Current and Future Prospective. *Biofuels Bioprod. Biorefining* **2019**, *13*, 1106–1118. [CrossRef]
67. Annamalai, N.; Al, H.; Nair, B.S.; Ahlam, A.; Azkawi, A.; Al, S. Enhanced Bioethanol Production from Waste Paper through Separate Hydrolysis and Fermentation. *Waste Biomass Valoriz.* **2020**, *11*, 121–131. [CrossRef]
68. Jha, P.; Singh, S.; Raghuram, M.; Nair, G.; Jobby, R.; Gupta, A.; Desai, N. Valorisation of Orange Peel: Supplement in Fermentation Media for Ethanol Production and Source of Limonene. *Environ. Sustain.* **2019**, *2*, 33–41. [CrossRef]
69. Jiménez-Castro, M.P.; Buller, L.S.; Sganzerla, W.G.; Forster-Carneiro, T. Bioenergy Production from Orange Industrial Waste: A Case Study. *Biofuels Bioprod. Biorefining* **2020**, *14*, 1239–1253. [CrossRef]
70. Mesa, L.; Martínez, Y.; Celia de Armas, A.; González, E. Ethanol Production from Sugarcane Straw Using Different Configurations of Fermentation and Techno-Economical Evaluation of the Best Schemes. *Renew. Energy* **2020**, *156*, 377–388. [CrossRef]
71. Salina, F.H.; Molina, F.B.; Gallego, A.G.; Palacios-Bereche, R. Fast Pyrolysis of Sugarcane Straw and Its Integration into the Conventional Ethanol Production Process through Pinch Analysis. *Energy* **2021**, *215*, 119066. [CrossRef]
72. Zhang, C.; Chen, H.; Pang, S.; Su, C.; Lv, M.; An, N.; Wang, K.; Cai, D.; Qin, P. Importance of Redefinition of Corn Stover Harvest Time to Enhancing Non-Food Bio-Ethanol Production. *Renew. Energy* **2020**, *146*, 1444–1450. [CrossRef]
73. Li, Q.; Qin, Y.; Liu, Y.; Liu, J.; Liu, Q.; Li, P.; Liu, L. Detoxification and Concentration of Corn Stover Hydrolysate and Its Fermentation for Ethanol Production. *Front. Chem. Sci. Eng.* **2019**, *13*, 140–151. [CrossRef]
74. da Siva Martins, L.H.; Komesu, A.; Neto, J.M.; de Oliveira, J.A.R.; Rabelo, S.C.; da Costa, A.C. Pretreatment of Sugarcane Bagasse by OX-B to Enhancing the Enzymatic Hydrolysis for Ethanol Fermentation. *J. Food Process Eng.* **2021**, *44*, 13579. [CrossRef]
75. Konde, K.S.; Nagarajan, S.; Kumar, V.; Patil, S.V.; Ranade, V.V. Sugarcane Bagasse Based Biorefineries in India: Potential and Challenges. *Sustain. Energy Fuels* **2021**, *5*, 52–78. [CrossRef]
76. Jin, X.; Song, J.; Liu, G. Bioethanol Production from Rice Straw through an Enzymatic Route Mediated by Enzymes Developed In-House from Aspergillus Fumigatus. *Energy* **2020**, *190*, 116395. [CrossRef]
77. Tajmirriahi, M.; Momayez, F.; Karimi, K. The Critical Impact of Rice Straw Extractives on Biogas and Bioethanol. *Bioresour. Technol.* **2021**, *319*, 124167. [CrossRef] [PubMed]
78. Zwirzitz, A.; Alteio, L.; Sulzenbacher, D.; Atanasoff, M.; Selg, M. Ethanol Production from Wheat Straw Hydrolysate by Issatchenkia Orientalis Isolated from Waste Cooking Oil. *J. Fungi* **2021**, *7*, 121. [CrossRef] [PubMed]
79. Hermosilla, E.; Schalchli, H.; Diez, M.C. Biodegradation Inducers to Enhance Wheat Straw Pretreatment by Gloeophyllum Trabeum to Second-Generation Ethanol Production. *Environ. Sci. Pollut. Res.* **2020**, *27*, 8467–8480. [CrossRef] [PubMed]
80. Fu, J.; Yan, X.; Jiang, D. Assessing the Sweet Sorghum—Based Ethanol Potential on Saline—Alkali Land with DSSAT Model and LCA Approach. *Biotechnol. Biofuels* **2021**, *14*, 44. [CrossRef] [PubMed]
81. Rakhmetova, S.O.; Vergun, O.M.; Blume, R.Y.; Bondarchuk, O.P.; Shymanska, O.V.; Tsygankov, S.P.; Yemets, A.I.; Blume, Y.B.; Rakhmetov, D.B. Ethanol Production Potential of Sweet Sorghum in North and Central Ukraine. *Open Agric. J.* **2020**, *14*, 321–338. [CrossRef]
82. Sukhang, S.; Choojit, S.; Reungpeerakul, T.; Sangwichien, C. Bioethanol Production from Oil Palm Empty Fruit Bunch with SSF and SHF Processes Using Kluyveromyces Marxianus Yeast. *Cellulose* **2020**, *27*, 301–314. [CrossRef]
83. Pangsang, N.; Rattanapan, U.; Thanapimmetha, A.; Srinopphakhun, P.; Liu, C.-G.; Zhao, X.-Q.; Bai, F.-W.; Sakdaronnarong, C. Chemical-Free Fractionation of Palm Empty Fruit Bunch and Palm Fiber by Hot-Compressed Water Technique for Ethanol Production. *Energy Rep.* **2019**, *5*, 337–348. [CrossRef]
84. Redondo-Gómez, C.; Quesada, M.R.; Astúa, S.V.; Zamora, J.P.M.; Lopretti, M.; Vega-Baudrit, J.R. Biorefinery of Biomass of Agro-Industrial Banana Waste to Obtain High-Value Biopolymers. *Molecules* **2020**, *25*, 3829. [CrossRef] [PubMed]
85. Utama, G.L.; Kurniawan, M.O.; Natiqoh, N.; Balia, R.L. Species Identification of Stress Resistance Yeasts Isolated from Banana Waste for Ethanol Production. *IOP Conf. Ser. Earth Environ. Sci.* **2019**, *306*, 12021. [CrossRef]
86. Karagoz, P.; Bill, R.M.; Ozkan, M. Lignocellulosic Ethanol Production: Evaluation of New Approaches, Cell Immobilization and Reactor Configurations. *Renew. Energy* **2019**, *143*, 741–752. [CrossRef]
87. Arora, A.; Nandal, P.; Singh, J.; Verma, M.L. Nanobiotechnological Advancements in Lignocellulosic Biomass Pretreatment. *Mater. Sci. Energy Technol.* **2020**, *3*, 308–318. [CrossRef]
88. Okolie, J.A.; Nanda, S.; Dalai, A.K.; Kozinski, J.A. Chemistry and Specialty Industrial Applications of Lignocellulosic Biomass. *Waste Biomass Valoriz.* **2020**, *12*, 2145–2169. [CrossRef]
89. Gopal, L.C.; Govindarajan, M.; Kavipriya, M.R.; Mahboob, S.; Al-ghanim, K.A.; Virik, P.; Ahmed, Z.; Al-mulhm, N.; Senthilkumaran, V.; Shankar, V. Science Optimization Strategies for Improved Biogas Production by Recycling of Waste through Response Surface Methodology and Artificial Neural Network: Sustainable Energy Perspective Research. *J. King Saud Univ.-Sci.* **2021**, *33*, 101241. [CrossRef]
90. Nishimura, H.; Kamiya, A.; Nagata, T.; Katahira, M.; Watanabe, T. Direct Evidence for α Ether Linkage between Lignin and Carbohydrates in Wood Cell Walls. *Sci. Rep.* **2018**, *8*, 6538. [CrossRef]
91. Santos, C.M.; Brito, P.L.; de Oliveira Santos, A.T.; de Araújo Pantoja, L.; da Costa, A.S.V.; dos Santos, A.S. Production of Lignocellulosic Ethanol from Waste Paper: Review on Production Technology. *Int. J. Dev. Res.* **2019**, *9*, 28383–28390.

92. Peinemann, J.C.; Pleissner, D. Continuous Pretreatment, Hydrolysis, and Fermentation of Organic Residues for the Production of Biochemicals. *Bioresour. Technol.* **2020**, *295*, 122256. [CrossRef]
93. Dey, P.; Pal, P.; Kevin, J.D.; Das, D.B. Lignocellulosic Bioethanol Production: Prospects of Emerging Membrane Technologies to Improve the Process—A Critical Review. *Rev. Chem. Eng.* **2020**, *36*, 333–367. [CrossRef]
94. Deng, L.; Li, J. Thread Rolling: An Efficient Mechanical Pretreatment for Corn Stover Saccharificatio. *Energies* **2021**, *14*, 542. [CrossRef]
95. Bhatia, L.; Garlapati, V.K.; Chandel, A.K. *Scalable Technologies for Lignocellulosic Biomass Processing into Cellulosic Ethanol*; Springer International Publishing: Cham, Swizerland, 2019; ISBN 9783030290696.
96. Cheah, W.Y.; Sankaran, R.; Show, P.L.; Ibrahim, T.N.B.T.; Chew, K.W.; Culaba, A.; Chang, J.S. Pretreatment Methods for Lignocellulosic Biofuels Production: Current Advances, Challenges and Future Prospects. *Biofuel Res. J.* **2020**, *7*, 1115–1127. [CrossRef]
97. Vasic, K.; Knez, Ž.; Leitgeb, M. Bioethanol Production by Enzymatic Hydrolysis from Different Lignocellulosic Sources. *Molecules* **2021**, *26*, 753. [CrossRef] [PubMed]
98. Aftab, M.N.; Iqbal, I.; Riaz, F.; Karadag, A.; Tabatabaei, M. Different Pretreatment Methods of Lignocellulosic Biomass for Use in Biofuel Production. In *Biomass for Bioenergy-Recent Trends and Future Challenges*; IntechOpen: London, UK, 2019; pp. 1–24. [CrossRef]
99. Shukla, A.; Kumar, D.; Girdhar, M.; Kumar, A.; Goyal, A.; Malik, T.; Mohan, A. Strategies of Pretreatment of Feedstocks for Optimized Bioethanol Production: Distinct and Integrated Approaches. *Biotechnol. Biofuels Bioprod.* **2023**, *16*, 44. [CrossRef]
100. Mustafa, A.H.; Rashid, S.S.; Rahim, M.H.A.; Roslan, R.; Musa, W.A.M.; Sikder, B.H.; Sasi, A.A. Enzymatic Pretreatment of Lignocellulosic Biomass: An Overview. *J. Chem. Eng. Ind. Biotechnol.* **2022**, *8*, 1–7. [CrossRef]
101. Sharma, S.; Tsai, M.-L.; Sharma, V.; Sun, P.-P.; Nargotra, P.; Bajaj, B.K.; Chen, C.-W.; Dong, C.-D. Environment Friendly Pretreatment Approaches for the Bioconversion of Lignocellulosic Biomass into Biofuels and Value-Added Products. *Environments* **2022**, *10*, 6. [CrossRef]
102. Klongklaew, A.; Unban, K.; Kalaimurugan, D.; Kanpiengjai, A.; Azaizeh, H.; Schroedter, L.; Venus, J.; Khanongnuch, C. Bioconversion of Dilute Acid Pretreated Corn Stover to L-Lactic Acid Using Co-Culture of Furfural Tolerant Enterococcus Mundtii WX1 and Lactobacillus Rhamnosus SCJ9. *Fermentation* **2023**, *9*, 112. [CrossRef]
103. Gong, J.-S.-Q.; Su, J.-E.; Cai, J.-Y.; Zou, L.; Chen, Y.; Jiang, Y.-L.; Hu, B.-B. Enhanced Enzymolysis and Bioethanol Yield from Tobacco Stem Waste Based on Mild Synergistic Pretreatment. *Front. Energy Res.* **2023**, *10*, 989393. [CrossRef]
104. Usino, D.O.; Sar, T.; Ylitervo, P.; Richards, T. Effect of Acid Pretreatment on the Primary Products of Biomass Fast Pyrolysis. *Energies* **2023**, *16*, 2377. [CrossRef]
105. Abdurrahman, A.; Richard, P.; Ibrahim Galadima, A.; Ubaida Muhammad, A.; Adamu, M. Dilute Sulphuric Acid Pre-Treatment for Efficient Production of Bioethanol from Sugarcane Bagasse Using Saccharomyces Cerevisiae. *J. Biotechnol.* **2022**, *1*, 56–65. [CrossRef]
106. Mazlan, N.A.; Samad, K.A.; Yussof, H.W.; Samah, R.A.; Jahim, J.M. Xylan Recovery from Dilute Nitric Acid Pretreated Oil Palm Frond Bagasse Using Fractional Factorial Design. *J. Oil Palm Res.* **2021**, *33*, 307–319. [CrossRef]
107. Woiciechowski, A.L.; Neto, C.J.D.; Vandenberghe, L.P.d.S.; Neto, D.P.d.C.; Sydney, A.C.N.; Letti, L.A.J.; Karp, S.G.; Torres, L.A.Z.; Soccol, C.R. Lignocellulosic Biomass: Acid and Alkaline Pretreatments and Their Effects on Biomass Recalcitrance—Conventional Processing and Recent Advances. *Bioresour. Technol.* **2020**, *304*, 122848. [CrossRef] [PubMed]
108. Eswari, A.P.; Ravi, Y.K.; Kavitha, S.; Banu, J.R. Recent Insight into Anaerobic Digestion of Lignocellulosic Biomass for Cost Effective Bioenergy Generation. *Adv. Electr. Eng. Electron. Energy* **2023**, *3*, 100119. [CrossRef]
109. Zhang, R.; Gao, H.; Wang, Y.; He, B.; Lu, J.; Zhu, W.; Peng, L.; Wang, Y. Challenges and Perspectives of Green-like Lignocellulose Pretreatments Selectable for Low-Cost Biofuels and High-Value Bioproduction. *Bioresour. Technol.* **2023**, *369*, 128315. [CrossRef]
110. Le Tan, N.T.; Dam, Q.P.; Mai, T.P.; Nguyen, Q.D. The Combination of Acidic and Alkaline Pretreatment for a Lignocellulose Material in Simultaneous Saccharification and Fermentation (SSF) Process. *Chem. Eng. Trans.* **2021**, *89*, 43–48. [CrossRef]
111. Siddique, M.; Mengal, A.N.; Khan, S.; Khan, L.A.; Kakar, E.K. Pretreatment of Lignocellulosic Biomass Conversion into Biofuel and Biochemical: A Comprehensive Review. *J. Biol. Med.* **2023**, *8*, 39–43. [CrossRef]
112. Li, P.; Yang, C.; Jiang, Z.; Jin, Y.; Wu, W. Lignocellulose Pretreatment by Deep Eutectic Solvents and Related Technologies: A Review. *J. Bioresour. Bioprod.* **2023**, *8*, 33–44. [CrossRef]
113. Tareen, A.K.; Punsuvon, V.; Parakulsuksatid, P. Investigation of Alkaline Hydrogen Peroxide Pretreatment to Enhance Enzymatic Hydrolysis and Phenolic Compounds of Oil Palm Trunk. *3 Biotech* **2020**, *10*, 179. [CrossRef]
114. Belay, J.B.; Habtu, N.G.; Ancha, V.R.; Hussen, A.S. Alkaline Hydrogen Peroxide Pretreatment of Cladodes of Cactus (Opuntia Ficus-Indica) for Biogas Production. *Heliyon* **2021**, *7*, e08002. [CrossRef]
115. Yang, L.; Ru, Y.; Xu, S.; Liu, T.; Tan, L. Features Correlated to Improved Enzymatic Digestibility of Corn Stover Subjected to Alkaline Hydrogen Peroxide Pretreatment. *Bioresour. Technol.* **2021**, *325*, 124688. [CrossRef]
116. Damaurai, J.; Preechakun, T.; Raita, M.; Champreda, V.; Laosiripojana, N. Investigation of Alkaline Hydrogen Peroxide in Aqueous Organic Solvent to Enhance Enzymatic Hydrolysis of Rice Straw. *BioEnergy Res.* **2021**, *14*, 122–134. [CrossRef]
117. Broda, M.; Yelle, D.J.; Serwanska, K. Bioethanol Production from Lignocellulosic Biomass—Challenges and Solutions. *Molecules* **2022**, *27*, 8717. [CrossRef]

118. Halim, F.N.B.A.; Taheri, A.; Yassin, Z.A.R.; Chia, K.F.; Goh, K.K.T.; Goh, S.M.; Du, J. Effects of Incorporating Alkaline Hydrogen Peroxide Treated Sugarcane Fibre on the Physical Properties and Glycemic Potency of White Bread. *Foods* **2023**, *12*, 1460. [CrossRef] [PubMed]
119. Chin, D.W.K.; Lim, S.; Pang, Y.L.; Lim, C.H.; Shuit, S.H.; Lee, K.M.; Chong, C.T. Effects of Organic Solvents on the Organosolv Pretreatment of Degraded Empty Fruit Bunch for Fractionation and Lignin Removal. *Sustainability* **2021**, *13*, 6757. [CrossRef]
120. Parchami, M.; Agnihotri, S.; Taherzadeh, M.J. Aqueous Ethanol Organosolv Process for the Valorization of Brewer's Spent Grain (BSG). *Bioresour. Technol.* **2022**, *362*, 127764. [CrossRef] [PubMed]
121. Adamcyk, J.; Beisl, S.; Friedl, A. High Temperature Lignin Separation for Improved Yields in Ethanol Organosolv Pre-Treatment. *Sustainability* **2023**, *15*, 3006. [CrossRef]
122. Chin, D.W.K.; Lim, S.; Pang, Y.L. Fundamental Review of Organosolv Pretreatment and Its Challenges in Emerging Consolidated Bioprocessing. *Biofuels Bioprod. Biorefining* **2020**, *14*, 808–829. [CrossRef]
123. Siacor, F.D.C.; Tabañag, I.D.F.; Lobarbio, C.F.Y.; Taboada, E.B. Effects of Aqueous Ethanol Concentration and Solid-to-Liquid Ratio in the Extraction of Organosolv Lignin from Mango (*Mangifera indica* L.) Seed Husk. *Sci. Technol. Asia* **2021**, *26*, 34–35. [CrossRef]
124. Ifeanyi-nze, F.O.; Omiyale, C.O. Insights into the Recent Advances in the Pretreatment of Biomass for Sustainable Bioenergy and Bio-Products Synthesis: Challenges and Future Directions. *Eur. J. Sustain. Dev. Res.* **2023**, *7*, em0209. [CrossRef]
125. Rabelo, S.C.; Nakasu, P.Y.S.; Scopel, E.; Araújo, M.F.; Cardoso, L.H.; da Costa, A.C. Organosolv Pretreatment for Biorefineries: Current Status, Perspectives, and Challenges. *Bioresour. Technol.* **2023**, *369*, 128331. [CrossRef]
126. Wang, R.; Yue, J.; Jiang, J.; Li, J.; Zhao, J.; Xia, H.; Wang, K.; Xu, J. Hydrothermal CO_2-Assisted Pretreatment of Wheat Straw for Hemicellulose Degradation Followed with Enzymatic Hydrolysis for Glucose Production. *Waste Biomass Valoriz.* **2020**, *12*, 1483–1492. [CrossRef]
127. Cano, M.E.; García-Martin, A.; Morales, P.C.; Wojtusik, M.; Santos, V.E.; Kovensky, J.; Ladero, M. Production of Oligosaccharides from Agrofood Wastes. *Fermentation* **2020**, *6*, 31. [CrossRef]
128. Nwachukwu, B.C.; Ayangbenro, A.S.; Babalola, O.O. Elucidating the Rhizosphere Associated Bacteria for Environmental Sustainability. *Agriculture* **2021**, *11*, 75. [CrossRef]
129. Parapouli, M.; Vasileiadis, A.; Afendra, A.S.; Hatziloukas, E. Saccharomyces Cerevisiae and Its Industrial Applications. *AIMS Microbiol.* **2020**, *6*, 1–31. [CrossRef]
130. Sharma, S.; Arora, A. Tracking Strategic Developments for Conferring Xylose Utilization / Fermentation by Saccharomyces Cerevisiae. *Ann. Microbiol.* **2020**, *70*, 50. [CrossRef]
131. Pacho, E.R.; Vaskan, P.; Gorgens, J.F.; Gnansounou, E. Process Design, Techno-Economic, and Life-Cycle Assessments of Selected Sugarcane-Based Biorefineries: A Case Study in the South African Context. In *Refining Biomass Residues for Sustainable Energy and Bioproducts*; Elsevier: Amsterdam, The Netherlands, 2020; pp. 567–597. [CrossRef]
132. Hernández, D.; Rebolledo-Leiva, R.; Fern, H.; Quinteros-Lama, H.; Cataldo, F.; Muñoz, E.; Tenreiro, C. Recovering Apple Agro-Industrial Waste for Bioethanol and Vinasse Joint Production: Screening the Potential of Chile. *Fermentation* **2021**, *7*, 203. [CrossRef]
133. Ma, S.; Dong, C.; Hu, X.; Xue, J.; Zhao, Y.; Wang, X. Techno-Economic Evaluation of a Combined Biomass Gasification-Solid Oxide Fuel Cell System for Ethanol Production via Syngas Fermentation. *Fuel* **2022**, *324*, 124395. [CrossRef]
134. Zentou, H.; Abidin, Z.Z.; Yunus, R.; Biak, D.R.A.; Korelskiy, D. Overview of Alternative Ethanol Removal Techniques for Enhancing Bioethanol Recovery from Fermentation Broth. *Processes* **2019**, *7*, 458. [CrossRef]
135. Do Thi, H.T.; Mizsey, P.; Toth, A.J. Separation of Alcohol-Water Mixtures by a Combination of Distillation, Hydrophilic and Organophilic Pervaporation Processes. *Membranes* **2020**, *10*, 345. [CrossRef]
136. Tareen, A.K.; Sultan, I.N.; Songprom, K.; Laemsak, N.; Sirisansaneeyakul, S.; Vanichsriratana, W.; Parakulsuksatid, P. Two-Step Pretreatment of Oil Palm Trunk for Ethanol Production by Thermotolerant Saccharomyces Cerevisiae SC90. *Bioresour. Technol.* **2021**, *320*, 124298. [CrossRef]
137. Morales, M.; Arvesen, A.; Cherubini, F. Integrated Process Simulation for Bioethanol Production: Effects of Varying Lignocellulosic Feedstocks on Technical Performance. *Bioresour. Technol.* **2021**, *328*, 124833. [CrossRef]
138. Cunha, J.T.; Romaní, A.; Inokuma, K.; Johansson, B.; Hasunuma, T.; Kondo, A.; Domingues, L. Consolidated Bioprocessing of Corn Cob - Derived Hemicellulose: Engineered Industrial Saccharomyces Cerevisiae as Efficient Whole Cell Biocatalysts. *Biotechnol. Biofuels* **2020**, *13*, 138. [CrossRef]
139. Baig, K.S. Interaction of Enzymes with Lignocellulosic Materials: Causes, Mechanism and Influencing Factors. *Bioresour. Bioprocess.* **2020**, *7*, 21. [CrossRef]
140. Marnoto, T.; Budiaman, I.G.S.; Hapsari, C.R.; Prakosa, R.A.Y.; Arifin, K. Dehydrating Ethanol Using a Ternary Solute and Extractive Batch Distillation. *Malays. J. Anal. Sci.* **2019**, *23*, 124–130. [CrossRef]
141. Nakamura, M.; Noguchi, K. Tolerant Mechanisms to O_2 Deficiency under Submergence Conditions in Plants. *J. Plant Res.* **2020**, *133*, 343–371. [CrossRef] [PubMed]
142. Galvanauskas, V.; Simutis, R.; Levišauskas, D.; Urniežius, R. Practical Solutions for Specific Growth Rate Control Systems in Industrial Bioreactors. *Processes* **2019**, *7*, 693. [CrossRef]
143. Gomes, D.; Cruz, M.; de Resende, M.; Ribeiro, E.; Teixeira, J.; Domingues, L. Very High Gravity Bioethanol Revisited: Main Challenges and Advances. *Fermentation* **2021**, *7*, 38. [CrossRef]

144. Veloso, I.I.K.; Rodrigues, K.C.S.; Sonego, J.L.S.; Cruz, A.J.G.; Badino, A.C. Fed-Batch Ethanol Fermentation at Low Temperature as a Way to Obtain Highly Concentrated Alcoholic Wines: Modeling and Optimization. *Biochem. Eng. J.* **2019**, *141*, 60–70. [CrossRef]
145. Knudsen, J.D.; Rønnow, B. Extended Fed-Batch Fermentation of a C5/C6 Optimised Yeast Strain on Wheat Straw Hydrolysate Using an Online Refractive Index Sensor to Measure the Relative Fermentation Rate. *Sci. Rep.* **2020**, *10*, 6705. [CrossRef]
146. Cruz, M.L.; de Resende, M.M.; Ribeiro, E.J. Improvement of Ethanol Production in Fed-Batch Fermentation Using a Mixture of Sugarcane Juice and Molasse under Very High-Gravity Conditions. *Bioprocess Biosyst. Eng.* **2021**, *44*, 617–625. [CrossRef]
147. Puligundla, P.; Smogrovicova, D.; Mok, C.; Obulam, V.S.R. A Review of Recent Advances in High Gravity Ethanol Fermentation. *Renew. Energy* **2019**, *133*, 1366–1379. [CrossRef]
148. Margono, M.; Kaavessina, M.; Mohd Zahari, M.A.K.; Hisyam, A. Continuous Bioethanol Production Using Uncontrolled Process in a Laboratory Scale of Integrated Aerobic-Anaerobic Baffled Reactor. *Period. Polytech. Chem. Eng.* **2020**, *64*, 172–178. [CrossRef]
149. Liu, Q.; Zhao, N.; Zou, Y.; Ying, H.; Liu, D.; Chen, Y. Feasibility Study on Long-Term Continuous Ethanol Production from Cassava Supernatant by Immobilized Yeast Cells in Packed Bed Reactor. *J. Microbiol. Biotechnol.* **2020**, *30*, 1227–1234. [CrossRef] [PubMed]
150. Degweker, G.; Lali, A. High Productivity Ethanol Fermentation of Glucose & Xylose Using Membrane Assisted Continuous Cell Recycle. *Sustain. Chem. Eng.* **2021**, *2*, 8–19. [CrossRef]
151. Benevenuti, C.; Branco, M.; Do Nascimento-Correa, M.; Botelho, A.; Ferreira, T.; Amaral, P. Residual Gas for Ethanol Production by Clostridium Carboxidivorans in a Dual Impeller Stirred Tank Bioreactor (STBR). *Fermentation* **2021**, *7*, 199. [CrossRef]
152. Pacheco, M.; Moura, P.; Silva, C. A Systematic Review of Syngas Bioconversion to Value-Added Products from 2012 to 2022. *Energies* **2023**, *16*, 3241. [CrossRef]
153. Rückel, A.; Hannemann, J.; Maierhofer, C.; Fuchs, A.; Weuster-Botz, D. Studies on Syngas Fermentation with Clostridium Carboxidivorans in Stirred-Tank Reactors with Defined Gas Impurities. *Front. Microbiol.* **2021**, *12*, 655390. [CrossRef]
154. Fernández-Blanco, C.; Robles-Iglesias, R.; Naveira-Pazos, C.; Veiga, M.C.; Kennes, C. Production of Biofuels from C1-Gases with Clostridium and Related Bacteria—Recent Advances. *Microb. Biotechnol.* **2023**, *16*, 726–741. [CrossRef]
155. Bäumler, M.; Schneider, M.; Ehrenreich, A.; Liebl, W.; Weuster-Botz, D. Synthetic Co-Culture of Autotrophic Clostridium Carboxidivorans and Chain Elongating Clostridium Kluyveri Monitored by Flow Cytometry. *Microb. Biotechnol.* **2022**, *15*, 1471–1485. [CrossRef]
156. Mann, M.; Effert, D.; Kottenhahn, P.; Hüser, A.; Philipps, G.; Jennewein, S.; Büchs, J. Impact of Different Trace Elements on Metabolic Routes during Heterotrophic Growth of C. Ljungdahlii Investigated through Online Measurement of the Carbon Dioxide Transfer Rate. *Biotechnol. Prog.* **2022**, *38*, e3263. [CrossRef]
157. Lakhssassi, N.; Baharlouei, A.; Meksem, J.; Hamilton-Brehm, S.D.; Lightfoot, D.A.; Meksem, K.; Liang, Y. EMS-Induced Mutagenesis of Clostridium Carboxidivorans for Increased Atmospheric CO_2 Reduction Efficiency and Solvent Production. *Microorganisms* **2020**, *8*, 1239. [CrossRef]
158. Pavan, M.; Reinmets, K.; Garg, S.; Mueller, A.P.; Marcellin, E.; Köpke, M.; Valgepea, K. Advances in Systems Metabolic Engineering of Autotrophic Carbon Oxide-Fixing Biocatalysts towards a Circular Economy. *Metab. Eng.* **2022**, *71*, 117–141. [CrossRef] [PubMed]
159. Vees, C.A.; Herwig, C.; Pflügl, S. Mixotrophic Co-Utilization of Glucose and Carbon Monoxide Boosts Ethanol and Butanol Productivity of Continuous Clostridium Carboxidivorans Cultures. *Bioresour. Technol.* **2022**, *353*, 127138. [CrossRef]
160. Liu, K.; Phillips, J.R.; Sun, X.; Mohammad, S.; Huhnke, R.L.; Atiyeh, H.K. Investigation and Modeling of Gas-Liquid Mass Transfer in a Sparged and Non-Sparged Continuous Stirred Tank Reactor with Potential Application in Syngas Fermentation. *Fermentation* **2019**, *5*, 75. [CrossRef]
161. Elisiário, M.P.; De Wever, H.; Van Hecke, W.; Noorman, H.; Straathof, A.J.J. Membrane Bioreactors for Syngas Permeation and Fermentation. *Crit. Rev. Biotechnol.* **2022**, *42*, 856–872. [CrossRef]
162. Puiman, L.; Abrahamson, B.; van der Lans, R.G.J.M.; Haringa, C.; Noorman, H.J.; Picioreanu, C. Alleviating Mass Transfer Limitations in Industrial External-Loop Syngas-to-Ethanol Fermentation. *Chem. Eng. Sci.* **2022**, *259*, 117770. [CrossRef]
163. Sajeev, E.; Shekher, S.; Ogbaga, C.C.; Desongu, K.S.; Gunes, B.; Okolie, J.A. Application of Nanoparticles in Bioreactors to Enhance Mass Transfer during Syngas Fermentation. *Encyclopedia* **2023**, *3*, 387–395. [CrossRef]
164. Sun, X.; Atiyeh, H.K.; Huhnke, R.L.; Tanner, R.S. Syngas Fermentation Process Development for Production of Biofuels and Chemicals: A Review. *Bioresour. Technol. Rep.* **2019**, *7*, 100279. [CrossRef]
165. Benevenuti, C.; Amaral, P.; Ferreira, T.; Seidl, P. Impacts of Syngas Composition on Anaerobic Fermentation. *Reactions* **2021**, *2*, 391–407. [CrossRef]
166. Prasoulas, G.; Gentikis, A.; Konti, A.; Kalantzi, S.; Kekos, D.; Mamma, D. Bioethanol Production from Food Waste Applying the Multienzyme System Produced On-Site by Fusarium Oxysporum F3 and Mixed Microbial Cultures. *Fermentation* **2020**, *6*, 39. [CrossRef]
167. Barahona, P.P.; Mayorga, B.B.; Martín-Gil, J.; Martín-Ramos, P.; Barriga, E.J.C. Cellulosic Ethanol: Improving Cost Efficiency by Coupling Semi-Continuous Fermentation and Simultaneous Saccharification Strategies. *Processes* **2020**, *8*, 1459. [CrossRef]
168. Hemansi; Kaushik, A.; Yadav, G.; Saini, J.K. Simultaneous Sacchari Fi Cation and Fermentation of Sequential Dilute Acid-Alkali Pretreated Cotton (*Gossypium hirsutum* L.) Stalk for Cellulosic Ethanol Production. *J. Chem. Technol. Biotechnol.* **2021**, *97*, 534–542. [CrossRef]

169. Yong, K.J.; Wu, T.Y. Second-Generation Bioenergy from Oilseed Crop Residues: Recent Technologies, Techno-Economic Assessments and Policies. *Energy Convers. Manag.* **2022**, *267*, 115869. [CrossRef]
170. David, A.N.; Sewsynker-Sukai, Y.; Sithole, B.; Kana, E.B.G. Development of a Green Liquor Dregs Pretreatment for Enhanced Glucose Recovery from Corn Cobs and Kinetic Assessment on Various Bioethanol Fermentation Types. *Fuel* **2020**, *274*, 117797. [CrossRef]
171. Tareen, A.K.; Punsuvon, V.; Sultan, I.N.; Khan, M.W.; Parakulsuksatid, P. Cellulase Addition and Pre-Hydrolysis Effect of High Solid Fed-Batch Simultaneous Saccharification and Ethanol Fermentation from a Combined Pretreated Oil Palm Trunk. *ACS Omega* **2021**, *6*, 26119–26129. [CrossRef]
172. Palacios, A.S.; Ilyina, A.; Ramos-gonzález, R.; Aguilar, C.N.; Martínez-hernández, J.L.; Segura-ceniceros, E.P.; Lizeth, M.; González, C.; Aguilar, M.; Ballesteros, M.; et al. Ethanol Production from Banana Peels at High Pretreated Substrate Loading: Comparison of Two Operational Strategies. *Biomass Convers. Biorefinery* **2019**, *11*, 1587–1596. [CrossRef]
173. Paschos, T.; Louloudi, A.; Papayannakos, N.; Kekos, D.; Mamma, D. Potential of Barley Straw for High Titer Bioethanol Production Applying Pre-Hydrolysis and Simultaneous Saccharification and Fermentation at High Solid Loading Pre-Hydrolysis and Simultaneous Saccharification and Fermentation at High. *Biofuels* **2020**, *13*, 467–473. [CrossRef]
174. Aparicio, E.; Rodríguez-Jasso, R.M.; Pinales-Marquez, C.D.; Loredo-Trevino, A.; Robledo-Olivo, A.; Aguilar, C.N.; Kostas, E.T.; Ruiz, H.A. High-Pressure Technology for Sargassum Spp Biomass Pretreatment and Fractionation in the Third Generation of Bioethanol Production. *Bioresour. Technol.* **2021**, *329*, 124935. [CrossRef]
175. Rodríguez-Martínez, B.; Coelho, E.; Gullon, B.; Yanez, R.; Domingues, L. Potato Peels Waste as a Sustainable Source for Biotechnological Production of Biofuels: Process Optimization. *Waste Manag.* **2023**, *155*, 320–328. [CrossRef]
176. Zhu, J.Q.; Zong, Q.J.; Li, W.C.; Chai, M.Z.; Xu, T.; Liu, H.; Fan, H.; Li, B.Z.; Yuan, Y.J. Temperature Profiled Simultaneous Saccharification and Co-Fermentation of Corn Stover Increases Ethanol Production at High Solid Loading. *Energy Convers. Manag.* **2020**, *205*, 112344. [CrossRef]
177. Tu, W.-L.; Ma, T.-Y.; Ou, C.-M.; Guo, G.-L.; Chao, Y. Simultaneous Saccharification and Co-Fermentation with a Thermotolerant Saccharomyces Cerevisiae to Produce Ethanol from Sugarcane Bagasse under High Temperature Conditions. *BioResources* **2021**, *16*, 1358–1372. [CrossRef]
178. Pinheiro, T.; Ying, K.; Lip, F.; Ríos, E.G.; Querol, A.; Teixeira, J. Differential Proteomic Analysis by SWATH - MS Unravels the Most Dominant Mechanisms Underlying Yeast Adaptation to Non-Optimal Temperatures under Anaerobic Conditions. *Sci. Rep.* **2020**, *10*, 22329. [CrossRef] [PubMed]
179. Fernandes, F.d.S.; de Souza, E.S.; Carneiro, L.M.; Silva, J.P.A.; de Souza, J.V.B.; Batista, J.d.S. Current Ethanol Production Requirements for the Yeast Saccharomyces Cerevisiae. *Int. J. Microbiol.* **2022**, *2022*, 7878830.
180. Phong, H.X.; Klanrit, P.; Thi, N.; Dung, P.; Thanonkeo, S. High-Temperature Ethanol Fermentation from Pineapple Waste Hydrolysate and Gene Expression Analysis of Thermotolerant Yeast Saccharomyces Cerevisiae. *Sci. Rep.* **2022**, *12*, 13965. [CrossRef]
181. Yang, P.; Wu, W.; Chen, J.; Jiang, S.; Zheng, Z.; Deng, Y.; Lu, J.; Wang, H.; Zhou, Y.; Geng, Y.; et al. Thermotolerance Improvement of Engineered Saccharomyces Cerevisiae ERG5 Delta ERG4 Delta ERG3 Delta, Molecular Mechanism, and Its Application in Corn Ethanol Production. *Biotechnol. Biofuels Bioprod.* **2023**, *16*, 66. [CrossRef]
182. Rahmadhani, N.; Astuti, R.I.; Meryandini, A. Ethanol Productivity of Ethanol-Tolerant Mutant Strain Pichia Kudriavzevii R-T3 in Monoculture and Co-Culture Fermentation with Saccharomyces Cerevisiae. *J. Biosci.* **2022**, *29*, 435–444. [CrossRef]
183. Van Nguyen, P.; Nguyen, K.H.V.; Nguyen, N.L.; Ho, X.T.T.; Truong, P.H.; Nguyen, K.C.T. Lychee-Derived, Thermotolerant Yeasts for Second-Generation Bioethanol Production. *Fermentation* **2022**, *8*, 515. [CrossRef]
184. Nuanpeng, S.; Thanonkeo, S.; Klanrit, P.; Yamada, M.; Thanonkeo, P. Optimization Conditions for Ethanol Production from Sweet Sorghum Juice by Thermotolerant Yeast Saccharomyces Cerevisiae: Using a Statistical Experimental Design. *Fermentation* **2023**, *9*, 450. [CrossRef]
185. Iyyappan, J.; Pravin, R.; Al-Ghanim, K.A.; Govindarajan, M.; Marcello, N.; Baskar, G. Dual Strategy for Bioconversion of Elephant Grass Biomass into Fermentable Sugars Using Trichoderma Reesei towards Bioethanol Production. *Bioresour. Technol.* **2023**, *374*, 128804. [CrossRef]
186. Oh, E.J.; Jin, Y. Engineering of Saccharomyces Cerevisiae for Efficient Fermentation of Cellulose. *FEMS Yeast Res.* **2020**, *20*, foz089. [CrossRef]
187. Singhania, R.R.; Patel, A.K.; Singh, A.; Haldar, D.; Soam, S.; Chen, C.-W.; Tsai, M.-L.; Dong, C.-D. Consolidated Bioprocessing of Lignocellulosic Biomass: Technological Advances and Challenges. *Bioresour. Technol.* **2022**, *354*, 127153. [CrossRef]
188. Banner, A.; Toogood, H.S.; Scrutton, N.S. Consolidated Bioprocessing: Synthetic Biology Routes to Fuels and Fine Chemicals. *Microorganisms* **2021**, *9*, 1079. [CrossRef]
189. Malherbe, S.J.M.; Cripwell, R.A.; Favaro, L.; Van Zyl, W.H.; Viljoen-bloom, M. Triticale and Sorghum as Feedstock for Bioethanol Production via Consolidated Bioprocessing. *Renew. Energy* **2023**, *206*, 498–505. [CrossRef]
190. Re, A.; Mazzoli, R. Current Progress on Engineering Microbial Strains and Consortia for Production of Cellulosic Butanol through Consolidated Bioprocessing. *Microb. Biotechnol.* **2023**, *16*, 238–261. [CrossRef]
191. Kavitha, S.; Gajendran, T.; Saranya, K.; Selvakumar, P.; Manivasagan, V. Study on Consolidated Bioprocessing of Pre-Treated Nannochloropsis Gaditana Biomass into Ethanol under Optimal Strategy. *Renew. Energy* **2021**, *172*, 440–452. [CrossRef]

192. Ghazali, M.F.S.M.; Mustafa, M.; Zainudin, N.A.I.M.; Aziz, N.A.A. Consolidated Bioethanol Production Using Trichoderma Asperellum B1581. *Jordan J. Biol. Sci.* **2022**, *15*, 621–627.
193. Lao, M.; Alfafara, C.; de Leon, R. Screening of Fusarium Moniliforme as Potential Fungus for Integrated Biodelignification and Consolidated Bioprocessing of Napier Grass for Bioethanol Production. *Catalysts* **2022**, *12*, 1204. [CrossRef]
194. Kavitha, S.; Gajendran, T.; Saranya, K.; Manivasagan, V. Bioconversion of Sargassum Wightii to Ethanol via Consolidated Bioprocessing Using Lachnoclostridium Phytofermentans KSM 1203. *Fuel* **2023**, *347*, 128465. [CrossRef]
195. Joshi, J.; Dhungana, P.; Prajapati, B.; Maharjan, R.; Poudyal, P.; Yadav, M.; Mainali, M.; Yadav, A.P.; Bhattarai, T.; Sreerama, L. Enhancement of Ethanol Production in Electrochemical Cell by Saccharomyces Cerevisiae (CDBT2) and Wickerhamomyces Anomalus (CDBT7). *Front. Energy Res.* **2019**, *7*, 1–11. [CrossRef]
196. Ruchala, J.; Kurylenko, O.O.; Dmytruk, K.V.; Sibirny, A.A. Construction of Advanced Producers of First- and Second-Generation Ethanol in Saccharomyces Cerevisiae and Selected Species of Non-Conventional Yeasts (Scheffersomyces Stipitis, Ogataea Polymorpha). *Ind. Microbiol. Biotechnol.* **2020**, *47*, 109–132. [CrossRef]
197. Kim, J.H.; Ryu, J.; Huh, I.Y.; Hong, S.K.; Kang, H.A.; Chang, Y.K. Ethanol Production from Galactose by a Newly Isolated Saccharomyces Cerevisiae KL17. *Bioprocess Biosyst. Eng.* **2014**, *37*, 1871–1878. [CrossRef]
198. Nachaiwieng, W.; Lumyong, S.; Yoshioka, K.; Watanabe, T.; Khanongnuch, C. Bioethanol Production from Rice Husk under Elevated Temperature Simultaneous Saccharification and Fermentation Using Kluyveromyces Marxianus CK8. *Biocatal. Agric. Biotechnol.* **2015**, *4*, 543–549. [CrossRef]
199. Saad, M.M.E.; Amani, F.A.H.; Keong, L.C. Optimization of Bioethanol Production Process Using Oil Palm Frond Juice as Substrate. *Malays. J. Microbiol.* **2016**, *12*, 308–314.
200. Li, Y.J.; Lu, Y.Y.; Zhang, Z.J.; Mei, S.; Tan, T.W.; Fan, L.H. Co-Fermentation of Cellulose and Sucrose/Xylose by Engineered Yeasts for Bioethanol Production. *Energy Fuels* **2017**, *31*, 4061–4067. [CrossRef]
201. Abdulla, R.; Derman, E.; Ravintaran, P.T.; Jambo, S.A. Fuel Ethanol Production from Papaya Waste Using Immobilized Saccharomyces Cerevisiae. *ASM Sci. J.* **2018**, *11*, 112–123.
202. Choojit, S.; Ruengpeerakul, T.; Sangwichien, C. Optimization of Acid Hydrolysis of Pineapple Leaf Residue and Bioconversion to Ethanol by Saccharomyces Cerevisiae. *Cell. Chem. Technol.* **2018**, *52*, 247–257.
203. Demiray, E.; Karatay, S.E.; Dönmez, G. Evaluation of Pomegranate Peel in Ethanol Production by Saccharomyces Cerevisiae and Pichia Stipitis. *Energy* **2018**, *159*, 988–994. [CrossRef]
204. Nuanpeng, S.; Thanonkeo, S.; Klanrit, P.; Thanonkeo, P. Ethanol Production from Sweet Sorghum by Saccharomyces Cerevisiae DBKKUY-53 Immobilized on Alginate-Loofah Matrices. *Braz. J. Microbiol.* **2018**, *49*, 140–150. [CrossRef]
205. Pandey, A.K.; Kumar, M.; Kumari, S.; Kumari, P.; Yusuf, F.; Jakeer, S.; Naz, S.; Chandna, P.; Bhatnagar, I.; Gaur, N.A. Evaluation of Divergent Yeast Genera for Fermentation-Associated Stresses and Identification of a Robust Sugarcane Distillery Waste Isolate Saccharomyces Cerevisiae NGY10 for Lignocellulosic Ethanol Production in SHF and SSF. *Biotechnol. Biofuels* **2019**, *12*, 40. [CrossRef]
206. Cripwell, R.A.; Rose, S.H.; Favaro, L.; Van Zyl, W.H. Construction of Industrial Saccharomyces Cerevisiae Strains for the Efficient Consolidated Bioprocessing of Raw Starch. *Biotechnol. Biofuels* **2019**, *12*, 201. [CrossRef]
207. Anu; Singh, B.; Kumar, A. Process Development for Sodium Carbonate Pretreatment and Enzymatic Saccharification of Rice Straw for Bioethanol Production. *Biomass Bioenergy* **2020**, *138*, 105574. [CrossRef]
208. Wu, R.; Chen, D.; Cao, S.; Lu, Z.; Huang, J.; Lu, Q.; Chen, Y.; Chen, X.; Guan, N.; Wei, Y.; et al. Enhanced Ethanol Production from Sugarcane Molasses by Industrially Engineered: Saccharomyces Cerevisiae via Replacement of the PHO4 Gene. *RSC Adv.* **2020**, *10*, 2267–2276. [CrossRef]
209. Aderibigbe, F.A.; Amosa, M.K.; Adejumo, A.L.; Mohammed, I.A.; Mustapha, S.I.; Saka, H.B.; Tijani, I.A.; Olufowora, F.O.; Bello, B.T.; Owolabi, R.U.; et al. Optimized Production of Bioethanol by Fermentation of Acid Hydrolyzed-Corn Stover Employing Saccharomyces Cerevisiae Yeast Strain. *J. Eng. Technol.* **2021**, *15*, 93–98.
210. Dhandayuthapani, K.; Sarumathi, V.; Selvakumar, P.; Temesgen, T.; Asaithambi, P.; Sivashanmugam, P. Study on the Ethanol Production from Hydrolysate Derived by Ultrasonic Pretreated Defatted Biomass of Chlorella Sorokiniana NITTS3. *Chem. Data Collect.* **2021**, *31*, 100641. [CrossRef]
211. Beigbeder, J.-B.; de Medeiros Dantas, J.M.; Lavoie, J.-M. Optimization of Yeast, Sugar and Nutrient Concentrations for High Ethanol Production Rate Using Industrial Sugar Beet Molasses and Response Surface Methodology. *Fermentation* **2021**, *7*, 86. [CrossRef]
212. Krajang, M.; Malairuang, K.; Sukna, J.; Rattanapradit, K.; Chamsart, S. Single-Step Ethanol Production from Raw Cassava Starch Using a Combination of Raw Starch Hydrolysis and Fermentation, Scale-up from 5-L Laboratory and 200-L Pilot Plant to 3000-L Industrial Fermenters. *Biotechnol. Biofuels* **2021**, *14*, 68. [CrossRef] [PubMed]
213. Hargono, H.; Jos, B.; Purwanto, P.; Sumardiono, S.; Zakaria, M. Bioethanol Purification from Fermentation of Suweg Starch Using Two Stage Distillation Method. *IOP Conf. Ser. Mater. Sci. Eng.* **2021**, *1053*, 012114. [CrossRef]
214. Koopraseytying, P.; Vanichsriratana, W.; Sirisansaneeyakul, S.; Laemsak, N.; Tareen, A.K.; Ullah, Z.; Parakulsuksatid, P.; Sultan, I.N. Ethanol Production through Optimized Alkaline Pretreated Elaeis Guineensis Frond Waste from Krabi Province, Thailand. *Fermentation* **2022**, *8*, 648. [CrossRef]

215. Tran, P.H.N.; Ko, J.K.; Gong, G.; Um, Y.; Lee, S.M. Improved Simultaneous Co-Fermentation of Glucose and Xylose by Saccharomyces Cerevisiae for Efficient Lignocellulosic Biorefinery. *Biotechnol. Biofuels* **2020**, *13*, 12. [CrossRef]
216. Lopez, P.C.; Abeykoon Udugama, I.; Thomsen, S.T.; Bayer, C.; Junicke, H.; Gernaey, K.V. Promoting the Co-Utilisation of Glucose and Xylose in Lignocellulosic Ethanol Fermentations Using a Data-Driven Feed-Back Controller. *Biotechnol. Biofuels* **2020**, *13*, 190. [CrossRef]
217. Sjulander, N.; Kikas, T. Origin, Impact and Control of Lignocellulosic Inhibitors in Bioethanol Production—A Review. *Energies* **2020**, *13*, 4751. [CrossRef]
218. Oktaviani, M.; Hermiati, E.; Thontowi, A.; Laksana, R.P.B.; Kholida, L.N.; Andriani, A.; Yopi; Mangunwardoyo, W. Production of Xylose, Glucose, and Other Products from Tropical Lignocellulose Biomass by Using Maleic Acid Pretreatment. *IOP Conf. Ser. Earth Environ. Sci.* **2019**, *251*, 012013. [CrossRef]
219. Galan-Mascaros, J.R. Photoelectrochemical Solar Fuels from Carbon Dioxide, Water and Sunlight. *Catal. Sci. Technol.* **2020**, *10*, 1967–1974. [CrossRef]
220. González-Garay, A.; Mac Dowell, N.; Shah, N. A Carbon Neutral Chemical Industry Powered by the Sun. *Discov. Chem. Eng.* **2021**, *1*, 2. [CrossRef]
221. Han, H.; Noh, Y.; Kim, Y.; Park, S.; Yoon, W.; Jang, D.; Choi, S.M.; Kim, W.B. Selective Electrochemical CO_2 Conversion to Multicarbon Alcohols on Highly Efficient N-Doped Porous Carbon-Supported Cu Catalysts. *Green Chem.* **2020**, *22*, 71–84. [CrossRef]
222. Yuan, C.Z.; Li, H.B.; Jiang, Y.F.; Liang, K.; Zhao, S.J.; Fang, X.X.; Ma, L.B.; Zhao, T.; Lin, C.; Xu, A.W. Tuning the Activity of N-Doped Carbon for CO_2 Reduction: Via in Situ Encapsulation of Nickel Nanoparticles into Nano-Hybrid Carbon Substrates. *Mater. Chem. A* **2019**, *7*, 6894–6900. [CrossRef]
223. Kou, W.; Zhang, Y.; Dong, J.; Mu, C.; Xu, L. Nickel-Nitrogen-Doped Three-Dimensional Ordered Macro-/Mesoporous Carbon as an Efficient Electrocatalyst for CO_2 Reduction to CO. *ACS Appl. Energy Mater.* **2020**, *3*, 1875–1882. [CrossRef]
224. Ramírez-Valencia, L.D.; Bailón-García, E.; Carrasco-Marín, F.; Pérez-Cadenas, A.F. From CO_2 to Value-Added Products: A Review about Carbon-Based Materials for Electro-Chemical CO_2 Conversion. *Catalysts* **2021**, *11*, 351. [CrossRef]
225. Liu, K.; Wang, J.; Shi, M.; Yan, J.; Jiang, Q. Simultaneous Achieving of High Faradaic Efficiency and CO Partial Current Density for CO_2 Reduction via Robust, Noble-Metal-Free Zn Nanosheets with Favorable Adsorption Energy. *Adv. Energy Mater.* **2019**, *9*, 1900276. [CrossRef]
226. Xu, H.; Rebollar, D.; He, H.; Chong, L.; Liu, Y.; Liu, C.; Sun, C.J.; Li, T.; Muntean, J.V.; Winans, R.E.; et al. Highly Selective Electrocatalytic CO_2 Reduction to Ethanol by Metallic Clusters Dynamically Formed from Atomically Dispersed Copper. *Nat. Energy* **2020**, *5*, 623–632. [CrossRef]
227. Park, S.; Wijaya, D.T.; Na, J.; Lee, C.W. Towards the Large-Scale Electrochemical Reduction of Carbon Dioxide. *Catalysts* **2021**, *11*, 253. [CrossRef]
228. Ramdin, M.; Morrison, A.R.T.; De Groen, M.; Van Haperen, R.; De Kler, R.; Van Den Broeke, L.J.P.; Martin Trusler, J.P.; De Jong, W.; Vlugt, T.J.H. High Pressure Electrochemical Reduction of CO_2 to Formic Acid/Formate: A Comparison between Bipolar Membranes and Cation Exchange Membranes. *Ind. Eng. Chem. Res.* **2019**, *58*, 1834–1847. [CrossRef] [PubMed]
229. Garg, S.; Li, M.; Weber, A.Z.; Ge, L.; Li, L.; Rudolph, V.; Wang, G.; Rufford, T.E. Advances and Challenges in Electrochemical CO_2 Reduction Processes: An Engineering and Design Perspective Looking beyond New Catalyst Materials. *J. Mater. Chem. A* **2020**, *8*, 1511–1544. [CrossRef]
230. Senocrate, A.; Battaglia, C. Electrochemical CO_2 Reduction at Room Temperature: Status and Perspectives. *J. Energy Storage* **2021**, *36*, 102373. [CrossRef]
231. Liu, X.; Schlexer, P.; Xiao, J.; Ji, Y.; Wang, L.; Sandberg, R.B.; Tang, M.; Brown, K.S.; Peng, H.; Ringe, S.; et al. PH Effects on the Electrochemical Reduction of CO_2 towards C_2 Products on Stepped Copper. *Nat. Commun.* **2019**, *10*, 32. [CrossRef] [PubMed]
232. Fan, L.; Xia, C.; Yang, F.; Wang, J.; Wang, H.; Lu, Y. Strategies in Catalysts and Electrolyzer Design for Electrochemical CO_2 Reduction toward C2+ Products. *Sci. Adv.* **2020**, *6*, eaay3111. [CrossRef] [PubMed]
233. Li, J.; Chang, X.; Zhang, H.; Malkani, A.S.; Cheng, M.J.; Xu, B.; Lu, Q. Electrokinetic and in Situ Spectroscopic Investigations of CO Electrochemical Reduction on Copper. *Nat. Commun.* **2021**, *12*, 3264. [CrossRef]
234. Shao, J.; Wang, C.; Shen, Y.; Shi, J.; Ding, D. Electrochemical Sensors and Biosensors for the Analysis of Tea Components: A Bibliometric Review. *Front. Chem.* **2022**, *9*, 818461. [CrossRef]
235. Li, Y.; Yuan, H.; Chen, Y.; Wei, X.; Sui, K.; Tan, Y. Application and Exploration of Nanofibrous Strategy in Electrode Design. *J. Mater. Sci. Technol.* **2021**, *74*, 189–202. [CrossRef]
236. Gao, D.; Arán-Ais, R.M.; Jeon, H.S.; Roldan Cuenya, B. Rational Catalyst and Electrolyte Design for CO_2 Electroreduction towards Multicarbon Products. *Nat. Catal.* **2019**, *2*, 198–210. [CrossRef]
237. Song, Y.; Chen, W.; Wei, W.; Sun, Y. Advances in Clean Fuel Ethanol Production from Electro-, Photo-and Photoelectro-Catalytic CO_2 Reduction. *Catalysts* **2020**, *10*, 1287. [CrossRef]
238. Sun, Y.; Dai, S. High-Entropy Materials for Catalysis: A New Frontier. *Sci. Adv.* **2021**, *7*, eabg1600. [CrossRef]
239. Nitopi, S.; Bertheussen, E.; Scott, S.B.; Liu, X.; Engstfeld, A.K.; Horch, S.; Seger, B.; Stephens, I.E.L.; Chan, K.; Hahn, C.; et al. Progress and Perspectives of Electrochemical CO_2 Reduction on Copper in Aqueous Electrolyte. *Chem. Rev.* **2019**, *119*, 7610–7672. [CrossRef] [PubMed]

240. Pei, Y.; Zhong, H.; Jin, F. A Brief Review of Electrocatalytic Reduction of CO_2—Materials, Reaction Conditions, and Devices. *Energy Sci. Eng.* **2021**, *9*, 1012–1032. [CrossRef]
241. Mustafa, A.; Lougou, B.G.; Shuai, Y.; Wang, Z.; Razzaq, S.; Zhao, J.; Tan, H. Theoretical Insights into the Factors Affecting the Electrochemical Reduction of CO_2. *Sustain. Energy Fuels* **2020**, *4*, 4352–4369. [CrossRef]
242. Liu, J.; Ma, J.; Zhang, Z.; Qin, Y.; Wang, Y.; Wang, Y.; Tan, R.; Duan, X.; Tian, T.Z.; Zhang, C.H.; et al. 2021 Roadmap: Electrocatalysts for Green Catalytic Processes. *J. Phys. Mater.* **2021**, *4*, 022004. [CrossRef]
243. Zhu, W.; Zhao, K.; Liu, S.; Liu, M.; Peng, F.; An, P.; Qin, B.; Zhou, H.; Li, H.; He, Z. Low-Overpotential Selective Reduction of CO_2 to Ethanol on Electrodeposited CuxAuy Nanowire Arrays. *Energy Chem.* **2019**, *37*, 176–182. [CrossRef]
244. Zhou, Y.; Yeo, B.S. Formation of C-C Bonds during Electrocatalytic CO_2 Reduction on Non-Copper Electrodes. *J. Mater. Chem. A* **2020**, *8*, 23162–23186. [CrossRef]
245. Chen, P.; Zhang, Y.; Zhou, Y.; Dong, F. Photoelectrocatalytic Carbon Dioxide Reduction: Fundamental, Advances and Challenges. *Nano Mater. Sci.* **2021**, *3*, 344–367. [CrossRef]
246. Zhang, X.; Guo, S.X.; Gandionco, K.A.; Bond, A.M.; Zhang, J. Electrocatalytic Carbon Dioxide Reduction: From Fundamental Principles to Catalyst Design. *Mater. Today Adv.* **2020**, *7*, 100074. [CrossRef]
247. da Gregorio, G.L.; Burdyny, T.; Loiudice, A.; Iyengar, P.; Smith, W.A.; Buonsanti, R. Facet-Dependent Selectivity of Cu Catalysts in Electrochemical CO_2 Reduction at Commercially Viable Current Densities. *ACS Catal.* **2020**, *10*, 4854–4862. [CrossRef] [PubMed]
248. Jung, H.; Lee, S.Y.; Lee, C.W.; Cho, M.K.; Won, D.H.; Kim, C.; Oh, H.S.; Min, B.K.; Hwang, Y.J. Electrochemical Fragmentation of Cu_2O Nanoparticles Enhancing Selective C-C Coupling from CO_2 Reduction Reaction. *J. Am. Chem. Soc.* **2019**, *141*, 4624–4633. [CrossRef] [PubMed]
249. Zhu, Q.; Sun, X.; Yang, D.; Ma, J.; Kang, X.; Zheng, L.; Zhang, J.; Wu, Z.; Han, B. Carbon Dioxide Electroreduction to C2 Products over Copper-Cuprous Oxide Derived from Electrosynthesized Copper Complex. *Nat. Commun.* **2019**, *10*, 3851. [CrossRef] [PubMed]
250. Ajmal, S.; Yang, Y.; Tahir, M.A.; Li, K.; Bacha, A.U.R.; Nabi, I.; Liu, Y.; Wang, T.; Zhang, L. Boosting C2 Products in Electrochemical CO_2 Reduction over Highly Dense Copper Nanoplates. *Catal. Sci. Technol.* **2020**, *10*, 4562–4570. [CrossRef]
251. Liu, G.; Lee, M.; Kwon, S.; Zeng, G.; Eichhorn, J.; Buckley, A.K.; Toste, F.D.; Goddard, W.A.; Toma, F.M. CO_2 Reduction on Pure Cu Produces Only H_2 after Subsurface O Is Depleted: Theory and Experiment. *Proc. Natl. Acad. Sci. USA* **2021**, *118*, 23. [CrossRef]
252. Bai, S.; Shao, Q.; Wang, P.; Dai, Q.; Wang, X.; Huang, X. Highly Active and Selective Hydrogenation of CO_2 to Ethanol by Ordered Pd-Cu Nanoparticles. *J. Am. Chem. Soc.* **2017**, *139*, 6827–6830. [CrossRef]
253. Jiwanti, P.K.; Natsui, K.; Nakata, K.; Einaga, Y. The Electrochemical Production of C2/C3 Species from Carbon Dioxide on Copper-Modified Boron-Doped Diamond Electrodes. *Electrochim. Acta* **2018**, *266*, 414–419. [CrossRef]
254. Li, F.; Li, Y.C.; Wang, Z.; Li, J.; Nam, D.H.; Lum, Y.; Luo, M.; Wang, X.; Ozden, A.; Hung, S.F.; et al. Cooperative CO_2-to-Ethanol Conversion via Enriched Intermediates at Molecule–Metal Catalyst Interfaces. *Nat. Catal.* **2020**, *3*, 75–82. [CrossRef]
255. Ting, L.R.L.; Piqué, O.; Lim, S.Y.; Tanhaei, M.; Calle-Vallejo, F.; Yeo, B.S. Enhancing CO_2 Electroreduction to Ethanol on Copper-Silver Composites by Opening an Alternative Catalytic Pathway. *ACS Catal.* **2020**, *10*, 4059–4069. [CrossRef]
256. Niu, D.; Wei, C.; Lu, Z.; Fang, Y.; Liu, B.; Sun, D.; Hao, X.; Pan, H.; Wang, G. Cu_2O-Ag Tandem Catalysts for Selective Electrochemical Reduction of CO_2 to C2 Products. *Molecules* **2021**, *26*, 2175. [CrossRef]
257. Ma, S.; Sadakiyo, M.; Luo, R.; Heima, M.; Yamauchi, M.; Kenis, P.J.A. One-Step Electrosynthesis of Ethylene and Ethanol from CO_2 in an Alkaline Electrolyzer. *J. Power Sources* **2016**, *301*, 219–228. [CrossRef]
258. Schmid, B.; Reller, C.; Neubauer, S.S.; Fleischer, M.; Dorta, R.; Schmid, G. Reactivity of Copper Electrodes towards Functional Groups and Small Molecules in the Context of CO_2 Electro-Reductions. *Catalysts* **2017**, *7*, 161. [CrossRef]
259. Zhuang, T.T.; Liang, Z.Q.; Seifitokaldani, A.; Li, Y.; De Luna, P.; Burdyny, T.; Che, F.; Meng, F.; Min, Y.; Quintero-Bermudez, R.; et al. Steering Post-C-C Coupling Selectivity Enables High Efficiency Electroreduction of Carbon Dioxide to Multi-Carbon Alcohols. *Nat. Catal.* **2018**, *1*, 421–428. [CrossRef]
260. Hoang, T.T.H.; Verma, S.; Ma, S.; Fister, T.T.; Timoshenko, J.; Frenkel, A.I.; Kenis, P.J.A.; Gewirth, A.A. Nanoporous Copper-Silver Alloys by Additive-Controlled Electrodeposition for the Selective Electroreduction of CO_2 to Ethylene and Ethanol. *J. Am. Chem. Soc.* **2018**, *140*, 5791–5797. [CrossRef]
261. Luo, M.; Wang, Z.; Li, Y.C.; Li, J.; Li, F.; Lum, Y.; Nam, D.H.; Chen, B.; Wicks, J.; Xu, A.; et al. Hydroxide Promotes Carbon Dioxide Electroreduction to Ethanol on Copper via Tuning of Adsorbed Hydrogen. *Nat. Commun.* **2019**, *10*, 5814. [CrossRef]
262. Zhang, J.; Luo, W.; Züttel, A. Self-Supported Copper-Based Gas Diffusion Electrodes for CO_2 Electrochemical Reduction. *J. Mater. Chem. A* **2019**, *7*, 26285–26292. [CrossRef]
263. Li, Y.C.; Wang, Z.; Yuan, T.; Nam, D.H.; Luo, M.; Wicks, J.; Chen, B.; Li, J.; Li, F.; De Arquer, F.P.G.; et al. Binding Site Diversity Promotes CO_2 Electroreduction to Ethanol. *J. Am. Chem. Soc.* **2019**, *141*, 8584–8591. [CrossRef] [PubMed]
264. Romero Cuellar, N.S.; Wiesner-Fleischer, K.; Fleischer, M.; Rucki, A.; Hinrichsen, O. Advantages of CO over CO_2 as Reactant for Electrochemical Reduction to Ethylene, Ethanol and n-Propanol on Gas Diffusion Electrodes at High Current Densities. *Electrochim. Acta* **2019**, *307*, 164–175. [CrossRef]
265. Wang, Y.; Wang, Z.; Dinh, C.T.; Li, J.; Ozden, A.; Golam Kibria, M.; Seifitokaldani, A.; Tan, C.S.; Gabardo, C.M.; Luo, M.; et al. Catalyst Synthesis under CO_2 Electroreduction Favours Faceting and Promotes Renewable Fuels Electrosynthesis. *Nat. Catal.* **2020**, *3*, 98–106. [CrossRef]

266. Wang, X.; Wang, Z.; García de Arquer, F.P.; Dinh, C.T.; Ozden, A.; Li, Y.C.; Nam, D.H.; Li, J.; Liu, Y.S.; Wicks, J.; et al. Efficient Electrically Powered CO$_2$-to-Ethanol via Suppression of Deoxygenation. *Nat. Energy* **2020**, *5*, 478–486. [CrossRef]
267. She, X.; Zhang, T.; Li, Z.; Li, H.; Xu, H.; Wu, J. Tandem Electrodes for Carbon Dioxide Reduction into C$_{2+}$ Products at Simultaneously High Production Efficiency and Rate. *Cell Rep. Phys. Sci.* **2020**, *1*, 100051. [CrossRef]
268. Tan, Y.C.; Lee, K.B.; Song, H.; Oh, J. Modulating Local CO$_2$ Concentration as a General Strategy for Enhancing C-C Coupling in CO$_2$ Electroreduction. *Joule* **2020**, *4*, 1104–1120. [CrossRef]
269. Yang, Q.; Liu, X.; Peng, W.; Zhao, Y.; Liu, Z.; Peng, M.; Lu, Y.R.; Chan, T.S.; Xu, X.; Tan, Y. Vanadium Oxide Integrated on Hierarchically Nanoporous Copper for Efficient Electroreduction of CO$_2$ to Ethanol. *J. Mater. Chem. A* **2021**, *9*, 3044–3051. [CrossRef]
270. Zhu, S.; Ren, X.; Li, X.; Niu, X.; Wang, M.; Xu, S.; Wang, Z.; Han, Y.; Wang, Q. Core-shell ZnO@Cu$_2$O as Catalyst to Enhance the Electrochemical Reduction of Carbon Dioxide to C2 Products. *Catalysts* **2021**, *11*, 535. [CrossRef]
271. Puring, K.J.; Siegmund, D.; Timm, J.; Möllenbruck, F.; Schemme, S.; Marschall, R.; Apfel, U. Electrochemical CO$_2$ Reduction: Tailoring Catalyst Layers in Gas Diffusion Electrodes. *Av. Sustain. Syst.* **2021**, *5*, 2000088. [CrossRef]
272. Xing, Z.; Hu, L.; Ripatti, D.S.; Hu, X.; Feng, X. Enhancing Carbon Dioxide Gas-Diffusion Electrolysis by Creating a Hydrophobic Catalyst Microenvironment. *Nat. Commun.* **2021**, *12*, 136. [CrossRef]
273. Mota, N.; Ordoñez, E.M.; Pawelec, B.; Fierro, J.L.G.; Navarro, R.M. Direct Synthesis of Dimethyl Ether from CO$_2$: Recent Advances in Bifunctional/Hybrid Catalytic Systems. *Catalysts* **2021**, *11*, 411. [CrossRef]
274. Dieterich, V.; Buttler, A.; Hanel, A.; Spliethof, H.; Fendt, S. Power-to-Liquid via Synthesis of Methanol, DME or Fischer–Tropsch-Fuels: A Review. *Energy Environ. Sci.* **2020**, *13*, 3207–3252. [CrossRef]
275. Makos, P.; Słupek, E.; Sobczak, J.; Zabrocki, D.; Hupka, J.; Rogala, A. Dimethyl Ether (DME) as Potential Environmental Friendly Fuel. *E3S Web Conf.* **2019**, *116*, 00048. [CrossRef]
276. Catizzone, E.; Freda, C.; Braccio, G.; Frusteri, F.; Bonura, G. Dimethyl Ether as Circular Hydrogen Carrier: Catalytic Aspects of Hydrogenation/Dehydrogenation Steps. *J. Energy Chem.* **2021**, *58*, 55–77. [CrossRef]
277. Bahari, N.A.; Wan Isahak, W.N.R.; Masdar, M.S.; Yaakob, Z. Clean Hydrogen Generation and Storage Strategies via CO$_2$ Utilization into Chemicals and Fuels: A Review. *Int. J. Energy Res.* **2019**, *43*, 5128–5150. [CrossRef]
278. Bao, J.; Yang, G.; Yoneyama, Y.; Tsubaki, N. Significant Advances in C1 Catalysis: Highly Efficient Catalysts and Catalytic Reactions. *ACS Catal.* **2019**, *9*, 3026–3053. [CrossRef]
279. Du, C.; Lu, P.; Tsubaki, N. Efficient and New Production Methods of Chemicals and Liquid Fuels by Carbon, Monoxide Hydrogenation. *ACS Omega* **2020**, *5*, 49–56. [CrossRef]
280. Zhang, X.; Zhang, G.; Song, C.; Guo, X. Catalytic Conversion of Carbon Dioxide to Methanol: Current Status and Future Perspective. *Front. Energy Res.* **2021**, *8*, 621119. [CrossRef]
281. Han, T.; Xu, H.; Liu, J.; Zhou, L.; Li, X.; Dong, J.; Ge, H. One-pass Conversion of Benzene and Syngas to Alkylbenzenes by Cu-ZnO-Al$_2$O$_3$ and ZSM-5 Relay. *Catal. Lett.* **2021**, *152*, 467–479. [CrossRef]
282. Yerga, R.M.N. Catalysts for Production and Conversion of Syngas. *Catalysts* **2021**, *11*, 752. [CrossRef]
283. Kamsuwan, T.; Krutpijit, C.; Praserthdam, S.; Phatanasri, S.; Jongsomjit, B.; Praserthdam, P. Comparative Study on the Effect of Different Copper Loading on Catalytic Behaviors and Activity of Cu/ZnO/Al$_2$O$_3$ Catalysts toward CO and CO$_2$ Hydrogenation. *Heliyon* **2021**, *7*, e07682. [CrossRef] [PubMed]
284. Zhong, J.; Yang, X.; Wu, Z. State of the Art and Perspectives in Heterogeneous Catalysis of CO$_2$ Hydrogenation. *Chem. Soc. Rev.* **2020**, *49*, 1385–1413. [CrossRef]
285. Liang, B.; Ma, J.; Su, X.; Yang, C.; Duan, H.; Zhou, H.; Deng, S.; Li, L.; Huang, Y. Investigation on Deactivation of Cu/ZnO/Al$_2$O$_3$ Catalyst for CO$_2$ Hydrogenation to Methanol. *Ind. Eng. Chem. Res.* **2019**, *58*, 9030–9037. [CrossRef]
286. Ren, S.; Fan, X.; Shang, Z.; Shoemaker, W.R.; Ma, L.; Wu, T.; Li, S.; Klinghoffer, N.B.; Yu, M.; Liang, X. Enhanced Catalytic Performance of Zr Modified CuO/ZnO/Al$_2$O$_3$ Catalyst for Methanol and DME Synthesis via CO$_2$ Hydrogenation. *J. CO$_2$ Util.* **2020**, *36*, 82–95. [CrossRef]
287. Divins, N.J.; Kordus, D.; Timoshenko, J.; Sinev, I.; Zegkinoglou, I.; Bergmann, A.; Chee, S.W.; Widrinna, S.; Karslıoğlu, O.; Mistry, H.; et al. Operando High-Pressure Investigation of Size-Controlled CuZn Catalysts for the Methanol Synthesis Reaction. *Nat. Commun.* **2021**, *12*, 1435. [CrossRef] [PubMed]
288. van Kampen, J.; Boon, J.; Vente, J.; van Sint Annaland, M. Sorption Enhanced Dimethyl Ether Synthesis under Industrially Relevant Conditions: Experimental Validation of Pressure Swing Regeneration. *React. Chem. Eng.* **2021**, *6*, 244–257. [CrossRef]
289. Peinado, C.; Liuzzi, D.; Retuerto, M.; Boon, J.; Peña, M.A.; Rojas, S. Study of Catalyst Bed Composition for the Direct Synthesis of Dimethyl Ether from CO$_2$-Rich Syngas. *Chem. Eng. J. Adv.* **2020**, *4*, 100039. [CrossRef]
290. de Oliveira Campos, B.L.; Delgado, K.H.; Wild, S.; Studt, F.; Pitter, S.; Sauer, J. Surface Reaction Kinetics of the Methanol Synthesis and the Water Gas Shift Reaction on Cu/ZnO/Al$_2$O$_3$. *React. Chem. Eng.* **2021**, *6*, 868–887. [CrossRef]
291. Styring, P.; Dowson, G.R.M.; Tozer, I.O. Synthetic Fuels Based on Dimethyl Ether as a Future Non-Fossil Fuel for Road Transport from Sustainable Feedstocks. *Front. Energy Res.* **2021**, *9*, 663331. [CrossRef]
292. Brunetti, A.; Migliori, M.; Cozza, D.; Catizzone, E.; Giordano, G.; Barbieri, G. Methanol Conversion to Dimethyl Ether in Catalytic Zeolite Membrane Reactors. *ACS Sustain. Chem. Eng.* **2020**, *8*, 10471–10479. [CrossRef]
293. Bizon, K.; Skrzypek-Markiewicz, K.; Continillo, G. Enhancement of the Direct Synthesis of Dimethyl Ether (DME) from Synthesis Gas by Macro-and Microstructuring of the Catalytic Bed. *Catalysts* **2020**, *10*, 852. [CrossRef]

294. Guffanti, S.; Visconti, C.G.; Groppi, G. Model Analysis of the Role of Kinetics, Adsorption Capacity, and Heat and Mass Transfer Effects in Sorption Enhanced Dimethyl Ether Synthesis. *Ind. Eng. Chem. Res.* **2021**, *60*, 6767–6783. [CrossRef]
295. Chen, T.Y.; Cao, C.; Chen, T.B.; Ding, X.; Huang, H.; Shen, L.; Cao, X.; Zhu, M.; Xu, J.; Gao, J.; et al. Unraveling Highly Tunable Selectivity in CO_2 Hydrogenation over Bimetallic In-Zr Oxide Catalysts. *ACS Catal.* **2019**, *9*, 8785–8797. [CrossRef]
296. Bonura, G.; Cannilla, C.; Frusteri, L.; Catizzone, E.; Todaro, S.; Migliori, M.; Giordano, G.; Frusteri, F. Interaction Effects between CuO-ZnO-ZrO_2 Methanol Phase and Zeolite Surface Affecting Stability of Hybrid Systems during One-Step CO_2 Hydrogenation to DME. *Catal. Today* **2020**, *345*, 175–182. [CrossRef]
297. Liu, C.; Kang, J.; Huang, Z.Q.; Song, Y.H.; Xiao, Y.S.; Song, J.; He, J.X.; Chang, C.R.; Ge, H.Q.; Wang, Y.; et al. Gallium Nitride Catalyzed the Direct Hydrogenation of Carbon Dioxide to Dimethyl Ether as Primary Product. *Nat. Commun.* **2021**, *12*, 2305. [CrossRef]
298. Tuygun, C.; Ipek, B. CO_2 Hydrogenation to Methanol and Dimethyl Ether at Atmospheric Pressure Using Cu-Ho-Ga/γ-Al_2O_3 and Cu-Ho-Ga/ZSM-5: Experimental Study and Thermodynamic Analysis. *Turk. J. Chem.* **2021**, *45*, 231–247. [CrossRef] [PubMed]
299. Wild, S.; Polierer, S.; Zevaco, T.A.; Guse, D.; Kind, M.; Pitter, S.; Delgado, K.H.; Sauer, J. Direct DME Synthesis on CZZ/H-FER from Variable CO_2/CO Syngas Feeds. *RSC Adv.* **2021**, *11*, 2556–2564. [CrossRef] [PubMed]
300. Esquius, J.R.; Bahruji, H.; Bowker, M.; Hutchings, G.J. Identification of C2-C5 Products from CO_2 Hydrogenation over PdZn/TiO_2-ZSM-5 Hybrid Catalysts. *Faraday Discuss.* **2021**, *230*, 52–67. [CrossRef]
301. Rodriguez-Vega, P.; Ateka, A.; Kumakiri, I.; Vicente, H.; Ereña, J.; Aguayo, A.T.; Bilbao, J. Experimental Implementation of a Catalytic Membrane Reactor for the Direct Synthesis of DME from H_2+CO/CO_2. *Chem. Eng. Sci.* **2021**, *234*, 116396. [CrossRef]
302. Liuzzi, D.; Peinado, C.; Peña, M.A.; Van Kampen, J.; Boon, J.; Rojas, S. Increasing Dimethyl Ether Production from Biomass-Derived Syngas: Via Sorption Enhanced Dimethyl Ether Synthesis. *Sustain. Energy Fuels* **2020**, *4*, 5674–5681. [CrossRef]
303. Du, C.; Hondo, E.; Chizema, L.G.; Wang, C.; Tong, M.; Xing, C.; Yang, R.; Lu, P.; Tsubaki, N. Developing Cu-MOR@SiO_2 Core—Shell Catalyst Microcapsules for Two-Stage Ethanol Direct Synthesis from DME and Syngas. *Ind. Eng. Chem. Res.* **2020**, *59*, 3293–3300. [CrossRef]
304. Du, C.; Hondo, E.; Gapu, L.; Hassan, R.; Chang, X.; Dai, L.; Ma, Q.; Lu, P.; Tsubaki, N. An Efficient Microcapsule Catalyst for One-Step Ethanol Synthesis from Dimethyl Ether and Syngas. *Fuel* **2021**, *283*, 118971. [CrossRef]
305. Feng, X.-B.; He, Z.-M.; Zhang, L.-Y.; Zhao, X.-Y.; Cao, J.-P. Facile Designing a Nanosheet HMOR Zeolite for Enhancing the Efficiency of Ethanol Synthesis from Dimethyl Ether and Syngas. *Int. J. Hydrogen Energy* **2022**, *47*, 9273–9282. [CrossRef]
306. Chen, X.; Chen, Y.; Song, C.; Ji, P.; Wang, N.; Wang, W.; Cui, L. Recent Advances in Supported Metal Catalysts and Oxide Catalysts for the Reverse Water Gas Shift Reaction. *Front. Chem.* **2020**, *8*, 709. [CrossRef]
307. González-Castaño, M.; Dorneanu, B.; Arellano-García, H. The Reverse Water Gas Shift Reaction: A Process Systems Engineering Perspective. *React. Chem. Eng.* **2021**, *6*, 954–976. [CrossRef]
308. Pearson, R.; Coe, A.; Paterson, J. Innovation in Fischer-Tropsch: A Sustainable Approach to Fuels Production. *Johnson Matthey Technol. Rev.* **2021**, *65*, 395–403. [CrossRef]
309. Aziz, M.A.A.; Setiabudi, H.D.; Teh, L.P.; Annuar, N.H.R.; Jalil, A.A. A Review of Heterogeneous Catalysts for Syngas Production via Dry Reforming. *J. Taiwan Inst. Chem. Eng.* **2019**, *101*, 139–158. [CrossRef]
310. Tan, K.B.; Zhan, G.; Sun, D.; Huang, J.; Li, Q. The Development of Bifunctional Catalysts for Carbon Dioxide Hydrogenation to Hydrocarbons via the Methanol Route: From Single Component to Integrated Components. *J. Mater. Chem. A* **2021**, *9*, 5197–5231. [CrossRef]
311. He, Z.; Cui, M.; Qian, Q.; Zhang, J.; Liu, H.; Han, B. Synthesis of Liquid Fuel via Direct Hydrogenation of CO_2. *Proc. Natl. Acad. Sci. USA* **2019**, *116*, 12654–12659. [CrossRef]
312. Gao, P.; Zhang, L.; Li, S.; Zhou, Z.; Sun, Y. Novel Heterogeneous Catalysts for CO_2 Hydrogenation to Liquid Fuels. *ACS Cent. Sci.* **2020**, *6*, 1657–1670. [CrossRef]
313. Zhang, S.; Liu, X.; Shao, Z.; Wang, H.; Sun, Y. Direct CO_2 Hydrogenation to Ethanol over Supported Co_2C Catalysts: Studies on Support Effects and Mechanism. *J. Catal.* **2020**, *382*, 86–96. [CrossRef]
314. Liu, H.; Wang, L. Highly Dispersed and Stable Ni/SBA-15 Catalyst for Reverse Water Gas Shift Reaction. *Crystals* **2021**, *11*, 790. [CrossRef]
315. Pauletto, G.; Galli, F.; Gaillardet, A.; Mocellin, P.; Patience, G.S. Techno Economic Analysis of a Micro Gas-to-Liquid Unit for Associated Natural Gas Conversion. *Renew. Sustain. Energy Rev.* **2021**, *150*, 111457. [CrossRef]
316. Vu, T.T.N.; Desgagnés, A.; Iliuta, M.C. Efficient Approaches to Overcome Challenges in Material Development for Conventional and Intensified CO_2 Catalytic Hydrogenation to CO, Methanol, and DME. *Appl. Catal. A Gen.* **2021**, *617*, 118119. [CrossRef]
317. Lv, C.; Xu, L.; Chen, M.; Cui, Y.; Wen, X.; Li, Y.; Wu, C.E.; Yang, B.; Miao, Z.; Hu, X.; et al. Recent Progresses in Constructing the Highly Efficient Ni Based Catalysts with Advanced Low-Temperature Activity toward CO_2 Methanation. *Front. Chem.* **2020**, *8*, 269. [CrossRef] [PubMed]
318. Xing, Y.; Ouyang, M.; Zhang, L.; Yang, M.; Wu, X.; Ran, R.; Weng, D.; Kang, F.; Si, Z. Single Atomic Pt on $SrTiO_3$ Catalyst in Reverse Water Gas Shift Reactions. *Catalysts* **2021**, *11*, 738. [CrossRef]
319. Azancot, L.; Bobadilla, L.F.; Centeno, M.A.; Odriozola, J.A. IR Spectroscopic Insights into the Coking-Resistance Effect of Potassium on Nickel-Based Catalyst during Dry Reforming of Methane. *Appl. Catal. B Environ.* **2021**, *285*, 119822. [CrossRef]
320. Cui, S.; Wang, X.; Wang, L.; Zheng, X. Enhanced Selectivity of the CO_2 Reverse Water-Gas Reaction over a Ni_2P/CeO_2 Catalyst. *Dalt. Trans.* **2021**, *50*, 5978–5987. [CrossRef]

321. Chou, C.Y.; Loiland, J.A.; Lobo, R.F. Reverse Water-Gas Shift Iron Catalyst Derived from Magnetite. *Catalysts* **2019**, *9*, 773. [CrossRef]
322. Zhang, Q.; Pastor-Pérez, L.; Gu, S.; Reina, T.R. Transition Metal Carbides (TMCS) Catalysts for Gas Phase CO_2 Upgrading Reactions: A Comprehensive Overview. *Catalysts* **2020**, *10*, 955. [CrossRef]
323. Pajares, A.; Prats, H.; Romero, A.; Viñes, F.; de la Piscina, P.R.; Sayós, R.; Homs, N.; Illas, F. Critical Effect of Carbon Vacancies on the Reverse Water Gas Shift Reaction over Vanadium Carbide Catalysts. *Appl. Catal. B Environ.* **2020**, *267*, 118719. [CrossRef]
324. Kuang, H.-Y.; Lin, Y.-X.; Li, X.-H.; Chen, J.-S. Chemical Fixation of CO_2 on Nanocarbons and Hybrids. *J. Mater. Chem. A* **2021**, *9*, 20857–20873. [CrossRef]
325. Chen, Y.; Wei, J.; Duyar, M.S.; Ordomsky, V.V.; Khodakov, A.Y.; Liu, J. Carbon-Based Catalysts for Fischer-Tropsch Synthesis. *Chem. Soc. Rev.* **2021**, *50*, 2337–2366. [CrossRef]
326. Zhang, P.; Han, F.; Yan, J.; Qiao, X.; Guan, Q.; Li, W. N-Doped Ordered Mesoporous Carbon (N-OMC) Confined Fe_3O_4-FeC_x Heterojunction for Efficient Conversion of CO_2 to Light Olefins. *Appl. Catal. B Environ.* **2021**, *299*, 120639. [CrossRef]
327. Li, Y.; Liang, G.; Wang, C.; Fang, Y.; Duan, H. Effect of Precipitated Precursor on the Catalytic Performance of Mesoporous Carbon Supported CuO-ZnO Catalysts. *Crystals* **2021**, *11*, 582. [CrossRef]
328. Valero-Romero, M.J.; Rodríguez-Cano, M.Á.; Palomo, J.; Rodríguez-Mirasol, J.; Cordero, T. Carbon-Based Materials as Catalyst Supports for Fischer–Tropsch Synthesis: A Review. *Front. Mater.* **2021**, *7*, 617432. [CrossRef]
329. González-Arias, J.; González-Castaño, M.; Sánchez, M.E.; Cara-Jiménez, J.; Arellano-García, H. Valorization of Biomass-Derived CO_2 Residues with Cu-MnOx Catalysts for RWGS Reaction. *Renew. Energy* **2022**, *182*, 443–451. [CrossRef]
330. Bahmanpour, A.M.; Signorile, M.; Kröcher, O. Recent Progress in Syngas Production via Catalytic CO_2 Hydrogenation Reaction. *Appl. Catal. B Environ.* **2021**, *295*, 120319. [CrossRef]
331. Elsernagawy, O.Y.H.; Hoadley, A.; Patel, J.; Bhatelia, T.; Lim, S.; Haque, N. Thermo-Economic Analysis of Reverse Water-Gas Shift Process with Different Temperatures for Green Methanol Production as a Hydrogen Carrier. *J. CO_2 Util.* **2020**, *41*, 101280. [CrossRef]
332. Okemoto, A.; Harada, M.R.; Ishizaka, T.; Hiyoshi, N.; Sato, K. Catalytic Performance of MoO_3/FAU Zeolite Catalysts Modified by Cu for Reverse Water Gas Shift Reaction. *Appl. Catal. A Gen.* **2020**, *592*, 117415. [CrossRef]
333. Zhang, Q.; Pastor-Pérez, L.; Jin, W.; Gu, S.; Reina, T.R. Understanding the Promoter Effect of Cu and Cs over Highly Effective β-Mo_2C Catalysts for the Reverse Water-Gas Shift Reaction. *Appl. Catal. B Environ.* **2019**, *244*, 889–898. [CrossRef]
334. Konsolakis, M.; Lykaki, M.; Stefa, S.; Carabineiro, S.A.C.; Varvoutis, G.; Papista, E.; Marnellos, G.E. CO_2 Hydrogenation over Nanoceria-Supported Transition Metal Catalysts: Role of Ceria Morphology (Nanorods versus Nanocubes) and Active Phase Nature (Co versus Cu). *Nanomaterials* **2019**, *9*, 1739. [CrossRef]
335. Zhang, Y.; Liang, L.; Chen, Z.; Wen, J.; Zhong, W.; Zou, S.; Fu, M.; Chen, L.; Ye, D. Highly Efficient Cu/CeO_2-Hollow Nanospheres Catalyst for the Reverse Water-Gas Shift Reaction: Investigation on the Role of Oxygen Vacancies through in Situ UV-Raman and DRIFTS. *Appl. Surf. Sci.* **2020**, *516*, 146035. [CrossRef]
336. Lykaki, M.; Stefa, S.; Carabineiro, S.A.C.; Soria, M.A.; Madeira, L.M.; Konsolakis, M. Shape Effects of Ceria Nanoparticles on the Water-Gas Shift Performance of CuOx/CeO_2 Catalysts. *Catalysts* **2021**, *11*, 753. [CrossRef]
337. Bahmanpour, A.M.; Héroguel, F.; Kılıç, M.; Baranowski, C.J.; Schouwink, P.; Röthlisberger, U.; Luterbacher, J.S.; Kröcher, O. Essential Role of Oxygen Vacancies of Cu-Al and Co-Al Spinel Oxides in Their Catalytic Activity for the Reverse Water Gas Shift Reaction. *Appl. Catal. B Environ.* **2020**, *266*, 118669. [CrossRef]
338. Su, X.; Yang, X.F.; Huang, Y.; Liu, B.; Zhang, T. Single-Atom Catalysis toward Efficient CO_2 Conversion to CO and Formate Products. *Acc. Chem. Res.* **2019**, *52*, 656–664. [CrossRef]
339. Liu, M.; Yi, Y.; Wang, L.; Guo, H.; Bogaerts, A. Hydrogenation of Carbon Dioxide to Value-Added Chemicals by Heterogeneous Catalysis and Plasma Catalysis. *Catalysts* **2019**, *9*, 275. [CrossRef]
340. Wang, L.; Guan, E.; Wang, Y.; Wang, L.; Gong, Z.; Cui, Y.; Meng, X.; Gates, B.C.; Xiao, F.S. Silica Accelerates the Selective Hydrogenation of CO_2 to Methanol on Cobalt Catalysts. *Nat. Commun.* **2020**, *11*, 1033. [CrossRef]
341. Ke, J.; Wang, Y.D.; Wang, C.M. First-Principles Microkinetic Simulations Revealing the Scaling Relations and Structure Sensitivity of CO_2 Hydrogenation to C1 & C2 Oxygenates on Pd Surfaces. *Catal. Sci. Technol.* **2021**, *11*, 4866–4881. [CrossRef]
342. Kang, J.; He, S.; Zhou, W.; Shen, Z.; Li, Y.; Chen, M.; Zhang, Q.; Wang, Y. Single-Pass Transformation of Syngas into Ethanol with High Selectivity by Triple Tandem Catalysis. *Nat. Commun.* **2020**, *11*, 827. [CrossRef] [PubMed]
343. Lou, Y.; Jiang, F.; Zhu, W.; Wang, L.; Yao, T.; Wang, S.; Yang, B.; Yang, B.; Zhu, Y.; Liu, X. CeO_2 Supported Pd Dimers Boosting CO_2 Hydrogenation to Ethanol. *Appl. Catal. B Environ.* **2021**, *291*, 120122. [CrossRef]
344. Ye, X.; Yang, C.; Pan, X.; Ma, J.; Zhang, Y.; Ren, Y.; Liu, X.; Li, L.; Huang, Y. Highly Selective Hydrogenation of CO_2 to Ethanol via Designed Bifunctional Ir1-In_2O_3 Single-Atom Catalyst. *J. Am. Chem. Soc.* **2020**, *142*, 19001–19005. [CrossRef] [PubMed]
345. Hasan, S.Z.; Ahmad, K.N.; Isahak, W.N.R.W.; Masdar, M.S.; Jahim, J.M. Synthesis of Low-Cost Catalyst NiO(111) for CO_2 Hydrogenation into Short-Chain Carboxylic Acids. *Int. J. Hydrogen Energy* **2020**, *45*, 22281–22290. [CrossRef]
346. Goryachev, A.; Pustovarenko, A.; Shterk, G.; Alhajri, N.S.; Jamal, A.; Albuali, M.; van Koppen, L.; Khan, I.S.; Russkikh, A.; Ramirez, A.; et al. A Multi-Parametric Catalyst Screening for CO_2 Hydrogenation to Ethanol. *ChemCatChem* **2021**, *13*, 3324–3332. [CrossRef]
347. Yang, C.; Mu, R.; Wang, G.; Song, J.; Tian, H.; Zhao, Z.J.; Gong, J. Hydroxyl-Mediated Ethanol Selectivity of CO_2 Hydrogenation. *Chem. Sci.* **2019**, *10*, 3161–3167. [CrossRef] [PubMed]

348. Wang, C.; Zhang, J.; Qin, G.; Wang, L.; Zuidema, E.; Yang, Q.; Dang, S.; Yang, C.; Xiao, J.; Meng, X.; et al. Direct Conversion of Syngas to Ethanol within Zeolite Crystals. *Chem* **2020**, *6*, 646–657. [CrossRef]
349. Kiss, J.; Sápi, A.; Tóth, M.; Kukovecz, Á.; Kónya, Z. Rh-Induced Support Transformation and Rh Incorporation in Titanate Structures and Their Influence on Catalytic Activity. *Catalysts* **2020**, *10*, 212. [CrossRef]
350. Ding, S.; Hülsey, M.J.; An, H.; He, Q.; Asakura, H.; Gao, M.; Hasegawa, J.; Tanaka, T.; Yan, N. Ionic Liquid-Stabilized Single-Atom Rh Catalyst against Leaching. *CCS Chem.* **2021**, *3*, 1814–1822. [CrossRef]
351. Wang, G.; Luo, R.; Yang, C.; Song, J.; Xiong, C.; Tian, H. Active Sites in CO_2 Hydrogenation over Confined VO_x-Rh Catalysts. *Sci. China Chem.* **2019**, *62*, 1710–1719. [CrossRef]
352. Ramirez, A.; Ould-Chikh, S.; Gevers, L.; Chowdhury, A.D.; Abou-Hamad, E.; Aguilar-Tapia, A.; Hazemann, J.L.; Wehbe, N.; Al Abdulghani, A.J.; Kozlov, S.M.; et al. Tandem Conversion of CO_2 to Valuable Hydrocarbons in Highly Concentrated Potassium Iron Catalysts. *ChemCatChem* **2019**, *11*, 2879–2886. [CrossRef]
353. Zeng, F.; Mebrahtu, C.; Xi, X.; Liao, L.; Ren, J.; Xie, J.; Heeres, H.J.; Palkovits, R. Catalysts Design for Higher Alcohols Synthesis by CO_2 Hydrogenation: Trends and Future Perspectives. *Appl. Catal. B Environ.* **2021**, *291*, 120073. [CrossRef]
354. Ojelade, O.A.; Zaman, S.F. A Review on Pd Based Catalysts for CO_2 Hydrogenation to Methanol: In-Depth Activity and DRIFTS Mechanistic Study. *Catal. Surv. Asia* **2020**, *24*, 11–37. [CrossRef]
355. Manrique, R.; Rodríguez-Pereira, J.; Rincón-Ortiz, S.A.; Bravo-Suárez, J.J.; Baldovino-Medrano, V.G.; Jiménez, R.; Karelovic, A. The Nature of the Active Sites of Pd-Ga Catalysts in the Hydrogenation of CO_2 to Methanol. *Catal. Sci. Technol.* **2020**, *10*, 6644–6658. [CrossRef]
356. Xu, D.; Ding, M.; Hong, X.; Liu, G.; Tsang, S.C.E. Selective C_{2+} Alcohol Synthesis from Direct CO_2 Hydrogenation over a Cs-Promoted Cu-Fe-Zn Catalyst. *ACS Catal.* **2020**, *10*, 5250–5260. [CrossRef]
357. Modak, A.; Bhanja, P.; Dutta, S.; Chowdhury, B.; Bhaumik, A. Catalytic Reduction of CO_2 into Fuels and Fine Chemicals. *Green Chem.* **2020**, *22*, 4002–4033. [CrossRef]
358. Mao, Z.; Gu, H.; Lin, X. Recent Advances of Pd/C-Catalyzed Reactions. *Catalysts* **2021**, *11*, 1078. [CrossRef]
359. Zheng, J.N.; An, K.; Wang, J.M.; Li, J.; Liu, Y. Direct Synthesis of Ethanol via CO_2 Hydrogenation over the Co/La-Ga-O Composite Oxide Catalyst. *Ranliao Huaxue Xuebao/J. Fuel Chem. Technol.* **2019**, *47*, 697–708. [CrossRef]
360. Wang, L.; He, S.; Wang, L.; Lei, Y.; Meng, X.; Xiao, F.S. Cobalt-Nickel Catalysts for Selective Hydrogenation of Carbon Ioxide into Ethanol. *ACS Catal.* **2019**, *9*, 11335–11340. [CrossRef]
361. Zhang, S.; Wu, Z.; Liu, X.; Shao, Z.; Xia, L.; Zhong, L.; Wang, H.; Sun, Y. Tuning the Interaction between Na and Co_2C to Promote Selective CO_2 Hydrogenation to Ethanol. *Appl. Catal. B Environ.* **2021**, *293*, 120207. [CrossRef]
362. Zhang, H.; Han, H.; Xiao, L.; Wu, W. Highly Selective Synthesis of Ethanol via CO_2 Hydrogenation over $CoMoC_x$ Catalysts. *ChemCatChem* **2021**, *13*, 3333–3339. [CrossRef]
363. Jiang, J.; Wen, C.; Tian, Z.; Wang, Y.; Zhai, Y.; Chen, L.; Li, Y.; Liu, Q.; Wang, C.; Ma, L. Manganese-Promoted Fe_3O_4 Microsphere for Efficient Conversion of CO_2 to Light Olefins. *Ind. Eng. Chem. Res.* **2020**, *59*, 2155–2162. [CrossRef]
364. Aitbekova, A.; Goodman, E.D.; Wu, L.; Boubnov, A.; Hoffman, A.S.; Genc, A.; Cheng, H.; Casalena, L.; Bare, S.R.; Cargnello, M. Engineering of Ruthenium–Iron Oxide Colloidal Heterostructures: Improved Yields in CO_2 Hydrogenation to Hydrocarbons. *Angew. Chem.-Int. Ed.* **2019**, *58*, 17451–17457. [CrossRef]
365. Li, Y.; Gao, W.; Peng, M.; Zhang, J.; Sun, J.; Xu, Y.; Hong, S.; Liu, X.; Wei, M.; et al. Interfacial Fe_5C_2-Cu Catalysts toward Low-Pressure Syngas Conversion to Long-Chain Alcohols. *Nat. Commun.* **2020**, *11*, 61. [CrossRef]
366. Xu, Y.; Zhai, P.; Deng, Y.; Xie, J.; Liu, X.; Wang, S.; Ma, D. Highly Selective Olefin Production from CO_2 Hydrogenation on Iron Catalysts: A Subtle Synergy between Manganese and Sodium Additives. *Angew. Chem.-Int. Ed.* **2020**, *59*, 21736–21744. [CrossRef]
367. Yao, R.; Wei, J.; Ge, Q.; Xu, J.; Han, Y.; Ma, Q.; Xu, H.; Sun, J. Monometallic Iron Catalysts with Synergistic Na and S for Higher Alcohols Synthesis via CO_2 Hydrogenation. *Appl. Catal. B Environ.* **2021**, *298*, 120556. [CrossRef]
368. Ye, X.; Ma, J.; Yu, W.; Pan, X.; Yang, C.; Wang, C.; Liu, Q.; Huang, Y. Construction of Bifunctional Single-Atom Catalysts on the Optimized β-Mo_2C Surface for Highly Selective Hydrogenation of CO_2 into Ethanol. *J. Energy Chem.* **2022**, *67*, 184–192. [CrossRef]
369. An, B.; Li, Z.; Song, Y.; Zhang, J.; Zeng, L.; Wang, C.; Lin, W. Cooperative Copper Centres in a Metal-Organic Framework for Selective Conversion of CO_2 to Ethanol. *Nat. Catal.* **2019**, *2*, 709–717. [CrossRef]
370. Shafer, W.D.; Jacobs, G.; Graham, U.M.; Hamdeh, H.H.; Davis, B.H. Increased CO_2 Hydrogenation to Liquid Products Using Promoted Iron Catalysts. *J. Catal.* **2019**, *369*, 239–248. [CrossRef]
371. Yang, C.; Liu, S.; Wang, Y.; Song, J.; Wang, G.; Wang, S.; Zhao, Z.J.; Mu, R.; Gong, J. The Interplay between Structure and Product Selectivity of CO_2 Hydrogenation. *Angew. Chem.-Int. Ed.* **2019**, *58*, 11242–11247. [CrossRef]
372. Liu, B.; Ouyang, B.; Zhang, Y.; Lv, K.; Li, Q.; Ding, Y.; Li, J. Effects of Mesoporous Structure and Pt Promoter on the Activity of Co-Based Catalysts in Low-Temperature CO_2 Hydrogenation for Higher Alcohol Synthesis. *J. Catal.* **2018**, *366*, 91–97. [CrossRef]
373. Wang, L.; Wang, L.; Zhang, J.; Liu, X.; Wang, H.; Zhang, W.; Yang, Q.; Ma, J.; Dong, X.; Yoo, S.J.; et al. Selective Hydrogenation of CO_2 to Ethanol over Cobalt Catalysts. *Angew. Chem.-Int. Ed.* **2018**, *57*, 6104–6108. [CrossRef]
374. Ouyang, B.; Xiong, S.; Zhang, Y.; Liu, B.; Li, J. The Study of Morphology Effect of Pt/Co_3O_4 Catalysts for Higher Alcohol Synthesis from CO_2 Hydrogenation. *Appl. Catal. A Gen.* **2017**, *543*, 189–195. [CrossRef]
375. Li, Z.; Ji, S.; Liu, Y.; Cao, X.; Tian, S.; Chen, Y.; Niu, Z.; Li, Y. Well-Defined Materials for Heterogeneous Catalysis: From Nanoparticles to Isolated Single-Atom Sites. *Chem. Rev.* **2020**, *120*, 623–682. [CrossRef]

376. Häusler, J.; Pasel, J.; Woltmann, F.; Everwand, A.; Meledina, M.; Valencia, H.; Lipińska-Chwałek, M.; Mayer, J.; Peters, R. Elucidating the Influence of the D-Band Center on the Synthesis of Isobutanol. *Catalysts* **2021**, *11*, 406. [CrossRef]
377. Tang, C.W.; Liu, C.H.; Wang, C.C.; Wang, C. Bin Electro-Oxidation of Methanol, Ethanol and Ethylene Glycol over Pt/TiO$_2$-C and PtSn/TiO$_2$-C Anodic Catalysts. *Int. J. Electrochem. Sci.* **2021**, *16*, 211045. [CrossRef]

Disclaimer/Publisher's Note: The statements, opinions and data contained in all publications are solely those of the individual author(s) and contributor(s) and not of MDPI and/or the editor(s). MDPI and/or the editor(s) disclaim responsibility for any injury to people or property resulting from any ideas, methods, instructions or products referred to in the content.

Review

Oxygenated Hydrocarbons from Catalytic Hydrogenation of Carbon Dioxide

Wan Nor Roslam Wan Isahak [1,*], Lina Mohammed Shaker [1] and Ahmed Al-Amiery [1,2]

[1] Department of Chemical and Process Engineering, Faculty of Engineering and Built Environment, Universiti Kebangsaan Malaysia (UKM), Bangi 43000, Malaysia
[2] Energy and Renewable Energies Technology Center, University of Technology, Baghdad 10001, Iraq
* Correspondence: wannorroslam@ukm.edu.my; Tel.: +60-038-9216424

Abstract: Once fundamental difficulties such as active sites and selectivity are fully resolved, metal-free catalysts such as 3D graphene or carbon nanotubes (CNT) are very cost-effective substitutes for the expensive noble metals used for catalyzing CO_2. A viable method for converting environmental wastes into useful energy storage or industrial wealth, and one which also addresses the environmental and energy problems brought on by emissions of CO_2, is CO_2 hydrogenation into hydrocarbon compounds. The creation of catalytic compounds and knowledge about the reaction mechanisms have received considerable attention. Numerous variables affect the catalytic process, including metal–support interaction, metal particle sizes, and promoters. CO_2 hydrogenation into different hydrocarbon compounds like lower olefins, alcoholic composites, long-chain hydrocarbon composites, and fuels, in addition to other categories, have been explained in previous studies. With respect to catalyst design, photocatalytic activity, and the reaction mechanism, recent advances in obtaining oxygenated hydrocarbons from CO_2 processing have been made both through experiments and through density functional theory (DFT) simulations. This review highlights the progress made in the use of three-dimensional (3D) nanomaterials and their compounds and methods for their synthesis in the process of hydrogenation of CO_2. Recent advances in catalytic performance and the conversion mechanism for CO_2 hydrogenation into hydrocarbons that have been made using both experiments and DFT simulations are also discussed. The development of 3D nanomaterials and metal catalysts supported on 3D nanomaterials is important for CO_2 conversion because of their stability and the ability to continuously support the catalytic processes, in addition to the ability to reduce CO_2 directly and hydrogenate it into oxygenated hydrocarbons.

Keywords: 3D nanomaterial; carbon nanotube; graphene; catalyst; hydrocarbon; oxygenated hydrocarbon

Citation: Isahak, W.N.R.W.; Shaker, L.M.; Al-Amiery, A. Oxygenated Hydrocarbons from Catalytic Hydrogenation of Carbon Dioxide. *Catalysts* 2023, *13*, 115. https://doi.org/10.3390/catal13010115

Academic Editors: Sagadevan Suresh and Is Fatimah

Received: 3 November 2022
Revised: 18 December 2022
Accepted: 19 December 2022
Published: 4 January 2023

Copyright: © 2023 by the authors. Licensee MDPI, Basel, Switzerland. This article is an open access article distributed under the terms and conditions of the Creative Commons Attribution (CC BY) license (https:// creativecommons.org/licenses/by/ 4.0/).

1. Introduction

A common "Janus" type of molecule is carbon dioxide (CO_2). Several issues, including the cost of the ligand and/or the base, remain unresolved when using ecologically desirable metals to reduce CO_2. Hydrocarbons with oxygenated functional groups [1], such as carbonylic (-CO-) [2] and alcoholic (-OH) groups, are known as oxygenated hydrocarbons (Oxy-HCs). Oxy-HCR has the potential to be a cleaner, more sustainable substitute for current fossil fuels. Steam reforming of mixtures of Oxy-HCs (Oxy-HCSR) is not thought to cause a net increase in atmospheric CO_2 because Oxy-HCs derived from biological/renewable resources are thought to be CO_2 neutral. In the semiconductor, precision machining, alcohol distillery, and biodiesel industries, oxy-HCs are typically obtained as waste byproducts [3]. Hydrocarbons are the principal constituents of petroleum and natural gas. They serve as fuels and lubricants as well as raw materials for the production of plastics, fibers, rubbers, solvents, explosives, and industrial chemicals. CO_2 is a stable compound and reactions with CO_2 are thus challenging. Nevertheless, there are various reaction pathways for CO_2 hydrogenation that are dependent on the nature of the catalyst, and a number of useful products can be obtained. Global climate change brought on by greenhouse gases has become a serious issue due to the continued use of fossil fuels, which has increased the amount of CO_2 in the atmosphere. Due to the continued increase in atmospheric CO_2 concentration (which exceeded 400 ppm in 2016), and its detrimental and potentially irreversible impact on the climate system, mitigation of CO_2 concentrations in the atmosphere is urgently needed [4]. Globally, there are plans and goals for this; the European Commission's goal is to achieve a reduction of 80–95% in greenhouse gas emissions by 2050 (compared to those of 1990) in order to achieve scientists' recommended reduction of at least 50% in global greenhouse gas emissions by 2050 [5]. China, Brazil, and Korea, among other important international partners of Europe, are tackling these problems by advancing the "low carbon economy" [6].

Currently, there are three approaches to minimize CO_2 emissions: by controlling CO_2 emissions, by capturing and storing CO_2, and by chemically converting and utilizing CO_2 [7]. Carbon storage is crucial for quickly reducing CO_2 emissions; however, it has the drawback of possible CO_2 leakage [8]. As a substitute for other carbon sources, CO_2 can be used to create feedstocks and value-added products that include carbon. In addition to providing a clean carbon supply for hydrogenation, using the CO_2 acquired through capture also helps to solve the leaking issue associated with CO_2 storage. The Sabatier reaction (CO_2 methanation) was therefore considered by the National Aeronautics and Space Administration (NASA) as a stage in recovering oxygen in closed-cycle life support systems [9]. It is possible to use even the CO_2 found in industrial exhaust gases directly as a feed for hydrogenation [10]. Therefore, it is essential and advantageous to make efficient use of renewable carbon resources in order to preserve the long-term and sustainable development of our civilization. Since CO_2 conversion needs energy input, pairing it with renewable energy would increase the sustainability and environmental friendliness of this technique. Electrocatalysis can be used to catalyze the reduction of CO_2 [11], as can photocatalysis [12] and thermal catalysis. Thermal catalysis stands out among these due to its quick kinetics and adaptable mixing of active ingredients. Being a very stable molecule, CO_2 requires energy just to be activated and then converted. The thermodynamics of the CO_2 conversion will be improved by the addition of a second material with a comparatively higher Gibbs energy. However, the fatal weakness of electrocatalysis and photocatalysis is low energy efficiency. To date, different types of metal-based electrocatalysts such as Au [13], Cu [14], Pd [15], Ag [16], Bi [17], Sn [18], and Co [19] have been intensively investigated in connection with electrochemical CO_2 reduction. Very recently, Chen et al. reported on the great importance of developing Au-based electrocatalysts with cost-effectiveness and high performance in order to commercialize CO_2 reduction technology [20]. Among the materials examined by Brouzgou et al. in 2016, reduced graphene oxide-based hybrid electrocatalysts exhibit both excellent activity and long-term stability [21]. They concluded that the development of the electrocatalyst by using materials with three-dimensional structures facilitates the electron and mass transfer process.

CO_2 is usually captured from high-concentration sources such as thermal power or chemical plants, steel mills, and cement factories. However, direct air capture (DAC) from the ambient air requires a separation unit to generate a concentrated CO_2 stream [22]. Consequently, CO_2 is hydrogenated with H_2 created using sustainable energy sources [23], and this is an exciting area of research that could yield chemicals and fuels as shown in Figure 1 [24]. CO_2 reacts over a catalyst with H_2 produced from water using renewable energy [25] to produce formic acid [26], lower olefins [27], methanol [28], and the higher alcohols [29], etc. In certain studies, the presence of H_2 was not detected, as reported by Sorcar et al. In 2019, in a study where researchers relied on natural sources, sunlight was used for a period of 6 h continuously (sustainable Joules) to recycle CO_2 into Joules-hydrocarbon fuel with a photoconversion efficiency of 1% and an efficiency of quantity estimated at 86%. The researchers reported on the use of Cu-Pt nanoparticles (Cu-Pt NPs) for the photoreduction of CO_2. From this process, methane and ethane resulted in the proportions of 3 mmol g^{-1} and 0.15 mmol g^{-1}, respectively [30]. The former problem has received considerable attention, and researchers have already made significant strides in water electrolysis to produce H_2 using electricity produced by solar, wind, or other renewable energy sources, as well as in water splitting using photocatalytic, photo-electrochemical, or other photochemical processes. Density Functional Theory (DFT) calculations and experimental studies of the CO_2 conversion mechanism and hydrocarbon chain formation have, however, received relatively little attention in reviews to date. This review highlights the progress of research into the use of nanomaterials with three-dimensional (3D) structures and their compounds, and methods for their synthesis, in the process of the hydrogenation of CO_2. Recent advances in catalytic performance and the conversion mechanism for CO_2 hydrogenation into hydrocarbons that have been made both through experiments and DFT simulations are also discussed. The development of 3D nanomaterials and metal catalysts supported on 3D nanomaterials is important for CO_2 conversion because of their stability and the ability to continuously support reverse transformation and Fischer–Tropsch catalysis (FT), in addition to the ability to reduce CO_2 directly and hydrogenate it into oxygenated hydrocarbons.

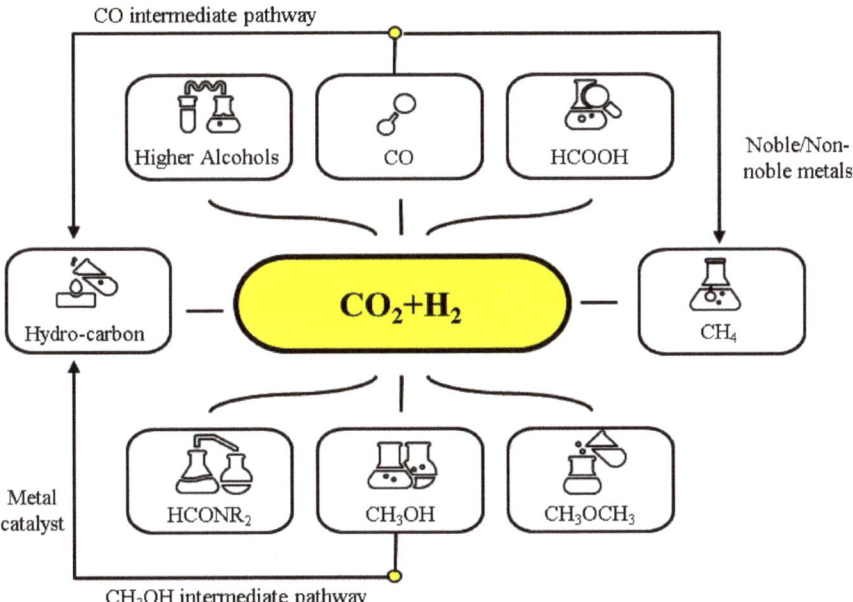

Figure 1. CO_2 to be hydrogenated to produce chemical compounds and fuels.

2. Synthesis of 3D-Structure Materials

Furthermore, different carbon or 2D-based material categories like single- or few-layer graphene are being employed for synthesizing 3D-based materials in the literature. That is, in addition to graphene, a variety of zero-, one-, and two-dimensional carbon compounds are also available, such as zero dimensional fullerene (0D) [31], 1D carbon nanotube (CNT) [32], graphene nanoribbon (GNR), carbon nanofiber (CNF) [33], and 2D transition disulfide (TMD) [34]. Due to the electrical properties being close to those of graphene, graphitic carbon nitride (g-C_3N_4) has been extensively exploited for 3D structure formation [35]. The discovery of fullerenes marked the beginning of research into carbon nanostructures (Figure 2a). The graphitization of nanodiamonds (ND) or monolithic structures of 0D materials like fullerenes or onion-like carbon are examples. 3DGMs have been created using fullerenes, which are C60 molecules [31]. The characteristics of C60 are entirely distinct from those of CNTs due to differences in size and shape. Consequently, the 3D structure of 0D materials (C60 molecules) has various preparation techniques and potential applications. One-dimensional CNTs have been employed much more to build 3D structures than 0D fullerenes. It is interesting to note that 3D CNT aerogels were first published before 3D graphene [36]. The creation of freestanding CNT aerogels was possible after the organic fabrication of a CNT suspension with an organogelator. For instance, some organic solvents, like chloroform, can gelate single-walled CNTs (SWCNTs) modified by ferrocene-grafted poly(p-phenyleneethynylene) to create sturdy 3D CNT aerogels [32]. C60 was converted into a 3D porous carbon by potassium hydroxide activation in ammonia by Zhu et al. in their work [31]. A 3D porous carbon can be created by activating C60 powder with KOH in an Ar flow, as is briefly illustrated in Figure 2b. A 3D porous carbon that has had N added to it can be produced if the KOH activation is carried out in an NH_3 atmosphere. Pyridinic and pyrrolic nitrogen are the two kinds of doped N atoms. Meso- and macropore volume in carbon results in the desired energy storage being significantly increased by N-doping scenarios. This is actually done by the graphitization of GO and ND films, such as the mesoporous graphite film prepared by Shi et al. in 2011 [37]. GO is reduced to RGO during the graphitizing process, while ND is transformed into carbon that resembles an onion. RGO sheets were sandwiched with carbon shaped like an onion, which not only stops graphene sheets from aggregating but also creates mesopores.

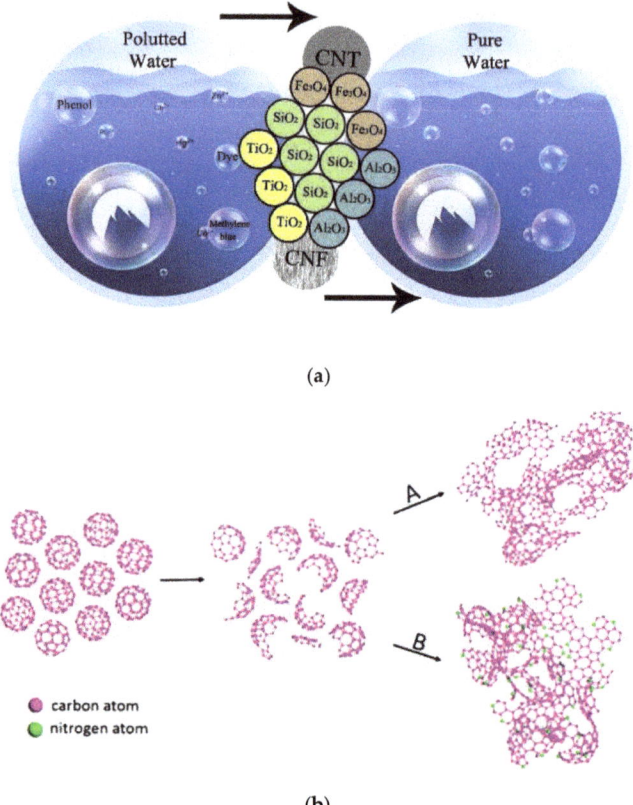

Figure 2. (a). Environmental uses of CNFs and CNTs, and (b) a diagram showing how KOH activates the C60 molecules. Route A depicts the typical activation carried out in an Ar flow, while Route B displays N-doping when NH3 flow is involved in the activation. Reprinted with permission from Ref. [31]. Copyright © 2016, John Wiley and Sons.

Surfactants are additionally employed to disperse the CNTs and produce the aerogels. In 2007, various quantities of SWCNTs were floated in the water with sodium dodecylbenzene sulfonate (from 5–13 mg mL^{-1}) by Yodh et al. [38]. The suspensions were allowed to be converted into elastic gels overnight before being dipped into poly(vinyl alcohol) (PVA) aqueous solutions at 90 °C. Using a chemical vapor deposition (CVD) approach, Gui et al. in 2010 reported injecting a ferrocene precursor solution into dichlorobenzene to design a macroscopic and monolithic multi-walled CNT (MWCNT) sponge [39]. The fact that the diameters and the lengths of these MWCNTs varied from 30–50 nm and 10–100 mm, respectively, indicates a thick sponge of many layers of CNTs. Many CNT piles continuously stack and reach a centimeter of thickness during the growing phase of CNT sponges. The constructive vertical alignment of CNTs up from the bottom to the upper surface was a promising method to construct a 3D CNT structure without the continuous-stacking growing process. One- to three-WCNT carpets can operate well on their own, especially when they are tightly coupled to high-quality graphene.

In 2010, Zhang et al. used the CVD approach to create CNT-pillared GO and RGO platelets, for which acetonitrile was used as a carbon source and nickel as a catalyst [40]. It is possible to customize the CNTs' alignment, density, and length. A technique to create carpets of covalently bound graphene and CNTs using a floating buffer layer was revealed by Zhu et al. in 2012 [41]. In this procedure, the deposition of iron (catalyst layer) and alumina (buffer layer) was achieved to coat the graphene in sequence by electron beam (e-beam) evaporation after graphene was first produced on the Cu foil. It should also be noted that hybrid graphene–CNT ohmic-linked carpets possess a high surface area without sacrificing their standalone features [42].

In the publications cited above, CNTs can only grow on the graphene surface where catalyst particles have been deposited via dip-coating or e-beam evaporation. This, in turn, is one of the major reasons for the difficulty of producing the higher loading of active components in these graphene–CNT hybrid materials, which is crucial for electrochemical devices with high energy density. A straightforward method for creating a 3D structure made of graphene foam (GF) and CNTs was devised by Liu et al. in 2014 [43]. Therein, a hydrothermal approach was used for loading the NPs catalyst over GF. This enhanced the active component (MnO_2) loading and allowed for significantly better CNT growth on the GF than was possible with dip-coating or e-beam deposition. More recently, Jin et al. in 2016 demonstrated how a 3D current collector could make a structure of covalent carbon bonds [44]. Covalent carbon–carbon bonds bind several micrometer-long bundles of CNTs into an ultrathin GF. The coupling of e-holes in the latter composite is enabled by carbon–carbon bonds, and such bonds facilitate the charges' transportation between out circuits and electrochemically active materials.

It is important to highlight that the GNR composites created by longitudinally unzipped MWCNTs may retain their proper structure and improved electrical conductivity, which are characteristics of both CNT and graphene. An in situ unzipped approach was also realized through the CNT sponge conversion directly into GNR aerogel by Peng et al. in 2014 [45]. In this instance, the walls were opened with $KMnO_4$ after the oxidative chemical fluid was in-filtered into the porous sponge in order to free the walls of the nanotubes from any kind of defects. Furthermore, organic compounds like pyrrole were used by Chen et al. in 2015 as nitrogen (N) source and reagent to fabricate ultralight, highly conductive, 3D N-doped GNR aerogels [46]. GNRs are unique in that they differ from graphene sheets in having higher length/width ratio and straight edges, as well as perfect surface regularity with few flaws on the substrate. This is achieved by using GNR aerogels doped with heteroatoms to get an enhanced electronic energy-gap modulation, boosting both the reactivity of the materials and their ability in electrocatalysis when utilized as oxygen reduction reaction (ORR) catalysts. The construction of 3D carbon material types frequently involves CNFs' self-assembly into a macroscopic structure [33]. Additionally, inexpensive components like bacterial cellulose (BC) and pitch may enable large-scale production [47]. For instance, in 2014, Chen et al. worked to develop a free-standing heteroatom-doped

CNF fabrication method [48]. Initially, BC was submerged for 10 h at normal conditions in H_3PO_4-H_3BO_3 liquid solution. A drying step was followed by thermal treating in N_2 atmosphere for the creation of 3D carbon CNF doping heteroatoms. Furthermore, such heteroatom co-doped 3D CNFs prepared from inexpensive raw materials have outstanding energy storage performance [49]. Additionally, the template-based method has also been widely employed to create 3D CNFs [50]. Using an inexpensive melamine sponge (MS) template, Zhu et al. in 2019 produced a macroscopic 3D porous graphite C_3N_4 structure from 2D graphite C_3N_4 by one-step thermal polymerization of urea [51]. This approach is practical for the production of 3D C_3N_4 structures due to high urea loading and the light weight and good water absorbability of MS. In this instance, the produced 3D C_3N_4 samples are readily shaped by blades. Without using strong acid, Wang et al. in 2017 created C_3N_4 aerogels using an aqueous sol-gel method [52]. First, C_3N_4 NPs were produced utilizing a salt molten technique along with temperature-induced condensation of melamine using potassium thiocyanate as the solvent. The C_3N_4 hydrogels could then be produced by the C_3N_4 NP sol solution self-assembling. C_3N_4 aerogels were made using a freeze-drying process. This approach is notable because of the ability to vary the produced size and mass through the process, and its affordability, as well as the assembly without cross-linking agents.

The production and characterization of monolithic, ultra-low-density TMD (WS_2 and MoS_2) aerogels were described by Worsley et al. in 2015 [34]. Thermal degradation of freeze-dried ammonium thio-molybdate (ATM) and ammonium thio-tungstate (ATT) solutions produces the monolithic WS_2 and MoS_2 aerogels, respectively. By merely altering the initial ATM and ATT concentrations, the densities of the pure TMD aerogels may be changed to correspond to 0.4 and 0.5% of the densities of single crystals of MoS_2 and WS_2, respectively. In 2019, Abu Zied and Alamry invented a new green synthesis method for producing 3D hierarchical Co_3O_4-C NPs [53]. The extract of basil leaves (BLE) was used as a low-cost source of carbon and green template. For use as catalysts in hydrogen generation via sodium borohydride hydrolysis, various Co_3O_4-C NPs have been tested. Research findings have shown both the presence of 3D porous hierarchical NPs and calcination temperature influence activity. Common features of 3D graphene, including superior mechanical strength, hierarchical porosity, large surface area, and perfect electrical conductivity, give such new materials considerable potential for applications related to catalysis, the environment, biomedicine, and most importantly, energy. Liu and Xu rapidly created a variety of 3D graphene compounds in 2019 by inventing a number of adaptable techniques [54]. The fact that graphene is a naturally occurring 2D polymer (2DP) has greatly sparked interest in the rational organic-chemical synthesis of novel 2DPs at the atomic or molecular level. The development of synthetic polymer chemistry can benefit greatly from the regulated synthesis of 2DPs with optimized molecules and superior ease of processing. Additionally, it demonstrates tremendous strength in the creation of unique polymer composites with desired characteristics and capabilities that are uncommon in traditional 1D polymers. Designing and making 2DPs that simultaneously incorporate a 2D conjugated plane, in-plane homogeneous microporosity, and electrochemically active groups is difficult yet important for the energy sector. On the other hand, due to the substantial effective surfaces, flexibility, and cycling stability, hierarchical 3D carbon nanoscale is a promising material variety for electrochemical energy applications. In order to build 3D carbon nanostructures made from carbon fibers (CFs) and electro-spun CNF (ECNFs), Alali et al. combined electrospinning with in situ CVD techniques in 2021, as illustrated in Figure 3 [55]. Ni/CNFs/ECNFs demonstrated satisfactory hydrogen evolution reaction (HER) activity in an alkaline medium with a low overpotential of 88 mV to give 10 mV cm^2 current density and Tafel slope of 170 mV dec^{-1}. This was based on the nano-nonwoven structures and forest-like growing nanostructured CNFs.

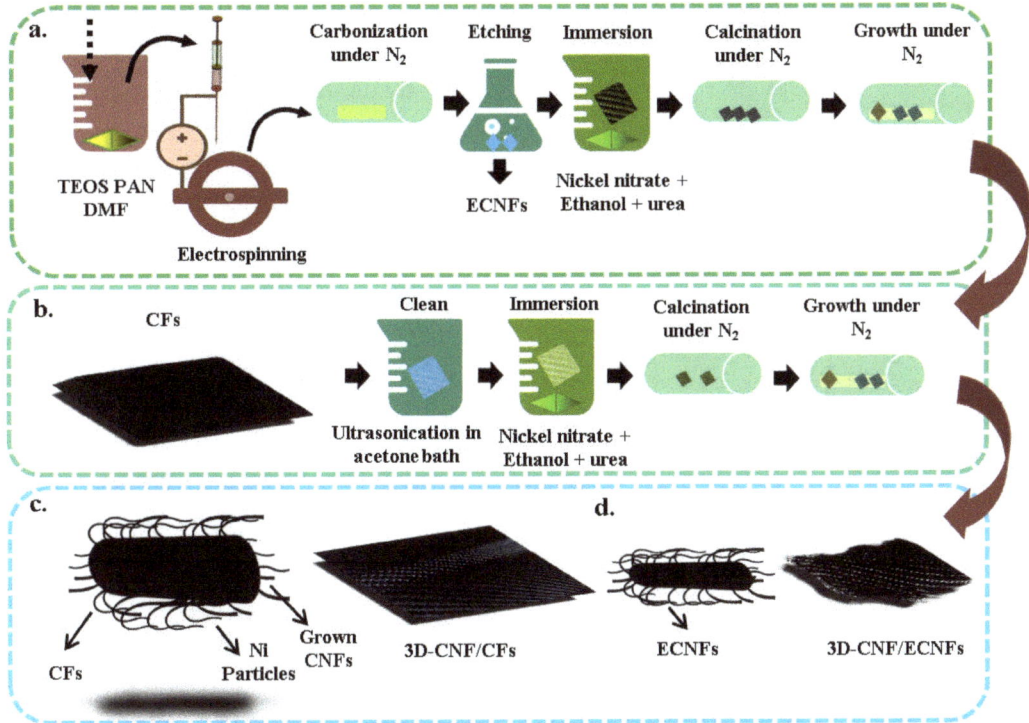

Figure 3. The steps resulting in the production of: (**a**) ECNFs and 3D-Ni/CNFs/ECNFs, (**b**) 3D-Ni/CNFs/CFs, (**c**) 3D-Ni/CNFs/CFs, and (**d**) 3D-Ni/CNFs/ECNFs. Adapted with permission from Ref. [55]. Copyright © 2021, Elsevier.

3. Graphene Production Methods
3.1. Chemical Vapor Deposition (CVD)-Based Methods

There are numerous CVD techniques that can be used today for synthesizing material compounds depending on graphene, as illustrated in Figure 4. These procedures can be categorized into seven major approaches based on the characteristics of process variables (temperature, pressure, nature of the precursor, gas flow state, deposition time, activation manner, and wall/substrate temperature) [56]. The procedures shown in Figure 5 accomplished by Arjmandi-Tash et al. in 2017 have developed the CVD-growth of graphene modalities that combine cold- and hot-wall reaction chambers [57]. Such a hybrid approach boosts growth quality to a level now comparable to other conventional CVD methods in hot-wall chambers while preserving the benefits of a cold-wall chamber, such as steady growth and high efficiency and maintaining power. Uniform monolayers of produced graphene were formed. Especially in comparison to graphene produced in cold-wall reaction chambers, charge transition experiments show a considerable increase in charge carrier mobility. Using a cold-wall CVD reactor, Alnuaimi et al. (2017) investigated the influence of graphene growth temperature and demonstrated that multilayer nucleation density is decreased under high temperatures [58]. The temperatures in that work ranged from 1000 to 1060 °C. Multilayer graphene was growing remarkably at a temperature of 1000 °C, but the nucleation rate was adversely affected at 1060 °C, so at lower growth temperatures, larger defect densities were detected. In 2019, Al-Hagri et al. created a single layer of vertically aligned graphene nano-sheet arrays (VAGNAs) with a high surface area on a Ge substrate at 625 °C using the radio frequency (RF) approach [59]. When evaluated as a surface-enhanced Raman spectroscopy (SERS) platform, the obtained graphene demonstrated detection performance reduced to 10^{-6} M of Rhodamine 6G (R6G). By adjusting H_2 (P_{H2}) and CH_4 (P_{CH4}) (P_{CH4}-P_{H2}) partial pressure ratio, Chen et al. in 2020 created single-crystalline hexagonal bi-layer graphene (BLG) in a single step with a controlled twist angle between the layers [60].

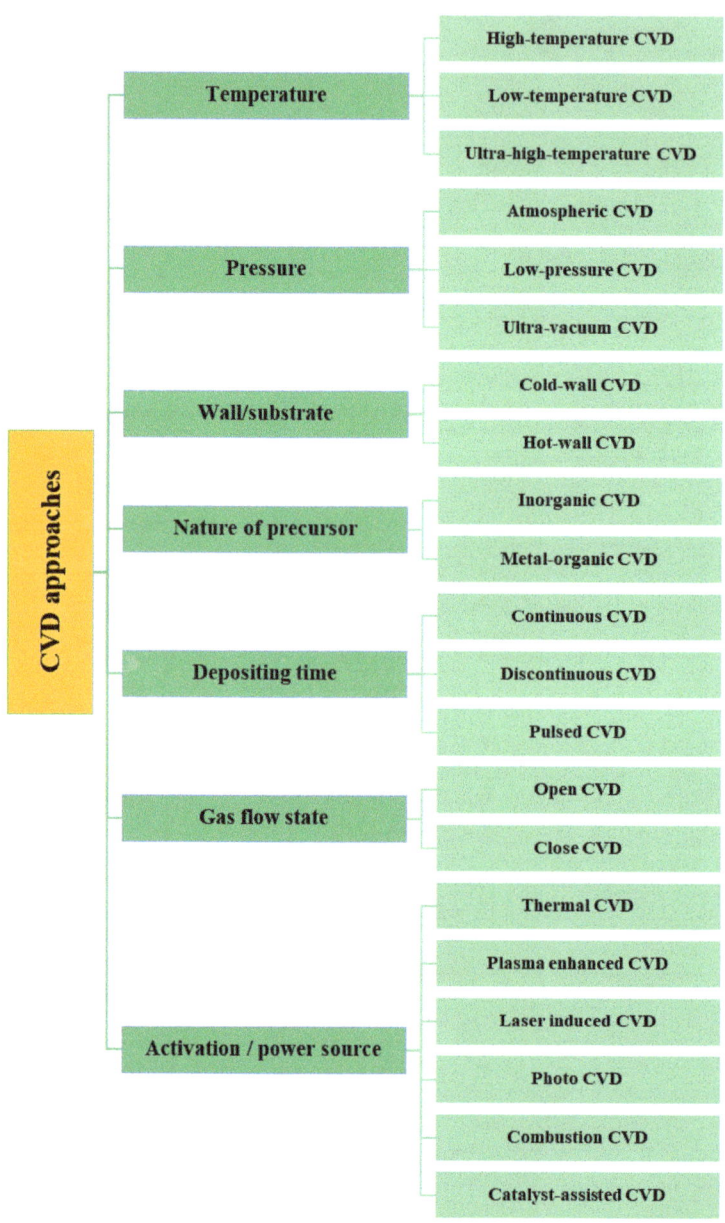

Figure 4. Summary of CVD methodologies.

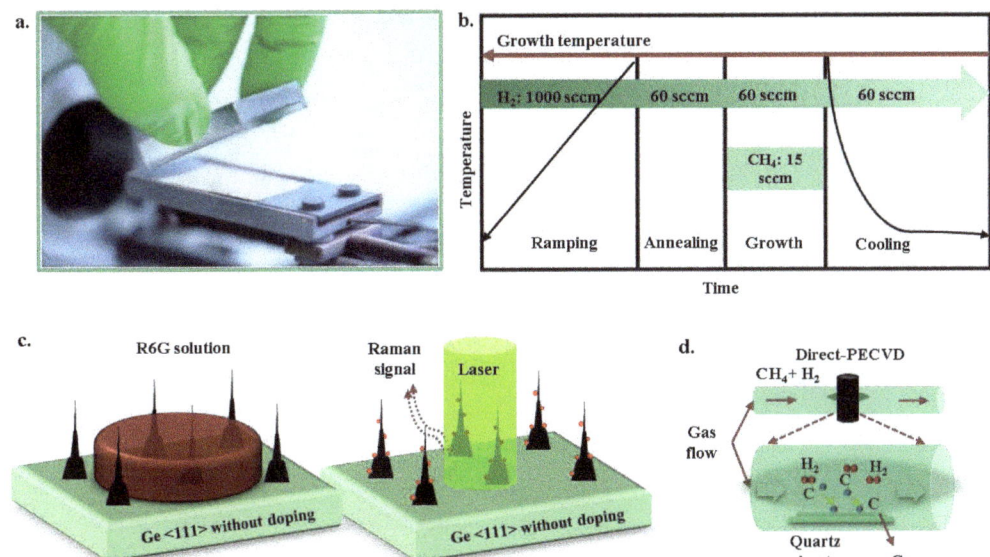

Figure 5. (**a**) Cold-wall chamber CVD growth of graphene. Reprinted with permission from Ref. [57]. Copyright © 2017, Elsevier. (**b**) The strategies that have been adopted during CVD-based graphene development. Adapted with permission from Ref. [58]. Copyright © 2017, RSC; (**c**) experimental process diagram, using VAGNAs as the SERS substrate. Adapted with permission from Ref. [59]. Copyright © 2022, Elsevier; and (**d**) production of twisted bilayer graphene with a controlled twist angle in sizes of mm and cm. Adapted with permission from Ref. [60]. Copyright © 2022, Elsevier.

3.2. Solution-Based Methods

Little if any specialist equipment is required for solution-based approaches, which assemble GO sheets onto 3D templates before chemically reducing GO to RGO (Figure 6a). In 2012, Sohn et al. succeeded in making 3D graphene capsules using spray pyrolysis with a mixture of GO-polystyrene colloidal particles [61]. An evaporation-induced capillary force was used for attaching the GO sheets to the polymer colloidal solution. The solvents used in a solution-based approach can have a significant impact on the nanostructure of the composites made of Li_2S-graphene. In 2014, Yan and his collaborators found a new methodology employing alumina fiber blanket (AFB) as a template for large-scale fabrication of microchannel-network graphene foams (mCNGFs) [62]. The procedure steps to prepare mCNGFs are depicted in Figure 6b. For effectively absorbing the GO suspension, a uniform GO solution had been used to immerse the AFB previously. AFB has a hydrophilic exterior due to the existence of hydroxyl groups on its surface. As a result, GO can readily bind to the carboxyl and hydroxyl groups on the surface of the AFB template. Additionally, capillary forces help to drive the GO to penetrate the AFB template and fill the unoccupied spaces between the alumina fibers. Following immersion in N_2 fluid, the GO-AFB composite was then freeze-dried, and the GO connected to the AFB template was converted to RGO via a thermal treatment under nitrogen atmosphere at 500 °C. Finally, HF was used to remove the AFB template to acquire the pure mCNGFs. In 2018, Shunxin et al. reported that anhydrous N-methylpyrrolidone (NMP) has excellent wettability characteristics with graphene and its functional groups have higher energies with Li_2S. NMP solvent was utilized in that study for Li_2S-graphene composite preparation [63]. The researchers claimed that the graphene surface had a good amount of uniformly distributed nano-Li_2S. Apparently, nano-Li_2S presence reduces the p-p interactions between graphene sheets, and the composite of Li_2S-graphene displays a honeycomb-like structures with a majority of micropores. The Li_2S-graphene composite showed better electrochemical performance in terms of high columbic efficiency, low potential barrier, highly energetic capacity, and a high-rate capability.

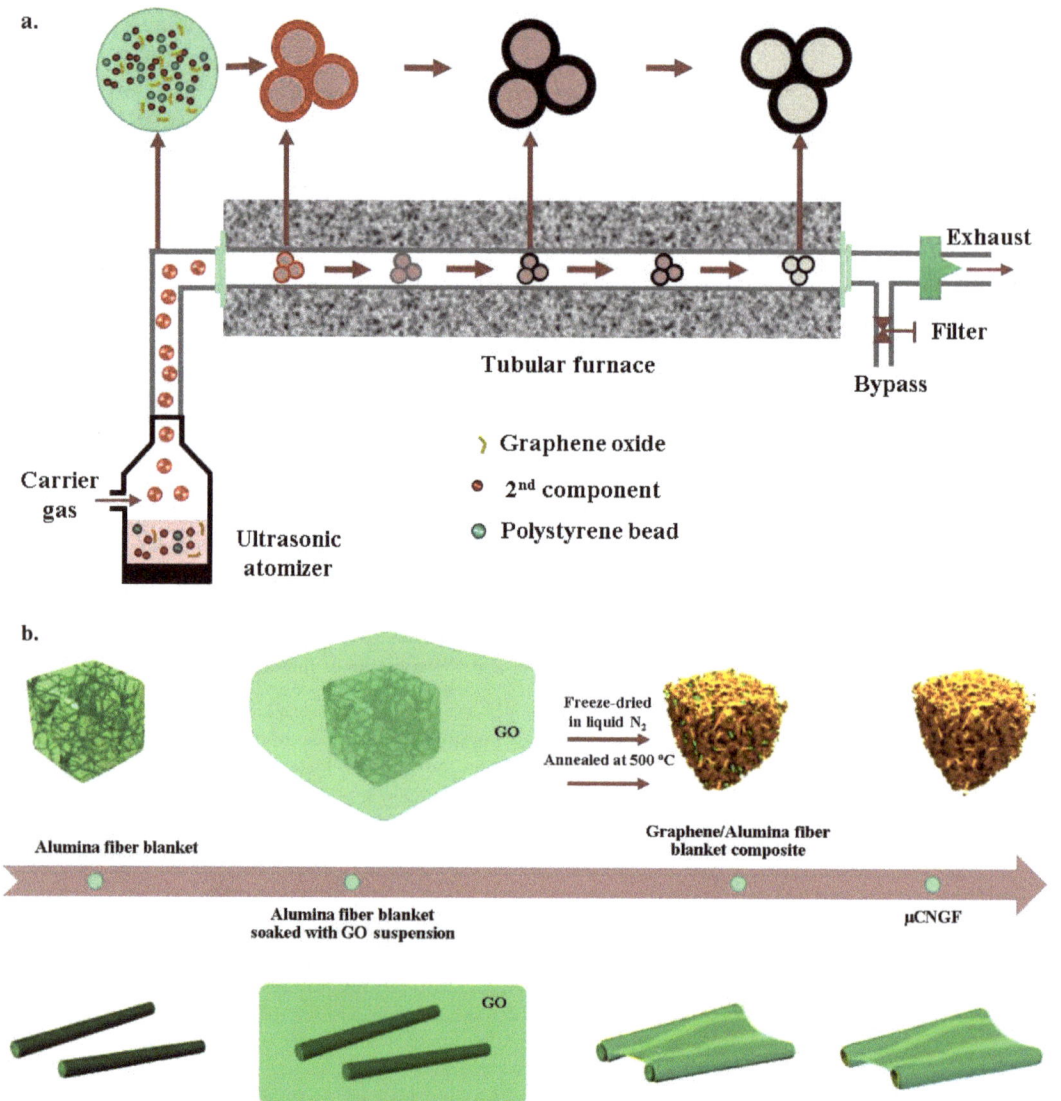

Figure 6. (a) rGO capsule production experimentally. Adapted with permission from Ref. [61]. Copyright © 2012, Royal Society of Chemistry. (b) The procedures involved in producing mCNGFs. Adapted with permission from Ref. [62]. Copyright © 2014, Royal Society of Chemistry.

3.3. Three-Dimensional (3D) Printing

An efficient and simple methodology to facilitate direct fabrication of 3D bulk objects is 3D printing. [64]. By carefully casting Ni and sucrose mixture onto a substrate, Sha et al. in 2017 reported the effective development of an automated metal powder 3D-printing approach for in situ synthesis of free-standing 3D GFs. They then used a commercial CO_2 laser to transform the Ni-sucrose mixture into 3D GFs, as clearly shown in Figure 7a [65].

This technology permits direct in situ 3D printing of GFs without the need for a rising furnace temperature or an extended growing phase. It blends powder metallurgy templating with 3D-printing techniques. The 3D-printed GFs exhibit multilayer, low density (0.015 g cm^{-3}), high quality, and high porosity rate (99.3%) for graphene characteristics. The GFs have an impressive storage modulus of 11 kPa, an electrical conductivity of 8.7 S cm^{-1}, and a high damping capacity of 0.06. By developing hybrid inks and printing schemes to enable mixed-dimensional hybrid printability, Tang et al. (2018) proposed a generalized 3D-printing methodology for graphene aerogels and graphene-based mixed-dimensional hybrid aerogels with complex architectures, overcoming the limitations of multicomponent inhomogeneity and harsh post-treatments for additives removal (Figure 7b) [66]. The 3D-printed hybrid aerogels were also shown to act as ultrathick electrodes in a microsupercapacitor that could withstand symmetrical compression while still demonstrating quasi-proportionally improved areal capacitances under heavy mass loading. The strong ion- and electron-transition routes offered by the 3D-printed, densely linked networks were responsible for the remarkable performance.

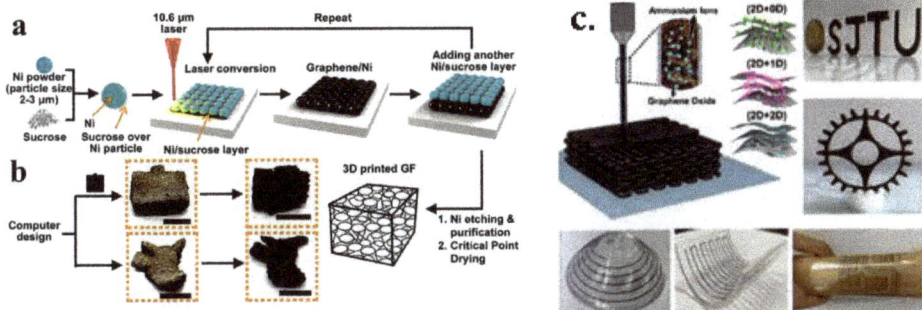

Figure 7. (**a**) Diagram of an in situ 3D GF synthesis method that simulates 3D printing. Reprinted with permission from Ref. [65]. Copyright © 2017, American Chemical Society, (**b**) images of 3D-printed GF taken before and after the Ni was dissolved. Reprinted with permission from Ref. [65]. Copyright © 2017, American Chemical Society. (**c**) Mixed-dimensional hybrid aerogels based on graphene. Reprinted with permission from Ref. [66]. Copyright © 2018, American Chemical Society.

3.4. Hydrothermal Method

Hydrothermal process flexibility increases the possibility of doping the graphene lattice with nitrogen or boron, for example, to prepare better quality 3DGMs. Specific additives, such as swelling and cross-linking agents, can be added to the GO dispersion forming these 3DGMs [67]. To produce extremely effective graphene–metal oxide-based hybrid supercapacitors, Bai et al. proposed in situ synthesis of 3D-graphene-MnO$_2$ foam composite in 2020 [68]. The 3D graphene-MnO$_2$ composite underwent in situ conformal development and exhibited excellent crystalline nature and low contact resistance, which increased the electrolyte performance at transporting charges. Relatively, the 3D conductive graphene foam allowed electrolyte ions to migrate across the MnO$_2$ surface quickly because of its porosity. In the supercapacitors, the 3D graphene-MnO$_2$ composite electrode demonstrated high specific capacitance (333.4 F g^{-1} at 0.2 A g^{-1}) and remarkable cycle stability in the absence of carbon black. This scientific method for creating a composite made of 3D graphene and MnO$_2$ offers a potential method for producing energy storage electrode materials to design high-performance supercapacitor devices. Pure 3D graphene is regarded as a suitable platform to load catalytic components, including metals, due to its low density, excellent electrical conductivity, exquisite flexibility, and high surface area [69], metal oxides [70], and metal sulfides [71]. In fact, using hydrothermal techniques, inorganic nanomaterials can grow in situ on the surface of 3D graphene.

3.5. In Situ Chemical Reduction

A 3D graphene architecture can be created using in situ self-assembling graphene fabricated by mild chemical reduction. The pore size, electrical conductivity, mechanical strength, and density of the 3D graphene preparation are all significantly affected by the choice of reducing agent. Different types of reducing agents have been studied, including hydrazine hydrate [72], metals [73], metal hydrides [73], phenolic compounds [74], and reduced sugars [75]. In 2011, Chen et al. created 3D graphene from a homogenous dispersion of Fe_3O_4 NPs in GO aqueous suspension. Additionally, they produced a 3D magnetic graphene-Fe_3O_4 aerogel during the reduction of GO to graphene [76]. This offers a useful approach for preparing 3D graphene-NP composites for a variety of uses, such as energy conversion and catalysis. For a superior and reasonably priced electro-catalyst, Kabtamu et al. reported a 3D annealed tungsten trioxide nanowire-graphene sheet (3D annealed WO_3 NWs/GS) foam in 2017 [77]. It was produced using in situ self-assembling graphene sheets that were prepared by mild chemical reduction, then freeze-dried, and finally annealed using vanadium redox flow battery (VRFB) electrodes (Figure 8). A 3D annealed WO_3 NWs/GS foam exhibited the desired electrocatalytic activity toward V^{2+}-V^{3+} and VO^{2+}-VO^{2+} redox couples. Charge–discharge tests further demonstrated that the 3D annealed WO_3 NWs/GS foam used in a single flow cell of a VRFB exhibited excellent energy efficiencies of 79.49 and 83.73% at current densities of 80 mA cm^{-2} and 40 mA cm^{-2}, respectively. These energy efficiencies are significantly higher than those of cells assembled with pristine graphite felt and 3D WO_3 NWs/GS foam with no specific heating process. It also does not appear to have degraded after 50 charge–discharge cycles. Such findings indicate that every WO_3 NW sample is firmly anchored to the GS and are essential for aiding the redox reactions of the vanadium redox couples, are attributable to the production of new W-O-C bonds. Additionally, WO_3 NWs/GS foam confirms the VRFBs electrochemical performance according to its 3D hierarchical porous structure after annealing, as is illustrated in Figure 8.

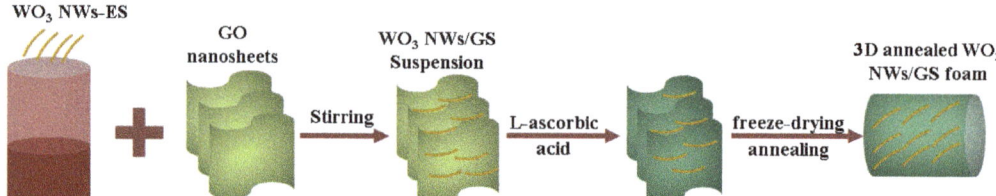

Figure 8. Graphical representation of 3D annealed WO_3 NWs-GS foam preparation. Adapted from Ref. [77]. Copyright © 2017, Royal Society of Chemistry.

3.6. Pyrolysis of Organic Precursors

Pyrolysis of organic precursor materials such as glucose [78], resorcinol-formaldehyde [79], phenol-furfural [80], and chitosan [81], is a quick and efficient way to manufacture 3DGMs on a large scale. A 3D porous graphite carbon was created by pyrolyzing a conjugated polymeric molecular precursor framework, according to To et al. in 2015 [82]. The obtained 3D porous graphite carbon had a record-high surface area (4073 m^2 g^{-1}), a sizable high porosity (2.26 cm^3), and outstanding electrical conductivity (300 S m^{-1}).

4. Uses of Graphene as a Catalyst

One of the carbon allotropes is called graphene, and it is composed of hexagons. Among the several elements of the carbon family, the use of graphene—a 2D single hexagonal carbon sheet is on the rise. Ever since the first manufacture of it was announced in 2004, the scientific and technical sectors have examined graphene in great detail [83]. Fullerene (0D), nanotube (1D), and graphite (3D) are some other allotropes, as seen in Figure 9. Reduced GO, graphene quantum dots, and GO are examples of graphene-based nanomaterials. Although some components of the graphene family have sp² and sp³ carbon atoms instead of the ideal sp² carbon atoms, this is because functional groups including hydroxyl group, carbonyl group, carboxyl group, and epoxy group have been added. GO, which is a single layer of graphite oxide, is typically created chemically by oxidizing graphite [84]. GO containing oxygen includes a variety of functional groups, as seen in Figure 9. With the exception of a small quantity of carbonyl, carboxyl, phenols, lactone, and quinones groups at the sheet's borders, these functional groups are primarily hydroxyl and epoxide groups in the basal planes [85]. GO possesses a wide variety of functional groups at its edges and basal planes, which enables it to be functionalized and exfoliated to produce well-dispersed fluids on distinct GO sheets in polar and non-polar fluids. As a result, it has many different applications, including nanocomposites [86], photocatalytic degradation [87], batteries [88], condensers [89], and sensors [90].

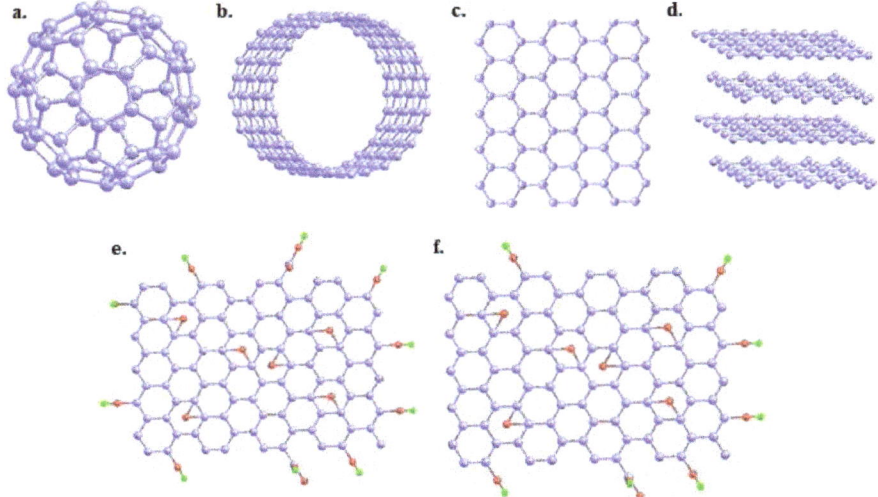

Figure 9. (a) 0D Fullerene, (b) 1D CNT, (c) 2D graphene, (d) 3D graphite structures, (e) GO, and (f) reduced GO (light purple—carbon; red—oxygen; and green—hydrogen).

Geim et al. isolated a single-graphite nanosheet layer in 2004 using a scotch tape peeling approach [91]. Because of the mentioned special characteristics in relation to chemical reactions, physical science, materials science, and mechanical applications, graphene and its compounds have received considerable attention ever since this significant advance. For example, graphene has a remarkable specific surface area of B2630 m² g^{-1}, B10,000 cm² V^{-1} s^{-1} carrier mobility, B5000 W m K^{-1} thermal conductivity at ambient temperatures, B97% optical transparency, strong chemical stability, and high mechanical strength with a B1.0 TPa Young's modulus [92].

Several synthesis techniques like CVD [93], physical exfoliation approach [94], graphitization, and chemical oxidizing cleavages [92], have been used for the purpose of 2D materials fabrication. Research on graphene in the areas of life sciences, energy applications, and environmental monitoring has advanced significantly as a result of simple techniques for graphene synthesis. Graphene is a versatile 2D building block that has been put together to form 1D fibers [95], 2D films [96], and 3D aerogels or hydrogels [97], all of which significantly broaden the scope of graphene applications and upcoming specific products. Due to the ability of 3D graphene to preserve the original properties of the material in the 2D phase, there have been extensive studies on the design and development of 3D graphene by assembling it from 2D graphene [65]. Based on these exciting characteristics, 3D graphene has indeed met some of the prerequisites to be regarded as an advanced catalyst or as a catalytic support. Meso-, micro-, and macropores are combined in 3D graphene-based materials (3DGMs) in such a way that the micro- and mesoporosity provide them with a high specific surface area while the macroporosity ensures accessibility to this surface, which is more advantageous for improving catalysis efficiency. It is important to remember that the confinement effect of catalytic elements within 3D graphene can stabilize effective regions through catalyzed reaction [98]. A distinctive benefit of these 3DGM monoliths is their integrated structure, thus making it simple to manipulate and collect when in use and eliminating any potential ecological concern brought on by the discharge of harmful graphene nanosheets [99]. The main functions and advantages of using 3DGM are illustrated in Figures 10 and 11, respectively. Numerous 3DGMs made of 2D materials with various distinct morphologies have been successfully produced and used as a result of the rapidly developing production approaches and evaluation procedures. Significantly, these 3DGMs have also been proven to perform admirably within catalyst regions. In addition to surface area, it is clear that the monolith contact angle will dictate how easily they can access electrolytes, which will ultimately impact the effectiveness of their catalysts. 3DGMs are stronger catalysts than 2DGMs because of their superior wettability. In catalytic reactions, 3DGMs appear to provide more benefits over 2DGMs. Unpolluted water shortages are a problem worldwide due to the rising demand for unpolluted water resources brought on by the rapid rise of industrialization, rising pollution emissions, and protracted droughts [100]. As a result, numerous approaches and solutions have been used to increase the amount of water resources that are readily available [101]. Long-term reusing of rural or municipal wastewater from treatment plants originating from agricultural and industrial operations can be accomplished with the use of chemical processes. Advanced oxidation processes (AOPs) are a category of water purification techniques based on the in situ generation of highly active transient species, like the reagents O_2 and OH; such reagents help in mineralizing the organics and disinfecting the harmful microorganisms in wastewater. Due to their potential to provide low-cost and extremely efficient platforms, photocatalytic techniques using semiconductor devices have been in the forefront of AOPs [102]. Using the hydrothermal approach, Qiu et al. (2014) embedded TiO_2 nanocrystals into high-porosity graphene aerogels (TiO_2-GAs) [103]. Substantial quantities of organic pollutants can be absorbed by TiO_2-GAs due to their higher surface area and hierarchical channel structures. Additionally, TiO_2-GA electron transmission facilitation and electrical conductivity can be improved by the addition of high conductivity GAs to the TiO_2 matrix. The researchers exploited this composite's excellent photocatalytic activity and its long-term stability for methyl orange (MO) degradation. The synergistic interfacial connections between TiO_2 nanocrystals and GAs, high conductivity, faceted features, and high elasticity were credited for these positive effects. Individual TiO_2-GA composites are considerably easier to separate from the liquid reaction medium due to their large dimensions, and they can be separated with just a pair of hand tweezers. According to Fan et al. in 2015, AgBr-GA composite is able to photoreduce CrVI and photooxidatively degrade MO [104]. As per the researchers, AgBr-Gas photocatalytic capability and observed stability resulted in the preservation of the AgBr-GAs quality and morphology even after numerous photocatalytic cycles (only 0.8% losses were reached after carrying out all the degradation cycles). In some traditional catalytic reactions, such as hydrolysis of ethyl acetate, the high mass transfer resistance in 3DGMs hinders the catalytic performance as compared to the 2D graphene. Although the defects from the formation process of 3D structure are favorable to some catalytic reactions, they lead to a reduction in the electrical conductivity of 3DGMs as compared to 2DGMs. The figures below list each function with its structural benefits, and also the advantages of 3DGMs in catalysis.

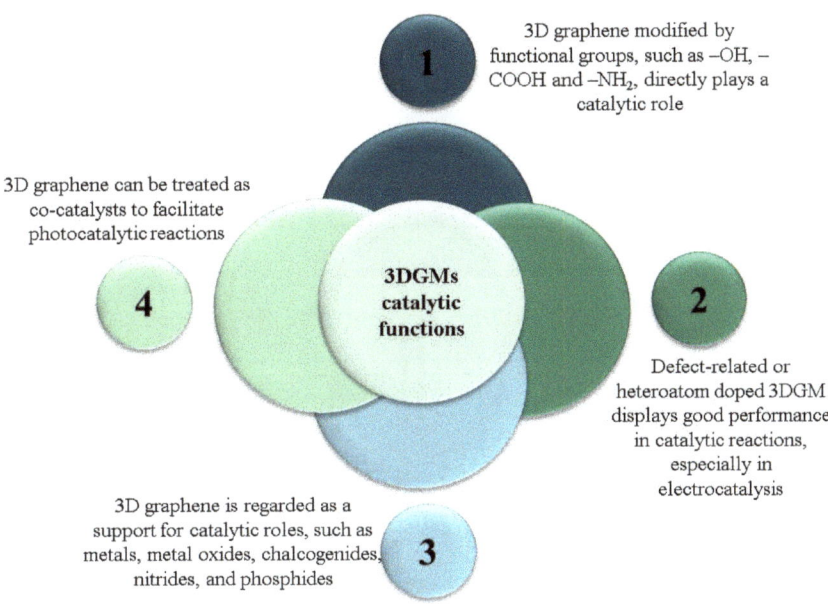

Figure 10. The main functions of 3DGMs in catalysis applications.

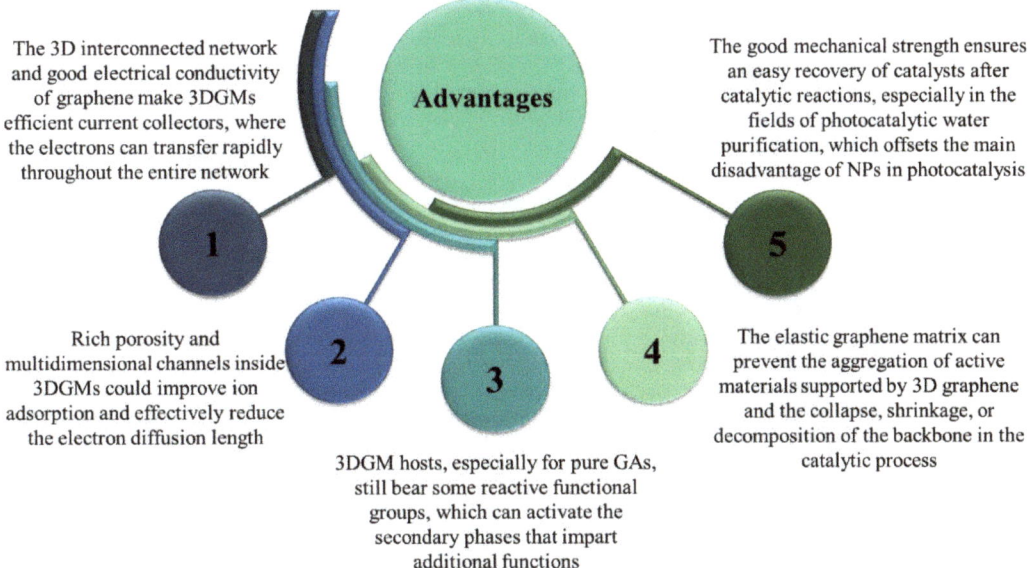

Figure 11. The main advantages of using 3DGMs in catalysis applications.

5. CNT Production Methods

Several methods in the literature have been developed to produce large quantities of CNTs via process gases or vacuum. In Figure 12, arc discharge, laser ablation, high-pressure carbon monoxide disproportionation, and chemical vapor deposition (CVD) are shown [33].

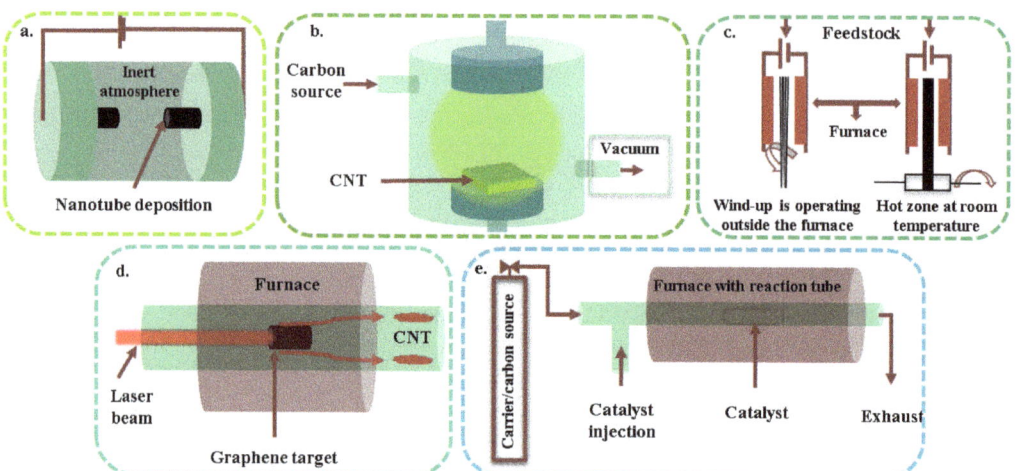

Figure 12. Various schematics for CNT fabrication techniques: (**a**) arc discharge apparatus, (**b**) parallel plate PECVD system, (**c**) direct aerosol spinning process, (**d**) laser ablation method, (**e**) thermal CVD system with a tube furnace.

5.1. Arc Discharge

In 1991, Iijima presented the creation of a novel class of finite carbon structures made of needle-like tubes [105]. Identical to how fullerene is made, the tubes were made utilizing an arc discharge evaporation process. The carbon electrode utilized for the direct current (dc) arc-discharge evaporation of carbon in a vessel filled with argon at a pressure of around 100 Torr was used to develop the carbon needles, which ranged in size from 4 to 30 nm in diameter and up to 1 mm in length. This process yielded both single-walled and multi-walled nanotubes with lengths of up to 50 μm and minimal structural flaws, with up to a 30% weight yield. Higher temperatures (over 1700 °C) are used in the arc-discharge method for CNT synthesis, which often results in expansion of CNTs with fewer structural flaws [106]. In another study, SWCNTs were produced by Zhao et al. in 2019 utilizing a modified arc-discharge furnace with buffer gas of 500 Torr helium at 600 °C [107]. Transmission electron microscopy (TEM), X-ray diffraction (XRD), and Raman spectroscopy were used to investigate the impact of the catalyst type on the generation of the SWCNTs. According to the research observations, the catalytic structure had a significant effect on the rate and purity of the final SWCNTs produced. At a catalyst loading of 3 wt%, Fe-Ni-Mg and Co-Ni powder catalysts displayed superior photocatalytic activity. The mean diameter of the SWCNTs was about 1.3 nm, and the soot production rate could reach 15 g/hr. Regarding the electrical energy storage and conversion devices application, Tepale-Cortés et al. in 2021 used the arc-discharge method to synthesize the required CNT structure by vaporizing the graphite rods while using Ni and a Ni/Y combination as catalysts. [108]. In a cylindrical glass reactor with a regulated Argon flow rate of 1.43 cm^3 min^{-1} and a chamber pressure of 39 kPa, CNTs were produced. Carbon powder had been gathered first from the reactor following chemical treatment with HCl solution at 1 M for metallic contaminants removal. SEM and TEM morphological characteristics revealed that MWCNTs have amorphous carbon particles stuck to their surfaces. FT-IR spectra showed bands at 1550 and 1200 cm^{-1} that corresponded to C=C bonds specific to CNT skeletons which were absent in the pristine graphite. UV-Vis was used to detect electromagnetic absorption, with peaks at 204 and 256 nm being associated with MWCNTs' sp^2 hybridization property.

5.2. Laser Ablation

Richard Smalley et al. in 1996 used a laser ablation method to vaporize graphite rods with a tiny quantity of cobalt and nickel at 1200 °C, yielding >70% high yields of SWNT [109]. The laser ablation technique requires bleeding an inert gas into a chamber, and laser pulses of pulsed-laser type supply a higher temperature for graphite evaporation. As the evaporated carbon particles condense on the reactor cooler surface, nanotubes will be developed there. A water-cooled surface should be used in the setup for collecting nanotubes. The tubes grow up until an excessive number of catalytic atoms accumulate at the nanotube end. The yield of the laser ablation approach is about 70% and it predominantly yields SWCNTs of a controlled size dictated by the reaction temperature. However, it costs much more than the CVD or arc-discharge methods. In 2017, Khashan et al. studied the antibacterial property of iron oxide NP-decorated CNTs that had been effectively created using liquid laser ablation with Nd:YAG pulses [110]. The composite NPs were observed using TEM, which revealed semi-spherical iron oxide NPs that were aggregated around rolled and unrolled graphene sheets. The existence of carbon and other iron oxide NP phases was demonstrated by the XRD spectra. Then, utilizing iron oxide-enhanced nutritional broth and nutrient agar procedures, the NPs' antibacterial activity was evaluated against various bacteria. Wu et al. in 2019 accomplished a patterning and ablation of CNT film using a femtosecond laser with various parameters [111]. Investigating the effects of laser pulse energy and pulse number on ablation holes led to the discovery of a 25 mJ/cm^2 ablation threshold. The pattern behavior groove was evaluated using Raman spectroscopy and SEM analysis. The outcomes demonstrated that the laser ablation removed the oligomer from the CNT film, which increased the Raman G band intensity. Once the pulse's energy was able to break the C-C bonds between distinct carbon atoms, the CNTs' ablation was brought about by the interaction of photon energy with laser-induced thermal elasticity. During laser cutting at higher energy, contaminants including amorphous carbon were discovered at and around the cut edge, and significant distortion and tensile stress formed on the CNT groove. In order to adorn CNTs with various concentrations of ZnO NPs on their tubular surface in only one step for catalytic degradation against methylene blue dye, Mostafa et al. in 2021 used the laser ablation procedure in liquid media [112]. By maximizing the laser ablation period, the number of decorated ZnO NPs was kept under control. The nanocomposite's structure was investigated using a variety of techniques, including optical, structural, and morphological analyses, which revealed that the interaction between ZnO and the CNTs had a different impact on the absorbance characteristic peak. It has been observed that if the ablation duration increasesd, the amount of ZnO coating on the CNTs increased. The researchers in this work demonstrated that the CNTs' presence in the composite significantly increased the photocatalytic performance when compared to pure ZnO.

5.3. Chemical Vapor Deposition (CVD)

Probably the most popular technique for creating CNTs is CVD [113]. A layer of metal catalyst particles, such as nickel, cobalt iron, or a mixture, is produced on a substrate during the CVD technique. Oxide reduction and oxide solid solutions are other methods for making metal NPs. The sizes of the metal NPs from which the nanotubes are to be generated will vary [114]. An annealed metal layer, patterned metal deposition, or plasma etching of a metal layer can all be used to manage this. In the CVD process, the reaction chamber is filled with a mixture of hydrocarbon gas, such as acetylene, methane, or ethylene, and nitrogen. When the hydrocarbons break down throughout the reaction at temperatures ranging from 700–900 °C and air pressure, nanotubes are constructed on the substrate. The direction of the electric field will be followed by the growing nanotube if a strong electric field is applied (plasma-enhanced CVD). Due to its low cost per unit and ability to produce nanotubes directly on a specified substrate, CVD holds the greatest promise for industrial-scale deposition [115]. The development of CNTs from liquefied petroleum gas (LPG) on a Fe_2O_3-Al_2O_3 precatalytic by using CVD procedure without hydrogen was reported by Duc et al. in 2019. The resulting MWCNTs had identical external tube diameters of 50 nm, but they had a less imperfect construction. The Fe_2O_3-Al_2O_3 precatalytic had been reduced to Fe-Al_2O_3 during the synthesis process utilizing the byproducts of LPG breakdown, according to the CNTs' growth mechanism, and a tip-growing process was proposed. The resulting CNTs were employed to adsorb the copper from liquids after being surface-modified with potassium permanganate in the acid medium. The adsorption data were evaluated using the Freundlich and Langmuir isotherm models, and the maximum adsorption capacity of Cu(II) was 163.7 mg g^{-1} [116]. First principles nonequilibrium quantum chemical molecular dynamics simulations of the breakdown of ferrocene (Fc) during floating catalyst CVD (FCCVD) were described by Mclean et al. in 2021 [117]. The key growth agents for the nucleation of carbon chains from Fc-derived species like cyclopentadienyl rings are produced when these species are dissociated into C_2H_x radicals and C atoms, according to their analysis of the effects of additional growth precursors like ethylene, methane, CO, and CO_2 on the Fc decomposition method. Without the need for an extra growth precursor, Fc degrades due to the spontaneous cleavage of the Fe-C and C-H bonds, allowing for the clustering of Fe atoms to create the floating catalysts. They described the two competing chemical routes that were present during the earliest stages of FCCVD—the growth of Fe NPs catalysts and the growth of carbonic chains—on the basis of these simulations. The latter can be facilitated in the presence of additional growth precursors, with the type of precursor dictating how these conflicting pathways are balanced. The latest CVD development of SWNTs from plastic polymers, such as low-density polyethylene (LDPE) and polypropylene (PP), was presented by Zhao et al. in 2022 [118]. Successfully developed cobalt catalysts supported by porous magnesia (Co-MgO), the porosity of which restricts the mobility of reduced Co NPs and facilitates the nucleation of small diameter SWNTs, was credited with the successful synthesis of SWNTs from the polymers. The method was also expanded to catalyze the creation of SWNTs from waste plastics such as food packaging film and melt-blown mask filters. This proof-of-concept development shows the potential of using plastic pollution as a feedstock to create valuable carbon nanomaterials.

5.4. Plasma Torch

A thermal plasma technique can also be used to create SWCNTs [119]. More CNTs can be continuously created using the plasma torch process. Since 2001, a number of academic institutions have studied how to make CNTs in plasma jet reactors using this method [120]. The procedure is inexpensive and ongoing. The atmospheric pressure plasma in a microwave's plasma torch, which takes the shape of an intense "flame", atomizes a gaseous combination of ethylene and argon. SWNT, metallic, and carbon NPs, as well as amorphous carbon, are all present in the flame's emissions. Using an induction thermal plasma technique with a plasma torch is another way to make SWCNTs. SWCNTs with various diameter distributions can be created. With this process, it is possible to manufacture two grammes of nanotube material each minute, which is more than with arc discharge or laser ablation [121]. An electron-rich poly(fluorene-co-carbazole) derivative is used to extract semiconducting species from the initial HiPCO or plasma-torch nanotube starting material, and then an electron-poor methylated poly(fluorene-co-pyridine) polymer is used to isolate the metallic species that are still present in the residue. Bodnaryk et al. reported this two-polymer system in 2018 [122]. The metallic species in the sample were two times more enriched using this process than they were from the raw starting material. These findings suggest that an efficient method for enriching metallic species is the use of polymers with low electron density. Assa et al. combined regioregular poly(3-hexylthiophene) (P3HT), 1-(3-methoxycarbonyl) propyl-1-phenyl[6,6]C61 (PCBM), and torch-plasma-grown SWCNTs as a hybrid photoactive layer for bulk heterojunction solar cell devices in 2019 [123]. Investigators showed that even when sputtering is done using a Cs^+ 2000 eV ion source, chemical information can be properly acquired by time-of-flight secondary ion mass spectrometry throughout the hybrid organic photoactive solar cell layers. The highest results were attained with 0.5 wt% SWCNT loads, resulting in a power conversion efficiency of 3.54% and an open-circuit voltage (VOC) of 660 mV. To better understand the charge-transfer mechanisms occurring at the P3HT:PCBM:SWCNT interfaces, Jsc was measured with respect to light intensity and exhibited a linear dependency (in the double logarithmic scale). This suggests that monomolecular recombination is more likely to be responsible for charge carrier losses at this optimal SWCNT concentration of 0.5 wt%. Finally, they reported that hybrid devices are able to significantly increase the exciton dissociation efficiency thanks to the fullerene's electron-accepting nature and the SWCNT's fast electron transit feature. In 2021, Gotthardt et al. demonstrated that 1,2,4,5-tetrakis(tetram-ethylguanidino)benzene (TTMGB), a guanidino-functionalized aromatic compound, is a successful n-dopant for field-effect transistors (FETs) with gold contacts and networks of semiconducting SWCNTs with small diameters and large band gaps [124]. After TTMGB treatment, the work functions of gold, palladium, and platinum were found to have decreased by about 1 eV, according to Kelvin probe measurements. In turn, gated four-point probe measurements revealed orders of magnitude lower contact resistances for electron injection into SWCNT networks. TTMGB treatment did not affect the electron transport or maximum mobilities in SWCNT networks at high carrier densities, according to measurement techniques that were temperature- and carrier concentration-dependent, but it significantly increased the subthreshold slope of nanotube FETs by removing shallow electron trap states.

6. Uses of CNT as a Catalyst

A variety of carbon bonds work to construct a new different structure of unique features. A layered structure with a weak out-of-plane van der Waals bond can be built by sp^2 hybridized carbon. The strong in-plane bonds play a major role in this purpose. A few to a few tens of concentric cylinders with regular periodic interlayer spacing locate around ordinary central hollow and made MWCNTs. The real-space analysis of multiwall nanotube images has shown a range of interlayer spacing (0.34 to 0.39 nm) [125]. It was discovered that CNTs had superior thermal transfer properties. For instance, it was discovered that CNTs had extraordinarily high axial thermal conductivities, around 2000 W/mK or more than 3000 W/mK for MWCNTs, and much higher for SWCNTs [126], and it was found that CNT-in-polymer and CNT-in-oil suspensions had massively enhanced thermal conductivity. Even the short CNTs agglomeration, randomly entangled with one another, have been employed in earlier studies [127]. Then, on ceramic spheres, large-scale CNT arrays with millimeter vertical alignment have been constructed [128]. High-speed shearing can easily spread them into fluffy CNTs. CNTs also demonstrated incredible catalytic uses [129]. Long CNTs (over 500 m) intercrossed Cu/Zn/Al/Zr catalyst (CD703) were produced in 2010 by dispersing CNT arrays of vertical alignment in Na_2CO_3 fluid and co-precipitating with metal nitrite. When comparing Cu/Zn/Al/Zr catalytic compound without CNTs, the space time yield (STY) of methanol on CD703 rose by 7 and 8%, respectively, to 0.94 and 0.28 g/(g_{cat} h) for CO and CO_2 hydrogenation. Additionally, dimethyl ether (DME) has been formed by one step CO and CO_2 hydrogenation with a STY of 0.90 and 0.077 g/(g_{cat} h) at 270 °C when paired with γ-Al_2O_3 catalyst and HZSM-5. A CD703 catalyst exhibited great action with production of methanol as a result of phase separation, ions dopant, valence compensating, hydrogen reversibly adsorbent and storage on CNTs promoting hydrogen spillover. Since CNTs have higher thermal conductivity, CD703 has better stability. It was thus revealed that using Cu/Zn/Al/Zr catalytic compounds for the synthesis of methanol and DME from CO/CO_2 hydrogenation was well-promoted by long CNTs [130]. Typically, bulk linked CNT constructions are used in the aerospace, automotive, robust electronics, and biomedical industries and have interesting properties [131]. It is still difficult to join CNTs at interconnects to form effective 3D constructions with a desired strength [132]. Spark plasma sintering (SPS) has been used under a range of pressure and temperature settings to synthesize bulk CNT linked structures. The interconnected 3D structures and the impact of processing conditions on structural damage to CNTs were examined with considerable detail using spectroscopic and microscopic techniques. Double-walled CNTs (DWCNTs) were produced in bulk by Guiderdoni et al., adopting SPS at 1100 °C and 100 MPa pressure [133]. According to reporting requirements, DWCNTs remained intact under those conditions. Extensive molecular dynamic simulations have been used to better study welding of CNTs that resulted in interconnected constructions. The Ozden team previously investigated how density and CNT structure are affected mechanically, electrically, and in terms of hydrophilicity (Figure 13) [134]. Al-Hakami et al. in 2013 investigated an approach to remove Escherichia coli (E. coli) bacteria from water using both naturally occurring CNTs and modified/functionalized CNTs containing 1-octadecanol groups (C18). As per their findings, E. coli was removed by CNT alone by 3–5%; however when paired with microwave radiation, unmodified CNT was able to remove up to 98% of bacteria. When CNT-C18 was employed in similar conditions, the bacteria had been removed by 100% [135]. Most textile wastewater is harmful and non-biodegradable. Semiconductor catalysts can be utilized to treat the environmental contamination. TiO_2 is a significant photocatalyst; unfortunately, TiO_2 has a limited spectrum of light sensitivity and poor efficiency. However, TiO_2 and CNT together can boost photocatalytic activity [136]. Using MWCNT and Ti as source materials, Ming-liang et al. synthesized a CNT-TiO_2 composite in 2009 to accelerate the photocatalytic oxidation of water contaminants [137]. The composite's photoactivity was assessed through the conversion of methylene blue in liquid phase under UV radiation. Researchers came to the conclusion that the CNT-TiO_2 composite's ability to remove methylene blue is facilitating the transfer of electrons between MWCNT and TiO_2, as well as MWCNT adsorption and TiO_2 photodegradation.

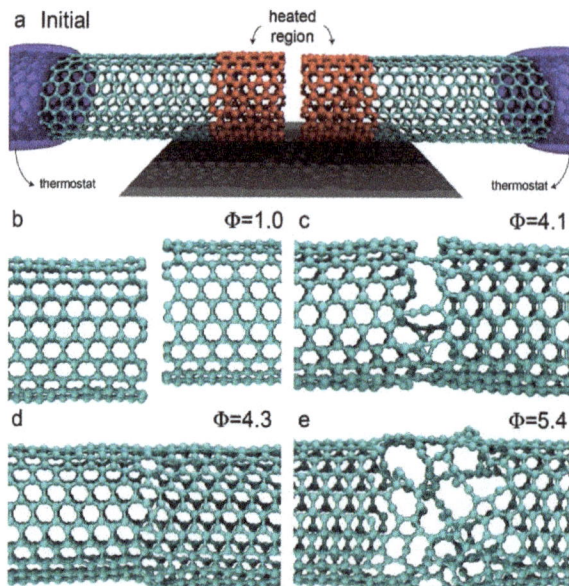

Figure 13. (a) Two CNTs are first arranged with their axes aligned at 180°. The atoms in red are heated from outside and are located at the contact tips of both tubes. (b–e). The tubes' final configuration following heating with heat fluxes of 1.0, 4.1, 4.3, and 5.4 kcal mol^{-1} fs^{-1}, respectively. Reprinted with permission from Ref. [134]. Copyright © 2016, John Wiley and Sons.

In 2016, Jauris et al. reported a relationship between SWCNTs and two artificial dyes (methylene blue and acridine orange) [138]. Because of π-π interactions' prevalence between each dye and the nanotubes, the researchers reached a conclusion that long-term configuration stability was where the dye is generally plano-parallel to the nanotube. SWCNT is a prospective adsorbent for dye degradation and could be employed commercially for treating wastewater. By increasing the nanotube's radius, the dye-nanotube binding energy increases. In order to prepare Au-TiO$_2$@CNT composite photocatalysts for photocatalytic gaseous styrene removal, Zhang W. et al. used a simple micro/nano-bubble approach [139]. High ternary-structure stability can be formed by reacting Au, and TiO$_2$ NPs coated onto CNTs can be efficiently facilitated by the micro/nano-bubbles. The response surface central composite design approach has been applied to examine Au-TiO$_2$@CNTs' photoactivity. Rapid development of a compact structure increased the photocatalytic degradation and mineralization of styrene over Au-TiO$_2$@CNTs dramatically as the reaction temperature increased. The increased photocatalytic mechanism of Au-TiO$_2$@CNTs was further disclosed through the examination of EPR, UV-vis DRS, electrochemical characteristics, and TPD-O$_2$. The further identification of free radicals revealed that oxidative radicals like hydroxyls and superoxides were closely related to the photocatalytic degradation and mineralization of styrene, which was primarily because of CNT and Au NP synergistic influence for increased activity through the photocatalysis process.

7. CO$_2$ Hydrogenation into Hydrocarbons and Oxygenated Hydrocarbons

Under certain conditions, it is thought that CO$_2$ catalytic hydrogenation with renewable hydrogen is an appropriate method for the chemical recycling of this hazardous and chemical resistance molecule into energy-carrying agents and chemical compounds. With a precise hydrogenation product, CO$_2$ can be hydrogenated into C$_1$ compounds like methane and methanol. It is more difficult to produce high (C$_{2+}$) hydrocarbons and oxygenates on a selective basis. Due to its higher volumetric energy density within a specific volume and compatibility with the current fuel infrastructure than C$_1$ compounds, such produced materials are desirable as entry platform chemicals and energy vectors [140]. The main challenge is integrating catalytic functions as effectively as possible for both the reductive and chain-growth stages [141]. The transformation of renewable energy also makes use of CO$_2$ as an energy carrier. Because renewable energy sources are intermittent by nature, there is presently a need for scalable storage [142]. Consequently, a much more practical and easier method for storing significant volumes of intermittent energy generated from renewable sources for longer durations is the generation of synthetic natural gas or liquid fuels. The power to gas (PtG) concept has received considerable attention, as seen in Figure 14 [143]. An alternate source of natural gas is produced when CO$_2$ combines with H$_2$, which is created by water electrolysis using renewable wind or solar energy. In Copenhagen, especially in 2016, a commercial-scale PtG project with 1.0 MW of capacity had been operating and successfully exploiting the transition to a sustainable energy system [144]. In the period 2003–2009, with capacities ranging from 25–6300 kW, there were five initiatives in Germany utilizing CO$_2$ methanation at pilot-plant or commercial scale [145]. The French chemist Paul Sabatier published his first study on CO$_2$ methanation in 1902 [146]. This age-old craft has gained fresh traction as a result of the growing need to combat global climate change and store excess renewable energy. The Sabatier reaction is an excellent method for converting CO$_2$ into chemical feedstocks and fuels, storing renewable energy sources like wind and solar energy, and efficiently converting biogas to biomethane [147]. CO$_2$ methanation is an endothermic reaction with higher equilibrium conversion between 25 and 400 °C [148]. By using the right catalysts, CO$_2$ methanation can achieve 99% CH$_4$ selectivity, avoid further product separation, and get around the challenges of dispersed product distribution. As a result, such a thermodynamic characteristic increases the importance of CO$_2$ methanation in terms of energy effectiveness and commercial viability.

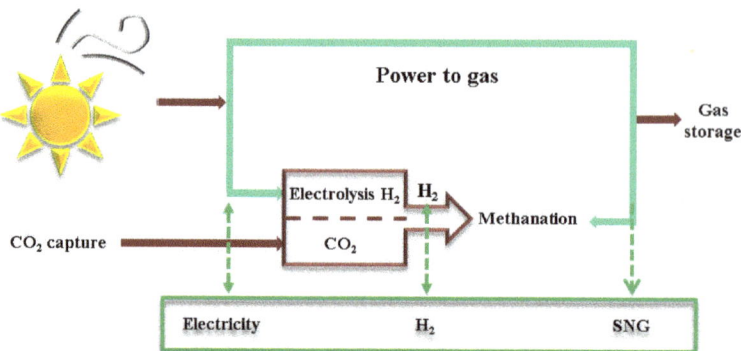

Figure 14. Chemical compounds and fuels produced sustainably using CO_2. At various temperatures, the equilibrium conversion of CO_2 to methane is plotted from the previous literature data.

In order to synthesize C_{2+} hydrocarbons, Fujiwara et al. (2015) investigated CO_2 hydrogenation over composite catalytic compounds made of Cu-Zn-Al oxide catalyst and HB zeolite via combining the production of methanol over Cu-Zn-Al oxide and the simultaneous conversion of methanol over HB zeolite [149]. The yield of C_{2+} hydrocarbons was low (0.5 C-mol%) and lower than that of oxygenated compounds when a non-modified zeolite was employed for the composite catalyst (methanol and dimethyl ether). For the conversion of dimethyl ether to C_{2+} hydrocarbons, the strong acid sites of zeolite were severely inactivated. The catalytic activity of the associated composite catalysts was significantly enhanced by the use of zeolites treated with 1,4-bis(hydroxydimethylsilyl) benzene to create C_{2+} hydrocarbons in yields of more than 7C-mol%. Under a pressure of 0.98 MPa, the best C_{2+} hydrocarbon production selectivity was approximately 12.6 C-mol%. Hydrophobic zeolites with water contact angles more than 130° were created by the disilane modifications. The disilane molecule was converted into a few condensed aromatics during CO_2 hydrogenation at 300 °C, although the hydrophobicity was maintained even after the reaction. The hydrophobic surface of the HB zeolite inhibits the deactivation of the strong acid sites, increasing catalytic activity. Even under low pressure situations, this enhanced composite catalyst will support the synthesis of C_{2+} hydrocarbons from CO_2.

In 2017, Zhang et al. suggested a procedure to create ethanol from paraformaldehyde, CO_2, and H_2 [150]. Under benign conditions, a ruthenium–cobalt (Ru-Co) bimetallic catalyst using LiI as the promoter in 1,3-dimethyl-2-imidazolidinone (DMI) may effectively speed up the process. Overall products had a selectivity of 50.9 C-mol% for ethanol, which was obviously higher than that of the disclosed methods. Additionally, the TOF for ethanol based on Ru metal reached a maximum of 17.9 h^{-1} as seen in Figure 15.

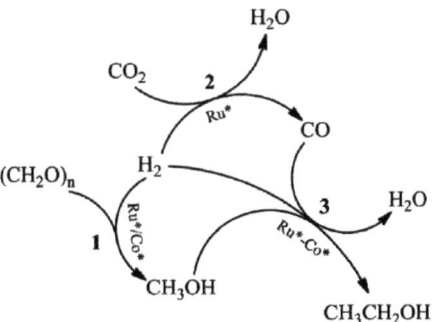

Figure 15. Synergy between the processes used to fabricate ethanol from paraformaldehyde, CO_2, and H_2. The active species of Ru and Co are represented, respectively, by Ru* and Co*. Reprinted with permission from Ref. [150]. Copyright © 2017, Royal Society of Chemistry.

Significant information about the involvement of oxygen in the electrochemical reduction of CO_2 on Cu electrodes was presented in 2017 by Le Duff and colleagues. They also regulated the surface structure and composition of Cu single crystal electrodes over time [151]. Since the pulse sequence may be controlled to ensure consistent beginning conditions for the reaction at every fraction of time and at a certain frequency, this was accomplished using pulsed voltammetry. Compared to the selective CO_2 reduction achieved using cyclic voltammetry [152], and chronoamperometric techniques [153], under alternating voltage, a wide range of oxygenated hydrocarbons was discovered. The coverage of oxygen species, which is reliant on surface structure and potential, was linked to product selectivity towards the synthesis of oxygenated hydrocarbon. A nanowire-like WSe_2-graphene catalyst was created by Ali and Oh in 2017 and examined for its ability to photocatalytically convert CO_2 into CH_3OH when exposed to UV–visible light. XRD, SEM, TEM, Raman, and XPS were used to further characterize the produced nanocomposite. Using gas chromatography (GCMS-QP2010SE), the photocurrent analysis was further evaluated for its photocatalytic reduction of CO_2. The sacrificial agent (Na_2S-Na_2SO_3) was added to WSe_2-graphene nanocomposite to further increase the photocatalytic effectiveness, and it was discovered that this improved the photocatalytic efficiency, with the methanol output reaching 5.0278 mol g^{-1} h^{-1} [154]. In a different study, Biswas et al. (2018) reported using ultrasonic techniques to create a WSe_2-Graphene-TiO_2 ternary nanocomposite [155]. According to estimates, the WSe_2-Graphene-TiO_2 band-gap is 1.62 eV, which is adequate for the photocatalytic degradation when exposed to UV–visible light. For the conversion of CO_2 to CH_3OH, the photocatalytic capability of nanocomposites was examined. After 48 h, WSe_2-G-TiO_2 with an optimal graphene loading of 8 wt% shown high photoactivity, yielding a total CH_3OH yield of 6.3262 mol g^{-1} h^{-1}. The gradual synergistic relationship between the WSe_2-TiO_2 and graphene components in the heterogeneous system is what causes this exceptional photoreduction activity. Ethylene could be produced from CO_2 electroreduction; however, present systems are constrained by low conversion efficiency, slow production rates, and unstable catalysts. In contrast to a reversible hydrogen electrode (RHE), Dinh et al. (2018) found that a copper electrocatalyst at an abrupt reaction interface in an alkaline electrolyte converts CO_2 to ethylene with a 70% faradaic efficiency (FE) at a potential of −0.55 volts [156]. The activation energy barriers for CO_2 reduction and carbon monoxide (CO)-CO coupling are lowered by hydroxyl ions on or near the copper surface, and as a result, ethylene evolution begins at −0.165 volts versus an RHE in 10 molar potassium hydroxide virtually concurrently with CO generation. By sandwiching the reaction interface between different hydrophobic and conductive supports, a polymer-based gas diffusion layer was introduced to increase operational stability while maintaining continuous ethylene selectivity over the first 150 operating hours. In 2019, Ma and Porosoff proposed reaction mechanisms by combining in situ characterization techniques with DFT calculations, identifying structure–property relationships for the zeolite support, strategizing methods to increase catalyst lifetime, and developing advanced synthesis techniques for depositing a metal-based active phase within a zeolite for highly active, selective, and stable tandem catalysts [157]. The critical research topics of reaction mechanism elucidation by combining in situ characterization methods with density functional theory calculations, identifying structure–property relationships for the zeolite support, developing advanced synthesis techniques for depositing a metal-based active phase within a zeolite for highly active, selective, and stable tandem cations, and strategizing methods to extend catalyst lifetime, are suggested as future research directions. An appealing method for storing such a renewable energy source in the form of chemical energy is the conversion of CO_2 into hydrocarbons using solar energy. A system that couples a photovoltaic (PV) cell to an electrochemical cell (EC) for CO_2 reduction can do this. Such a system should have minimum energy losses related to the catalysts at the anode and cathode, as well as the electrolyzer device, in order to be advantageous and usable. It should also use inexpensive and easily processed solar cells. All of these factors were taken into account by Huan et al. in 2019 when setting up a reference PV-EC system for CO_2 reduction to hydrocarbons [158]. Combined with a fairly priced, state-of-the-art perovskite photovoltaic minimodule, this system sets a standard for a low-cost, all-earth-abundant PV-EC system with a solar-to-hydrocarbon efficiency of 2.3%. In 2019, Wu et al. demonstrated cobalt phthalocyanine (CoPc) catalysis for the six-electron reduction of CO_2 to methanol with considerable activity and selectivity when immobilized on CNTs [159]. They discovered that the conversion produces methanol with FE > 40% and a partial current density exceeding 10 mm/cm^2 at −0.94 volts with respect to the reversible hydrogen electrode in a near-neutral electrolyte. CO serves as an intermediary in a special domino mechanism that moves the conversion along. By adding amino substituents that donate electrons to the phthalocyanine ring, it is possible to prevent the harmful reduction of the phthalocyanine ligand from having a negative effect on the catalytic activity. With significant activity, selectivity, and stable performance for at least 12 h, the enhanced molecule-based electrocatalyst converts CO_2 to methanol.

A novel multifunctional catalyst made of Fe_2O_3 encapsulated in K_2CO_3 was introduced by Ramirez et al. in 2019 and has the ability to use a tandem process to convert CO_2 into olefins [160]. The authors established that, unlike the conventional systems in FT processes, very large K loadings are essential to activate CO_2 via the well-known "potassium carbonate mechanism." While utilizing CO_2 as a feedstock, the suggested catalytic process proved to be just as productive as currently used commercial synthesis gas-based techniques. By employing Cu-doped $MgAl_2O_4$ ($Mg_{1-x}Cu_xAl_2O_4$) and a straightforward deposition–reduction process, Tada et al. in 2020 investigated the synthesis of Cu NPs. The following three Cu^{2+} species were present in $Mg_{1-x}Cu_xAl_2O_4$ [161]: short O-Cu octahedrally coordinated $[CuO_6]_s$, elongated O-Cu octahedrally coordinated $[CuO_6]_{el}$, and tetrahedrally coordinated $[CuO_4]_t$. The first two were discovered in $Mg_{1-x}Cu_xAl_2O_4$ of the inverse-spinel type, and the third was discovered in $Mg_{1-x}Cu_xAl_2O_4$ of the normal-spinel type. In addition, they made it clear that their percentage is related to Cu loading by concentrating on the variation in the reducibility of the Cu^{2+} species. $Mg_{1-x}Cu_xAl_2O_4$ predominantly comprised the $[CuO_6]_s$ species at low Cu loading (x < 0.3). In contrast, the fraction of the $[CuO_6]_{el}$ and $[CuO_4]_t$ species rose with high Cu loading (x ≥ 0.3). Notably, the H_2-reduced $Mg_{0.8}Cu_{0.2}Al_2O_4$ (x = 0.2) catalyst showed the best photocatalytic activity among the synthesized catalysts because it had the most exposed metallic Cu sites. Therefore, the formation of metallic Cu NPs on metal oxides depends on the H_2 reduction of $[CuO_6]_s$.

In 2021, Tada et al., suggested bifunctional tandem $ZnZrO_x$ catalysts for the hydrogenation of CO_2 to methanol along with a number of solid acid catalysts (for subsequent methanol conversion to light olefins) [162]. Researchers used zeolites and silicoaluminophosphates with a variety of topologies, including MOR, FER, MFI, BEA, CHA, and ERI, as solid acid catalysts. A study using ammonia adsorption revealed that they likewise displayed the equivalent acid characteristics. Lower olefins were being synthesized in one step using the tandem catalysts, whereas with ZnZrO$_x$ alone, no hydrocarbons could be produced. There appears to be no relationship between product yields and acid strength, at least according to the reaction test and ammonia adsorption. The product selectivity is influenced by the pore sizes and the channel dimensionality of the zeolites; zeolites with small pores, like MOR, SAPO-34, and ERI, are promising, whereas zeolites with bigger pores, like MFI, generate heavier hydrocarbons. The outcomes offer fresh perspective on the creation of creative catalysts for CO_2 usage. A low-temperature atmospheric surface dielectric barrier discharge reactor that converts biogas into liquid chemicals was introduced by Rahmani et al. in 2021 [163]. The effect of steam on the conversion of methane and CO_2 was investigated, as well as the distribution of products in relation to a given energy input based on the operational circumstances. The authors reported conversion rates of 44% for CH_4 and 22% for CO_2. When steam was introduced at the in-feed, the lowest energy cost of 26 eV/molecule was attained. For liquid hydrocarbons, a selectivity of 3% was attained. The transformation of biogas ($CH_4 + CO_2$) resulted in the production of more than 12 compounds. At ambient temperature, the most prevalent oxygenated hydrocarbon liquids were acetone, methanol, ethanol, and isopropanol. H_2, CO, C_2H_4, and C_2H_6 were the major gases produced.

In order to explain an alternative approach for the chemical CO_2 reduction reaction, Islam et al. in 2021 subjected up to 3% CO_2-saturated pure water, NaCl, and artificial seawater solutions to high-power ultrasound (488 kHz ultrasonic plate transducer) [164]. The converted CO_2 products under ultrasonic settings were discovered to be mostly CH_4, C_2H_4, and C_2H_6, as well as a significant amount of CO that was later converted into CH_4. The analysis revealed that adding molecular H_2 to the CO_2 conversion process is essential, and that raising the hydrogen concentration boosted hydrocarbon yields. However, it was found that the overall conversion decreased at higher hydrogen concentrations because hydrogen, a diatomic gas, is known to reduce cavitation activity in liquids. Additionally, it was discovered that the maximum hydrocarbon yields (nearly 5%) were achieved with 1.0 M NaCl solutions saturated with 2% CO_2 + 98% H_2, and that increasing salt concentrations further decreased the yield of hydrocarbons due to the combined physical and chemical effects of ultrasound. By diluting the flue gas with hydrogen, it was demonstrated that the CO_2 present in a synthetic industrial flue gas (86.74% N_2, 13.5% CO_2, 0.2% O_2, and 600 ppm of CO) could be transformed into hydrocarbons. Additionally, it has been demonstrated that the conversion process can be carried out in ambient circumstances, i.e., at room temperature and pressure, without the use of catalysts, when low-frequency, high-power ultrasound is used. Tian et al. in 2022 newly manufactured In_2O_3, MnO-In_2O_3, and MgO-In_2O_3 catalysts using the co-precipitation method, and they looked into the hydrogenation of CO_2 to methanol [165]. The ability of In_2O_3 to absorb CO_2 was significantly improved by the addition of Mn and Mg oxides. The CO_2 adsorption capacity and the changing trend of methanol selectivity were consistent. As opposed to In_2O_3, the methanol selectivity of MnO-In_2O_3 and MgO-In_2O_3 catalysts is higher. ODP, or oxidative dehydrogenation of propane with CO_2, is a promising solution for efficient CO_2 usage. In a new study published in 2022, Chernyak et al. included various C-materials for the first time as supports for Cr-based catalysts of CO_2-assisted ODP [166]. A commercially available activated carbon was evaluated alongside CNTs, jellyfish-like graphene nanoflakes, and their oxidized and N-doped derivatives. The oxidized CNT- and pure GNF-supported catalysts showed the highest activity and a propylene yield of up to 25%. Raman spectroscopy was used to confirm that these two catalysts were stable throughout tests against disintegration and particle sintering. The oxidized CNT- and pristine graphene nanoflakes-supported catalysts' high activity and durability were explained by their macro- and mesoporosity, which improve reagent and product diffusion, as well as by the highest surface graphitization degree, which was validated by XPS. These catalysts performed significantly better than the catalyst supported by activated carbon. As a result, CNTs and graphene nanoflakes are suitable supports for CO_2-ODP catalytic compounds. Several main catalysts used in the hydrogenation of CO_2 to hydrocarbon were summarized in Table 1.

Table 1. Some of the main catalysts used in the hydrogenation of CO_2 to hydrocarbon products.

Catalysts	Metal/Metal-free	Preparation Method	Process Type	Conversion	Selectivity
Cu-Zn-Al oxide and HB zeolite	Metal	Co-precipitation method	The production of C_{2+} hydrocarbons by CO_2 hydrogenation.	2.8%	12.6 C-mol%
Ru-Co	Metal	-	The production of ethanol from paraformaldehyde, CO_2, and H_2	-	50.9 C-mol%
WSe_2-graphene	Metal	Ultra-sonication method	Photocatalytic reduction of CO_2 into CH_3OH	5.0278 μmol g^{-1} h^{-1}.	-
WSe_2-graphene-TiO_2	Hybrid	Ultra-sonication method	CO_2 reduction to CH_3OH	6.3262 μmol g^{-1} h^{-1}	-
hydroxide-mediated Cu	Metal	Hydroxide-mediated abrupt reaction interface	CO_2 conversion to ethylene	70%	65%
CoPc/CNT	Hybrid	CO_2 reduction to methanol	Dispersion process	40%	-
Fe_2O_3@K_2CO_3	Metal	CO_2 conversion to olefins	Mortar mixing	40%	60%

Although the technique of hydrogenating carbon dioxide to produce methanol is inexpensive and environmentally benign, nevertheless, because of the increased stability and inertness of CO_2, this reaction is thermodynamically constrained. The reaction below reveals the process of methanol production.

$$CO_2 + 3H_2 \leftrightarrow CH_3OH + H_2O, \Delta H_{25\ °C} = -49.5\ kJ \cdot mol^{-1}$$

The reaction is thermodynamically advantageous at low temperatures and high pressure, according to the Le Chatelier's principle. The enhanced thermal stability and chemical inertness of CO_2 means that this method of methanol synthesis requires a high temperature to proceed. Indeed, at elevated reaction temperatures (for example, higher than 240 °C), CO_2 activation and subsequent methanol production are facilitated. However, the higher temperature procedure strongly conflicts with the reaction's thermodynamics. Furthermore, when the reaction is conducted at a higher temperature, undesired byproducts including higher alcohols and hydrocarbons are formed. Similar to this, because the CO_2 hydrogenation process produces methanol, which is a molecular reducing reaction, it is thermodynamically more advantageous at high pressure. By employing various catalysts, various reaction pressure sizes have been suggested for the best CO_2 conversion [140].

8. Mechanism of Conversion

Alternative catalytic routes for CO_2 fixing into chemicals have the disadvantage of not being connected to the well-established array of technologies for the conversion of syngas mixtures (CO + H_2) into synthetic liquid fuels and platform chemicals. Hydrogenation routes, on the other hand, have the advantage of being connected to these technologies and could, therefore, first achieve an on-purpose CO_2-recycling industrial application. CO_2 hydrogenation is being used to create a variety of different chemicals [141]. CH_4 and CH_3OH, which seem to be important examples of compounds that could be produced by established or industrial-scale technologies like the methanation and methanol synthesis processes, have received the majority of research attention. Another C_1 product that has already received attention is the derivative dimethyl ether, which would also be formally a C_1 product. Nevertheless, due to a number of factors, high (C_{2+}) hydrocarbons and alcohols, whose production is essentially thermochemically more favorable than that of methanol, may be more desirable products [161]. If somehow the major aim of the hydrogenation process should be applied to produce energy carriers, then lengthening the products' carbon chains will result in ever-higher volumetric energy densities. By directly injecting short-chain alkanes (C_2–C_4) into gas distribution networks, it is possible to raise the calorific value of natural gas or biogas. Furthermore, average volumetric concentration limits for ethane, propane, and butane in conventional pipeline specifications are around 10, 5, and 2%, respectively [167]. Under ambient conditions, the liquid higher alkanes (C_{5+}) [168], are quite desirable as predecessors of, e.g., jet fuels, because they combine higher energy densities with complete compliance to current liquid-fuel distribution and end-use infrastructures. Base chemical compounds with a chain length between C_2 and C_4 and considered light olefins [169], for the creation of polymers, are an illustration. Regarding oxygenated products, C_{2+} alcohols have higher energy densities than methanol and are less poisonous and corrosive. As a result, they are better suited for use in existing internal combustion engines as blending fractions or even as pure fuels. In addition, strong alcohols like ethanol and butanol are great precursors for short-chain (C_2–C_4) olefins [170], which, when manufactured very selectively through alcohol dehydration under relatively uncomplicated reaction conditions, can be added to the value chains already in place. Overviews on the catalytic hydrogenation of CO_2 have been published in earlier surveys of the literature [171]. Processes for the manufacturing of C_1 products, particularly methane and methanol and its derivatives, have undergone more concentrated changes [172].

Attention should be paid to the opportunities and intrinsic kinetic constraints, sometimes intimately connected to thermodynamic bounds, that exist for certain catalytic systems and which must be taken into account in the creation of new catalysts and procedures [173].

Due to its capacity to generate clean hydrocarbons like CO, CH_4, and C_{2+} products, carbon dioxide hydrogenation is a potential reaction that is being studied. Through thermal, photocatalytic, and photothermal catalysis, hydrogenation reactions can move forward. First, CO_2 will be hydrogenated in the Reverse Water Gas Shift (RWGS) reaction (Equation (1)) to yield CO. Then, CH_4 can be created by either directly methanizing CO_2 through the Sabatier reaction (Equation (3)) or further hydrogenating CO (Equation (2)). Both the CO and CO_2 methanation processes are exothermic and take place at temperatures between 200 and 500 °C. Other than that, the Fischer–Tropsch process can be used to create paraffins (alkanes) and olefins (alkenes), which are both significant and pricey chemical feedstocks. Additionally, because they are volume-reducing reactions, a rise in pressure favors the production of CO and hydrocarbon products. This is because, in accordance with Le Chatelier's principle, the equilibrium will change as pressure rises to favor the side of the reaction that contains less moles of gas. Most significantly, as the pressure of the reactor was raised from ambient pressure to 4 bar, the CO_2 methanation activity over RuF catalyst was improved. Additionally, it was noted that the reduction in apparent activation energy caused by the pressure rise helped the methanation reaction progress [174].

$$CO_2 + H_2 \leftrightarrow CO + H_2O \tag{1}$$

$$CO + 3H_2 \leftrightarrow CH_4 + H_2O \tag{2}$$

$$CO_2 + 4H_2 \leftrightarrow CH_4 + 2H_2O \tag{3}$$

$$(2n+1)H_2 + nCO \rightarrow C_nH_{2n} + nH_2O \tag{4}$$

$$2nH_2 + nCO \rightarrow C_nH_{2n} + nH_2O \tag{5}$$

Examples of Conversion Mechanisms

Liu et al. examined the advancement of CO_2 direct hydrogenation to value-added compounds such as CO [175], CH_3OH [154], CH_4 [176], DME [177], higher hydrocarbons [178], and olefins [27]. There has also been a summary of heterogeneous catalysis, plasma catalyst supports, and CO_2 hydrogenation research activities [179–181]. A method to increase the oxygen vacancies of nickel-based catalysts for CO_2 methanation was reported by Zhu et al. in 2021 [182–184]. At low reaction temperatures (300 °C), a Y_2O_3-promoted NiO-CeO_2 catalyst was developed and found to have a remarkable methanation activity that is up to three times higher than NiO-CeO_2 and six times higher than NiO-Y_2O_3. The addition of Y_2O_3 to CeO_2 significantly speeds up the production of surface oxygen vacancies during the reaction, as researchers showed both theoretically and experimentally. That study also demonstrated that these regions support CO_2 dissociation directly, which is kinetically more advantageous than associative routes. As a result, it significantly increased the activity of CO_2 methanation. In the feed stream including methane and traces of H_2S, Gac et al. in 2021 discussed CO_2 methanation over ceria- and alumina-supported nickel catalysts in [185]. With a packed-bed reactor, stability tests conducted for 20 h at 350 and 600 °C revealed the catalysts' great resistance to sintering processes. At 350 °C, a higher conversion has been seen for the nickel catalyst assisted by ceria. According to a thermodynamic analysis, under certain reaction conditions, the CO_2 present in biogas can be transformed to methane without carbon production. The decrease in CO_2 conversion and increase in CH_4 selectivity were caused by the addition of CH_4 to the CO_2–H_2 feed stream. When trace amounts of H_2S were added to the feed stream, CO_2 conversion and CH_4 selectivity quickly decreased. Al_2O_3-supported catalysts were shown to be more durable (20%) than CeO_2-based catalysts. Heterogeneous nanocatalytic compounds' atomic usage and activity are typically improved as their size is decreased in a variety of catalytic reactions. This method has, however, been less successful for Cu-based electrocatalysts in the reduction of CO_2 to multi-carbon (C_{2+}) products because of the excessively strong intermediate binding to small-sized (15 nm) Cu NPs. Here, Chang et al. in 2022 effectively added pyrenyl-graphdiyne (Pyr-GDY) to ultrafine (2 nm) Cu NPs to give them a greatly increased selectivity for CO_2-to-C_{2+} conversion [186]. By adjusting the catalyst d-band center, Pyr-GDY would be aided in reducing the excessively tight binding between adsorbed H* and CO* intermediates on Cu NPs as well as maintain the ultrafine Cu NPs due to a higher affinity between alkyne moieties and Cu NPs. In comparison to support-free Cu NPs of C_{2+} 20% FE, CNT-supported Cu NPs of 18% C_{2+} FE, GO-supported Cu NPs of 8% C_{2+} FE, and other reported ultrafine Cu NPs, the resultant Pyr-GDY-Cu catalytic composite gave up to 74% FE for C_{2+} products. Their findings highlight the crucial role graphdiyne plays in the selectivity of Cu-catalyzed CO_2 electroreduction and highlight the potential of ultrafine Cu NP catalysts to transform CO_2 into a product with added value (C_{2+}). Kattel et al. (2016) used DFT calculations to identify the mechanism of CO_2 hydrogenation at the metal-oxide contacts [187]. In experiments on PtCo-TiO_2 and PtCo-ZrO_2, *HCOO and *HOCO were both shown to be reaction intermediates, but *CH_3O was only found on PtCo-ZrO_2.

9. Theoretical Studies of CO_2 Conversion

DFT calculations were performed by Kumari et al. in 2016 to study the mechanisms of CO_2 reduction to CO and the hydrogenation of CO_2 to methanol on both the stoichiometric and reduced $CeO_2(110)$ surfaces [188]. It was found that CO_2 dissociates to CO through interaction with the oxygen vacancy on the reduced ceria surface, and the produced CO can be further hydrogenated to methanol. In 2016, Cheng et al. investigated the conversion of CO_2 to methanol on the reduced $CeO_2(110)$ surface by performing DFT calculations corrected by on-site Coulomb interaction (DFT + U) and microkinetic analysis [189]. They also found that the HCOO route is the dominant pathway for methanol formation on the reduced $CeO_2(110)$. DFT studies show that the energy required to hydrogenate CO to *CHO on PtCo-TiO_2 is substantially lower than that required to desorb it. The result was a selective generation of CO since the chemisorbed CO preferred energetic desorption over the subsequent hydrogenation. Mostly on PtCo-ZrO$_2$ catalysts, however, CO generation was hampered, and CH$_4$ was produced instead. Chai and Guo demonstrated in 2016 that the interaction of N-doping and curvature can successfully control the activity and selectivity of graphene-CNT catalysts using both DFT and ab initio molecular dynamic calculations [190]. For graphitic N-doped graphene edges, as opposed to the un-doped equivalent, the CO$_2$ activation barrier can be adjusted to 0.58 eV. While the (6, 0) CNT with a high degree of curvature is efficient for both CH$_3$OH and HCHO synthesis, the graphene catalyst without curvature demonstrated great selectivity for CO-HCOOH generation. Curvature played a significant role in adjusting the overpotential for a particular product, e.g., for the synthesis of CO, from 1.5 to 0.02 V, and for CH$_3$OH, from 1.29 to 0.49 V. Thus, as demonstrated here for CO$_2$ reduction, graphene-CNT nanostructures provide significant scope and flexibility for effective tuning of catalytic efficiency and selectivity. Green chemistry is a fascinating field that deals with chemicals like ethanol produced from CO$_2$ transformation. Li et al. in 2019 investigated the mechanism of thermal catalytic hydrogenation of CO$_2$ to methanol on reduced CeO$_2$(100) by using DFT calculations [191]. They found that CO$_2$ was hydrogenated via the HCOO route rather than the COOH route. These results then indicate that oxygen vacancies on the reduced CeO$_2$ surface are crucial to the conversion of CO$_2$ to CH$_3$OH.

The combined experimental and density functional theory (DFT) study of Liu et al. in 2020 reported that the morphology control of CeO$_2$ nano-catalysts is important for methanol synthesis [192]. They also proposed that methanol was likely generated via the so-called formate (HCOO) pathway where the adsorbed CO$_2$ is firstly hydrogenated to the HCOO* species. In 2020, Coufourier et al. created a catalytic system that was both effective and affordable [193]. The CO$_2$ reductions, hydrogenocarbonate, and carbonate in pure water were described for use in an iron catalyst system that is very effective, has higher stability, is free of phosphine, and is simple to produce. Carbonic derivative hydrogenation occurs in good yields with good catalytic performance in just the existence of the bifunctional cyclopentadienone iron tricarbonyl. For the hydrogenation of CO$_2$, hydrogenocarbonate, and carbonate into formate in pure water, turnover numbers (TON) of up to 3343, 4234, and 40, respectively, have been obtained. Cao et al. in 2021 coupled the DFT calculations with micro-kinetic modeling [194]. The lack of interactions between adsorbed formate and intermediates, which would understate the rate of CO$_2$ pathway by possessing a too high formate coverage, was shown to be the cause of the discrepancy between the investigational rate and the earlier simulated predictions in the literature. The researchers demonstrated that CO$_2$ hydrogenation dominates for pure Cu catalysts, which is consistent with results, when adsorbate–adsorbate interactions, particularly the generated H bond, were considered. In particular, it has been found that the adsorbed HCOOH* can hydrogenate in a new transition state that is already being stable by hydrogen bonds. In an earlier study in 2022, Rasteiro et al. applied DFT calculations to analyze the impact of alloy–support synergy on the catalytic performance of Ni$_5$Ga$_3$ supported by SiO$_2$, CeO$_2$, and ZrO$_2$ [195]. The most promising catalyst, Ni$_5$Ga$_3$-ZrO$_2$, had a reaction mechanism that the researchers proposed according to DFT results. In 2022, Kovalskii et al. examined Au-h-BN(O) and Pt-h-BN(O) nanohybrids in CO oxidation and CO$_2$ hydrogenation reactions, and on the basis of DFT calculations, postulated potential catalytic reaction pathways [196]. Oxygen-related chemical reactions were accelerated by a charge density distribution at the Pt-h-BN interface via increasing oxygen absorption. In order to elucidate the mechanism underlying the catalysis of selective hydrogenation of CO$_2$ to methanol, Wang et al. in 2022 performed extensive DFT calculations corrected by on-site Coulomb interaction (DFT + U) to investigate the H$_2$ dissociation and the reaction between the active H species and CO$_2$ on the pristine and Cu-doped CeO$_2$(111) (denoted as Cu/CeO$_2$(111)) surfaces [197]. Their calculations evidenced that the heterolytic H$_2$ dissociation for hydride generation can more readily occur on the Cu/CeO$_2$(111) surface than on the pristine CeO$_2$(111) surface. They also found that the Cu dopant can facilitate the formation of surface oxygen vacancies, further promoting the generation of hydride species. Moreover, the adsorption of CO$_2$

and the hydrogenation of CO_2 to HCOO* can be greatly promoted on the $Cu/CeO_2(111)$ surface with hydride species, which can lead to high activity and selectivity toward CO_2 hydrogenation into methanol.

10. Preparation and Approximate Cost of CNTs

Table 2 describes the numerous purification steps and the mass variations during each purification stage. Based on this, each step in the purification table is clearly intended to remove some contaminants in order to produce pure CNTs [198]. The definition of "yield" varies considerably from author to author; some claim that yield is based on removing only metallic particles, while others claim that it is based on removing everything extraneous other than the CNTs. The second reason is thought to constitute the foundation for the yield in the current experiment. As a result, although having what seem to be lower yields, the current technique of synthesis has a more cost-effective yield because it is less expensive [199].

Table 2. Purification stages and each stage's effective mass variation.

Method	Set up	Purity	Cost in USD
Conventional arc discharge in vacuum	TIG power source, inert atmosphere, metal cabinet with water cooling system, automated process, and chemical purification	80–95 wt%	15 USD/gm
Chemical vapor deposition (CVD)	Furnace, inert atmosphere, metal catalyst	95%	40 USD/gm
Laser ablation	Laser source, furnace, inert atmosphere, metal catalyst–graphite composite	20–80 wt%	Due to the high capital cost of the laser and the lower quantity of CNT after final purification, this method is not commercially viable.
Floating catalyst method	Tubular reactor, quartz tube, thermocouples, inert gas	70–90 wt%	It requires a complicated set up. The cost of aromatic hydrocarbons is very high (Benzene: 44 USD/10 g).
Cyclic oxidation	Plant materials, ceramic reactor	No reports on purity	Even though the source materials are cheap, pre-treatment and heating takes longer duration in a high pressure vacuum chamber. Yield details are not available.
EDM process	Plasma sputtering unit, microelectric discharge apparatus, metal catalyst	No reports on purity	It requires costly equipment such as plasma-sputtering unit and microelectric discharge unit. Yield details are not mentioned.
Combustion process	Bunsen burner, liquefied butane, metal catalyst	No reports on purity	This method is simple but the yield seems to be much less compared to other methods (in mg).
Simplified arc discharge in air	Manual metal arc welding machine and chemical purification	75–80 wt%	3 USD/gm

11. CNTs as Catalysts

For catalytic reactions, carbon nanostructures offer a catalytic support framework that demonstrates good adhesion, metal particle stability at high temperatures, and relative chemical inertness [129]. In the past, FT catalysis has been carried out using carbon-based catalysts, including carbon nanotubes. C_{2+} hydrocarbons can now be formed with good selectivity using carbon-based catalysts [149]. Due to the well-stressed and superior graphical nature of the curved support, metal particles placed on carbon nanotubes behave differently from those deposited on flat non-carbon nanosupports. It has been demonstrated that hydrogen spillage from conductive nanoparticles on stents is superior to that from non-conductive analogues. Where there is a physical route for hydrogen to get from the NP to the support surface, bridging happens [189]. As the hydrogen travels to the surface supporting the nanotube, it interacts with the nanoparticles and needs to be stabilized. In the absence of a physical bridge, hydrogen species created by the nanoparticles cannot be transferred to the support surface. The intrinsic capacity of the substrates to support hydrogen species is ruled out in the event of poor transport from the nanoparticles to the surface. Less hydrogen on the nanotube's surface results from decreased hydrogen leakage, which hinders the catalyst's capacity to reduce carbon dioxide or carbon dioxide throughout the reaction [200].

With many more potential uses in the future, CNTs have already found commercial success in the domains of energy storage (such as consumer lithium-ion batteries), coatings and films (such as fouling-resistant paints for ship hulls), and composite materials (such as enhanced wind turbine blades). They are also desired as support materials for catalytic transition metal nanoparticles in the field of chemical catalysis. The high surface area, tunable structure (e.g., diameter, porosity, and surface composition), and excellent chemical and thermal stability of CNTs have all been recognized to make them particularly desirable as catalyst supports. Their surface area, which ranges from 400 to 900 and 200 to 400 m^2 g^{-1} for SWCNTs and MWCNTs, respectively, is desirable because it enables the deposition and dispersion of catalytic metal nanoparticles with a high surface area-to-volume ratio. An sp^2 hybridized network of carbon atoms that forms a tube with a nanometer-scale diameter and often a high aspect ratio is known as a CNT. Due to their many attractive characteristics, CNTs have attracted considerable research attention for uses in everything from construction to electronics, catalysis, and beyond [201].

12. Conclusions

CO_2 is a significant greenhouse gas that, due to its growing atmospheric concentrations, is thought to be the primary cause of both global warming and climate change. As a result, global attention has shifted to CO_2 concentration reduction. The chemical transformation of CO_2 produces carbon compounds that can be used as precursors for the production of chemicals and fuels. The conversion of CO_2 into fuels and chemicals presents options for reducing the rising CO_2 buildup because CO_2 is both a renewable and ecofriendly source of carbon and it can be used as a C_1 building block for valuable chemicals. Studies have shown that increased concentrations of carbon dioxide increase photosynthesis, spurring plant growth. While rising carbon dioxide concentrations in the air can be beneficial for plants, they are also the chief culprit in climate change. There are only three sources of renewable carbon: renewable carbon from the recycling of already existing plastics (mechanical and chemical recycling), renewable carbon obtained from all types of biomass, and renewable carbon from direct CO_2 utilization of fossil point sources (while they still exist), as well as from permanently biogenous point sources and direct air capture. All three sources are essential for a complete transition to renewable carbon, and all of them, in equal shares, should be used by industry, supported by politicians, and accepted by the population. In a sustainable chemical industry, bulk chemicals will primarily rely on chemical CO_2 utilization through methane, methanol, and naphtha, while specialty chemicals and complex molecules will more likely be produced from biomass (and CO_2 fermentation). At the same time, mechanical and chemical recycling will reduce the overall need for additional renewable carbon. Whereas traditional recycling re-uses products and materials, the use of biomass and direct CO_2 utilization is tantamount to a recycling process, which also constitutes part of an extended circular economy. The hydrogenation of CO_2 is a practical and efficient procedure in this respect, as has already been mentioned in the review. One of the main issues in developing an exergonic CO_2 conversion reaction is thermodynamically unfavorable CO_2 thermochemical properties. This gas is in the highest oxidation state of carbon, which results in its relatively low standard enthalpy of formation. That is the reason for this molecule being one of the main products of combustion reactions. Different technical directions and targeted research methods on the logical design of catalysts, reactor optimization, and investigation of reaction mechanisms have been proposed to overcome the limits on conversion and selectivity. In fact, oxides have been found to be able to overcome the limits on CO_2 conversion into oxygenated hydrocarbons by means of the supported metallic catalysts at moderate temperatures and pressures. Furthermore, the preparation of bifunctional catalysts combining metal oxides and zeolites has demonstrated an effective way to control the product selectivity for the conversion. Graphene and CNT have been extensively studied by academics over the past 15 years as 3D nanostructured materials for catalytic applications because of their impressive chemical and physical properties. There have been numerous reviews of the heterogeneous catalytic hydrogenation of CO_2 that can be categorized by the methods used, such as thermal, electrochemical, and photochemical hydrogenation, as well as by the homogeneous and heterogeneous catalysts used, or by the resulting product distributions or catalysts used. Despite the difficulties, the transformation of CO_2 to value-added chemicals still receives great attention worldwide because of its significance for providing sustainable alternatives to solve urgent issues such as those of energy and the environment. Recent years have seen the emergence of experimental and computational technologies for more efficient search and design of catalysts and other materials. Experimental technologies are increasingly being employed for the rapid discovery of novel catalysts and materials. On the other hand, a similar array of computational technologies, including high-throughput and automated computational simulations and reaction modeling, coupled with machine learning algorithms, have also started to enable the theoretical understanding and prediction of new catalysts.

Author Contributions: Conceptualization, W.N.R.W.I. and L.M.S.; validation, W.N.R.W.I.; formal analysis, L.M.S.; investigation, A.A.-A. and L.M.S.; resources, W.N.R.W.I.; data curation, L.M.S.; writing—original draft preparation, L.M.S.; writing—review and editing, A.A.-A.; supervision, W.N.R.W.I. and A.A.-A.; project administration, W.N.R.W.I. and A.A.-A.; funding acquisition, W.N.R.W.I. All authors have read and agreed to the published version of the manuscript.

Funding: This research was funded by Ministry of Higher Education Malaysia FRGS/1/2020/TK0/UKM/02/31.

Acknowledgments: The authors thank Universiti Kebangsaan Malaysia (Malaysia) and the University of Technology (Iraq) for supporting this work.

Conflicts of Interest: The authors declare no conflict of interest.

References

1. Ma, R.; Xu, B.; Zhang, X. Catalytic partial oxidation (CPOX) of natural gas and renewable hydrocarbons/oxygenated hydrocarbons—A review. *Catal. Today* **2019**, *338*, 18–30. [CrossRef]
2. Hasan, S.Z.; Ahmad, K.N.; Isahak, W.N.R.W.; Pudukudy, M.; Masdar, M.S.; Jahim, J.M. Synthesis, Characterisation and Catalytic Activity of NiO supported Al_2O_3 for CO_2 Hydrogenation to Carboxylic Acids: Influence of Catalyst Structure. *IOP Conf. Ser. Earth Environ. Sci.* **2019**, *268*, 012079. [CrossRef]
3. Palmeri, N.; Chiodo, V.; Freni, S.; Frusteri, F.; Bart, J.; Cavallaro, S. Hydrogen from oxygenated solvents by steam reforming on Ni/Al_2O_3 catalyst. *Int. J. Hydrogen Energy* **2008**, *33*, 6627–6634. [CrossRef]
4. Kahn, B. Earth's CO_2 Passes the 400 PPM Threshold—Maybe Permanently. *Sci. Am.* **2016**. Available online: https://www.scientificamerican.com/article/earth-s-co2-passes-the-400-ppm-threshold-maybe-permanently/ (accessed on 1 November 2022).
5. Capros, P.; Tasios, N.; De Vita, A.; Mantzos, L.; Paroussos, L. Model-based analysis of decarbonising the EU economy in the time horizon to 2050. *Energy Strat. Rev.* **2012**, *1*, 76–84. [CrossRef]
6. European Commission. *A Roadmap for Moving to a Competitive Low Carbon Economy in 2050. COM(2011) 112 Final*; European Commission: Brussels, Belgium, 2011; Volume 34, pp. 1–34.

7. Dimitriou, I.; García-Gutiérrez, P.; Elder, R.H.; Cuéllar-Franca, R.M.; Azapagic, A.; Allen, R.W.K. Carbon dioxide utilisation for production of transport fuels: Process and economic analysis. *Energy Environ. Sci.* **2015**, *8*, 1775–1789. [CrossRef]
8. Ma, X.; Wang, X.; Song, C. "Molecular Basket" Sorbents for Separation of CO_2 and H_2S from Various Gas Streams. *J. Am. Chem. Soc.* **2009**, *131*, 5777–5783. [CrossRef]
9. Du, G.; Lim, S.; Yang, Y.; Wang, C.; Pfefferle, L.; Haller, G.L. Methanation of carbon dioxide on Ni-incorporated MCM-41 catalysts: The influence of catalyst pretreatment and study of steady-state reaction. *J. Catal.* **2007**, *249*, 370–379. [CrossRef]
10. Duyar, M.S.; Treviño, M.A.A.; Farrauto, R.J. Dual function materials for CO_2 capture and conversion using renewable H_2. *Appl. Catal. B Environ.* **2015**, *168–169*, 370–376. [CrossRef]
11. Lee, C.-Y.; Zhao, Y.; Wang, C.; Mitchell, D.R.G.; Wallace, G.G. Rapid formation of self-organised Ag nanosheets with high efficiency and selectivity in CO_2 electroreduction to CO. *Sustain. Energy Fuels* **2017**, *1*, 1023–1027. [CrossRef]
12. Li, K.; Peng, B.; Peng, T. Recent Advances in Heterogeneous Photocatalytic CO_2 Conversion to Solar Fuels. *ACS Catal.* **2016**, *6*, 7485–7527. [CrossRef]
13. Welch, A.J.; DuChene, J.S.; Tagliabue, G.; Davoyan, A.R.; Cheng, W.-H.; Atwater, H.A. Nanoporous Gold as a Highly Selective and Active Carbon Dioxide Reduction Catalyst. *ACS Appl. Energy Mater.* **2019**, *2*, 164–170. [CrossRef]
14. Xie, H.; Wang, T.; Liang, J.; Li, Q.; Sun, S. Cu-based nanocatalysts for electrochemical reduction of CO_2. *Nano Today* **2018**, *21*, 41–54. [CrossRef]
15. Huang, H.; Jia, H.; Liu, Z.; Gao, P.; Zhao, J.; Luo, Z.; Yang, J.; Zeng, J. Understanding of Strain Effects in the Electrochemical Reduction of CO_2: Using Pd Nanostructures as an Ideal Platform. *Angew. Chem. Int. Ed.* **2017**, *56*, 3594–3598. [CrossRef]
16. Ma, M.; Trześniewski, B.J.; Xie, J.; Smith, W.A. Selective and Efficient Reduction of Carbon Dioxide to Carbon Monoxide on Oxide-Derived Nanostructured Silver Electrocatalysts. *Angew. Chem. Int. Ed.* **2016**, *55*, 9748–9752. [CrossRef]
17. Su, P.; Xu, W.; Qiu, Y.; Zhang, T.; Li, X.; Zhang, H. Ultrathin Bismuth Nanosheets as a Highly Efficient CO_2 Reduction Electrocatalyst. *Chemsuschem* **2018**, *11*, 848–853. [CrossRef]
18. Li, Q.; Fu, J.; Zhu, W.; Chen, Z.; Shen, B.; Wu, L.; Xi, Z.; Wang, T.; Lu, G.; Zhu, J.-J.; et al. Tuning Sn-Catalysis for Electrochemical Reduction of CO_2 to CO via the Core/Shell Cu/SnO_2 Structure. *J. Am. Chem. Soc.* **2017**, *139*, 4290–4293. [CrossRef]
19. Hu, X.-M.; Rønne, M.H.; Pedersen, S.U.; Skrydstrup, T.; Daasbjerg, K. Enhanced Catalytic Activity of Cobalt Porphyrin in CO_2 Electroreduction upon Immobilization on Carbon Materials. *Angew. Chem. Int. Ed.* **2017**, *56*, 6468–6472. [CrossRef]
20. Chen, Q.; Tsiakaras, P.; Shen, P. Electrochemical Reduction of Carbon Dioxide: Recent Advances on Au-Based Nanocatalysts. *Catalysts* **2022**, *12*, 1348. [CrossRef]
21. Brouzgou, A.; Song, S.; Liang, Z.-X.; Tsiakaras, P. Non-Precious Electrocatalysts for Oxygen Reduction Reaction in Alkaline Media: Latest Achievements on Novel Carbon Materials. *Catalysts* **2016**, *6*, 159. [CrossRef]
22. Alaba, P.A.; Abbas, A.; Daud, W.M.W. Insight into catalytic reduction of CO_2: Catalysis and reactor design. *J. Clean. Prod.* **2017**, *140*, 1298–1312. [CrossRef]
23. Centi, G.; Perathoner, S. CO_2-based energy vectors for the storage of solar energy. *Greenh. Gases Sci. Technol.* **2011**, *1*, 21–35. [CrossRef]
24. Porosoff, M.D.; Yan, B.; Chen, J.G. Catalytic reduction of CO_2 by H_2 for synthesis of CO, methanol and hydrocarbons: Challenges and opportunities. *Energy Environ. Sci.* **2016**, *9*, 62–73. [CrossRef]
25. Gao, P.; Li, S.; Bu, X.; Dang, S.; Liu, Z.; Wang, H.; Zhong, L.; Qiu, M.; Yang, C.; Cai, J.; et al. Direct conversion of CO_2 into liquid fuels with high selectivity over a bifunctional catalyst. *Nat. Chem.* **2017**, *9*, 1019–1024. [CrossRef] [PubMed]
26. Song, H.; Zhang, N.; Zhong, C.; Liu, Z.; Xiao, M.; Gai, H. Hydrogenation of CO_2 into formic acid using a palladium catalyst on chitin. *New J. Chem.* **2017**, *41*, 9170–9177. [CrossRef]
27. Visconti, C.G.; Martinelli, M.; Falbo, L.; Infantes-Molina, A.; Lietti, L.; Forzatti, P.; Iaquaniello, G.; Palo, E.; Picutti, B.; Brignoli, F. CO_2 hydrogenation to lower olefins on a high surface area K-promoted bulk Fe-catalyst. *Appl. Catal. B Environ.* **2017**, *200*, 530–542. [CrossRef]
28. Larmier, K.; Liao, W.-C.; Tada, S.; Lam, E.; Verel, R.; Bansode, A.; Urakawa, A.; Comas-Vives, A.; Copéret, C. CO_2-to-Methanol Hydrogenation on Zirconia-Supported Copper Nanoparticles: Reaction Intermediates and the Role of the Metal-Support Interface. *Angew. Chem. Int. Ed.* **2017**, *56*, 2318–2323. [CrossRef]
29. Bai, S.; Shao, Q.; Wang, P.; Dai, Q.; Wang, X.; Huang, X. Highly Active and Selective Hydrogenation of CO_2 to Ethanol by Ordered Pd–Cu Nanoparticles. *J. Am. Chem. Soc.* **2017**, *139*, 6827–6830. [CrossRef]
30. Sorcar, S.; Hwang, Y.; Lee, J.; Kim, H.; Grimes, K.M.; Grimes, C.A.; Jung, J.-W.; Cho, C.-H.; Majima, T.; Hoffmann, M.R.; et al. CO_2, water, and sunlight to hydrocarbon fuels: A sustained sunlight to fuel (Joule-to-Joule) photoconversion efficiency of 1%. *Energy Environ. Sci.* **2019**, *12*, 2685–2696. [CrossRef]
31. Tan, Z.; Ni, K.; Chen, G.; Zeng, W.; Tao, Z.; Ikram, M.; Zhang, Q.; Wang, H.; Sun, L.; Zhu, X.; et al. Incorporating Pyrrolic and Pyridinic Nitrogen into a Porous Carbon made from C60 Molecules to Obtain Superior Energy Storage. *Adv. Mater.* **2017**, *29*, 160341. [CrossRef]
32. Chen, J.; Xue, C.; Ramasubramaniam, R.; Liu, H. A new method for the preparation of stable carbon nanotube organogels. *Carbon* **2006**, *44*, 2142–2146. [CrossRef]
33. Navrotskaya, A.G.; Aleksandrova, D.D.; Krivoshapkina, E.F.; Sillanpää, M.; Krivoshapkin, P.V. Hybrid Materials Based on Carbon Nanotubes and Nanofibers for Environmental Applications. *Front. Chem.* **2020**, *8*, 546. [CrossRef]

34. Worsley, M.A.; Shin, S.J.; Merrill, M.D.; Lenhardt, J.; Nelson, A.J.; Woo, L.Y.; Gash, A.E.; Baumann, T.F.; Orme, C.A. Ultralow Density, Monolithic WS$_2$, MoS$_2$, and MoS$_2$/Graphene Aerogels. *ACS Nano* **2015**, *9*, 4698–4705. [CrossRef]
35. Mamba, G.; Mishra, A.K. Graphitic carbon nitride (g-C$_3$N$_4$) nanocomposites: A new and exciting generation of visible light driven photocatalysts for environmental pollution remediation. *Appl. Catal. B Environ.* **2016**, *198*, 347–377. [CrossRef]
36. Kovtyukhova, N.I.; Mallouk, T.E.; Pan, A.L.; Dickey, E.C. Individual Single-Walled Nanotubes and Hydrogels Made by Oxidative Exfoliation of Carbon Nanotube Ropes. *J. Am. Chem. Soc.* **2003**, *125*, 9761–9769. [CrossRef]
37. Sun, Y.; Wu, Q.; Xu, Y.; Bai, H.; Li, C.; Shi, G. Highly conductive and flexible mesoporous graphitic films prepared by graphitizing the composites of graphene oxide and nanodiamond. *J. Mater. Chem.* **2011**, *21*, 7154–7160. [CrossRef]
38. Bryning, M.B.; Milkie, D.E.; Islam, M.F.; Hough, L.A.; Kikkawa, J.M.; Yodh, A.G. Carbon Nanotube Aerogels. *Adv. Mater.* **2007**, *19*, 661–664. [CrossRef]
39. Gui, X.; Wei, J.; Wang, K.; Cao, A.; Zhu, H.; Jia, Y.; Shu, Q.; Wu, D. Carbon Nanotube Sponges. *Adv. Mater.* **2010**, *22*, 617–621. [CrossRef]
40. Zhang, L.L.; Xiong, Z.; Zhao, X.S. Pillaring Chemically Exfoliated Graphene Oxide with Carbon Nanotubes for Photocatalytic Degradation of Dyes under Visible Light Irradiation. *ACS Nano* **2010**, *4*, 7030–7036. [CrossRef]
41. Zhu, Y.; Li, L.; Zhang, C.; Casillas, G.; Sun, Z.; Yan, Z.; Ruan, G.; Peng, Z.; Raji, A.-R.; Kittrell, C.; et al. A seamless three-dimensional carbon nanotube graphene hybrid material. *Nat. Commun.* **2012**, *3*, 1225. [CrossRef]
42. Tripathi, M.; Valentini, L.; Rong, Y.; Bon, S.B.; Pantano, M.F.; Speranza, G.; Guarino, R.; Novel, D.; Iacob, E.; Liu, W.; et al. Free-Standing Graphene Oxide and Carbon Nanotube Hybrid Papers with Enhanced Electrical and Mechanical Performance and Their Synergy in Polymer Laminates. *Int. J. Mol. Sci.* **2020**, *21*, 8585. [CrossRef] [PubMed]
43. Liu, J.; Zhang, L.; Bin Wu, H.; Lin, J.; Shen, Z.; Lou, X.W. High-performance flexible asymmetric supercapacitors based on a new graphene foam/carbon nanotube hybrid film. *Energy Environ. Sci.* **2014**, *7*, 3709–3719. [CrossRef]
44. Jin, S.; Xin, S.; Wang, L.; Du, Z.; Cao, L.; Chen, J.; Kong, X.; Gong, M.; Lu, J.; Zhu, Y.; et al. Covalently Connected Carbon Nanostructures for Current Collectors in Both the Cathode and Anode of Li-S Batteries. *Adv. Mater.* **2016**, *28*, 9094–9102. [CrossRef]
45. Peng, Q.; Li, Y.; He, X.; Gui, X.; Shang, Y.; Wang, C.; Wang, C.; Zhao, W.; Du, S.; Shi, E.; et al. Graphene Nanoribbon Aerogels Unzipped from Carbon Nanotube Sponges. *Adv. Mater.* **2014**, *26*, 3241–3247. [CrossRef]
46. Chen, L.; Du, R.; Zhu, J.; Mao, Y.; Xue, C.; Zhang, N.; Hou, Y.; Zhang, J.; Yi, T. Three-Dimensional Nitrogen-Doped Graphene Nanoribbons Aerogel as a Highly Efficient Catalyst for the Oxygen Reduction Reaction. *Small* **2015**, *11*, 1423–1429. [CrossRef] [PubMed]
47. Chen, L.-F.; Zhang, X.-D.; Liang, H.-W.; Kong, M.; Guan, Q.-F.; Chen, P.; Wu, Z.-Y.; Yu, S.-H. Synthesis of Nitrogen-Doped Porous Carbon Nanofibers as an Efficient Electrode Material for Supercapacitors. *ACS Nano* **2012**, *6*, 7092–7102. [CrossRef] [PubMed]
48. Chen, L.-F.; Huang, Z.-H.; Liang, H.-W.; Gao, H.-L.; Yu, S.-H. Three-Dimensional Heteroatom-Doped Carbon Nanofiber Networks Derived from Bacterial Cellulose for Supercapacitors. *Adv. Funct. Mater.* **2014**, *24*, 5104–5111. [CrossRef]
49. Yan, X.; You, H.; Liu, W.; Wang, X.; Wu, D. Free-Standing and Heteroatoms-Doped Carbon Nanofiber Networks as a Binder-Free Flexible Electrode for High-Performance Supercapacitors. *Nanomaterials* **2019**, *9*, 1189. [CrossRef]
50. He, Z.; Yang, Y.; Liu, J.-W.; Yu, S.-H. Emerging tellurium nanostructures: Controllable synthesis and their applications. *Chem. Soc. Rev.* **2017**, *46*, 2732–2753. [CrossRef]
51. Luo, W.; Chen, X.; Wei, Z.; Liu, D.; Yao, W.; Zhu, Y. Three-dimensional network structure assembled by g-C$_3$N$_4$ nanorods for improving visible-light photocatalytic performance. *Appl. Catal. B Environ.* **2019**, *255*, 117761. [CrossRef]
52. Ou, H.; Yang, P.; Lin, L.; Anpo, M.; Wang, X. Carbon Nitride Aerogels for the Photoredox Conversion of Water. *Angew. Chem. Int. Ed.* **2017**, *56*, 10905–10910. [CrossRef] [PubMed]
53. Abu-Zied, B.M.; Alamry, K.A. Green synthesis of 3D hierarchical nanostructured Co$_3$O$_4$/carbon catalysts for the application in sodium borohydride hydrolysis. *J. Alloy. Compd.* **2019**, *798*, 820–831. [CrossRef]
54. Liu, J.J.; Xu, Y.X. Three-dimensional Graphene-based Composites and Two-dimensional Polymers: Synthesis and Application in Energy Storage and Conversion. *Acta Polym. Sin.* **2019**, *50*, 219–232.
55. Alali, K.T.; Yu, J.; Moharram, D.; Liu, Q.; Chen, R.; Zhu, J.; Li, R.; Liu, P.; Liu, J.; Wang, J. In situ construction of 3-dimensional hierarchical carbon nanostructure; investigation of the synthesis parameters and hydrogen evolution reaction performance. *Carbon* **2021**, *178*, 48–57. [CrossRef]
56. Ferrari, A.C.; Bonaccorso, F.; Fal'Ko, V.; Novoselov, K.S.; Roche, S.; Bøggild, P.; Borini, S.; Koppens, F.H.L.; Palermo, V.; Pugno, N.; et al. Science and technology roadmap for graphene, related two-dimensional crystals, and hybrid systems. *Nanoscale* **2015**, *7*, 4598–4810. [CrossRef]
57. Arjmandi-Tash, H.; Lebedev, N.; van Deursen, P.M.; Aarts, J.; Schneider, G.F. Hybrid cold and hot-wall reaction chamber for the rapid synthesis of uniform graphene. *Carbon* **2017**, *118*, 438–442. [CrossRef]
58. Alnuaimi, A.; Almansouri, I.; Saadat, I.; Nayfeh, A. Toward fast growth of large area high quality graphene using a cold-wall CVD reactor. *RSC Adv.* **2017**, *7*, 51951–51957. [CrossRef]
59. Al-Hagri, A.; Li, R.; Rajput, N.S.; Lu, X.; Cong, S.; Sloyan, K.; Almahri, M.A.; Tamalampudi, S.R.; Chiesa, M.; Al Ghaferi, A. Direct growth of single-layer terminated vertical graphene array on germanium by plasma enhanced chemical vapor deposition. *Carbon* **2019**, *155*, 320–325. [CrossRef]

60. Chen, Y.-C.; Lin, W.-H.; Tseng, W.-S.; Chen, C.-C.; Rossman, G.; Chen, C.-D.; Wu, Y.-S.; Yeh, N.-C. Direct growth of mm-size twisted bilayer graphene by plasma-enhanced chemical vapor deposition. *Carbon* **2020**, *156*, 212–224. [CrossRef]
61. Sohn, K.; Na, Y.J.; Chang, H.; Roh, K.-M.; Jang, H.D.; Huang, J. Oil absorbing graphene capsules by capillary molding. *Chem. Commun.* **2012**, *48*, 5968–5970. [CrossRef]
62. Yan, J.; Ding, Y.; Hu, C.; Cheng, H.; Chen, N.; Feng, Z.; Zhang, Z.; Qu, L. Preparation of multifunctional microchannel-network graphene foams. *J. Mater. Chem. A* **2014**, *2*, 16786–16792. [CrossRef]
63. Shunxin, J. A Simple Solution-Based Method to Prepare Honeycomb-Like Li2S/Graphene Composite for Lithium-Sulfur Batteries. *Int. J. Electrochem. Sci.* **2018**, *13*, 3407–3419. [CrossRef]
64. Moura, D.; Pereira, R.F.; Gonçalves, I.C. Recent advances on bioprinting of hydrogels containing carbon materials. *Mater. Today Chem.* **2022**, *23*, 100617. [CrossRef]
65. Sha, J.; Li, Y.; Salvatierra, R.V.; Wang, T.; Dong, P.; Ji, Y.; Lee, S.-K.; Zhang, C.; Zhang, J.; Smith, R.H.; et al. Three-Dimensional Printed Graphene Foams. *ACS Nano* **2017**, *11*, 6860–6867. [CrossRef] [PubMed]
66. Tang, X.; Zhou, H.; Cai, Z.; Cheng, D.; He, P.; Xie, P.; Zhang, D.; Fan, T. Generalized 3D Printing of Graphene-Based Mixed-Dimensional Hybrid Aerogels. *ACS Nano* **2018**, *12*, 3502–3511. [CrossRef]
67. Yu, Z.-Y.; Chen, L.-F.; Song, L.-T.; Zhu, Y.-W.; Ji, H.-X.; Yu, S.-H. Free-standing boron and oxygen co-doped carbon nanofiber films for large volumetric capacitance and high rate capability supercapacitors. *Nano Energy* **2015**, *15*, 235–243. [CrossRef]
68. Bai, X.-L.; Gao, Y.-L.; Gao, Z.-Y.; Ma, J.-Y.; Tong, X.-L.; Sun, H.-B.; Wang, J.A. Supercapacitor performance of 3D-graphene/MnO2 foam synthesized via the combination of chemical vapor deposition with hydrothermal method. *Appl. Phys. Lett.* **2020**, *117*, 183901. [CrossRef]
69. Li, J.; Yang, S.; Jiao, P.; Peng, Q.; Yin, W.; Yuan, Y.; Lu, H.; He, X.; Li, Y. Three-dimensional macroassembly of hybrid C@CoFe nanoparticles/reduced graphene oxide nanosheets towards multifunctional foam. *Carbon* **2020**, *157*, 427–436. [CrossRef]
70. Liu, W.; Cai, J.; Li, Z. Self-Assembly of Semiconductor Nanoparticles/Reduced Graphene Oxide (RGO) Composite Aerogels for Enhanced Photocatalytic Performance and Facile Recycling in Aqueous Photocatalysis. *ACS Sustain. Chem. Eng.* **2015**, *3*, 277–282. [CrossRef]
71. Zhang, M.; Song, Z.; Liu, H.; Wang, A.; Shao, S. MoO2 coated few layers of MoS2 and FeS2 nanoflower decorated S-doped graphene interoverlapped network for high-energy asymmetric supercapacitor. *J. Colloid Interface Sci.* **2021**, *584*, 418–428. [CrossRef] [PubMed]
72. Park, S.; An, J.; Potts, J.R.; Velamakanni, A.; Murali, S.; Ruoff, R.S. Hydrazine-reduction of graphite- and graphene oxide. *Carbon* **2011**, *49*, 3019–3023. [CrossRef]
73. Yan, L.; Zhou, M.; Pang, X.; Gao, K. One-Step in Situ Synthesis of Reduced Graphene Oxide/Zn–Al Layered Double Hydroxide Film for Enhanced Corrosion Protection of Magnesium Alloys. *Langmuir* **2019**, *35*, 6312–6320. [CrossRef]
74. Shandilya, P.; Mittal, D.; Sudhaik, A.; Soni, M.; Raizada, P.; Saini, A.; Singh, P. GdVO4 modified fluorine doped graphene nanosheets as dispersed photocatalyst for mitigation of phenolic compounds in aqueous environment and bacterial disinfection. *Sep. Purif. Technol.* **2019**, *210*, 804–816. [CrossRef]
75. Santos, F.C.U.; Paim, L.L.; da Silva, J.L.; Stradiotto, N.R. Electrochemical determination of total reducing sugars from bioethanol production using glassy carbon electrode modified with graphene oxide containing copper nanoparticles. *Fuel* **2016**, *163*, 112–121. [CrossRef]
76. Chen, W.; Li, S.; Chen, C.; Yan, L. Self-Assembly and Embedding of Nanoparticles by In Situ Reduced Graphene for Preparation of a 3D Graphene/Nanoparticle Aerogel. *Adv. Mater.* **2011**, *23*, 5679–5683. [CrossRef] [PubMed]
77. Kabtamu, D.M.; Chang, Y.-C.; Lin, G.-Y.; Bayeh, A.W.; Chen, J.-Y.; Wondimu, T.H.; Wang, C.-H. Three-dimensional annealed WO3 nanowire/graphene foam as an electrocatalytic material for all vanadium redox flow batteries. *Sustain. Energy Fuels* **2017**, *1*, 2091–2100. [CrossRef]
78. Cai, L.; Lin, Z.; Wang, M.; Pan, F.; Chen, J.; Wang, Y.; Shen, X.; Chai, Y. Improved interfacial H2O supply by surface hydroxyl groups for enhanced alkaline hydrogen evolution. *J. Mater. Chem. A* **2017**, *5*, 24091–24097. [CrossRef]
79. Worsley, M.A.; Pauzauskie, P.J.; Olson, T.Y.; Biener, J.; Satcher, J.J.H.; Baumann, T.F. Synthesis of Graphene Aerogel with High Electrical Conductivity. *J. Am. Chem. Soc.* **2010**, *132*, 14067–14069. [CrossRef]
80. Zhang, L.; Li, H.; Lai, X.; Wu, W.; Zeng, X. Hindered phenol functionalized graphene oxide for natural rubber. *Mater. Lett.* **2018**, *210*, 239–242. [CrossRef]
81. Bustos-Ramírez, K.; Martínez-Hernández, A.L.; Martínez-Barrera, G.; De Icaza, M.; Castaño, V.M.; Velasco-Santos, C. Covalently Bonded Chitosan on Graphene Oxide via Redox Reaction. *Materials* **2013**, *6*, 911–926. [CrossRef]
82. To, J.W.F.; Chen, Z.; Yao, H.; He, J.; Kim, K.; Chou, H.-H.; Pan, L.; Wilcox, J.; Cui, Y.; Bao, Z. Ultrahigh Surface Area Three-Dimensional Porous Graphitic Carbon from Conjugated Polymeric Molecular Framework. *ACS Central Sci.* **2015**, *1*, 68–76. [CrossRef] [PubMed]
83. Fang, B.; Chang, D.; Xu, Z.; Gao, C. A Review on Graphene Fibers: Expectations, Advances, and Prospects. *Adv. Mater.* **2020**, *32*, e1902664. [CrossRef] [PubMed]
84. Hummers, W.S., Jr.; Offeman, R.E. Preparation of Graphitic Oxide. *J. Am. Chem. Soc.* **1958**, *80*, 1339. [CrossRef]
85. Lawal, A.T. Graphene-based nano composites and their applications. A review. *Biosens. Bioelectron.* **2019**, *141*, 111384. [CrossRef] [PubMed]

86. Zhao, W.; Liu, H.; Meng, N.; Jian, M.; Wang, H.; Zhang, X. Graphene oxide incorporated thin film nanocomposite membrane at low concentration monomers. *J. Membr. Sci.* **2018**, *565*, 380–389. [CrossRef]
87. Akyüz, D.; Koca, A. Photocatalytic hydrogen production with reduced graphene oxide (RGO)-CdZnS nano-composites synthesized by solvothermal decomposition of dimethyl sulfoxide as the sulfur source. *J. Photochem. Photobiol. A Chem.* **2018**, *364*, 625–634. [CrossRef]
88. Kumar, N.; Rodriguez, J.R.; Pol, V.G.; Sen, A. Facile synthesis of 2D graphene oxide sheet enveloping ultrafine 1D $LiMn_2O_4$ as interconnected framework to enhance cathodic property for Li-ion battery. *Appl. Surf. Sci.* **2019**, *463*, 132–140. [CrossRef]
89. Aghazadeh, M. One-step electrophoretic/electrochemical synthesis of reduced graphene oxide/manganese oxide (RGO-Mn_3O_4) nanocomposite and study of its capacitive performance. *Anal. Bioanal. Electrochem.* **2018**, *10*, 961–973.
90. Jasmi, F.; Azeman, N.H.; Bakar, A.A.A.; Zan, M.S.D.; Badri, K.H.; Su'Ait, M.S. Ionic Conductive Polyurethane-Graphene Nanocomposite for Performance Enhancement of Optical Fiber Bragg Grating Temperature Sensor. *IEEE Access* **2018**, *6*, 47355–47363. [CrossRef]
91. Novoselov, K.S.; Geim, A.K.; Morozov, S.V.; Jiang, D.; Zhang, Y.; Dubonos, S.V.; Grigorieva, I.V.; Firsov, A.A. Electric Field Effect in Atomically Thin Carbon Films Supplementary. *Science* **2004**, *5*, 105–110. [CrossRef]
92. Qiu, B.; Li, Q.; Shen, B.; Xing, M.; Zhang, J. Stöber-like method to synthesize ultradispersed Fe_3O_4 nanoparticles on graphene with excellent Photo-Fenton reaction and high-performance lithium storage. *Appl. Catal. B Environ.* **2016**, *183*, 216–223. [CrossRef]
93. Wang, Y.; Zheng, Y.; Xu, X.; Dubuisson, E.; Bao, Q.; Lu, J.; Loh, K.P. Electrochemical Delamination of CVD-Grown Graphene Film: Toward the Recyclable Use of Copper Catalyst. *ACS Nano* **2011**, *5*, 9927–9933. [CrossRef] [PubMed]
94. Geim, A.K.; Novoselov, K.S. The rise of graphene. *Nat. Mater.* **2007**, *6*, 183–191. [CrossRef] [PubMed]
95. Dong, Z.; Jiang, C.; Cheng, H.; Zhao, Y.; Shi, G.; Jiang, L.; Qu, L. Facile Fabrication of Light, Flexible and Multifunctional Graphene Fibers. *Adv. Mater.* **2012**, *24*, 1856–1861. [CrossRef] [PubMed]
96. Kim, K.S.; Zhao, Y.; Jang, H.; Lee, S.Y.; Kim, J.M.; Kim, K.S.; Ahn, J.-H.; Kim, P.; Choi, J.-Y.; Hong, B.H. Large-scale pattern growth of graphene films for stretchable transparent electrodes. *Nature* **2009**, *457*, 706–710. [CrossRef] [PubMed]
97. Qiu, B.; Deng, Y.; Du, M.; Xing, M.; Zhang, J. Ultradispersed Cobalt Ferrite Nanoparticles Assembled in Graphene Aerogel for Continuous Photo-Fenton Reaction and Enhanced Lithium Storage Performance. *Sci. Rep.* **2016**, *6*, 29099. [CrossRef] [PubMed]
98. Chen, S.; Duan, J.; Ran, J.; Jaroniec, M.; Qiao, S.Z. N-doped graphene film-confined nickel nanoparticles as a highly efficient three-dimensional oxygen evolution electrocatalyst. *Energy Environ. Sci.* **2013**, *6*, 3693–3699. [CrossRef]
99. Shen, Y.; Fang, Q.; Chen, B. Environmental Applications of Three-Dimensional Graphene-Based Macrostructures: Adsorption, Transformation, and Detection. *Environ. Sci. Technol.* **2015**, *49*, 67–84. [CrossRef]
100. Qiu, B.; Deng, Y.; Li, Q.; Shen, B.; Xing, M.; Zhang, J. Rational Design of a Unique Ternary Structure for Highly Photocatalytic Nitrobenzene Reduction. *J. Phys. Chem. C* **2016**, *120*, 12125–12131. [CrossRef]
101. Belver, C.; Bedia, J.; Gómez-Avilés, A.; Peñas-Garzón, M.; Rodriguez, J.J. Semiconductor Photocatalysis for Water Purification. *Nanoscale Mater. Water Purif.* **2018**, *22*, 581–651.
102. Zhang, M.; Luo, W.; Wei, Z.; Jiang, W.; Liu, D.; Zhu, Y. Separation free C_3N_4/SiO_2 hybrid hydrogels as high active photocatalysts for TOC removal. *Appl. Catal. B Environ.* **2016**, *194*, 105–110. [CrossRef]
103. Qiu, B.; Xing, M.; Zhang, J. Mesoporous TiO_2 Nanocrystals Grown in Situ on Graphene Aerogels for High Photocatalysis and Lithium-Ion Batteries. *J. Am. Chem. Soc.* **2014**, *136*, 5852–5855. [CrossRef] [PubMed]
104. Fan, Y.; Ma, W.; Han, D.; Gan, S.; Dong, X.; Niu, L. Convenient Recycling of 3D AgX/Graphene Aerogels (X = Br, Cl) for Efficient Photocatalytic Degradation of Water Pollutants. *Adv. Mater.* **2015**, *27*, 3767–3773. [CrossRef] [PubMed]
105. Iijima, S. Helical microtubules of graphitic carbon. *Nature* **1991**, *354*, 56–58. [CrossRef]
106. Sun, D.; Hong, R.; Liu, J.; Wang, F.; Wang, Y. Preparation of carbon nanomaterials using two-group arc discharge plasma. *Chem. Eng. J.* **2016**, *303*, 217–230. [CrossRef]
107. Zhao, X.; Zhao, T.; Peng, X.; Hu, J.; Yang, W. Catalyst effect on the preparation of single-walled carbon nanotubes by a modified arc discharge. *Full-Nanotub. Carbon Nanostruct.* **2019**, *27*, 52–57. [CrossRef]
108. Tepale-Cortés, A.; Moreno-Saavedra, H.; Hernández-Tenorio, C.; Rojas-Ramírez, T.; Illescas, J. Multi-walled Carbon Nanotubes Synthesis by Arc Discharge Method in a Glass Chamber. *J. Mex. Chem. Soc.* **2021**, *65*, 480–490. [CrossRef]
109. Smalley, R.E. Discovering the Fullerenes (Nobel Lecture). *Angew. Chem. (Int. Ed. Engl.)* **1997**, *36*, 1594–1601. [CrossRef]
110. Khashan, K.S.; Sulaiman, G.; Mahdi, R. Preparation of iron oxide nanoparticles-decorated carbon nanotube using laser ablation in liquid and their antimicrobial activity. *Artif. Cells Nanomed. Biotechnol.* **2017**, *45*, 1699–1709. [CrossRef]
111. Wu, X.; Yin, H.; Li, Q. Ablation and Patterning of Carbon Nanotube Film by Femtosecond Laser Irradiation. *Appl. Sci.* **2019**, *9*, 3045. [CrossRef]
112. Mostafa, A.M.; Mwafy, E.A.; Toghan, A. ZnO nanoparticles decorated carbon nanotubes via pulsed laser ablation method for degradation of methylene blue dyes. *Colloids Surfaces A Physicochem. Eng. Asp.* **2021**, *627*, 127204. [CrossRef]
113. Jourdain, V.; Bichara, C. Current understanding of the growth of carbon nanotubes in catalytic chemical vapour deposition. *Carbon* **2013**, *58*, 2–39. [CrossRef]
114. Razak, S.A.; Nordin, N.N.; Sulaiman, M.A.; Yusoff, M.; Masri, M.N. A Brief Review on Recent Development of Carbon Nanotubes by Chemical Vapor Deposition. *J. Trop. Resour. Sustain. Sci.* **2021**, *4*, 68–71. [CrossRef]
115. Gspann, T.S.; Juckes, S.M.; Niven, J.F.; Johnson, M.B.; Elliott, J.A.; White, M.A.; Windle, A.H. High thermal conductivities of carbon nanotube films and micro-fibres and their dependence on morphology. *Carbon* **2017**, *114*, 160–168. [CrossRef]

116. Duc Vu Quyen, N.; Khieu, D.Q.; Tuyen, T.N.; Tin, D.X.; Thi Hoang Diem, B. Carbon Nanotubes: Synthesis via Chemical Vapour Deposition without Hydrogen, Surface Modification, and Application. *J. Chem.* **2019**, *2019*, 4260153. [CrossRef]
117. McLean, B.; Kauppinen, E.I.; Page, A.J. Initial competing chemical pathways during floating catalyst chemical vapor deposition carbon nanotube growth. *J. Appl. Phys.* **2021**, *129*, 044302. [CrossRef]
118. Zhao, N.; Wu, Q.; Zhang, X.; Yang, T.; Li, D.; Zhang, X.; Ma, C.; Liu, R.; Xin, L.; He, M. Chemical vapor deposition growth of single-walled carbon nanotubes from plastic polymers. *Carbon* **2022**, *187*, 29–34. [CrossRef]
119. Asinovsky, E.I.; Amirov, R.H.; Isakaev, E.K.; Kiselev, V.I. Thermal plasma torch for synthesis of carbon nanotubes. *High Temp. Mater. Process.* **2006**, *10*, 197–206. [CrossRef]
120. Hong, Y.C.; Uhm, H.S. Production of carbon nanotubes by microwave plasma torch at atmospheric pressure. *Phys. Plasmas* **2005**, *12*, 053504. [CrossRef]
121. Amirov, R.H.; Isakaev, E.K.; Shavelkina, M.B.; Shatalova, T. Synthesis of carbon nanotubes by high current divergent anode-channel plasma torch. *J. Phys. Conf. Ser.* **2014**, *550*, 012023. [CrossRef]
122. Bodnaryk, W.J.; Fong, D.; Adronov, A. Enrichment of Metallic Carbon Nanotubes Using a Two-Polymer Extraction Method. *ACS Omega* **2018**, *3*, 16238–16245. [CrossRef]
123. Aïssa, B.; Ali, A.; Bentouaf, A.; Khan, W.; Zakaria, Y.; Mahmoud, K.A.; Ali, K.; Muhammad, N.M.; Mansour, S.A. Influence of single-walled carbon nanotubes induced exciton dissociation improvement on hybrid organic photovoltaic devices. *J. Appl. Phys.* **2019**, *126*, 113101. [CrossRef]
124. Gotthardt, J.M.; Schneider, S.; Brohmann, M.; Leingang, S.; Sauter, E.; Zharnikov, M.; Himmel, H.-J.; Zaumseil, J. Molecular n-Doping of Large- and Small-Diameter Carbon Nanotube Field-Effect Transistors with Tetrakis(tetramethylguanidino)benzene. *ACS Appl. Electron. Mater.* **2021**, *3*, 804–812. [CrossRef]
125. Ajayan, P.M.; Ebbesen, T.W. Nanometre-size tubes of carbon. *Rep. Prog. Phys.* **1997**, *60*, 1025–1062. [CrossRef]
126. Ruoff, R.S.; Lorents, D.C. Mechanical and thermal properties of carbon nanotubes. *Carbon* **1995**, *33*, 925–930. [CrossRef]
127. Dong, X.; Zhang, H.-B.; Lin, G.-D.; Yuan, Y.-Z.; Tsai, K. Highly Active CNT-Promoted Cu–ZnO–Al$_2$O$_3$ Catalyst for Methanol Synthesis from H$_2$/CO/CO$_2$. *Catal. Lett.* **2003**, *85*, 237–246. [CrossRef]
128. Huang, J.; Zhang, Q.; Zhao, M.; Wei, F. A review of the large-scale production of carbon nanotubes: The practice of nanoscale process engineering. *Chin. Sci. Bull.* **2012**, *57*, 157–166. [CrossRef]
129. Schnorr, J.M.; Swager, T.M. Emerging Applications of Carbon Nanotubes. *Chem. Mater.* **2011**, *23*, 646–657. [CrossRef]
130. Zhang, Q.; Zuo, Y.-Z.; Han, M.-H.; Wang, J.-F.; Jin, Y.; Wei, F. Long carbon nanotubes intercrossed Cu/Zn/Al/Zr catalyst for CO/CO$_2$ hydrogenation to methanol/dimethyl ether. *Catal. Today* **2010**, *150*, 55–60. [CrossRef]
131. Ozden, S.; Tiwary, C.S.; Hart, A.H.C.; Chipara, A.C.; Romero-Aburto, R.; Rodrigues, M.-T.F.; Taha-Tijerina, J.; Vajtai, R.; Ajayan, P.M. Density Variant Carbon Nanotube Interconnected Solids. *Adv. Mater.* **2015**, *27*, 1842–1850. [CrossRef]
132. Ozden, S.; Narayanan, T.N.; Tiwary, C.S.; Dong, P.; Hart, A.H.C.; Vajtai, R.; Ajayan, P.M. 3D Macroporous Solids from Chemically Cross-linked Carbon Nanotubes. *Small* **2015**, *11*, 688–693. [CrossRef] [PubMed]
133. Guiderdoni, C.; Estournes, C.; Peigney, A.; Weibel, A.; Turq, V.; Laurent, C. The preparation of double-walled carbon nanotube/Cu composites by spark plasma sintering, and their hardness and friction properties. *Carbon* **2011**, *49*, 4535–4543. [CrossRef]
134. Ozden, S.; Brunetto, G.; Karthiselva, N.S.; Galvão, D.S.; Roy, A.; Bakshi, S.R.; Tiwary, C.S.; Ajayan, P.M. Controlled 3D Carbon Nanotube Structures by Plasma Welding. *Adv. Mater. Interfaces* **2016**, *3*, 1500755. [CrossRef]
135. Al-Hakami, S.M.; Khalil, A.B.; Laoui, T.; Atieh, M.A. Fast Disinfection of *Escherichia coli* Bacteria Using Carbon Nanotubes Interaction with Microwave Radiation. *Bioinorg. Chem. Appl.* **2013**, *2013*, 458943. [CrossRef] [PubMed]
136. Khalid, N.; Majid, A.; Tahir, M.B.; Niaz, N.; Khalid, S. Carbonaceous-TiO$_2$ nanomaterials for photocatalytic degradation of pollutants: A review. *Ceram. Int.* **2017**, *43*, 14552–14571. [CrossRef]
137. Chen, M.-L.; Zhang, F.-J.; Oh, W.-C. Synthesis, characterization, and photocatalytic analysis of CNT/TiO$_2$ composites derived from MWCNTs and titanium sources. *New Carbon Mater.* **2009**, *24*, 159–166. [CrossRef]
138. Jauris, I.M.; Fagan, S.B.; Adebayo, M.A.; Machado, F.M. Adsorption of acridine orange and methylene blue synthetic dyes and anthracene on single wall carbon nanotubes: A first principle approach. *Comput. Theor. Chem.* **2016**, *1076*, 42–50. [CrossRef]
139. Zhang, W.; Li, G.; Liu, H.; Chen, J.; Ma, S.; An, T. Micro/nano-bubble assisted synthesis of Au/TiO$_2$@CNTs composite photocatalyst for photocatalytic degradation of gaseous styrene and its enhanced catalytic mechanism. *Environ. Sci. Nano* **2019**, *6*, 948–958. [CrossRef]
140. Wang, W.; Wang, S.; Ma, X.; Gong, J. Recent advances in catalytic hydrogenation of carbon dioxide. *Chem. Soc. Rev.* **2011**, *40*, 3703–3727. [CrossRef]
141. Wang, D.; Xie, Z.; Porosoff, M.D.; Chen, J.G. Recent advances in carbon dioxide hydrogenation to produce olefins and aromatics. *Chem* **2021**, *7*, 2277–2311. [CrossRef]
142. Duyar, M.; Ramachandran, A.; Wang, C.; Farrauto, R.J. Kinetics of CO$_2$ methanation over Ru/γ-Al$_2$O$_3$ and implications for renewable energy storage applications. *J. CO2 Util.* **2015**, *12*, 27–33. [CrossRef]
143. Mutz, B.; Carvalho, H.W.; Mangold, S.; Kleist, W.; Grunwaldt, J.-D. Methanation of CO$_2$: Structural response of a Ni-based catalyst under fluctuating reaction conditions unraveled by operando spectroscopy. *J. Catal.* **2015**, *327*, 48–53. [CrossRef]
144. Younas, M.; Kong, L.L.; Bashir, M.J.K.; Nadeem, H.; Shehzad, A.; Sethupathi, S. Recent Advancements, Fundamental Challenges, and Opportunities in Catalytic Methanation of CO$_2$. *Energy Fuels* **2016**, *30*, 8815–8831. [CrossRef]

145. Rönsch, S.; Schneider, J.; Matthischke, S.; Schlüter, M.; Götz, M.; Lefebvre, J.; Prabhakaran, P.; Bajohr, S. Review on methanation—From fundamentals to current projects. *Fuel* **2016**, *166*, 276–296. [CrossRef]
146. Potocnik, P. *Natural Gas*; BoD–Books on Demand; IntechOpen: London, UK, 2010.
147. Thampi, K.R.; Kiwi, J.; Grätzel, M. Methanation and photo-methanation of carbon dioxide at room temperature and atmospheric pressure. *Nature* **1987**, *327*, 506–508. [CrossRef]
148. Koschany, F.; Schlereth, D.; Hinrichsen, O. On the kinetics of the methanation of carbon dioxide on coprecipitated NiAl(O)$_x$. *Appl. Catal. B Environ.* **2016**, *181*, 504–516. [CrossRef]
149. Fujiwara, M.; Satake, T.; Shiokawa, K.; Sakurai, H. CO_2 hydrogenation for C_{2+} hydrocarbon synthesis over composite catalyst using surface modified HB zeolite. *Appl. Catal. B Environ.* **2015**, *179*, 37–43. [CrossRef]
150. Zhang, J.; Qian, Q.; Cui, M.; Chen, C.; Liu, S.; Han, B. Synthesis of ethanol from paraformaldehyde, CO_2 and H_2. *Green Chem.* **2017**, *19*, 4396–4401. [CrossRef]
151. Le Duff, C.S.; Lawrence, M.J.; Rodriguez, P. Role of the Adsorbed Oxygen Species in the Selective Electrochemical Reduction of CO_2 to Alcohols and Carbonyls on Copper Electrodes. *Angew. Chem.* **2017**, *129*, 13099–13104. [CrossRef]
152. Climent, V.; Feliu, J.M. Cyclic Voltammetry. In *Encyclopedia of Interfacial Chemistry: Surface Science and Electrochemistry*; Elsevier: Amsterdam, The Netherlands, 2018; pp. 48–74.
153. Ezenarro, J.J.; Párraga-Niño, N.; Sabrià, M.; Del Campo, F.; Muñoz-Pascual, F.-X.; Mas, J.; Uria, N. Rapid Detection of *Legionella pneumophila* in Drinking Water, Based on Filter Immunoassay and Chronoamperometric Measurement. *Biosensors* **2020**, *10*, 102. [CrossRef]
154. Ali, A.; Oh, W.-C. Preparation of Nanowire like WSe_2-Graphene Nanocomposite for Photocatalytic Reduction of CO_2 into CH_3OH with the Presence of Sacrificial Agents. *Sci. Rep.* **2017**, *7*, 1867. [CrossRef] [PubMed]
155. Biswas, R.U.D.; Ali, A.; Cho, K.Y.; Oh, W.-C. Novel synthesis of WSe_2-Graphene-TiO_2 ternary nanocomposite via ultrasonic technics for high photocatalytic reduction of CO_2 into CH_3OH. *Ultrason. Sonochem.* **2018**, *42*, 738–746. [CrossRef]
156. Dinh, C.-T.; Burdyny, T.; Kibria, M.G.; Seifitokaldani, A.; Gabardo, C.M.; de Arquer, F.P.G.; Kiani, A.; Edwards, J.P.; De Luna, P.; Bushuyev, O.S.; et al. CO_2 electroreduction to ethylene via hydroxide-mediated copper catalysis at an abrupt interface. *Science* **2018**, *360*, 783–787. [CrossRef] [PubMed]
157. Ma, Z.; Porosoff, M.D. Development of Tandem Catalysts for CO_2 Hydrogenation to Olefins. *ACS Catal.* **2019**, *9*, 2639–2656. [CrossRef]
158. Huan, T.N.; Corte, D.A.D.; Lamaison, S.; Karapinar, D.; Lutz, L.; Menguy, N.; Foldyna, M.; Turren-Cruz, S.-H.; Hagfeldt, A.; Bella, F.; et al. Low-cost high-efficiency system for solar-driven conversion of CO_2 to hydrocarbons. *Proc. Natl. Acad. Sci. USA* **2019**, *116*, 9735–9740. [CrossRef] [PubMed]
159. Wu, Y.; Jiang, Z.; Lu, X.; Liang, Y.; Wang, H. Domino electroreduction of CO_2 to methanol on a molecular catalyst. *Nature* **2019**, *575*, 639–642. [CrossRef]
160. Ramirez, A.; Ould-Chikh, S.; Gevers, L.; Chowdhury, A.D.; Abou-Hamad, E.; Aguilar-Tapia, A.; Hazemann, J.; Wehbe, N.; Al Abdulghani, A.J.; Kozlov, S.M.; et al. Tandem Conversion of CO_2 to Valuable Hydrocarbons in Highly Concentrated Potassium Iron Catalysts. *Chemcatchem* **2019**, *11*, 2879–2886. [CrossRef]
161. Tada, S.; Otsuka, F.; Fujiwara, K.; Moularas, C.; Deligiannakis, Y.; Kinoshita, Y.; Uchida, S.; Honma, T.; Nishijima, M.; Kikuchi, R. Development of CO_2-to-Methanol Hydrogenation Catalyst by Focusing on the Coordination Structure of the Cu Species in Spinel-Type Oxide $Mg_{1-x}Cu_xAl_2O_4$. *ACS Catal.* **2020**, *10*, 15186–15194. [CrossRef]
162. Tada, S.; Kinoshita, H.; Ochiai, N.; Chokkalingam, A.; Hu, P.; Yamauchi, N.; Kobayashi, Y.; Iyoki, K. Search for solid acid catalysts aiming at the development of bifunctional tandem catalysts for the one-pass synthesis of lower olefins via CO_2 hydrogenation. *Int. J. Hydrogen Energy* **2021**, *46*, 36721–36730. [CrossRef]
163. Rahmani, A.; Aubert, X.; Fagnon, N.; Nikravech, M. Liquid oxygenated hydrocarbons produced during reforming of CH_4 and CO_2 in a surface dielectric barrier discharge: Effects of steam on conversion and products distribution. *J. Appl. Phys.* **2021**, *129*, 193304. [CrossRef]
164. Islam, H.; Burheim, O.S.; Hihn, J.-Y.; Pollet, B. Sonochemical conversion of CO_2 into hydrocarbons: The Sabatier reaction at ambient conditions. *Ultrason. Sonochem.* **2021**, *73*, 105474. [CrossRef] [PubMed]
165. Tian, G.; Wu, Y.; Wu, S.; Huang, S.; Gao, J. Influence of Mn and Mg oxides on the performance of In_2O_3 catalysts for CO_2 hydrogenation to methanol. *Chem. Phys. Lett.* **2022**, *786*, 139173. [CrossRef]
166. Chernyak, S.A.; Kustov, A.L.; Stolbov, D.N.; Tedeeva, M.A.; Isaikina, O.Y.; Maslakov, K.I.; Usol'Tseva, N.V.; Savilov, S.V. Chromium catalysts supported on carbon nanotubes and graphene nanoflakes for CO_2-assisted oxidative dehydrogenation of propane. *Appl. Surf. Sci.* **2022**, *578*, 152099. [CrossRef]
167. Davis, B.H. Fischer–Tropsch synthesis: Current mechanism and futuristic needs. *Fuel Process. Technol.* **2001**, *71*, 157–166. [CrossRef]
168. Wang, H.; Li, J. Microporous Metal–Organic Frameworks for Adsorptive Separation of C5–C6 Alkane Isomers. *Acc. Chem. Res.* **2019**, *52*, 1968–1978. [CrossRef]
169. Golubev, K.; Batova, T.; Kolesnichenko, N.; Maximov, A. Synthesis of C2–C4 olefins from methanol as a product of methane partial oxidation over zeolite catalyst. *Catal. Commun.* **2019**, *129*, 105744. [CrossRef]
170. Zhang, Z.; Liu, Y.; Jia, L.; Sun, C.; Chen, B.; Liu, R.; Tan, Y.; Tu, W. Effects of the reducing gas atmosphere on performance of FeCeNa catalyst for the hydrogenation of CO_2 to olefins. *Chem. Eng. J.* **2022**, *428*, 131388. [CrossRef]

171. Saeidi, S.; Amin, N.A.S.; Rahimpour, M.R. Hydrogenation of CO_2 to value-added products—A review and potential future developments. *J. CO2 Util.* **2014**, *5*, 66–81. [CrossRef]
172. Aziz, M.A.A.; Jalil, A.A.; Triwahyono, S.; Ahmad, A. CO_2 methanation over heterogeneous catalysts: Recent progress and future prospects. *Green Chem.* **2015**, *17*, 2647–2663. [CrossRef]
173. Reimer, J.; Müller, S.; De Boni, E.; Vogel, F. Hydrogen-enhanced catalytic hydrothermal gasification of biomass. *Biomass-Convers. Biorefinery* **2017**, *7*, 511–519. [CrossRef]
174. Navarro-Jaén, S.; Navarro, J.C.; Bobadilla, L.F.; Centeno, M.A.; Laguna, O.H.; Odriozola, J.A. Size-tailored Ru nanoparticles deposited over γ-Al_2O_3 for the CO_2 methanation reaction. *Appl. Surf. Sci.* **2019**, *483*, 750–761. [CrossRef]
175. Jin, S.; Hao, Z.; Zhang, K.; Yan, Z.; Chen, J. Advances and Challenges for the Electrochemical Reduction of CO_2 to CO: From Fundamentals to Industrialization. *Angew. Chem. Int. Ed.* **2021**, *60*, 20627–20648. [CrossRef] [PubMed]
176. Fu, S.; Angelidaki, I.; Zhang, Y. In situ Biogas Upgrading by CO_2-to-CH_4 Bioconversion. *Trends Biotechnol.* **2021**, *39*, 336–347. [CrossRef] [PubMed]
177. Miletto, I.; Catizzone, E.; Bonura, G.; Ivaldi, C.; Migliori, M.; Gianotti, E.; Marchese, L.; Frusteri, F.; Giordano, G. In Situ FT-IR Characterization of CuZnZr/Ferrierite Hybrid Catalysts for One-Pot CO_2-to-DME Conversion. *Materials* **2018**, *11*, 2275. [CrossRef]
178. Boreriboon, N.; Jiang, X.; Song, C.; Prasassarakich, P. Higher Hydrocarbons Synthesis from CO_2 Hydrogenation over K- and La-Promoted Fe–Cu/TiO_2 Catalysts. *Top. Catal.* **2018**, *61*, 1551–1562. [CrossRef]
179. Liu, M.; Yi, Y.; Wang, L.; Guo, H.; Bogaerts, A. Hydrogenation of Carbon Dioxide to Value-Added Plasma Catalysis. *Catalysts* **2019**, *9*, 275. [CrossRef]
180. Ahmad, K.N.; Anuar, S.A.; Isahak, W.N.R.W.; Rosli, M.I.; Yarmo, M.A. Influences of calcination atmosphere on nickel catalyst supported on mesoporous graphitic carbon nitride thin sheets for CO methanation. *ACS Appl. Mater. Interfaces* **2020**, *12*, 7102–7113. [CrossRef]
181. Ahmad, K.N.; Isahak, W.N.R.W.; Rosli, M.I.; Yusop, M.R.; Kassim, M.B.; Yarmo, M.A. Rare earth metal doped nickel catalysts supported on exfoliated graphitic carbon nitride for highly selective CO and CO_2 methanation. *Appl. Surf. Sci.* **2022**, *571*, 151321. [CrossRef]
182. Bahari, N.A.; Isahak, W.N.R.W.; Masdar, M.S.; Yaakob, Z. Clean hydrogen generation and storage strategies via CO_2 utilization into chemicals and fuels: A review. *Int. J. Energy Res.* **2020**, *43*, 5128–5150. [CrossRef]
183. Hasan, S.Z.; Ahmad, K.N.; Isahak, W.N.R.W.; Masdar, M.S.; Jahim, J.M. Synthesis of low-cost catalyst NiO (111) for CO_2 hydrogenation into short-chain carboxylic acids. *Int. J. Hydrog. Energy* **2020**, *45*, 22281–22290. [CrossRef]
184. Zhu, M.; Tian, P.; Cao, X.; Chen, J.; Pu, T.; Shi, B.; Xu, J.; Moon, J.; Wu, Z.; Han, Y.-F. Vacancy engineering of the nickel-based catalysts for enhanced CO_2 methanation. *Appl. Catal. B Environ.* **2021**, *282*, 119561. [CrossRef]
185. Gac, W.; Zawadzki, W.; Rotko, M.; Greluk, M.; Słowik, G.; Pennemann, H.; Neuberg, S.; Zapf, R.; Kolb, G. Direct Conversion of Carbon Dioxide to Methane over Ceria- and Alumina-Supported Nickel Catalysts for Biogas Valorization. *Chempluschem* **2021**, *86*, 889–903. [CrossRef] [PubMed]
186. Chang, Y.-B.; Zhang, C.; Lu, X.-L.; Zhang, W.; Lu, T.-B. Graphdiyne enables ultrafine Cu nanoparticles to selectively reduce CO_2 to C_{2+} products. *Nano Res.* **2022**, *15*, 195–201. [CrossRef]
187. Kattel, S.; Yan, B.; Yang, Y.; Chen, J.G.; Liu, P. Optimizing Binding Energies of Key Intermediates for CO_2 Hydrogenation to Methanol over Oxide-Supported Copper. *J. Am. Chem. Soc.* **2016**, *138*, 12440–12450. [CrossRef] [PubMed]
188. Kumari, N.; Haider, M.A.; Agarwal, M.; Sinha, N.; Basu, S. Role of Reduced CeO_2(110) Surface for CO_2 Reduction to CO and Methanol. *J. Phys. Chem. C* **2016**, *120*, 16626–16635. [CrossRef]
189. Cheng, Z.; Lo, C.S. Mechanistic and microkinetic analysis of CO_2 hydrogenation on ceria. *Phys. Chem. Chem. Phys.* **2016**, *18*, 7987–7996. [CrossRef]
190. Chai, G.-L.; Guo, Z.-X. Highly effective sites and selectivity of nitrogen-doped graphene/CNT catalysts for CO_2 electrochemical reduction. *Chem. Sci.* **2016**, *7*, 1268–1275. [CrossRef]
191. Zhang, W.; Ma, X.-L.; Xiao, H.; Lei, M.; Li, J. Mechanistic Investigations on Thermal Hydrogenation of CO_2 to Methanol by Nanostructured CeO_2(100): The Crystal-Plane Effect on Catalytic Reactivity. *J. Phys. Chem. C* **2019**, *123*, 11763–11771. [CrossRef]
192. Jiang, F.; Wang, S.; Liu, B.; Liu, J.; Wang, L.; Xiao, Y.; Xu, Y.; Liu, X. Insights into the Influence of CeO_2 Crystal Facet on CO_2 Hydrogenation to Methanol over Pd/CeO_2 Catalysts. *ACS Catal.* **2020**, *10*, 11493–11509. [CrossRef]
193. Coufourier, S.; Gaillard, Q.G.; Lohier, J.-F.; Poater, A.; Gaillard, S.; Renaud, J.-L. Hydrogenation of CO_2, Hydrogenocarbonate, and Carbonate to Formate in Water using Phosphine Free Bifunctional Iron Complexes. *ACS Catal.* **2020**, *10*, 2108–2116. [CrossRef]
194. Cao, A.; Wang, Z.; Li, H.; Elnabawy, A.O.; Nørskov, J.K. New insights on CO and CO_2 hydrogenation for methanol synthesis: The key role of adsorbate-adsorbate interactions on Cu and the highly active MgO-Cu interface. *J. Catal.* **2021**, *400*, 325–331. [CrossRef]
195. Rasteiro, L.F.; De Sousa, R.A.; Vieira, L.H.; Ocampo-Restrepo, V.K.; Verga, L.G.; Assaf, J.M.; Da Silva, J.L.; Assaf, E.M. Insights into the alloy-support synergistic effects for the CO_2 hydrogenation towards methanol on oxide-supported Ni_5Ga_3 catalysts: An experimental and DFT study. *Appl. Catal. B Environ.* **2022**, *302*, 120842. [CrossRef]
196. Kovalskii, A.M.; Volkov, I.N.; Evdokimenko, N.D.; Tkachenko, O.P.; Leybo, D.V.; Chepkasov, I.V.; Popov, Z.I.; Matveev, A.T.; Manakhov, A.; Permyakova, E.S.; et al. Hexagonal BN- and BNO-supported Au and Pt nanocatalysts in carbon monoxide oxidation and carbon dioxide hydrogenation reactions. *Appl. Catal. B Environ.* **2022**, *303*, 120891. [CrossRef]

197. Wang, Z.-Q.; Liu, H.-H.; Wu, X.-P.; Hu, P.; Gong, X.-Q. Hydride Generation on the Cu-Doped CeO$_2$(111) Surface and Its Role in CO$_2$ Hydrogenation Reactions. *Catalysts* **2022**, *12*, 963. [CrossRef]
198. Patel, J.; Parikh, S.; Patel, S.; Patel, R.; Patel, P. Carbon Nanotube (CNTs): Structure, Synthesis, Purification, Functionalisation, Pharmacology, Toxicology, Biodegradation and Application as Nanomedicine and Biosensor. *J. Pharm. Sci. Med. Res.* **2021**, *1*, 17–44. [CrossRef]
199. Saravanan, M.; Babu, S.; Sivaprasad, K.; Jagannatham, M. Techno-economics of carbon nanotubes produced by open air arc discharge method. *Int. J. Eng. Sci. Technol.* **2010**, *2*, 100–108. [CrossRef]
200. O'Byrne, J.P.; Owen, R.E.; Minett, D.R.; Pascu, S.I.; Plucinski, P.K.; Jones, M.D.; Mattia, D. High CO$_2$ and CO conversion to hydrocarbons using bridged Fe nanoparticles on carbon nanotubes. *Catal. Sci. Technol.* **2013**, *3*, 1202–1207. [CrossRef]
201. Serp, P.; Corrias, M.; Kalck, P. Carbon nanotubes and nanofibers in catalysis. *Appl. Catal. A Gen.* **2003**, *253*, 337–358. [CrossRef]

Disclaimer/Publisher's Note: The statements, opinions and data contained in all publications are solely those of the individual author(s) and contributor(s) and not of MDPI and/or the editor(s). MDPI and/or the editor(s) disclaim responsibility for any injury to people or property resulting from any ideas, methods, instructions or products referred to in the content.

Article

Mango Seed-Derived Hybrid Composites and Sodium Alginate Beads for the Efficient Uptake of 2,4,6-Trichlorophenol from Simulated Wastewater

Asma Jabeen [1], Urooj Kamran [2,3], Saima Noreen [1], Soo-Jin Park [2,*] and Haq Nawaz Bhatti [1,*]

1. Department of Chemistry, University of Agriculture, Faisalabad 38000, Pakistan
2. Department of Chemistry, Inha University, 100 Inharo, Incheon 22212, Korea
3. Department of Mechanical Engineering, College of Engineering, Kyung Hee University, Yongin 445-701, Korea
* Correspondence: sjpark@inha.ac.kr (S.-J.P.); hnbhatti2005@yahoo.com (H.N.B.); Tel./Fax: +92-333-6528455 (H.N.B.)

Abstract: In this study, mango seed shell (MS)-based hybrid composite and composite beads (FeCl$_3$-NaBH$_4$/MS and Na-Alginate/MS) were designed. Batch and column experimental analyses were performed for the uptake of 2,4,6-trichlorophenol (2,4,6-TCP) from wastewater. The physicochemical characteristics of both composites were also examined. From the batch adsorption experiments, the best adsorption capacities of 28.77 mg/g and 27.42 mg/g were observed in basic media (pH 9–10) at 308 K for FeCl$_3$-NaBH$_4$/MS and 333 K for Na-Alginate/MS with 25 mg/L of 2,4,6-TCP concentration for 120 min. The rate of reaction was satisfactorily followed by the pseudo-second-order kinetics. Equilibrium models revealed that the mechanism of reaction followed the Langmuir isotherm. The thermodynamic study also indicated that the nature of the reaction was exothermic and spontaneous with both adsorbents. Desorption experiments were also carried out to investigate the reliability and reusability of the composites. Furthermore, the efficiency of the adsorbents was checked in the presence of different electrolytes and heavy metals. From the batch experimental study, the FeCl$_3$-NaBH$_4$/MS composite proved to be the best adsorbent for the removal of the 2,4,6-TCP pollutant, hence it is further selected for fixed-bed column experimentation. The column study data were analyzed using the BDST and Thomas models and the as-selected FeCl$_3$-NaBH$_4$/MS hybrid composites showed satisfactory results for the fixed-bed adsorption of the 2,4,6-TPC contaminants.

Keywords: mango seed shell; hybrid composite; alginate; 2,4,6-trichlorophenol; adsorption

Citation: Jabeen, A.; Kamran, U.; Noreen, S.; Park, S.-J.; Bhatti, H.N. Mango Seed-Derived Hybrid Composites and Sodium Alginate Beads for the Efficient Uptake of 2,4,6-Trichlorophenol from Simulated Wastewater. *Catalysts* **2022**, *12*, 972. https://doi.org/10.3390/catal12090972

Academic Editor: Sagadevan Suresh

Received: 25 July 2022
Accepted: 25 August 2022
Published: 30 August 2022

Publisher's Note: MDPI stays neutral with regard to jurisdictional claims in published maps and institutional affiliations.

Copyright: © 2022 by the authors. Licensee MDPI, Basel, Switzerland. This article is an open access article distributed under the terms and conditions of the Creative Commons Attribution (CC BY) license (https://creativecommons.org/licenses/by/4.0/).

1. Introduction

Contamination by volatile phenolic compounds threatens the use of water resources. Phenolic compounds with an unpleasant odor and a half-life span of 2–72 days cause extremely toxic effects in water [1–3]. Phenolic compounds are widely used in various products such as pharmaceuticals, plastics manufacturing, petrochemicals, oil refineries, pesticide/insecticide units, leather, paper, paint, wood, and other chemical manufacturing processes [4,5]. The wastewater discharged from the manufacture of these products contains many toxic phenolic compounds which are considered important to treat before discharge into water reservoirs [6]. Due to the toxicity of these phenolic pollutants, the U.S. Environmental protection agency lists the most phenolic pollutants as hazardous to human health and other living organisms [7,8]. Therefore, the discharge or removal of these hazardous pollutants is highly significant. The World Health Organization (W.H.O.) recommends the permissible concentration of phenolics to be about 0.001 mg/L in potable water and less than 1 mg/L in industrial wastewater for the safe discharge of polluted water into the environment [9,10].

Until now, various methods have been used for the elimination of phenolic contaminants from wastewater including biological, chemical, and electrochemical treatments and adsorption, photochemical oxidation, and catalytic reduction [11–13]. Among these, adsorption is considered the most convenient, eco-friendly, and cost-effective technique for the treatment of wastewater [14,15]. Adsorption is also trending due to its ease of operation, flexibility and lowest production of harmful by-products [16,17]. Various adsorbing materials or catalysts have been employed for the treatment of wastewater to eliminate various phenolic pollutants and other contaminants such as metal oxides, minerals, zeolites, polymers, activated carbons, and carbon nanotubes [18–21]. The use of polymers and biopolymers as adsorbents has become a well-established purification method, providing good efficiency to eliminate various toxic pollutants from water in a short time [22,23]. Biomass-based polymeric materials, such as rice husks, sugarcane bagasse, cotton sticks, coconut shells, leaves and barks of different plats, waste teas, seeds of plants, and seed shells of different fruits, provide a highly polymeric and porous structure that favors the adsorption mechanism [24–28].

Recently, the use of activated carbon, magnetically active materials, nanomaterials, chemically modified composites, and hybrid materials has gained more attention from researchers due to their modified structures and extensive adsorption properties [29–31]. The chemical and polymeric blending and grafting of materials are trending due to their excellent performance in wastewater treatment [32]. The synthesis of composites and modification of polymeric materials have also been used to resolve the disposal issue of waste slug, which is mostly produced during the treatment of water with simple or native biomasses as adsorbents [33–36].

Hence, this work is focused on the fabrication of an $FeCl_3$-$NaBH_4$-modified mango seed shell (MS)-based hybrid composite ($FeCl_3$-$NaBH_4$/MS) and sodium alginate-modified mango seed shell (MS)-based composite beads (Na-Alginate/MS) for the adsorptive removal of 2,4,6-trichlorophenol from aqueous media. The adsorption properties and structural changes on the surface of the polymeric structure of the adsorbents were determined by FTIR, TGA, and SEM analytical techniques. The experimental work was performed in batch and column modes for the optimization of various parameters. The reusability and recovery of adsorbents after adsorption experiments is also a concerning process for researchers; therefore, desorption experiments were also performed to check the reusability and percentage of recovery of both adsorbents ($FeCl_3$-$NaBH_4$/MS and Na-Alginate/MS) after experimentation.

2. Results and Discussion
2.1. Functional, Thermal, and Morphological Analysis of Composite Materials

FTIR spectra for both adsorbents ($FeCl_3$-$NaBH_4$/MS and Na-Alginate/MS) before and after adsorption for the comparative study are given in Figure 1a,b. The examination of the different functional groups shows that there were a different number of groups present such as a broad peak observed at 1030 cm^{-1} associated with C-O stretching vibration in primary and secondary alcohols some other peaks observed at 1575–1735 cm^{-1} mostly for the aromatic bending of the C=C groups associated with the aromatic ring, and some medium-intensity peaks observed at 1420 cm^{-1}, indicating the presence of O-H bending associated with phenol [37–39]. Similarly, a strong and broad peak was observed at 3330 cm^{-1} for the O-H stretching in the aromatic ring [40,41]. Some bands' intensities were found to be decreased after the adsorption spectra, indicating the decline in the number of functional groups after the adsorption of the 2,4,6-TCP contaminants.

Thermo-gravimetric analyses were performed to determine the thermal stability of both composites at an increasing temperature rate of 10 °C/min. As illustrated in Figure 1c,d, the decompositions of the composites' structures due to the elimination of volatile compounds occurred at high temperatures. A very small reduction in weight was observed at the initial 150 min of the temperature stage due to the removal of the loosely bound moisture content from the surface of the composites. The Na-Alginate/MS composite beads have a high moisture content compared to the $FeCl_3$-$NaBH_4$/MS hybrid composite and it experiences a consistent decrease in weight loss from the start of the reaction. The initial stages of weight loss were observed due to the elimination of the lignin and hemicellulose functional groups from the structure of the composites after 230 °C [42]. Another prominent stage of weight loss of about 80% was observed in the case of FeCl$_3$-NaBH$_4$/MS just after 350–400 °C due to the elimination of aromatic derivatives from the base structure (Figure 1c) [43], but in the case of the Na-Alginate/MS composite beads, only a 60% weight loss was observed at this point and the composition of Na-Alginate/MS showed better structural stability up to 560 °C (Figure 1d). In addition, a >80% weight loss was found after 400 °C in the case of FeCl$_3$-NaBH$_4$/MS and about an 80% weight loss was observed after 600 °C in the case of Na-Alginate/MS. However, the remaining backbone structure with 20% weight remained stable even after 800 °C.

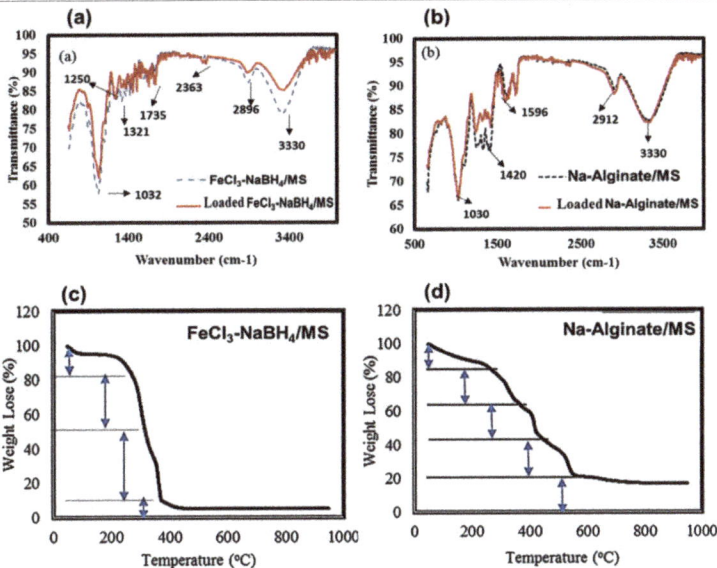

Figure 1. FTIR spectra of (**a**) FeCl$_3$-NaBH$_4$/MS hybrid composites and (**b**) Na-Alginate/MS composite beads, both for loaded and unloaded 2,4,6-TCP, and TGA thermograms for (**c**) FeCl$_3$-NaBH$_4$/MS hybrid composites and (**d**) Na-Alginate/MS composite beads.

The surface morphological structural characteristics were determined by SEM-EDX analysis for both adsorbents (FeCl$_3$-NaBH$_4$/MS and Na-Alginate/MS). Figure 2a,b present the rough and porous structure of the adsorbents that facilitates the adsorption of pollutant molecules. The rough and porous surface of the adsorbents is related to their efficiency in the uptake of pollutants and the porous surface is usually capable of enhancing adsorption capacity [30]. In addition, the presence of different functional groups on the surface (as determined by FTIR) also played an important role in the adsorption performance of the 2,4,6-TCP molecules on the surface of the adsorbents.

2.2. Batch Experimental Study
2.2.1. Influence of pH

In the batch adsorption study, the pH is considered an important factor that influences the capacity of the adsorbent [44,45]. The current study was carried out to check the efficiency of adsorbents in acidic and basic media. To determine the dependency of the adsorption reaction on the pH, the reactions were performed at a pH range of 2–10 with both adsorbents. It was observed from the q_e values that were obtained at the end of each experiment, that the adsorption capacities of 2,4,6-TCP were more favorable in the basic media. As 2,4,6-TCP was founded as a neutral molecule at pH 6.15, the anionic form of 2,4,6-TCP on the increasing pH influenced by the electrostatic force between the negatively charged TCP anion and the adsorbent molecules (FeCl$_3$-NaBH$_4$/MS and Na-Alginate/MS composites) results in an increase in adsorption capacity up to pH 10 (Figure 3a). Further increases in the pH cause the aggregation of pollutant molecules that affects the capacities of the adsorbents [46,47]. Olu-Owolabi and coworkers carried out the adsorption of 2,4,6-T on KAC and PCK and found an increase in adsorption capacity with increasing adsorbent doses up to pH 9 [48]. The variability of the pH was found to be high for PCK compared to KAC at a pH between 6 and 10. This could be due to the increase in electrostatic attraction between 2,4,6-T and KAC in this pH region, which enhanced the adsorption capacity, although the PCK showed less adsorption efficiency in the acidic pH region due to the limited functional groups on the surface.

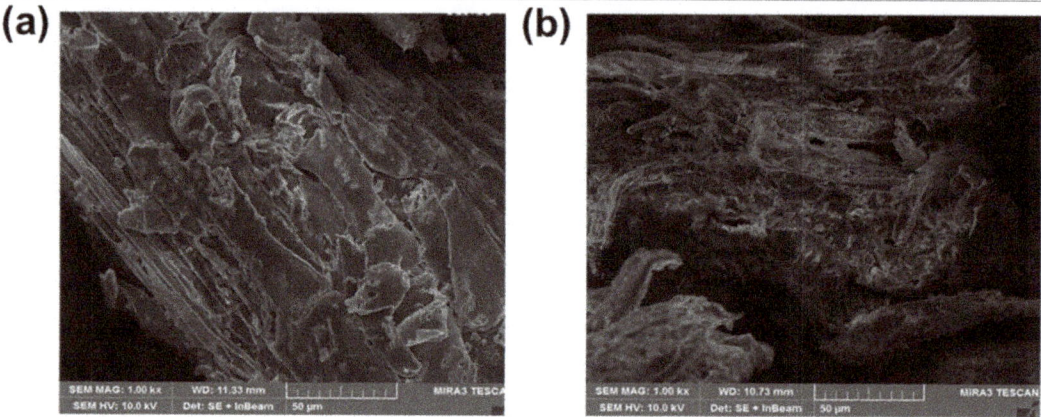

Figure 2. SEM micrographs of (**a**) FeCl$_3$-NaBH$_4$/MS hybrid composites and (**b**) Na-Alginate/MS composite beads.

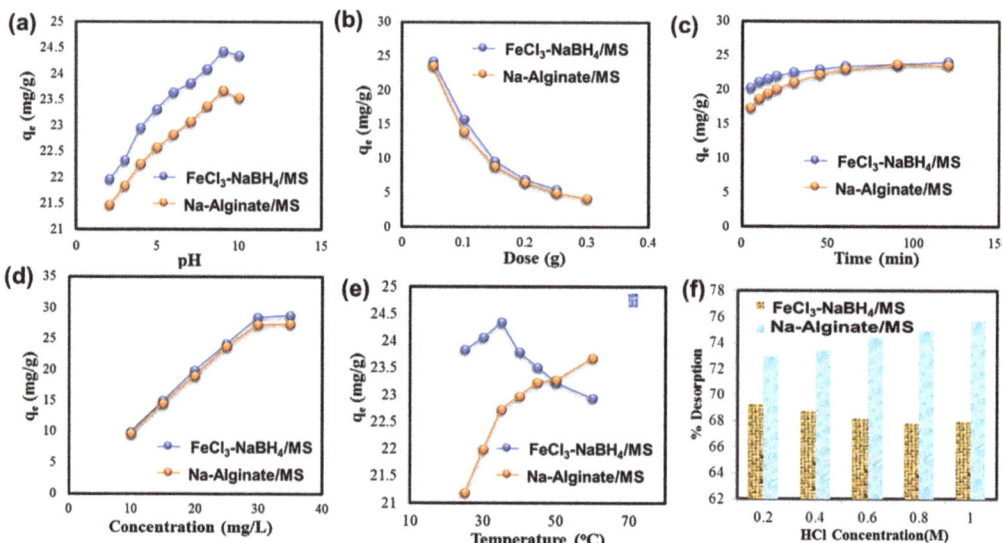

Figure 3. Influence of (**a**) pH, (**b**) dose, (**c**) contact time, (**d**) 2,4,6-TCP initial concentration, and (**e**) temperature on 2,4,6-TCP adsorption capacities of adsorbents and (**f**) 2,4,6-TCP contaminant desorption efficiency of adsorbents in desorbing agents.

2.2.2. Influence of Adsorbent Dosage

The increase in the adsorbent dose provides the opportunity for the pollutants to attach to the maximum available active sites of the adsorbent. The effect of increasing the adsorbent dose for the adsorption of 2,4,6-TCP onto $FeCl_3$-$NaBH_4$/MS and Na-Alginate/MS composites was determined in this study. It was found that by increasing the amount of the adsorbent from 0.1 to 0.4 g, the adsorption capacities were decreased as shown in Figure 3b. As there were a high number of active sites for the adsorption of the pollutant molecules for the available concentration of the 2,4,6-TCP molecules, these molecules occupied all available sites and on further increasing the dose [49,50], the number of active sites increased but as the concentration was fixed, this led to decreased q_e values [51]. A similar phenomenon was observed previously [52] for phenolic compounds, where a decrease in sorption capacity (q_e) with increasing adsorbent doses was considered a reason for the unsaturation of active binding sites in an adsorption phenomenon. The same trend was also reported by El-Naas and coworkers for the adsorption of phenol onto date-pit activated carbon [53]. They found that the phenol concentration decreased with the increasing adsorbent dose, which provided more adsorption sites for phenol and increased the adsorption rate. However, at a specific level, the available active sites were more than the available phenol molecules, which resulted in a decline in adsorption capacity.

2.2.3. Influence of Contact Time and Kinetic Study

The rate of the adsorption reaction between the adsorbents and 2,4,6-TCP molecules was investigated under various time intervals. The adsorption capacity of 2,4,6-TCP at a concentration of 25 mg/L with 0.05 mg of composites was determined. As shown in Figure 3c, at the start of the reaction for both adsorbents, a rapid increase in the rate of adsorption was observed (in the first 30–60 min of reaction) because all of the surface sites became occupied by the available pollutant molecules; after saturation of the active sites (up to 60 min), the adsorption process slowed down due to the hindrance of the occupying pollutant molecules and stopped at a point where maximum adsorption occurred [54–56]. The adsorption kinetic models, pseudo-first-order and pseudo-second-order, and the intraparticle model of diffusion were applied to predict the adsorption of the 2,4,6-TCP contaminants onto the $FeCl_3$-$NaBH_4$/MS and Na-Alginate/MS composites. Linear plots of ln ($q_e - q_t$) vs time were obtained to determine the rate constant [8]. The linear equation (Equation (2)) derived from Equation (1) can be used to calculate the constant values for the pseudo-first-order reaction

$$\frac{dq}{dt} = k_1(q_e - q_t) \tag{1}$$

$$\log(q_e - q_t) = \log q_e - \frac{k_1}{2.303}t \tag{2}$$

In the above equations, k_1 (min^{-1}) is the first-order rate constant, q_e and q_t (mg/g) are the equilibrium adsorption capacities for 2,4,6-TCP, and t (min) is the adsorption time [57]. Similarly, the differential and linear form of pseudo-second-order reactions can be written as Equations (3) and (4):

$$\frac{dt}{q_t} = k_2(q_e - q_t)^2 \tag{3}$$

$$\frac{t}{q_t} = \frac{1}{k_2 q_e^2} + \frac{t}{q_t} \tag{4}$$

In the above equations, q_e and q_t (mg/g) are the adsorption capacities at the equilibrium and time (t), whereas, k_2 is used to represent the rate constant of the pseudo-second-order reaction.

The intraparticle diffusion model was used to determine the diffusion of the phenolic molecules into the inner pores of the adsorbents. This model shows the interaction of molecules towards the binding adsorption sites and influences the biosorption efficiency of adsorbents [58]. The equation used to calculate the values of the intraparticle diffusion constants is given below:

$$q_t = k_{pi}t^{1/2} + C_i \quad (5)$$

The values for the kinetic rate constants (k) and correlation coefficients (R^2) are given in Table 1 and the kinetic linear plots are illustrated in Figure 4a–c. If the values of R^2 were found close to unity ($\geq$0.999), they showed the best fitness of the respective kinetic models. In the current study, the R^2 values of the pseudo-second-order reactions found as the maximum (0.999 for both adsorbents), as well as the closeness of calculated and experimental q_e values also showing the best fitness of the pseudo-second-order kinetic model (Table 1).

Table 1. Kinetic constant parameters for the adsorption of 2,4,6-TCP onto FeCl$_3$-NaBH$_4$/MS and Na-Alginate/MS composites.

Kinetic Model	FeCl$_3$-NaBH$_4$/MS	Na-Alginate/MS
Pseudo 1st order		
q_e exp. (mg/g)	24.11	23.75
q_e cal. (mg/g)	−0.363	−0.091
K_1 (min^{-1})	0.014	0.028
R^2	0.632	0.922
Pseudo 2nd order		
q_e exp. (mg/g)	24.11	23.75
q_e cal. (mg/g)	24.39	24.93
K_2 (g mg^{-1} min^{-1})	0.022	0.012
R^2	0.999	0.999
Intraparticle diffusion		
K_{pi}	0.411	0.735
R^2	0.923	0.920
C_i	19.99	16.64

2.2.4. Influence of Initial Concentration and Equilibrium Studies

The number of pollutants highly influenced the adsorption capacities of the adsorbents. In order to investigate this influence, various concentrations (10, 15, 20, 25, 30, and 35 mg/L) of 2,4,6-TCP were investigated at fixed amounts of adsorbents (0.05 g). As shown in Figure 3d, it was found that with the increasing concentration of the pollutant, the adsorption capacities increased up to the saturation of all adsorption sites and this saturation was achieved at a 30 mg/L concentration of 2,4,6-TCP. The mechanism followed by the adsorption reaction was determined by applying various theoretical equilibrium models such as Langmuir, Freundlich, Temkin, Harkins-Jura (H-J), and Doubinin-Radushkevich (D-R) to the experimental data. The optimized contact time that allowed us to study the mechanism of the reaction was 120 min for each experiment.

The linear form of the Langmuir isotherm was used, which is based on the formation of a monolayer between the finite number of adsorbate-adsorbent identical sites having a homogeneous distribution of energy overall binding sites [59]. Equation (6) presents the linear form of the Langmuir isotherm:

$$\frac{C_e}{q_e} = \frac{1}{q_m b} + \frac{C_e}{q_m} \quad (6)$$

where q_m and q_e (mg/g) are the maximum and equilibrium adsorption capacities, respectively, and b (L/mg) and C_e (mg/L) are the binding energies and equilibrium concentrations. The Freundlich isotherm provided the idea for the heterogeneity of surface binding sites and favored the multilayer adsorption mechanism [60]. The linear form presented in Equation (7) was used to calculate the values of the Freundlich parameters:

$$\log q_e = \log k_F + \frac{1}{n} \log C_e \tag{7}$$

In the above equation, k_F is the Freundlich constant, and q_e (mg/g) and C_e (mg/g) are the adsorption capacities and concentrations at equilibrium. Another equilibrium isothermal model named Temkin expresses homogeneous energy distribution on the active sites and heat of the adsorption [61]. Equation (8) presents the linear form of the Temkin isotherm:

$$q_e = B \ln A + B \ln C_e \tag{8}$$

Both A and B are Temkin's constants. A plot between q_e and $\ln C_e$ (equilibrium adsorption capacities and concentrations) was drawn to calculate the values of the Temkin's constants (A and B). Another equilibrium model the H-J isotherm based on the multilayer adsorption phenomena on the heterogeneous distribution of energy sites was also used [62]. The linear form of this model is given in Equation (9) and was used to calculate the values of the H-J constants (A and B):

$$\frac{1}{q_e^2} = \left(\frac{B}{A}\right) - \left(\frac{1}{A}\right) \log C_e \tag{9}$$

The D-R model rejected the idea of the Langmuir model (homogeneous energy distribution) and used a linear equation to estimate the apparent free-energy binding sites on the adsorbent surfaces [63]. The linear form of the D-R model as given in Equations (10) and (11) helps to calculate the biosorption energy (β) and Polanyi potential (ε) of the D-R isotherm:

$$\ln q_e = \ln q_m - \beta \varepsilon^2 \tag{10}$$

$$\varepsilon = RT \ln\left(1 + \frac{1}{C_e}\right) \tag{11}$$

In Equation (11), R is the general gas constant (8.314 J·mol^{-1}K^{-1}) and T (K) is the absolute temperature. Similarly, the value of mean free energy (E) can be calculated by Equation (12):

$$E = \frac{1}{\sqrt{2\beta}} \tag{12}$$

The Elovich model explains the multilayer adsorption phenomenon of the pollutants on the exponentially increasing active binding sites on the surface of the adsorbents. The linear form of the Elovich model [64,65] is presented in Equation (13):

$$\ln \frac{q_e}{C_e} = \ln K_E q_{max} - \frac{q_e}{q_{max}} \tag{13}$$

where, K_E presents the Elovich constant (L/mg), C_e is the equilibrium 2,4,6-TCP concentration, and q_m shows the maximum calculated adsorption capacity (mg/g) from the Elovich isotherm. The values of K_E, q_{max}, and the regression coefficient R^2 were calculated by plotting a graph, $\ln (q_e/C_e)$ vs. q_e (Table 2).

The calculated and experimental values for all isotherms along with their correlation values are given in Table 2. By comparing the regression values of all isothermal models, we determined the fitness of the best model in order to find the mechanism that was being followed by the batch experiment at various concentrations of pollutants. The R^2 values of the Langmuir model showed the best fitness of this model on the experimental data with very high R^2 values (0.999 and 0.996) for both adsorbents (FeCl$_3$-NaBH$_4$/MS and Na-Alginate/MS) compared to the other isothermal models. Furthermore, the fitness of the Langmuir model can be verified by comparing the calculated and experimental q_e values.

In addition, the D-R model also showed good fitness with a high R^2 value (0.945) for only the FeCl$_3$-NaBH$_4$/MS hybrid composite.

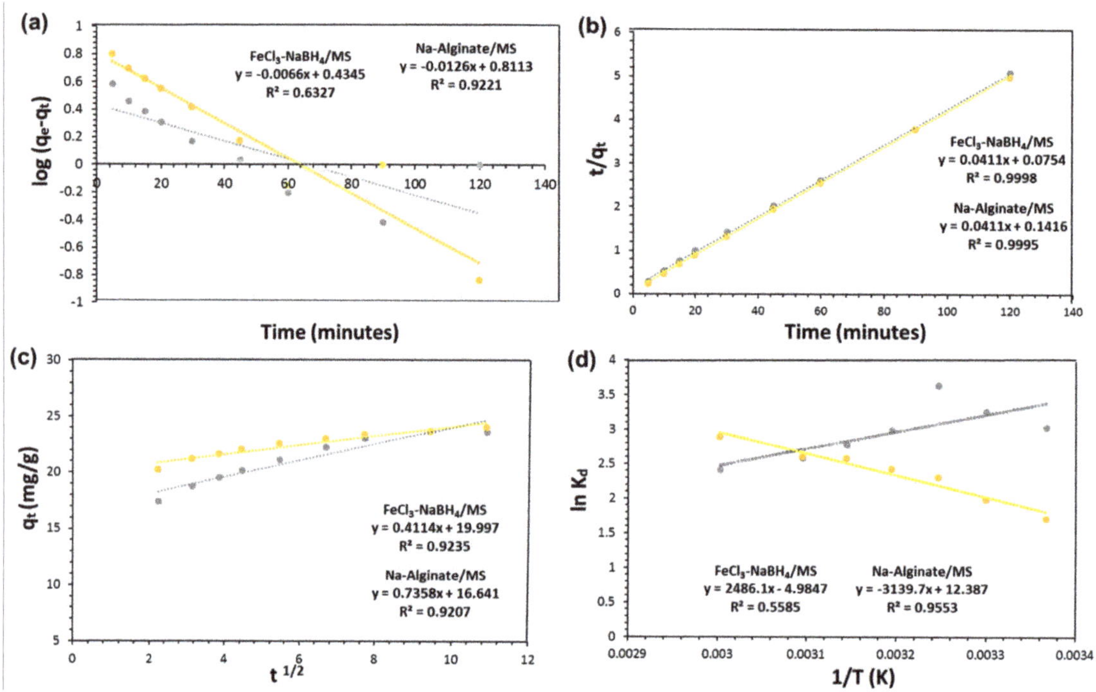

Figure 4. Kinetic plots for 2,4,6-TCP adsorption (**a**) Pseudo 1st order, (**b**) Pseudo 2nd order, (**c**) Intraparticle diffusion, and (**d**) adsorption thermodynamics of 2,4,6-TCP above FeCl$_3$-NaBH$_4$/MS and Na-Alginate/MS composites.

2.2.5. Influence of Temperature and Thermodynamics Studies

To determine the effect of increasing temperature on the uptake of 2,4,6-TCP onto the FeCl$_3$-NaBH$_4$/MS and Na-Alginate/MS composites, experiments were performed at different temperatures ranging from 297 to 333 K. It was found that the adsorption capacities of the FeCl$_3$-NaBH$_4$/MS hybrid composite decreased after 308 K, but increased for the Na-Alginate/MS composite bead up to 323 K, showing the stability of the active sites for the adsorption of 2,4,6-TCP molecules at this temperature (Figure 3e). The values for the various thermodynamic parameters (enthalpy "ΔH", entropy "ΔS", and Gibbs free energy "ΔG") were calculated to determine the nature of the chemical reaction that occurred on the surface of the adsorbents [66] using Equations (14)–(16):

$$\Delta G° = \Delta H° - T\Delta S° \quad (14)$$

$$\Delta G° = -RT \ln K_d \quad (15)$$

In the above equation, R is a gas constant (8.314 J·mol^{-1}K^{-1}), T (K) is the temperature, and K_d is a thermodynamics constant. The value of K_d can be determined using Equation (16):

$$\ln(K_d) = \frac{\Delta S°}{R} - \frac{\Delta H°}{R} \times \frac{1}{T} \quad (16)$$

As shown in Table 3, in the case of the 2,4,6-TCP adsorption performance on the Na-Alginate/MS composite beads, the decreasing values of ΔG on increasing temperatures up to 333 K were noticed due to the spontaneous nature of the reaction. The positive values of ΔH show the endothermic nature of the reaction as the heat evolved and the entropy (ΔS) of the system increased [37,40]. On the other hand, in the case of the FeCl$_3$-NaBH$_4$/MS hybrid composite, the adsorption only favorable at temperature up to 308 K, hence the values of ΔG first decreased then increased after 308 K (Figure 4d). The value of ΔH was negative, showing the exothermic nature of the reaction; as the heat contents decreased the value of ΔS also decreased compared to the value observed in the case of adsorption with the Na-Alginate/MS composite beads.

Table 2. Equilibrium isothermal constant parameters of 2,4,6-TCP onto FeCl$_3$-NaBH$_4$/MS and Na-Alginate/MS composites.

Isothermal Models	Composites	
Langmuir	FeCl$_3$-NaBH$_4$/MS	Na-Alginate/MS
K_L	0.998	0.013
b	0.0005	2.119
q_{max} Cal. (mg/g)	28.99	29.49
q_{max} Exp. (mg/g)	28.77	27.42
R^2	0.999	0.996
Freundlich		
K_F (mg/g)	0.144	0.096
n	7.018	3.114
R^2	0.771	0.786
D-R		
q_m (mg/g)	8.1×10^{11}	4.2×10^{11}
β 10^4 (mol^2/kL^2)	0.00001	0.0001
E	235.71	70.71
R^2	0.863	0.945
Temkin		
A	9.284	3.23
B	2.73	5.85
R^2	0.852	0.848
Harkins–Jura		
A	0.001	0.109
B	0.267	0.226
R^2	0.503	0.775
Elovich		
K_E (L/mg)	12.838	1.4898
q_{max} Cal. (mg/g)	3.911	11.299
q_{max} Exp. (mg/g)	28.77	27.42
R^2	0.784	0.647

Table 3. Thermodynamic parameters for the adsorption of 2,4,6-TCP onto FeCl$_3$-NaBH$_4$/MS and Na-Alginate/MS composites.

	FeCl$_3$-NaBH$_4$/MS			Na-Alginate/MS		
Temp. (K)	$\Delta G°$ (kJ·mol^{-1})	$\Delta H°$ (kJ·mol^{-1})	$\Delta S°$ (kJ/mol·K)	$\Delta G°$ (kJ·mol^{-1})	$\Delta H°$ (kJ·mol^{-1})	$\Delta S°$ (kJ/mol·K)
297	−7.49			−4.24		
303	−8.20			−5.01		
308	−9.31			−5.91		
313	−7.77	−20.74	0.042	−6.33	26.19	0.103
318	−7.33			−6.83		
323	−6.94			−6.997		
333	−6.68			−8.01		

2.3. Desorption Study

Desorption is an important parameter to check the efficiency of adsorbents for a number of repeated cycles [67]. This study was carried out in the presence of different molar concentrations (0.2, 0.4, 0.6, 0.8, and 1 M) of HCl solutions. The percentage of pollutant desorbed was calculated using the percent desorption formula in Equation (17):

$$Desorption\% = \frac{\text{Amount of pollutant desorbed}\left(\frac{mg}{g}\right)}{\text{Amount of pollutant sorbed}} \times 100 \quad (17)$$

For the desorption analysis of the 2,4,6-TCP pollutants, the molar solutions of HCl were used for both adsorbents because the best adsorption capacities in batch experiments were observed in basic media. From Figure 3f, it was found that on increasing the HCl concentration up to 1 M, the 2,4,6-TCP desorption capacities were increased [16,24] and up to 75% of Na-Alginate/MS and 69% of $FeCl_3$-$NaBH_4$/MS were successfully recovered.

2.4. Influence of Electrolytes and Heavy Metals

The effect of different electrolytes and heavy metals on the adsorption capacity of both adsorbents was investigated. In these experiments, the effects of different electrolytes, such as Na^+, K^+, Al^{3+}, Mg^{2+}, and Ca^{2+}, and heavy metals, such as Cr Cu, Co, Pb, and Cd, at fixed 2,4,6-TCP concentrations (25 mg/L) were examined. The amounts of adsorbents were also kept optimum (0.05 g) and the reactions were performed at ambient temperatures for 120 min. It was observed that the presence of electrolytes Ca^{2+} and Al^{3+} slightly fluctuated the adsorption capacities of both adsorbents compared to the control values (Figure 5a). Similarly, the presence of heavy metals caused a prominent decrease in the adsorption capacities with the $FeCl_3$-$NaBH_4$/MS hybrid composite but a slight reduction with the Na-Alginate/MS composite beads as shown in Figure 5b. This small fluctuation was due to the aggregation of metal ions in the solution phase that resisted the attachment of the pollutant molecules on the surface of the adsorbents but was enhanced in the presence of electrolytes [30,37].

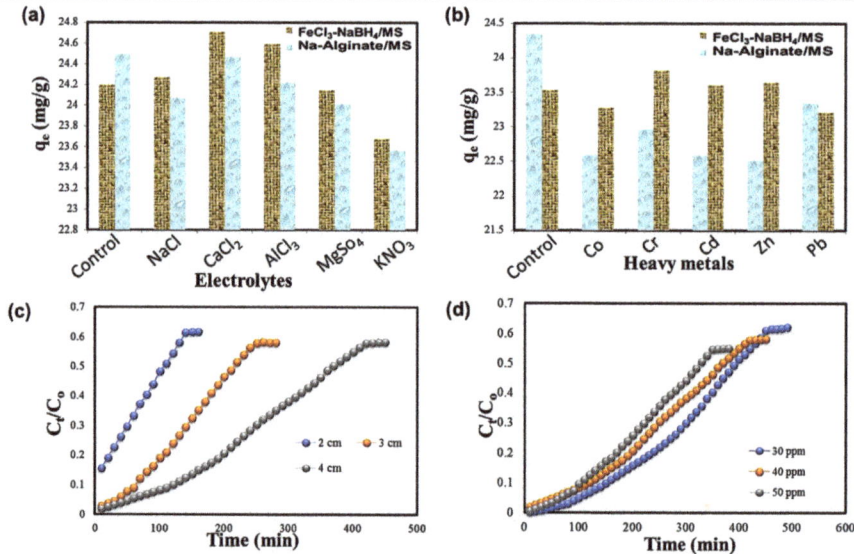

Figure 5. (a) Effect of different electrolytes, (b) effect of various heavy metals towards adsorption performance of 2,4,6-TCP onto $FeCl_3$-$NaBH_4$/MS and Na-Alginate/MS composites, and (c,d) effect of bed height on the uptake of 2,4,6-TCP by the $FeCl_3$-$NaBH_4$/MS hybrid composite.

2.5. Column Study

The efficiency of the selected adsorbent (FeCl$_3$-NaBH$_4$/MS) based on the best adsorption power was also investigated in column study experiments. The effects of different process parameters, such as bed height and pollutant concentration, were investigated at a fixed flow rate of 1.8 mL/min. The salient feature of continued study experimentation is its breakthrough time, which helps to investigate the effectiveness of the adsorbent's bed height and the renaissance time of the column [68]. The breakthrough time (Q_{50}) was calculated by Equation (18) using the values of breakthrough time (BT), flow rate, the concentration of 2,4,6-TCP (C_i), and the mass of the FeCl$_3$-NaBH$_4$/MS hybrid composite in the column bed:

$$\text{BT capacity}(Q_{50\%}) = \frac{\text{BT time(at 50\%)} \times \text{flowrate} \times C_i}{\text{Mass of adsorbent in bed}(g)} \quad (18)$$

The calculated values of breakthrough time and capacities are given in Table 4 along with bed height concentrations and flow rates.

Table 4. Breakthrough point and capacities at various conditions for 2,4,6-TCP uptake by FeCl$_3$-NaBH$_4$/MS hybrid composite.

Inlet Conc. (ppm)	Bed Height (cm)	Flow Rate (mL·min^{-1})	Breakthrough Point (50%) (min)	Biosorption Capacity (mg/g)
30	4	1.8	400	14.4
40	2	1.8	100	14.4
40	3	1.8	210	15.12
40	4	1.8	360	17.28
50	4	1.8	300	18

2.5.1. Influence of Bed Height

The effect of the column study parameter bed height was determined using different amounts of the FeCl$_3$-NaBH$_4$/MS adsorbent at a fixed-inlet 2,4,6-TCP concentration (40 mg/L) with a fixed flow rate (1.8 mL/min). As seen in Figure 5c, the adsorption capacity of FeCl$_3$-NaBH$_4$/MS increased on increasing the bed height from 2 to 4cm; the maximum recorded adsorption capacity of FeCl$_3$-NaBH$_4$/MS was 17.28 mg/g at 360 min. It was also found that the shifting of the breakthrough time toward higher values was due to the availability of an increased number of vacant sites with the increasing bed height of the column [45,69]. This increasing number of active sites favored the uptake of more pollutant (2,4,6-TCP) molecules by intra-particulate diffusion. A higher breakthrough time means a longer stay for the pollutants in the column and as a result, more effluent was being treated. The calculated values of the breakthrough time are given in Table 4.

2.5.2. Influence of 2,4,6-TCP Initial Dye Concentration

The effect of another parameter pollutant of the 2,4,6-TCP concentrations was also investigated in the column study experiments. Measurements of different concentrations of 30, 40, and 50 mg/L of 2,4,6-TCP were taken by keeping the bed height and flow rate constant, at 3 cm and 1.8 mL/L, respectively (Figure 5d). It was concluded that on the increasing concentration of 2,4,6-TCP, the adsorption capacity was also increased from 14.4 mg/g (with 30 mg/L) to 18 mg/g (with 50 mg/L of pollutant concentration). The decrease in the breakthrough time was noticed due to the presence of a high concentration gradient at an increased concentration of 2,4,6-TCP as it increased the rate of the reaction [70,71]. The available active sites saturated faster with the available concentration of 2,4,6-TCP as this concentration was enough to saturate the active sites by reducing the breakthrough time. This effect steepened the curve by reducing the BT volume due to the availability of weak driving forces for the mass transfer from the bulk to the surfaces of the adsorbents [72].

2.5.3. Models for Column Study

Two kinetic models including the Thomas and BDST (bed-depth service time) models were applied to the experimental data in the column study of the two parameters in order to investigate the performance of the adsorbent in the column.

The Thomas model is based on the concept of the Langmuir isotherm and second-order reversible kinetics at equilibrium followed by the driving force present between the pollutants and adsorbent molecules. The second-order kinetics (reversible) and Langmuir isotherm can be used under favorable or unfavorable conditions [73]. The linear form of the Thomas model used for the calculations of different constants is given in Equation (19),

$$\ln\left(\frac{C_o}{C_t} - 1\right) = \frac{K_{Th} \times q_o \times W}{Q} - K_{Th} \times C_o \times t \tag{19}$$

In the above equation, K_{Th} (mL/min.mg) is a rate constant of the Thomas model, the C_o (mg/L) for the inlet and C_t (mg/L) for the outlet are the 2,4,6-TCP concentrations, q_o (mg/g) is for the 2,4,6-TCP uptake at equilibrium, W (g) is the mass of the FeCl$_3$-NaBH$_4$/MS adsorbent, t (min) is the time and Q (mL/min) is the flow rate. The calculated values of the parameters are presented in Table 5. The higher calculated values of R^2 and close values of the adsorption capacities (observed values with the calculated values) showed the fitness of the Thomas model on the experimental data. It was also observed that with increasing concentrations, the rate of the inter-phase mass transfer decreased [48,74].

Table 5. Thomas model parameters for the adsorption of 2,4,6-TCP onto FeCl$_3$-NaBH$_4$/MS hybrid composite.

Inlet Conc. (mg/L)	Bed Height (cm)	Flow Rate (mL/min)	K_{Th} (mL/min) × 10^3	q_e cal. (mg/g)	q_e exp. (mg/g)	R^2
30	4	1.8	0.00037	14.27	14.4	0.924
40	2	1.8	0.00035	14.85	14.4	0.980
40	3	1.8	0.00035	15.95	15.12	0.954
40	4	1.8	0.00025	17.14	17.28	0.982
50	4	1.8	0.00024	18.92	18	0.917

The BDST model was used to investigate the relationship between the amount of adsorbent in the bed (bed-depth Z) and the adsorbent efficiency (service time t) and was applied to the experimental values obtained from the column experimental studies (effect of initial 2,4,6-TCP concentration and bed height). The service time in this model can be defined as "this is the maximum time at which adsorbent remains able to remove pollutant from solution before regeneration is required" or "the time taken by the column bed to attain the breakthrough point" [19,75,76]. The calculations of the different parameters of this model were carried out using the linear form given in Equation (20):

$$t = \frac{N_o Z}{C_o U} - \frac{1}{K_a C_o} \ln\left(\frac{C_o}{C_b} - 1\right) \tag{20}$$

In the above equation, C_o and C_b (mg/L) are the initial and BT (breakthrough) concentrations of 2,4,6-TCP, U is the linear velocity, K_a is the rate constant of the BDST model, N_o and Z (cm) are the biosorption capacity and bed height, respectively [71]. The values of the parameters for the BDST model were calculated using the values of the slope and intercept of the graph and are presented in Table 6.

$$t = aZ \tag{21}$$

where,

$$Slope = a = \frac{N_o}{C_o U} \tag{22}$$

Table 6. BDST model parameters for the adsorption of 2,4,6-TCP onto $FeCl_3$-$NaBH_4$/MS hybrid composite.

C_t/C_o	A	B	K_a (L·mg^{-1}·min^{-1}) × 10^4	N_o (mg·L^{-1}) × 10^{-4}	R^2
0.2	85	−151.6	0.00024	1938	0.998
0.4	100	−110	0.00012	2280	0.970
0.6	145	−168.3	0.00004	3306	0.990

And

$$Intercept = b = \frac{1}{K_a C_o} \ln\left(\frac{C_o}{C_b} - 1\right) \quad (23)$$

The values of R^2 obtained at different bed heights illustrate the fitness of the BDST model on the experimental data of the column study.

3. Synthetic Methodologies

3.1. Chemical Reagents and Biomass

Phenolic pollutants named 2,4,6-trichlorophenol (2,4,6-TCP) were supplied by AGROSOL Ltd., Karachi, Pakista. All reagents and chemicals sodium borohydride ($NaBH_4$), iron(III) chloride hexahydrate ($FeCl_3·6H_2O$), ethanol, sodium alginate, polyvinyl alcohol (PVA), calcium chloride ($CaCl_2$), boric acid, NaCl solution (saline), NaOH, and HCl used throughout the experiments were of analytical grade and purchased from Sigma-Aldrich (Saint Louis, MO, USA). The mango seed shell biomass was purchased from the local market in the city of Faisalabad, Pakistan.

3.2. Preparation of Composite and Composite Beads

Mango seed shell (MS) biomass-based $FeCl_3$-$NaBH_4$/MS hybrid composite and Na-Alginate/MS composite beads were prepared following the reported methodology [54,77]. In order to prepare $FeCl_3$-$NaBH_4$/MS hybrid composites, first, 2 g of washed and ground MS biomass (particle size 300 µm) was soaked in 0.1 M of HCl solution (200 mL) for 1 h and 30 min. Then, the MS biomass was washed several times with deionized water and oven dried at 60 °C. The dried MS biomass was added to a 0.05 M $FeCl_3·6H_2O$ solution at a 1:1 ratio and stirred at 250 rpm on a magnetic stirrer. Then, 10 mL of 0.53 M ($NaBH_4$) was poured into the mixture and it was further stirred at 150 rpm. Products (precipitates) were filtered, washed with ethanol, and dried in an oven at 50 °C overnight.

Similarly, the MS biomass-based sodium alginate composite beads (Na-Alginate/MS) were synthesized by preparing a solution of PVA (3.5 g) and sodium alginate (1 g) in deionized water. This solution was stirred at 50 °C for 30 min, and then 1.5 g of biomass was added and it was further stirred for 1 h under the same experimental conditions. Then, 0.1 M of $CaCl_2$ (500 mL) and boric acid were prepared. The mixture of biomass-PVA and Na-Alginate was poured drop-wise into this solution for the formation of the Na-Alginate/MS composite beads. The as-prepared composite beads were separated from the solution after 24 h, washed with saline solution several times, and stored in saline solution at 5 °C.

3.3. Characterization of Composites

FTIR Cary-630 (Fourier transform infrared spectrometer) (Agilent Technologies, Clara, CA 95051, USA) was used to determine the functional group modifications that occurred on the surface of the MS biomass after the formation of the composites; spectra were recorded in the IR region 4000–500 cm^{-1} before and after the adsorption. The surface textures and morphological properties of the adsorbents were determined by a JMT-300 SEM-EDS (scanning electron microscopy paired with energy dispersive X-ray) (JEOL Ltd. Akishima, Tokyo, Japan). Surface information was obtained at 1.0 kx–25.0 kx magnifications and 11 mm focus distances. Similarly, a TGA/DSA Axxx multi-analyzer (Metter Toledo, Columbus, OH, USA) was used to analyze the thermal stability of the composites. Approximately 40–70 mg

(300 µm size) samples were taken into the sampler and heated at a temperature range of 10–900 °C with an increasing temperature rate of 10 °C/min under a nitrogen atmosphere.

3.4. Preparation of 2,4,6-TCP Stock Solution

A stock solution of 2,4,6-TCP was prepared by dissolving 1 g per 1000 mL of deionized water. Further diluted solutions of low concentrations of 10–25 mg/L (25 ppm) were prepared from this stock solution for the batch experiments.

3.5. Experimental Batch Study

Batch experimental work was performed to check the effect of the different batch-adsorption parameters on the adsorption behavior of both adsorbents (FeCl$_3$-NaBH$_4$/MS and Na-Alginate/MS). The effect of the pH was examined at a pH range of 2–9; contact time experiments were performed at different time intervals such as 5, 10, 15, 20, 30, 45, 60, 90, and 120 min; the effects of the pollutant concentrations were checked at a range of 2,4,6-TCP initial concentrations of 10, 25, 30, 40, and 50 mg/L; and the temperature effect experiment was carried out at room temperature (about 297 K), 303, 308, 313, 318, 323, and 333 K. Similarly, the effect of different adsorbent doses of 0.05, 0.1, 0.15, 0.2, 0.25, and 0.3 g were investigated for both adsorbents. Other effects such as the presence of different electrolytes and heavy metals were also examined in the batch study experiments. Different studies were carried out for the adsorption of the phenolic compounds in the batch experiments. The adsorption capacity values of some reported adsorbents [55,78–80] for the adsorption of the phenolic compounds are given in Table 7. The process diagram for the batch adsorption reaction is presented in Figure 6a.

Table 7. Comparison of adsorption capacities reported in current work and adsorption capacities of other adsorbents reported by various researchers.

Pollutants	Adsorbent	Initial Concentration (mg/L)	Adsorption Capacity (mg/g)	Reference
2,6-dichlorophenol	Modified plantain peel adsorbents	50–100	15.9	[54]
Para-Chlorophenol	Magnetic powdered activated carbon	100–150	66.5	[78]
2,4,6-Trichlorophenol	Modified polypropylene hollow fiber composites	40–100	66.49	[79]
2,4,6-Trichlorophenol	chitosan/fly-ash-based magnetic composites	100–150	68.89	[80]
Bisphenol A	chitosan/fly-ash-based magnetic composites	50–150	31.92	[80]
2,4,6-Trichlorophenol	Fe^{3+}- and Fe^{2+}-enriched magnetic composites	35–50	31.27	[55]
2,4,6-Trichlorophenol	FeCl$_3$-NaBH$_4$/MS	10–35	28.77	This Study
2,4,6-Trichlorophenol	Na-Alginate/MS	10–35	27.42	This Study

3.6. Desorption Study

The efficiency of the adsorbents depends on the reliability and reusability of the adsorbents at different cycles. The percentage of desorption of the adsorbents after the adsorption experiments is directly related to the efficiency of the adsorbents. Desorption experiments were performed using different HCl molar concentrations (as desorbing agents) for the recovery of the composites.

3.7. Column Study

The column study was carried out in a glass column with a diameter of 1.2 cm and a path length of 42 cm. The effects of the different column study parameters, such as bed height and the concentration of 2,4,6-TCP, were investigated at a fixed flow rate of 1.8 mL/min. The inlet pollutant concentrations (30, 40, and 50 mg/L) and bed heights (1.5, 2.5, and 3.5 g) of the adsorbents at 1, 2, and 3 cm column lengths were investigated with the FeCl$_3$-NaBH$_4$/MS hybrid composites. A bit of glass wool was used to hold the adsorbent inside the column. All the experiments were carried out at room temperature and with a solution at pH 9, which was obtained from the batch study experiments. The outlet concentrations after adsorption were analyzed at 10 min intervals of time by a Uv-Vis spectrophotometer to calculate the C_i/C_o ratios. The process diagram for the fixed-bed column experiments is given in Figure 6b.

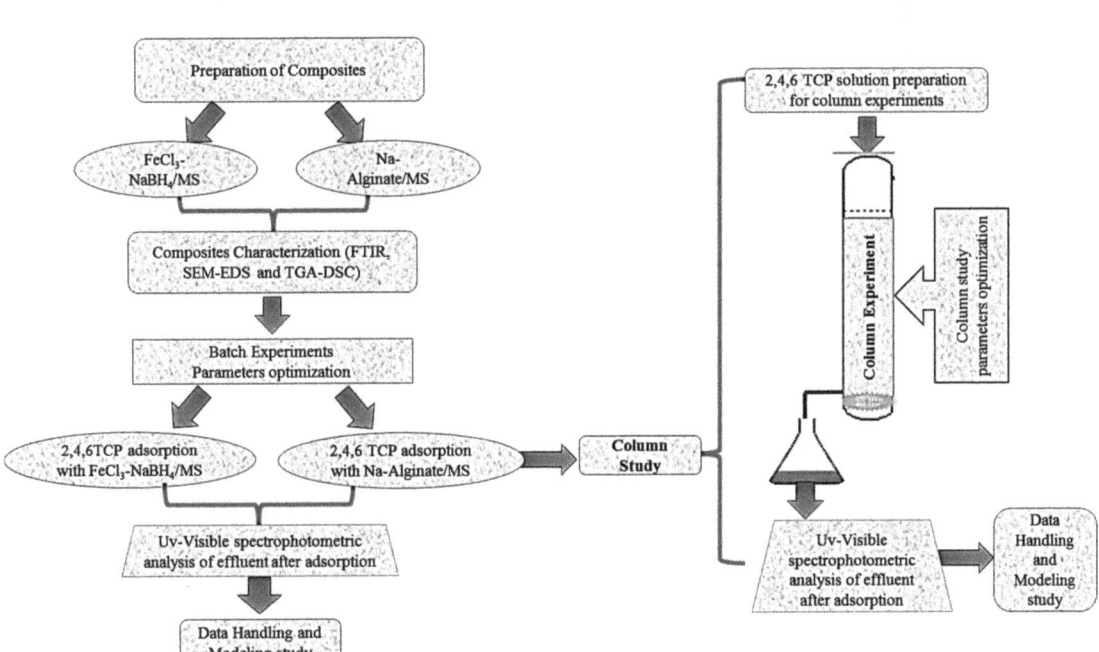

Figure 6. Flow sheet diagram for (**a**) Batch study experiments and (**b**) column study experiments.

4. Conclusions

In summary, mango seed shell biomass-based hybrid composite and composite beads (FeCl$_3$-NaBH$_4$/MS and Na-Alginate/MS) were successfully investigated for the removal of 2,4,6-TCP pollutants from wastewater. Both composites were considered efficient and low-cost adsorption materials for the removal of phenolics from wastewater. The batch experiments illustrate that the best working adsorption conditions were pH: alkaline 9–10, dose: 0.05 g, adsorption equilibrium time: 120 min at temperatures of 308 K (for FeCl$_3$-NaBH$_4$/MS) and 333 K (for Na-Alginate/MS), and 2,4,6-TCP initial concentration: 25 mg/L. The results obtained in the batch experiments were also tested by applying kinetics, equilibrium, and thermodynamic models and the best-fitting models were the pseudo-second-order and Langmuir for both the kinetics and equilibrium data. The thermodynamic models showed the exothermic, endothermic, and spontaneous or non-spontaneous nature of the chemical reactions. The best-working adsorbent (FeCl$_3$-NaBH$_4$/MS) was selected for column study experiments and investigated by applying the column study parameters, bed height, and increasing initial concentration parameters. The experimental data were also best fitted with the column study models (Thomas and BDST) with high R^2 values. From the results, it can be concluded that the as-designed FeCl$_3$-NaBH$_4$/MS hybrid composite and Na-Alginate/MS composite beads can be considered efficient adsorbents for the adsorptive removal of 2,4,6-TCP or other phenolic compounds from wastewater at the commercial level.

Author Contributions: Conceptualization, A.J.; investigation, A.J.; writing—original draft preparation, A.J.; writing—review, A.J.; writing—review & editing, U.K. and S.N.; Help and Support in data handling, S.-J.P.; supervision, H.N.B.; All authors have read and agreed to the published version of the manuscript.

Funding: This work did not receive any external funding and this research work was conducted in the Department of Chemistry, University of Agriculture, Faisalabad, Pakistan; and this work was also supported by Department of Chemistry, Inha University, Incheon, South Korea, and the Department of Mechanical Engineering, Kyung Hee University, Yongin, South Korea These institutes provided all the facilities to conduct this research and modeling work.

Data Availability Statement: Not applicable.

Acknowledgments: The authors are thankful to the Department of Chemistry, University of Agriculture, Faisalabad, Pakistan for providing some of the research facilities. In addition to this, the authors are also thankful to the Polymer-Carbon Nanomaterials Laboratory, the Department of Chemistry, Inha University, Incheon, South Korea, and the Department of Mechanical Engineering, Kyung Hee University, Yongin, South Korea, for providing some of the instrumental analytical facilities and the modeling for this work.

Conflicts of Interest: The authors declare no conflict of interest.

References

1. Devi, P.; Saroha, A.K. Simultaneous adsorption and dechlorination of pentachlorophenol from effluent by Ni–ZVI magnetic biochar composites synthesized from paper mill sludge. *Chem. Eng. J.* **2015**, *271*, 195–203. [CrossRef]
2. Wang, L.; Gan, K.; Lu, D.; Zhang, J. Hydrophilic Fe_3O_4 @C for High-Capacity Adsorption of 2,4-Dichlorophenol. *Eur. J. Inorg. Chem.* **2016**, *2016*, 890–896. [CrossRef]
3. Shi, W.; Ren, H.; Huang, X.; Li, M.; Tang, Y.; Guo, F. Low cost red mud modified graphitic carbon nitride for the removal of organic pollutants in wastewater by the synergistic effect of adsorption and photocatalysis. *Sep. Purif. Technol.* **2019**, *237*, 116477. [CrossRef]
4. Pei, Z.; Li, L.; Sun, L.; Zhang, S.; Shan, X.-Q.; Yang, S.; Wen, B. Adsorption characteristics of 1,2,4-trichlorobenzene, 2,4,6-trichlorophenol, 2-naphthol and naphthalene on graphene and graphene oxide. *Carbon* **2013**, *51*, 156–163. [CrossRef]
5. Liu, H.; Ruan, X.; Zhao, D.; Fan, X.; Feng, T. Enhanced Adsorption of 2,4-Dichlorophenol by Nanoscale Zero-Valent Iron Loaded on Bentonite and Modified with a Cationic Surfactant. *Ind. Eng. Chem. Res.* **2016**, *56*, 191–197. [CrossRef]
6. Hasan, Z.; Jhung, S.H. Removal of hazardous organics from water using metal-organic frameworks (MOFs): Plausible mechanisms for selective adsorptions. *J. Hazard. Mater.* **2015**, *283*, 329–339. [CrossRef]
7. Heo, J.; Yoon, Y.; Lee, G.; Kim, Y.; Han, J.; Park, C.M. Enhanced adsorption of bisphenol A and sulfamethoxazole by a novel magnetic CuZnFe2O4–biochar composite. *Bioresour. Technol.* **2019**, *281*, 179–187. [CrossRef]
8. Ahmad, N.; Al-Fatesh, A.S.; Wahab, R.; Alam, M.; Fakeeha, A.H. Synthesis of silver nanoparticles decorated on reduced graphene oxide nanosheets and their electrochemical sensing towards hazardous 4-nitrophenol. *J. Mater. Sci. Mater. Electron.* **2020**, *31*, 11927–11937. [CrossRef]
9. Kwon, J.; Lee, B. Bisphenol A adsorption using reduced graphene oxide prepared by physical and chemical reduction methods. *Chem. Eng. Res. Des.* **2015**, *104*, 519–529. [CrossRef]
10. Jin, M.-Y.; Lin, Y.; Liao, Y.; Tan, C.-H.; Wang, R. Development of highly-efficient ZIF-8@PDMS/PVDF nanofibrous composite membrane for phenol removal in aqueous-aqueous membrane extractive process. *J. Membr. Sci.* **2018**, *568*, 121–133. [CrossRef]
11. Chen, X.H.; Shan, Z.J.; Zhai, H.L. QSAR models for predicting the toxicity of halogenated phenols to *Tetrahymena*. *Toxicol. Environ. Chem.* **2016**, *99*, 273–284. [CrossRef]
12. Gomri, M.; Abderrazak, H.; Chabbah, T.; Souissi, R.; Saint-Martin, P.; Casabianca, H.; Chatti, S.; Mercier, R.; Errachid, A.; Hammami, M.; et al. Adsorption characteristics of aromatic pollutants and their halogenated derivatives on bio-based poly (ether-pyridine)s. *J. Environ. Chem. Eng.* **2020**, *8*, 104333. [CrossRef]
13. Pham, T.D.; Bui, T.T.; Truong, T.T.T.; Hoang, T.H.; Le, T.S.; Duong, V.D.; Yamaguchi, A.; Kobayashi, M.; Adachi, Y. Adsorption characteristics of beta-lactam cefixime onto nanosilica fabricated from rice HUSK with surface modification by polyelectrolyte. *J. Mol. Liq.* **2019**, *298*, 111981. [CrossRef]
14. Kamran, U.; Park, S.-J. Microwave-assisted acid functionalized carbon nanofibers decorated with Mn doped TNTs nanocomposites: Efficient contenders for lithium adsorption and recovery from aqueous media. *J. Ind. Eng. Chem.* **2020**, *92*, 263–277. [CrossRef]
15. Kamran, U.; Park, S.-J. Functionalized titanate nanotubes for efficient lithium adsorption and recovery from aqueous media. *J. Solid State Chem.* **2019**, *283*, 121157. [CrossRef]
16. Shankar, A.; Kongot, M.; Saini, V.K.; Kumar, A. Removal of pentachlorophenol pesticide from aqueous solutions using modified chitosan. *Arab. J. Chem.* **2020**, *13*, 1821–1830. [CrossRef]
17. Naganathan, K.K.; Faizal, A.N.M.; Zaini, M.A.A.; Ali, A. Adsorptive removal of Bisphenol a from aqueous solution using activated carbon from coffee residue. *Mater. Today: Proc.* **2021**, *47*, 1307–1312. [CrossRef]

18. Zhu, H.; Li, Z.; Yang, J. A novel composite hydrogel for adsorption and photocatalytic degradation of bisphenol A by visible light irradiation. *Chem. Eng. J.* **2018**, *334*, 1679–1690. [CrossRef]
19. Batra, S.; Datta, D.; Beesabathuni, N.S.; Kanjolia, N.; Saha, S. Adsorption of Bisphenol-A from aqueous solution using amberlite XAD-7 impregnated with aliquat 336: Batch, column, and design studies. *Process Saf. Environ. Prot.* **2018**, *122*, 232–246. [CrossRef]
20. Men, X.; Guo, Q.; Meng, B.; Ren, S.; Shen, B. Adsorption of bisphenol A in aqueous solution by composite bentonite with organic moity. *Microporous Mesoporous Mater.* **2020**, *308*, 110450. [CrossRef]
21. Kamran, U.; Heo, Y.-J.; Lee, J.W.; Park, S.-J. Chemically modified activated carbon decorated with MnO_2 nanocomposites for improving lithium adsorption and recovery from aqueous media. *J. Alloy. Compd.* **2019**, *794*, 425–434. [CrossRef]
22. Duan, F.; Chen, C.; Zhao, X.; Yang, Y.; Liu, X.; Qin, Y. Water-compatible surface molecularly imprinted polymers with synergy of bi-functional monomers for enhanced selective adsorption of bisphenol A from aqueous solution. *Environ. Sci. Nano* **2016**, *3*, 213–222. [CrossRef]
23. Enyoh, C.E.; Isiuku, B.O. 2,4,6-Trichlorophenol (TCP) removal from aqueous solution using Canna indica L.: Kinetic, isotherm and Thermodynamic studies. *Chem. Ecol.* **2020**, *37*, 64–82. [CrossRef]
24. Sahnoun, S.; Boutahala, M.; Zaghouane-Boudiaf, H.; Zerroual, L. Trichlorophenol removal from aqueous solutions by modified halloysite: Kinetic and equilibrium studies. *DESALINATION Water Treat.* **2015**, *57*, 15941–15951. [CrossRef]
25. Obinna, I.B.; Ebere, E.C. A review: Water pollution by heavy metal and organic pollutants: Brief review of sources, effects and progress on remediation with aquatic plants. *Anal. Methods Environ. Chem. J.* **2019**, *2*, 5–38. [CrossRef]
26. Kamran, U.; Park, S.-J. MnO_2-decorated biochar composites of coconut shell and rice husk: An efficient lithium ions adsorption-desorption performance in aqueous media. *Chemosphere* **2020**, *260*, 127500. [CrossRef]
27. Kamran, U.; Bhatti, H.N.; Noreen, S.; Tahir, M.A.; Park, S.-J. Chemically modified sugarcane bagasse-based biocomposites for efficient removal of acid red 1 dye: Kinetics, isotherms, thermodynamics, and desorption studies. *Chemosphere* **2021**, *291*, 132796. [CrossRef]
28. Kamran, U.; Bhatti, H.N.; Iqbal, M.; Jamil, S.; Zahid, M. Biogenic synthesis, characterization and investigation of photocatalytic and antimicrobial activity of manganese nanoparticles synthesized from Cinnamomum verum bark extract. *J. Mol. Struct.* **2018**, *1179*, 532–539. [CrossRef]
29. Tan, I.A.W.; Ahmad, A.L.; Hameed, B.H. Adsorption isotherms, kinetics, thermodynamics and desorption studies of 2,4,6-trichlorophenol on oil palm empty fruit bunch-based activated carbon. *J. Hazard. Mater.* **2009**, *164*, 473–482. [CrossRef]
30. Mubarik, S.; Saeed, A.; Athar, M.; Iqbal, M. Characterization and mechanism of the adsorptive removal of 2,4,6-trichlorophenol by biochar prepared from sugarcane baggase. *J. Ind. Eng. Chem.* **2016**, *33*, 115–121. [CrossRef]
31. Lisowski, P.; Colmenares, J.C.; Mašek, O.; Lisowski, W.; Lisovytskiy, D.; Kamińska, A.; Łomot, D. Dual Functionality of TiO_2/Biochar Hybrid Materials: Photocatalytic Phenol Degradation in the Liquid Phase and Selective Oxidation of Methanol in the Gas Phase. *ACS Sustain. Chem. Eng.* **2017**, *5*, 6274–6287. [CrossRef]
32. Gundogdu, A.; Duran, C.; Senturk, H.B.; Soylak, M.; Ozdes, D.; Serencam, H.; Imamoglu, M. Adsorption of Phenol from Aqueous Solution on a Low-Cost Activated Carbon Produced from Tea Industry Waste: Equilibrium, Kinetic, and Thermodynamic Study. *J. Chem. Eng. Data* **2012**, *57*, 2733–2743. [CrossRef]
33. Tao, X.; Zhou, G.; Zhuang, X.; Cheng, B.; Li, X.; Li, H. Solution blowing of activated carbon nanofibers for phenol adsorption. *RSC Adv.* **2014**, *5*, 5801–5808. [CrossRef]
34. Diao, Z.-H.; Xu, X.-R.; Chen, H.; Jiang, D.; Yang, Y.-X.; Kong, L.-J.; Sun, Y.-X.; Hu, Y.-X.; Hao, Q.-W.; Liu, L. Simultaneous removal of Cr (VI) and phenol by persulfate activated with bentonite-supported nanoscale zero-valent iron: Reactivity and mechanism. *J. Hazard. Mater.* **2016**, *316*, 186–193. [CrossRef]
35. Lunagariya, J.; Chabhadiya, K.; Pathak, P.; Mashru, D. Application of Taguchi method in activated carbon adsorption process of phenol removal from ceramic gasifier wastewater. *Environ. Challenges* **2022**, *6*, 100450. [CrossRef]
36. Kamran, U.; Bhatti, H.N.; Iqbal, M.; Nazir, A. Green synthesis of metal nanoparticles and their applications in different fields: A review. *Z. Phys. Chem.* **2019**, *233*, 1325–1349. [CrossRef]
37. Giraldo, L.; Moreno-Piraján, J.C. Study of adsorption of phenol on activated carbons obtained from eggshells. *J. Anal. Appl. Pyrolysis* **2014**, *106*, 41–47. [CrossRef]
38. Sagbas, S.; Kantar, C.; Sahiner, N. Preparation of Poly(Humic Acid) Particles and Their Use in Toxic Organo-Phenolic Compound Removal from Aqueous Environments. *Water, Air, Soil Pollut.* **2013**, *225*, 1–10. [CrossRef]
39. Tamang, M.; Paul, K.K. Adsorptive treatment of phenol from aqueous solution using chitosan/calcined eggshell adsorbent: Optimization of preparation process using Taguchi statistical analysis. *J. Indian Chem. Soc.* **2022**, *99*, 100251. [CrossRef]
40. Issabayeva, G.; Hang, S.Y.; Wong, M.C.; Aroua, M.K. A review on the adsorption of phenols from wastewater onto diverse groups of adsorbents. *Rev. Chem. Eng.* **2017**, *34*, 855–873. [CrossRef]
41. Kazachenko, A.S.; Vasilieva, N.Y.; Fetisova, O.Y.; Sychev, V.V.; Elsuf'Ev, E.V.; Malyar, Y.N.; Issaoui, N.; Miroshnikova, A.V.; Borovkova, V.S.; Kazachenko, A.S.; et al. New reactions of betulin with sulfamic acid and ammonium sulfamate in the presence of solid catalysts. *Biomass Convers. Biorefinery* **2022**, 1–12. [CrossRef]
42. Fathy, M.; Selim, H.; Shahawy, A.E.L. Chitosan/MCM-48 nanocomposite as a potential adsorbent for removing phenol from aqueous solution. *RSC Adv.* **2020**, *10*, 23417–23430. [CrossRef]

43. Salari, M.; Dehghani, M.H.; Azari, A.; Motevalli, M.D.; Shabanloo, A.; Ali, I. High performance removal of phenol from aqueous solution by magnetic chitosan based on response surface methodology and genetic algorithm. *J. Mol. Liq.* **2019**, *285*, 146–157. [CrossRef]
44. Kamran, U.; Park, S.-J. Hybrid biochar supported transition metal doped MnO$_2$ composites: Efficient contenders for lithium adsorption and recovery from aqueous solutions. *Desalination* **2021**, *522*, 115387. [CrossRef]
45. Soni, U.; Bajpai, J.; Singh, S.K.; Bajpai, A. Evaluation of chitosan-carbon based biocomposite for efficient removal of phenols from aqueous solutions. *J. Water Process Eng.* **2017**, *16*, 56–63. [CrossRef]
46. Anirudhan, T.; Ramachandran, M. Removal of 2,4,6-trichlorophenol from water and petroleum refinery industry effluents by surfactant-modified bentonite. *J. Water Process Eng.* **2014**, *1*, 46–53. [CrossRef]
47. Hussain, A.; Dubey, S.K.; Kumar, V. Kinetic study for aerobic treatment of phenolic wastewater. *Water Resour. Ind.* **2015**, *11*, 81–90. [CrossRef]
48. Olu-Owolabi, B.I.; Alabi, A.H.; Diagboya, P.N.; Unuabonah, E.I.; Düring, R.-A. Adsorptive removal of 2,4,6-trichlorophenol in aqueous solution using calcined kaolinite-biomass composites. *J. Environ. Manag.* **2017**, *192*, 94–99. [CrossRef]
49. Loh, C.H.; Zhang, Y.; Goh, S.; Wang, R.; Fane, A.G. Composite hollow fiber membranes with different poly(dimethylsiloxane) intrusions into substrate for phenol removal via extractive membrane bioreactor. *J. Membr. Sci.* **2015**, *500*, 236–244. [CrossRef]
50. Sierra, J.D.M.; Wang, W.; Cerqueda-Garcia, D.; Oosterkamp, M.J.; Spanjers, H.; van Lier, J.B. Temperature susceptibility of a mesophilic anaerobic membrane bioreactor treating saline phenol-containing wastewater. *Chemosphere* **2018**, *213*, 92–102. [CrossRef]
51. Daramola, M.; Sadare, O.; Oluwasina, O.; Iyuke, S. Synthesis and Application of Functionalized Carbon Nanotube Infused Polymer Membrane (fCNT/PSF/PVA) for Treatment of Phenol-Containing Wastewater. *J. Membr. Sci. Res.* **2019**, *5*, 310–316. [CrossRef]
52. Ahmaruzzaman, M.; Gayatri, S.L. Activated Tea Waste as a Potential Low-Cost Adsorbent for the Removal of p-Nitrophenol from Wastewater. *J. Chem. Eng. Data* **2010**, *55*, 4614–4623. [CrossRef]
53. El-Naas, M.; Al-Zuhair, S.; Abu-Alhaija, M. Removal of phenol from petroleum refinery wastewater through adsorption on date-pit activated carbon. *Chem. Eng. J.* **2010**, *162*, 997–1005. [CrossRef]
54. Agarry, S.E.; Owabor, C.N.; Ajani, A.O. Modified plantain peel as cellulose-based low-cost adsorbent for the removal of 2,6-dichlorophenol from aqueous solution: Adsorption isotherms, kinetic modeling, and thermodynamic studies. *Chem. Eng. Commun.* **2013**, *200*, 1121–1147. [CrossRef]
55. Jabeen, A.; Bhatti, H.N.; Noreen, S.; Gaffar, A. Adsorptive removal of 2,4,6-trichloro-phenol from wastewater by mango seed shell and its magnetic composites: Batch and column study. *Int. J. Environ. Anal. Chem.* **2021**, 1–21. [CrossRef]
56. Hai, L.; Zhang, T.; Zhang, X.; Zhang, G.; Li, B.; Jiang, S.; Ma, X. Catalytic hydroxylation of phenol to dihydroxybenzene by Fe(II) complex in aqueous phase at ambient temperature. *Catal. Commun.* **2017**, *101*, 93–97. [CrossRef]
57. Kim, Y.-H.; Lee, B.; Choo, K.-H.; Choi, S.-J. Adsorption characteristics of phenolic and amino organic compounds on nano-structured silicas functionalized with phenyl groups. *Microporous Mesoporous Mater.* **2014**, *185*, 121–129. [CrossRef]
58. Kesavan, G.; Nataraj, N.; Chen, S.-M.; Lin, L.-H. Hydrothermal synthesis of NiFe$_2$O$_4$ nanoparticles as an efficient electrocatalyst for the electrochemical detection of bisphenol A. *New J. Chem.* **2020**, *44*, 7698–7707. [CrossRef]
59. Zhang, J.; Fang, P.; Yang, L.; Zhang, J.; Wang, X. Rapid Method for the Separation and Recovery of Endocrine-Disrupting Compound Bisphenol AP from Wastewater. *Langmuir* **2013**, *29*, 3968–3975. [CrossRef]
60. Freundlich, H.M.F. Über die adsorption in lösungen. *Z. Phys. Chem.* **1907**, *57U*, 385–470. [CrossRef]
61. Temkin, M.J.; Pyzhev, V. Recent modifications to Langmuir isotherms. *Acta Phys.Chim. Sin.* **1940**, *12*, 217–222.
62. Rahman, M.M.; Marwani, H.M.; Algethami, F.K.; Asiri, A.M.; Hameed, S.A.; Alhogbi, B. Ultra-sensitive p-nitrophenol sensing performances based on various Ag 2 O conjugated carbon material composites. *Environ. Nanotechnology, Monit. Manag.* **2017**, *8*, 73–82. [CrossRef]
63. Dubinin, M.M.; Raduskhevich, L.V. Proceedings of the academy of sciences of the USSR. *Phys. Chem.* **1947**, *55*, 327–329.
64. Hamdaoui, O.; Naffrechoux, E. Modeling of adsorption isotherms of phenol and chlorophenols onto granular activated carbon: Part I. Two-parameter models and equations allowing determination of thermodynamic parameters. *J. Hazard. Mater.* **2007**, *147*, 381–394. [CrossRef]
65. López-Luna, J.; Ramírez-Montes, L.E.; Martinez-Vargas, S.; Martínez, A.I.; Mijangos-Ricardez, O.F.; González-Chávez, M.D.C.A.; Carrillo-González, R.; Solís-Domínguez, F.A.; Cuevas-Díaz, M.D.C.; Vázquez-Hipólito, V. Linear and nonlinear kinetic and isotherm adsorption models for arsenic removal by manganese ferrite nanoparticles. *SN Appl. Sci.* **2019**, *1*, 1–19. [CrossRef]
66. Goel, J.; Kadirvelu, K.; Rajagopal, C.; Garg, V.K. Removal of lead(II) by adsorption using treated granular activated carbon: Batch and column studies. *J. Hazard. Mater.* **2005**, *125*, 211–220. [CrossRef]
67. Kamran, U.; Heo, Y.-J.; Min, B.-G.; In, I.; Park, S.-J. Effect of nickel ion doping in MnO$_2$/reduced graphene oxide nanocomposites for lithium adsorption and recovery from aqueous media. *RSC Adv.* **2020**, *10*, 9245–9257. [CrossRef]
68. Juang, R.-S.; Lin, S.-H.; Tsao, K.-H. Sorption of phenols from water in column systems using surfactant-modified montmorillonite. *J. Colloid Interface Sci.* **2003**, *269*, 46–52. [CrossRef]
69. Baek, K.W.; Song, S.H.; Kang, S.H.; Rhee, Y.W.; Lee, C.S.; Lee, B.J.; Hudson, S.; Hwang, T.S. Adsorption kinetics of boron by anion exchange resin in packed column bed. *J. Ind. Eng. Chem.* **2007**, *13*, 452–456.

70. Yousef, R.I.; El-Eswed, B. The effect of pH on the adsorption of phenol and chlorophenols onto natural zeolite. *Colloids Surfaces A: Physicochem. Eng. Asp.* **2009**, *334*, 92–99. [CrossRef]
71. Karunarathne, H.; Amarasinghe, B. Fixed Bed Adsorption Column Studies for the Removal of Aqueous Phenol from Activated Carbon Prepared from Sugarcane Bagasse. *Energy Procedia* **2013**, *34*, 83–90. [CrossRef]
72. Vijayaraghavan, K.; Jegan, J.; Palanivelu, K.; Velan, M. Removal of nickel(II) ions from aqueous solution using crab shell particles in a packed bed up-flow column. *J. Hazard. Mater.* **2004**, *113*, 223–230. [CrossRef] [PubMed]
73. Thomas, H.C. Heterogeneous Ion Exchange in a Flowing System. *J. Am. Chem. Soc.* **1944**, *66*, 1664–1666. [CrossRef]
74. Rao, K.; Anand, S.; Venkateswarlu, P. Modeling the kinetics of Cd(II) adsorption on Syzygium cumini L leaf powder in a fixed bed mini column. *J. Ind. Eng. Chem.* **2011**, *17*, 174–181. [CrossRef]
75. Oladipo, A.A.; Gazi, M. Fixed-bed column sorption of borate onto pomegranate seed powder-PVA beads: A response surface methodology approach. *Toxicol. Environ. Chem.* **2014**, *96*, 837–848. [CrossRef]
76. Cruz-Olivares, J.; Perez-Alonso, C.; Barrera-Díaz, C.; Ureña-Nuñez, F.; Chaparro-Mercado, M.; Bilyeu, B. Modeling of lead (II) biosorption by residue of allspice in a fixed-bed column. *Chem. Eng. J.* **2013**, *228*, 21–27. [CrossRef]
77. Jerold, M.; Vasantharaj, K.; Joseph, D.; Sivasubramanian, V. Fabrication of hybrid biosorbent nanoscale zero-valent iron-*Sargassum swartzii* biocomposite for the removal of crystal violet from aqueous solution. *Int. J. Phytoremediation* **2016**, *19*, 214–224. [CrossRef]
78. Kakavandi, B.; Jahangiri-Rad, M.; Rafiee, M.; Esfahani, A.R.; Babaei, A.A. Development of response surface methodology for optimization of phenol and p-chlorophenol adsorption on magnetic recoverable carbon. *Microporous Mesoporous Mater.* **2016**, *231*, 192–206. [CrossRef]
79. Motsa, M.M.; Thwala, J.M.; Msagati, T.A.M.; Mamba, B.B. Adsorption of 2,4,6-Trichlorophenol and ortho-Nitrophenol from Aqueous Media Using Surfactant-Modified Clinoptilolite–Polypropylene Hollow Fibre Composites. *Water, Air, Soil Pollut.* **2011**, *223*, 1555–1569. [CrossRef]
80. Pan, J.; Yao, H.; Li, X.; Wang, B.; Huo, P.; Xu, W.; Ou, H.; Yan, Y. Synthesis of chitosan/γ-Fe_2O_3/fly-ash-cenospheres composites for the fast removal of bisphenol A and 2,4,6-trichlorophenol from aqueous solutions. *J. Hazard. Mater.* **2011**, *190*, 276–284. [CrossRef]

Article

Production of 1,3-Butadiene from Ethanol Using Treated Zr-Based Catalyst

Adama A. Bojang and Ho-Shing Wu *

Department of Chemical Engineering and Materials Science, Yuan Ze University, 135 Yuan Tung Road, Chung Li, Taoyuan 32003, Taiwan; zazafj1990@gmail.com
* Correspondence: cehswu@saturn.yzu.edu.tw; Tel.: +886-3-4638800-2564

Abstract: The conversion of ethanol to 1,3-butadiene was carried out using a treated Zr-based catalyst at a temperature of 350–400 °C with different weight hourly space velocities in a fixed bed reactor. The catalysts used are commercial, but they underwent pretreatment. The commercial catalysts used were ZrO_2, $Zr(OH)_2$, 2% $CaO\text{-}ZrO_2$, 30% $TiO_2\text{-}ZrO_2$, 50% $CeO_2\text{-}ZrO_2$ and 10% $SiO_2\text{-}ZrO_2$ in their modified or treated form. The characterizations of the catalysts were carried out using XRD, XPS, and TGA. The results indicated that ethanol conversion, yield, and selectivity of 1,3-butadiene operated weight hourly space velocity of 2.5 h^{-1} using 10% $SiO_2\text{-}ZrO_2$ were 95%, 80%, and 85%, respectively, at 350 °C. Using 50% $CeO_2\text{-}ZrO_2$ converted 70% ethanol with a 1,3-butadiene yield of 65%. The best Zr-based catalyst was 10% $SiO_2\text{-}ZrO_2$ as it gives a steady 1,3-butadiene yield, the Si-composition with ZrO_2 gives a good catalytic pour of the catalyst-bed structure; hence, the life span was good. Using 30% $TiO_2\text{-}ZrO_2$ has an ethanol conversion of 70% with a 1,3-butadiene yield of 43%.

Keywords: 1,3-butadiene; ethanol; zirconium; acetaldehyde; ethylene

Citation: Bojang, A.A.; Wu, H.-S. Production of 1,3-Butadiene from Ethanol Using Treated Zr-Based Catalyst. *Catalysts* 2022, 12, 766. https://doi.org/10.3390/catal12070766

Academic Editors: Sagadevan Suresh and Is Fatimah

Received: 31 May 2022
Accepted: 8 July 2022
Published: 11 July 2022

Publisher's Note: MDPI stays neutral with regard to jurisdictional claims in published maps and institutional affiliations.

Copyright: © 2022 by the authors. Licensee MDPI, Basel, Switzerland. This article is an open access article distributed under the terms and conditions of the Creative Commons Attribution (CC BY) license (https://creativecommons.org/licenses/by/4.0/).

1. Introduction

The chemical 1,3-butadiene is very important in the manufacturing of isomers in rubber industries. It can be used in the large-scale production of organic chemicals in synthetic chemical reactions, e.g., the Diels–Alder reaction. Since it belongs to the lower hydrocarbon group, its production is mostly from condensation reactions in the form of cracking. Since the butadiene vintage hinges principally on the type of biomass or the reactant in the condensation cracker, butadiene fabrication is vulnerable to market unpredictability or styles in the gasoline business, notably the evolving routine of gas, which might lead to butadiene deficiencies. The insufficiency of greenhouse gas-emitting fuel reserves is another long-running issue with the present hydrocarbon production methodology, in terms of profitable and environmental property. These staples have recently revived an interest in the century-old, heterogeneous chemical process transformation of alcohol to hydrocarbon, within which vaporized alcohol is primarily reworked to hydrocarbon (butadiene).

The ethanol to butadiene production process can occur in two basic steps: one-step and two-step processes. The Lebedev process often used a one-step process, by which ethanol is processed in the gas-phase form and later is converted to butadiene [1]. The Ostromislensky process is also known as the two-step process. In this process, ethanol is in a ration with acetaldehyde in a gas phase in which the dehydrogenation of ethanol is ideal [2]. The one-step process typically favors the production of acetaldehyde with little butadiene yield. Meanwhile, acetaldehyde can be recycled in industries to produce butadiene in further reaction synthesis with a specifically designed catalyst.

Consequently, even in the one-step process, it is indispensable to test feeds containing acetaldehyde to assess the comportment of the catalytic system in acetaldehyde-containing mixtures, nevertheless, in lower quantities compared to the two-step process [3–7]. Since ancient times, humans have used sugar fermentation to produce ethanol, one of the earliest of several biotechnologies [8]. Bioethanol made from the microbial fermentation of the biomass was changed from the sugar into ethanol and the same happens with petrochemical raw materials to produce ethanol, which can be produced from fermentation processes using renewable substrates such as glucose, starch, and others. The dehydration of ethanol can replace steam-cracking fossil fuels to produce ethylene and 1,3-butadiene [5,6,9]. In addition, with the progress of bioethanol technology and the popularization of bioethanol industrial equipment, the production of butadiene from bioethanol has become a sustainable green chemical route for the supply of butadiene.

Generally, ethanol dehydration to butadiene or ethylene is conducted using solid acidic or basic catalysts. The Zr-based catalyst was its reactivity and stability over the Mg-based catalyst since those are the catalysts in our preliminary study. Also, treating Zr with oxalic acid gives well-coordinated acidic and basic surroundings if necessary. ZrO_2, Zr/Si, and CaO/Zr-containing catalysts are usually used at a temperature of 300–425 °C, at standard pressure [10–13]. It was suggested that the catalyst activity could be correlated with the number of strong Brønsted acid sites in the catalyst. Table 1 highlights the use of Zr-containing catalyst for 1,3-butadiene production from ethanol. Modifying the Cu/Zr-Si catalyst with lanthanum increases their activity in aqueous ethanol conversion into 1,3-butadien [14]. Sushkevich et al. discuss the use of Zr-based catalyst in combination with zeolites beta catalyst which was obtained using synthetic modification methods, the result yielded a high selectivity of 1,3-butadiene (74 mol%), and the acidic nature of the catalyst was a key determinant for the general reaction synthesis [15,16].

The novelty of this study gives a general understanding of the catalytic performance of treated-Zr-based catalytic systems when using aqueous ethanol in an ethanol to butadiene process. This study examines the effect of a treated-Zr-based catalyst in the production of 1,3-butadiene. It also gives the relevant information on oxalic acid used in the treatment of each sample of the commercial catalyst [13,17]. The reason for selecting the Zr-based catalyst was its reactivity and stability over the Mg-based catalyst, since those are the catalysts at our disposal. Also, treating Zr with oxalic acid gives well-coordinated acidic and basic surroundings if necessary. This study aims to improve the catalytic activity of the commercial catalyst by treating it with oxalic acid, NaOH, and other essential acidic compounds. As the results show, the treated commercial catalyst yielded a better 1,3-butadiene production than the untreated. Also, the presence of oxalic acid in combination with NaOH will provide a good desilication process and increase the catalyst's acidic properties.

Table 1. Zr-containing catalyst for 1,3-butadiene production from ethanol.

Catalyst	T (°C)	WHSV (h^{-1})	TOS (h)	X_{EtOH} (%)	$Y_{1,3-BD}$ (%)	Refs.
4.9% Cu/MCF + 2.7% Zr/MCF	425	3.7	10	92	64.4	[18]
3.7% Ag/Zr/BEA	350	1.2–3.7	3	-	-	[15,19]
3% ZrO_2/SiO_2	350	1.8	-	45.4	31.6	[20,21]
3000 ppm Na/$Zn_1Zr_{10}O_n$	400	6.2	-	54.4	15.2	[22]
ZrO_2	350	2.5	14	50	28	This study
$Zr(OH)_2$	350	2.5	14	30	19	
2%CaO-ZrO_2	350	2.5	14	79	30	
10% SiO_2-ZrO_2	350	2.5	15	95	80	
30% TiO_2-ZrO_2	350	2.5	14	70	43	
50% CeO_2-ZrO_2	350	2.5	14	70	65	

X: ethanol conversion Y: 1,3 butadiene selectivity, WHSV: weight hourly space velocity, TOS: time on stream, T: temperature.

2. Results and Discussion

2.1. Crystalline Properties of the Catalyst

The catalytic characterizations of ZrO_2, $Zr(OH)_2$, 10% SiO_2-ZrO_2, and 2% CaO-ZrO_2 using XRD are shown in Figure 1. For ZrO_2, Figure 1c shows a greater crystallinity pattern (2θ = 25–28°), decreasing between 60 and 80°. The XRD pattern shows the crystalline phase nature at a diffraction peak of 2θ values of 28.3°, 32.6°, 38.7°, 50.2°, and 59.9°, corresponding to monoclinic ZrO$_2$. Generally, ZrO$_2$ has nanostructures of the monoclinic phase, possesses nanograins, and has a low strain, as shown on the XRD pattern. This finding means the thermal modification process helps to explain the phenomenon of crystallinity along the lower degree, which gives more incredible lactic structure and acidity. Zr(OH)$_2$ in Figure 1b shows less crystallinity than ZrO$_2$. The XRD pattern shows that the amorphous Zr(OH)$_2$ was less crystalline after calcination than ZrO$_2$. Thus, this dynamic character of zirconium hydroxide-to-oxide thermal evolution is also in concordance with other authors [23]. The powered content of Zr(OH)$_2$ is high, and the intensity of the peak has a significant peak (2θ = 20–30°), which is the essential nature of the catalyst.

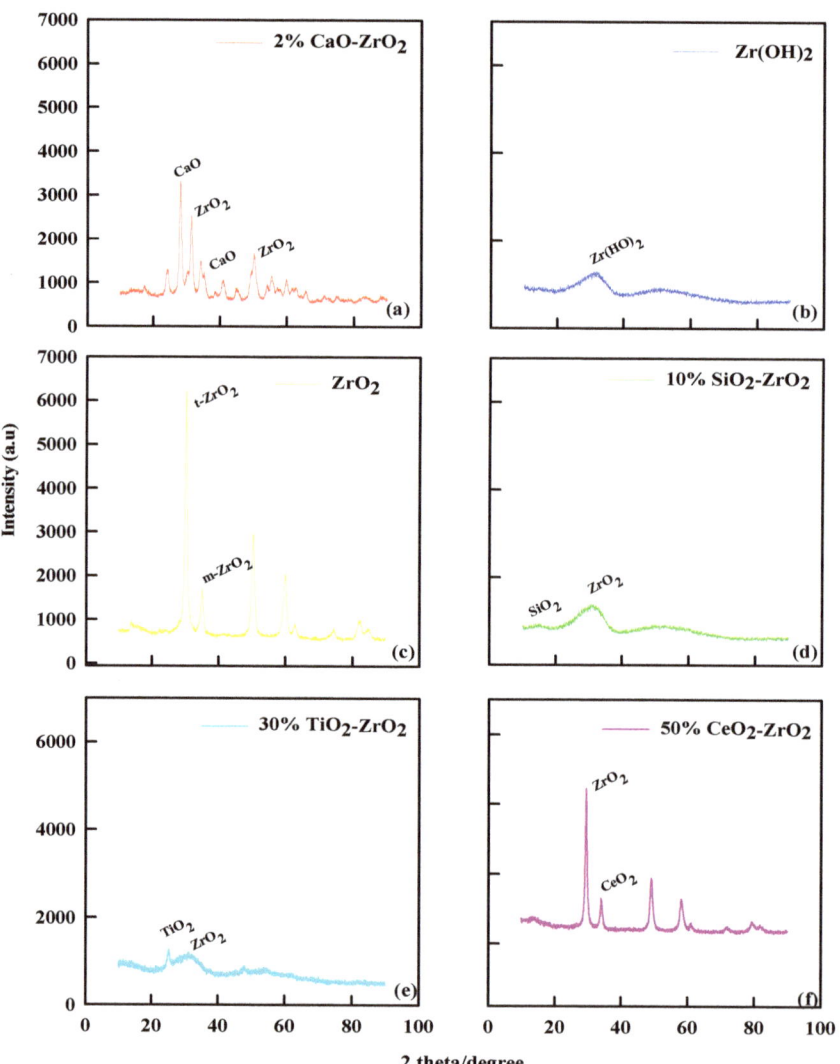

Figure 1. XRD pattern of (**a**) 2% CaO-ZrO$_2$, (**b**) Zr(OH)$_2$, (**c**) ZrO$_2$, (**d**) 10% SiO$_2$-ZrO$_2$, (**e**) 30% TiO$_2$-ZrO$_2$, and (**f**) 50% CeO$_2$-ZrO$_2$.

The XRD pattern for 10% SiO$_2$-ZrO$_2$ shows a similar trend to Zr(OH)$_2$. The XRD pattern in Figure 1d shows that when the catalyst was calcined at 700 °C, a change in morphology of the crystal structure of ZrO_2 is typical for the diffraction peak around 10° (2θ), which indicates mesoporous materials. The single diffused peak at 20–30° indicates that 10% SiO_2-ZrO_2 catalyst maintains an amorphous structure at 700 °C calcination. The crystallization of ZrO_2 in the 10% SiO_2-ZrO_2 catalyst was prevented by 10% SiO_2 during the calcining process because Si-O-Zr bonds in 10% SiO_2-ZrO_2 xerogels retard the crystal growth and phase transition [24]. Likewise, for 2% CaO-ZrO_2, as shown in Figure 1a, the XRD pattern shows that the catalyst has either a cubic or a tetragonal phase, which also indicates that CaO was not amorphous nor well dispersed on the surface of the zirconia. The presence of CaO shows a greater degree of crystal pattern and increases the angular lattice point of interaction [25].

For 30% TiO_2-ZrO_2, as shown in Figure 1e, the XRD pattern of pure ZrO_2 was a monoclinic phase, and pure TiO_2 showed an anatase phase, but the mixed oxide of 30% TiO_2-ZrO_2 was found to be X-ray amorphous, as Zou and Lin stated [26]. The XRD pattern of the mixed oxide of 30% TiO_2-ZrO_2 exhibited poor crystallinity with the tetragonal phase of ZrO_2. Generally, under the same catalytic preparation with other commercial catalysts, pure ZrO_2 has a monoclinic phase, but with 30% TiO_2, the component ZrO_2 was stabilized as the tetragonal phase. For 50% CeO_2-ZrO_2, Figure 1f shows that the XRD pattern is highlighted by a peak that appears to be ZrO_2. It was remarkable that the XRD peak positions and lattice parameters of Ce/ZrO_2 were continuously shifted from CeO_2 to ZrO_2 depending on the ratio of Ce/Zr in the precursor solution, as explained by Vegard's law [27]. Thus, from the results, the XRD pattern of 50% CeO_2-ZrO_2 shows that the crystal sizes would be small and homogeneously well dispersed at the XRD detection level. The XRD pattern for 50% CeO_2-ZrO_2 exhibited a mixed profile of cubic CeO_2 and tetragonal ZrO_2 phase.

Generally, the crystallinity of the catalyst was calculated using the XRD data. The equation is as follows:

$$\text{Crystallinity} = \frac{\text{Area of crystalline peaks}}{\text{Area of all peak(crytalline + Amorphous)}} \times 100 \quad (1)$$

Therefore, the crystallinity of ZrO_2 was 78.3%, showing distinct peaks and hence its crystalline nature. $Zr(OH)_2$ has an amorphous morphology in which its crystallinity was 20.4%; The 2% CaO-ZrO_2 catalyst has a crystallinity of 87.3%, which can also be identified by the many distinct peaks and also the appearance of CaO structures. The 50% CeO_2-ZrO_2 catalyst has a crystallinity of 74.3%. The 30% TiO_2-ZrO_2 catalyst has 30.2% crystallinity, while SiO_2-ZrO_2 has a crystallinity of 28.3%.

2.2. XPS Analytical Profile of the Catalyst

The XPS analyses of catalysts show the mass surface concentration of carbon, oxygen, zirconium, and silicon (C 1s, O 1s, Zr 3d, and Si 2p) (Figure 2). Table 2 lists the percentage of surface oxygen and carbon as about 47.3% and 28.0% for ZrO_2; 53.2% and 23% for 2% CaO-ZrO_2; 53.4% and 19% for 10% SiO_2-ZrO_2; 49.6% and 29% for $Zr(OH)_2$; 54% and 22.8% for 30% TiO_2-ZrO_2 and 51.4% and 24.8% for 50% CeO_2-ZrO_2, respectively. The silicon level is shown as 6.8% in 10% SiO_2-ZrO_2. Meanwhile, 10% SiO_2-ZrO_2 has increased oxygen content compared to all the others. This effect explains the method of catalytic preparation for the commercial catalysts using the treated methods. The carbon content varies from 25% for ZrO_2, which was the highest, to 19% for 10% SiO_2-ZrO_2. Since all the catalysts have some amount of Zr, this effect influences ethanol conversion in various degrees, from 25% for ZrO_2 to the lowest 20.9% for 10% SiO_2-ZrO_2.

Comparing the spectra lines shows the band energies for Zr 3d and Si 2p. Different binding energies are associated with each species, between 140 to 200 eV for Zr 3d, 105 to 100 eV for Si 2p, Ti 2p has a binding energy of 459 eV, and Ce 3d has a binding energy of 884.5 eV. The decrease in the nature of the peak level for Si 2p shows the desilication of the silicon group in the catalytic preparation methods. It indicated the performance of the catalyst during the reaction. The peak, shifting from the maximum binding energies, indicated oxidized silicon elements' general direct proportional ratio.

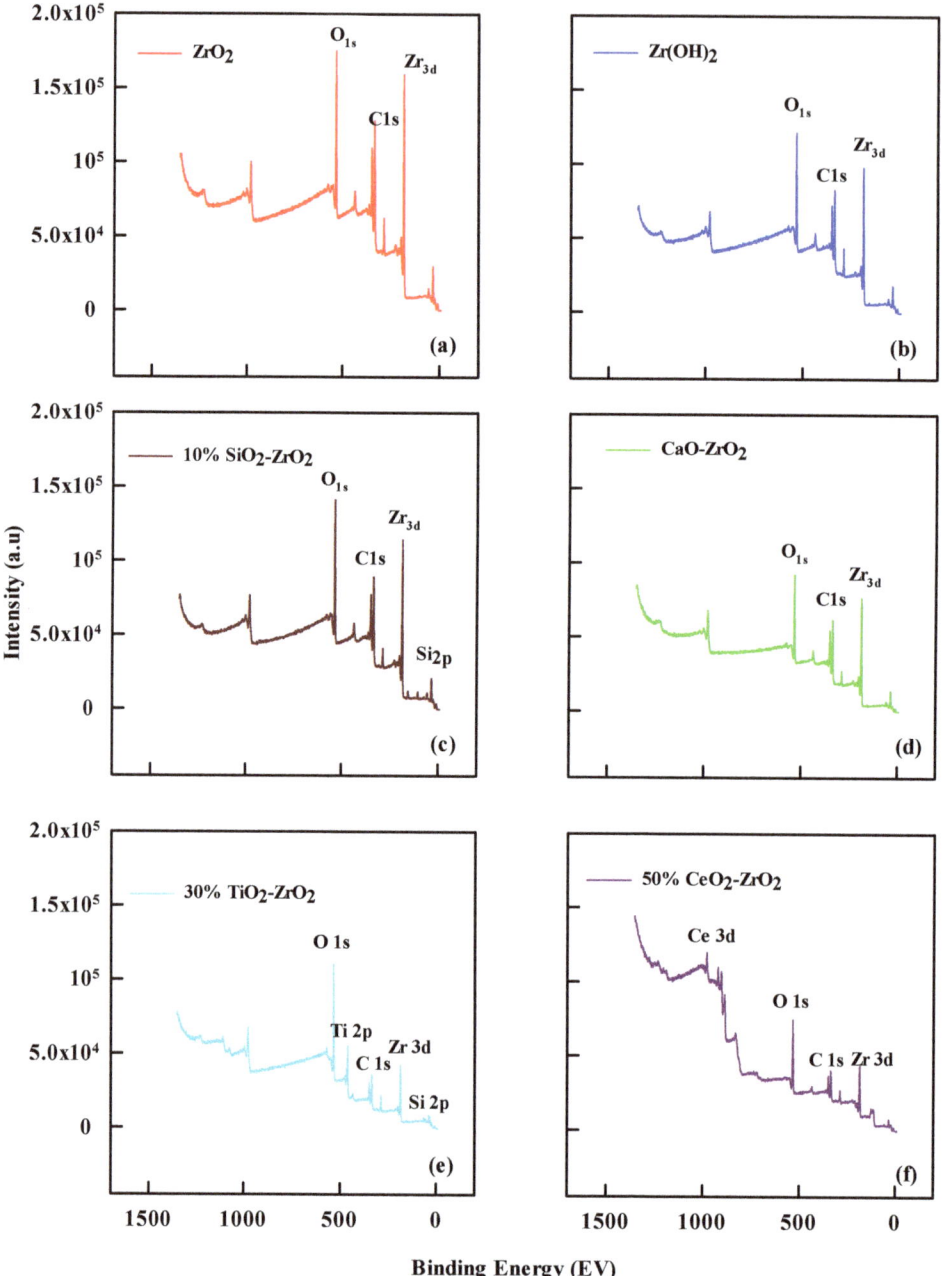

Figure 2. XPS profile for (**a**) ZrO_2, (**b**) 10% SiO_2-ZrO_2, (**c**) $Zr(OH)_2$ (**d**) 2% CaO-ZrO_2 (**e**) 30% TiO_2-ZrO_2, and (**f**) 50% CeO_2-ZrO_2.

Table 2. Weight percentage (%) of the elements by XPS analysis.

Catalyst	O 1s (%)	C 1s (%)	Zr 3d (%)	Si 2p (%)	Ca 2p (%)	Ti 2p (%)	Ce 3d (%)
ZrO_2	47.3	27.9	24.9	-			
2% CaO-ZrO_2	50.3	23.0	23.8	-	2.9		
10% SiO_2-ZrO_2	53.4	19.0	20.9	6.8			
$Zr(OH)_2$	49.6	29.0	21.4	-			
30% TiO_2-ZrO_2	54	22.8	10.7	2.3		10.3	
50% CeO_2-ZrO_2	51.4	24.8	13.4				10.4

2.3. BET Profiles of the Catalyst

Figure 3 shows the relative pore-diameter distribution of the catalyst before and after the reaction. The characterization of the pore volume was investigated by N_2 adsorption-desorption methods. Tables 3 and 4 list the surface area and pore volume of the catalyst before and after the reaction. For the ZrO_2 sample, the pore volume (V_{meso}) of 0.0012 cm^3/g increased to 0.0014 cm^3/g after the reaction. The surface area of ZrO_2 (S_{BET}) was 98 m^2/g before the reaction. It decreased to 78 m^2/g after the reaction. $Zr(OH)_2$ has a pore area of 50 m^2/g and micro-pore volume of 0.004 cm^3/g, which went down to 32 m^2/g and 0.002 cm^3/g after the reaction. The 10% SiO_2-ZrO_2 catalyst has a surface area of 90 m^2/g and micro-pore volume of 0.0172 cm^3/g, decreasing to 88 m^2/g and 0.142 cm^3/g after the reaction. The 2% CaO_2-ZrO_2 catalyst has a surface area of 195 m^2/g and micro-pore volume of 0.006 cm^3/g, down to 100 m^2/g and 0.004 cm^3/g. For 30% TiO_2-ZrO_2, a surface area of 293 m^2/g and micro-pore volume of 0.0093 cm^3/g decreased to 212 m^2/g and 0.0087 cm^3/g after the reaction.

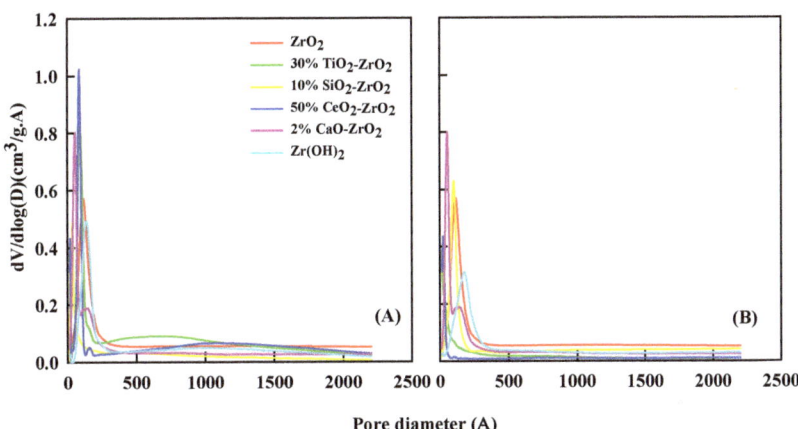

Figure 3. Pore-size distribution of treated catalyst (**A**) before reaction and (**B**) after reaction.

Table 3. Catalyst particle size and BET surface area before reaction.

Catalyst	Surface Area (m^2/g)			Pore Volume (cm^3/g)	
	S_{BET} [a]	S_{micro} [b]	S_{ext} [b]	V_{micro} [b]	V_{meso} [c]
ZrO_2	98	-	116	-	0.0012
$Zr(OH)_2$	50	5.5	45	0.0004	0.0002
2% CaO-ZrO_2	195	3.4	191	0.0006	0.0004
10% SiO_2-ZrO_2	90	7.4	83	0.0172	0.0013
30% TiO_2-ZrO_2	293	43	250	0.0093	0.0063
50% CeO_2-ZrO_2	205	-	218	-	0.0044

[a] BET method, [b] t-Plot method, [c] BJH method (adsorption branch).

Table 4. Catalyst particle size and BET surface area after the reaction.

Catalyst	Surface Area (m²/g)			Pore Volume (cm³/g)	
	S_{BET} [a]	S_{micro} [b]	S_{ext} [b]	V_{micro} [b]	V_{meso} [c]
ZrO_2	78	-	89	-	0.0014
$Zr(OH)_2$	32	3.2	16	0.0002	0.0005
2% CaO-ZrO_2	100	1.6	78	0.0004	0.0008
10% SiO_2-ZrO_2	88	6.8	100	0.0142	0.0015
30% TiO_2-ZrO_2	212	12.8	145	0.0087	0.0093
50% CeO_2-ZrO_2	178	-	164	-	0.0056

[a] BET method, [b] t-Plot method, [c] BJH method (adsorption branch).

The higher loss of micro-pore area in catalysts suggests that the coke precursors or coke produced during the reaction procedure tend to deposit in the newly created micropores. The deposition of coke could be due to the formation of ash. If the catalyst is exposed to a high temperature for a long time, the ash or coke deposition can affect the reactivity, decreasing the pore size or altering it, as suggested in previous literature [28,29]. Consequently, the newly created micro-pores may quarter the part of coke deposition, sinking the materialization of coke deposition in its inherent micro-pores to some extent. The coke deposition causes the decrease of S_{micro} and V_{micro}. The strait obstruction by coke deposition over the prepared catalyst limits the access of reactants/intermediates to the core active sites on the catalyst. The surface area and pore volume size will distress the catalyst for ethanol to 1,3-butadiene reaction. The coke deposition effect will be amplified when the surface area and pore volume are small. When the coke effect is strong, the catalyst activity will be decreased. Because the coke effect will block the surface area and pore volume, the catalyst will be less active.

2.4. Measurement of Catalyst Deposition and Regeneration Using TGA

For the coke decomposition of the catalyst after reaction, TGA analysis was performed on ZrO_2, $Zr(OH)_2$, CaO-ZrO_2, and 10% SiO_2-ZrO_2. As shown in Figure 4, the total weight loss follows a trend up to 800 °C. The weight loss of the catalysts is given as follows: ZrO_2 has a weight loss of 4 wt%; $Zr(OH)_2$ has a weight loss of 19 wt%; 2% CaO-ZrO_2 has a weight loss of 2 wt%, and 10% SiO_2-ZrO_2 has a weight loss of 9 wt%.

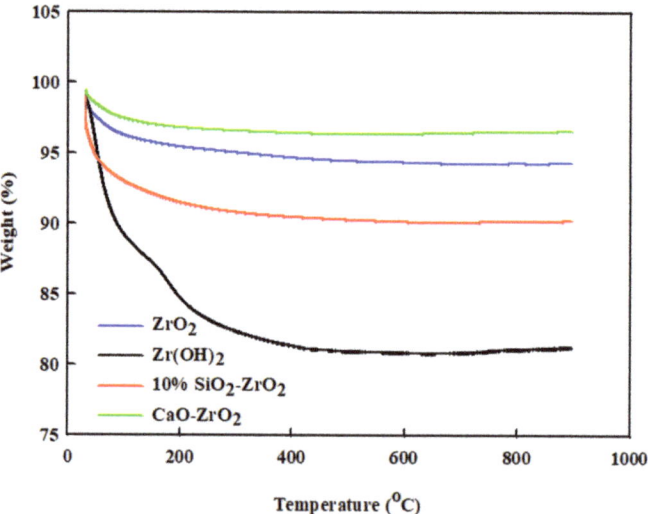

Figure 4. TGA-profile analysis for ZrO_2, $Zr(OH)_2$, CaO-ZrO_2 and 10% SiO_2-ZrO_2.

Generally, if the catalyst loses weight at a lower temperature (i.e., below 170 °C), it is normally associated with the effect of water absorption, since the catalysts are porous in many aspects. The weight loss is due to the burning of the catalyst, which could lead to the accumulation of coke at 250–800 °C in the reactor. $Zr(OH)_2$ shows a more significant amount of coke decomposition than all the others. The 2% CaO-ZrO_2 catalyst has more thermal stability compared to other catalytic systems. Hence, it has low coke formation. The TGA profile also shows some regenerative nature in all the catalysts (ZrO_2, $Zr(OH)_2$, 2% CaO-ZrO_2, and 10% SiO_2-ZrO_2) because the weight loss remains mostly constant for each of them at 200 °C. The TGA profiles also help to show that the ZrO_2-containing catalyst has a good regeneration characteristic and thermal stability. A TGA sample for each catalyst was taken, and some present a similar pattern: 30% TiO_2-ZrO_2 has a similar curve to 2% CaO-ZrO_2, and 50% CeO-ZrO_2 has a similar curve to ZrO_2. This finding demonstrates that they have similar thermo-stability but different relativities, see Figure S2 in Supplementary Materials.

2.5. Effect of Weight Hourly Space Velocity (WHSV)

Ethanol dehydration using catalysts dramatically affects the yield of 1,3-butadiene at different WHSV. Space velocity is an essential factor in catalytic activity and explains the concentration profiles [30]. The production rate of 1,3-butadiene is directly proportional to ethanol's WHSV [31]. Mass transfer increases are effective, resulting from increased space velocity and gas velocity [9].

2.5.1. Effect of Catalyst at WHSV = 2.5 h^{-1}

Figure 5 and Table 5 show the performance of different pretreated commercial catalysts at different temperatures at WHSV = 2.5 h^{-1}. The ethanol dehydration is incomplete at lower temperatures; thus, ethanol conversion was low. The ethanol conversion was 40% at 250 °C for ZrO_2, while the ethanol conversion increased to 60% at 400 °C. The same trend can be observed for $Zr(OH)_2$, which shows 10% ethanol conversion at 250 °C and 22% at 350 °C, but at 400 °C, the ethanol conversion drops to 8%. This finding could be explained by the coke decomposition of the $Zr(OH)_2$. For 10% SiO_2-ZrO_2, ethanol conversion was 50% at 250 °C, while ethanol conversions were 90% and 90% at 300 and 350 °C, respectively. CaO-ZrO_2 shows an ethanol conversion of 40% at 250 °C and 79% at 300 °C, respectively.

The product yields follow a different trajectory at WHSV of 2.5 h^{-1}, ZrO_2 catalytic performance shows a 25% yield of 1,3-butadiene at 250 °C and 30% yield at 300 and 350 °C, respectively, and at 400 °C, it drops to 20%. Moreover, ZrO_2 gave better acetaldehyde 20% yield at 300 and 350 °C, although it decreased to 14% when the temperature was 400 °C. The ethylene yield was 6% at 250 °C and 10% at 350 °C. This result was due to the structural morphology of the catalyst and its crystalline nature, which will effectively favor more of an acidic reaction, and this effect will resist any new active sites to be developed on the catalyst unless further modification takes over. For $Zr(OH)_2$, 1,3-butadiene yield is 18% at 250 °C, and shows a little increase at 350 °C (20%) and at 400 °C (22%). The acetaldehyde yield was 10% at 250 °C, and it eventually increased to 14% at 350 °C and 400 °C, respectively. The ethylene yield was 24% at 350 °C.

The 2% CaO-ZrO_2 catalyst favors an ethanol conversion of 40% at 250 °C and 80% at 300 and 350 °C, respectively, which shows a steady-state mechanism. Ethanol conversion decreases to 60% at 400 °C. It generates a 1,3-butadiene yield of 30% at 350 °C and an acetaldehyde yield of 15% at 300 and 350 °C. It gives an ethylene yield of 30% from 250 to 350 °C. This result is in concordance with [30] for ethylene production. The 10% SiO_2-ZrO_2 catalyst gives an ethanol conversion of 50%, 90%, 90% and 87% at 250, 300, 350 and 400 °C, respectively. The highest 1,3-butadiene yield was 80% at 350 and 400 °C. The combination of SiO_2 and ZrO_2 tends to favor more ethylene production (9.8% at 300 °C) than acetaldehyde.

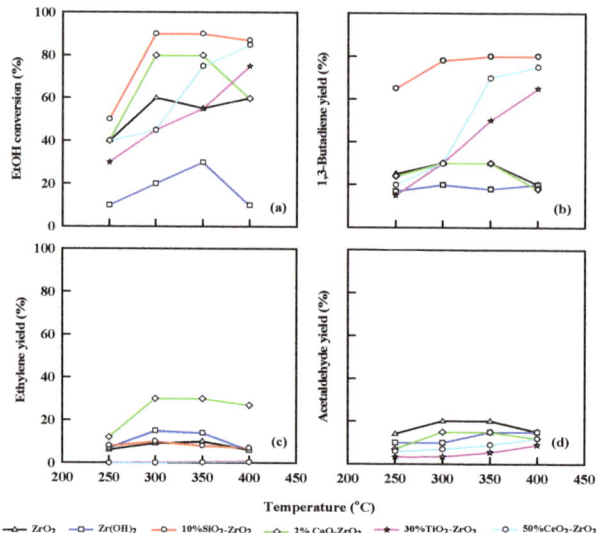

Figure 5. Plot of ethanol conversion (**a**), yields of 1,3-butadiene (**b**), ethylene (**c**), and acetaldehyde (**d**) vs. reaction temperature for different pretreated commercial catalysts. WHSV: 2.5 h^{-1}, catalyst: 0.4 g, N$_2$: 30 mL/min.

Table 5. STY of 1,3-butadiene using different temperatures and pretreated catalysts.

Catalyst	T (°C)	WHSV (h^{-1})	
		1.25	2.5
ZrO$_2$	250	256	456
	300	328	528
	350	343	503
	400	87.6	97.6
Zr(OH)$_2$	250	79.7	99.7
	300	189	199
	350	284	204
	400	278	308
2% CaO-ZrO$_2$	250	348	231
	300	457	346
	350	468	408
	400	85.6	89.3
10% SiO$_2$-ZrO$_2$	250	549	449
	300	557	589
	350	566	578
	400	580	558
30% TiO$_2$-ZrO$_2$	250	87.2	98.7
	300	212	412
	350	329	423
	400	399	489
50% CeO$_2$-ZrO$_2$	250	99.7	249
	300	254	348
	350	444	459
	400	522	579

Catalyst: 0.4 g, T (Temperature), X (Conversion), Y (Yield), STY (Space time yield) (gKg^{-1} h^{-1}).

For ZrO$_2$, the 1,3-butadiene yield was 23% at 250 °C, and the space-time yield (STY) was 456 gKg^{-1} h^{-1}. The STY decreased when the temperature was 400 °C (97.6 gKg^{-1} h^{-1}). Zr(OH)$_2$ has a STY of 308 gKg^{-1} h^{-1} when the temperature was increased from 350 to 400 °C, and the 1,3-butadiene yield attained was 20%. CaO-ZrO$_2$ shows a steady yield of 1,3-butadiene at 250 to 350 °C with a 30% yield, and the STY increases from 346 to 408 gKg^{-1} h^{-1}. The 10% SiO$_2$-ZrO$_2$ catalyst has 80% 1,3-butadiene yield, and the STY increased from 449 gKg^{-1} h^{-1} at 250 °C to 578 gKg^{-1} h^{-1} at 350 °C. At 400 °C, 30% TiO$_2$ combined with ZrO$_2$ realized an ethanol conversion of 79% with 65% 1,3-butadiene yield. The 1,3-butadiene yield increases with an increase in temperature from 250–400 °C, and the STY increases from 98.7 to 489 gKg^{-1} h^{-1} as temperature increases from 250 to 400 °C. The same trend was observed for 50% CeO$_2$-ZrO$_2$ with 90% ethanol conversion at 400 °C and 2.5 h^{-1} WHSV. The 1,3-butadiene yield increases with an increase in temperature. The highest yield of 1,3-butadiene for CeO$_2$-ZrO$_2$ was 79% at 400 °C. An increase in temperature also increases STY for CeO$_2$-ZrO$_2$ (249–579 gKg^{-1} h^{-1}), as shown in Table 5.

2.5.2. Effect of Catalyst at WHSV = 1.25 h^{-1}

Figure 6 and Table 5 show the different modified commercial catalysts' performances at different temperatures and WHSV of 1.25 h^{-1}. For ZrO$_2$, the modified catalyst, in this instance, converted ethanol to a maximum of 60% at 350 °C. The 1,3-butadiene yield was 30% at 350 °C, acetaldehyde yield was 20% at 350 °C, and ethylene yield was 9.7% at 350 °C. Zr(OH)$_2$ generates an ethanol conversion of 39% at 350 °C, and 1,3-butadiene yield of 20% at 300 °C. The ethylene yield was 10%, while acetaldehyde was 14% at 350 °C. The 2% CaO-ZrO$_2$ catalyst produces an ethanol conversion of 80% at 350 °C and a 1,3-butadiene yield of 30% at 350 °C. The acetaldehyde yield was 14% at 300 °C. Surprisingly, it yields more ethylene (30%) at 350 °C than acetaldehyde. This effect was due to the presence of CaO, which allows the catalyst to behave with a more amphiprotic nature. The 10% SiO$_2$-ZrO$_2$ catalyst gives an ethanol conversion of 90% at 300 °C and 80% yield of 1,3-butadiene at 300 to 400 °C, respectively. It also gives an ethylene yield of 8.9% at 350 °C. Acetaldehyde yield was less noticeable during this catalytic reaction.

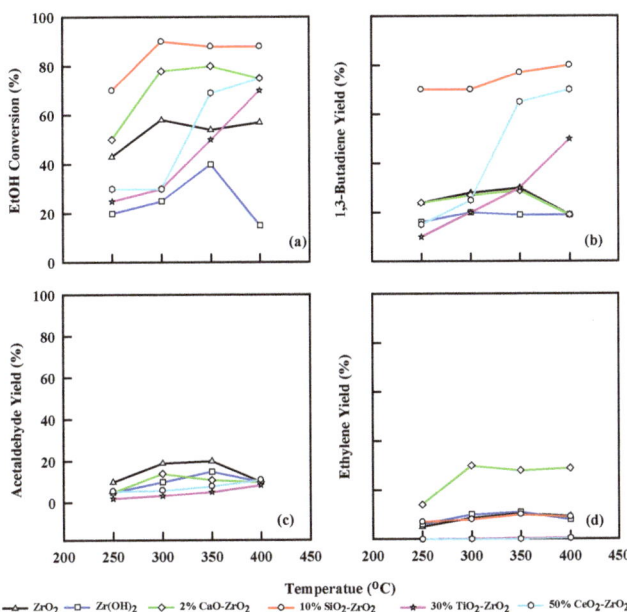

Figure 6. Plot of ethanol conversion (**a**), yields of 1,3-butadiene (**b**), acetaldehyde (**c**), and ethylene (**d**) vs. reaction temperature for different modified commercial catalysts. WHSV: 1.25 h^{-1}, catalyst: 0.4 g, N$_2$: 30 mL/min.

When WHSV is 1.25 h^{-1}, ZrO$_2$ has a 1,3-butadiene yield of 21% and STY of 256 gkg^{-1}h^{-1} at 250 °C. The yield of 1,3-butadiene increases by 28% at 350 °C and a STY of 343 gkg^{-1}h^{-1}.

When the temperature increases to 400 °C, the STY decreases to 87.6 gkg^{-1}h^{-1}, and the 1,3-butadiene yield drops to 15%. Using Zr(OH)$_2$ generated 20.8% butadiene yield at 400 °C and 278 gkg^{-1}h^{-1} STY, compared to ZrO$_2$, which yielded less. CaO-ZrO$_2$ has a 1,3-butadiene yield of 39% at a STY of 468 gkg^{-1}h^{-1} at 350 °C, but at 400 °C, the STY is decreased to 85.6 gkg^{-1}h^{-1}. The 10% SiO$_2$-ZrO$_2$ catalyst has a good 1,3-butadiene yield of 70% at 350 °C and STY of 580 gkg^{-1}h^{-1}.

The ethanol conversion at WHSV = 1.25 h^{-1} for TiO$_2$-ZrO$_2$ shows a similar trend to WHSV of 2.5 h^{-1}, but there was a decrease in the total ethanol conversion, which was 65% at 400 °C compared to 2.5 h^{-1}, which was 79% at 400 °C. However, the general increase in ethanol conversion was realized as the temperature increased from 250–400 °C. The 1,3-butadiene yield was 50% at 400 °C, the highest for TiO$_2$ catalyst at 1.25 h^{-1}. Acetaldehyde and ethylene yields were 8% and 0.2%, respectively, at 400 °C. The STY was increased from 87.2 gkg^{-1}h^{-1} at 250 °C to 399 gkg^{-1}h^{-1} at 400 °C for TiO$_2$-ZrO$_2$. CeO$_2$-ZrO$_2$ has an ethanol conversion of 79% and 1,3-butadiene yield of 65% at 400 °C. The acetaldehyde and ethylene yields were 12% and 0.3% at 350 °C. Also, an increase in STY was observed from 250 °C–400 °C (99.7–52 gkg^{-1}h^{-1}) (Table 5) for CeO$_2$-ZrO$_2$.

2.5.3. Effect of Catalyst at WHSV = 6.0 h^{-1}

Figure 7 illustrates the performance of different modified catalysts for different temperatures at WHSV of 6.0 h^{-1}. Likewise, ZrO$_2$ produces an ethanol conversion of 59% at 350 °C and a 1,3-butadiene yield of 30% at the same temperature. The acetaldehyde yield was 30% at 300 °C, while the ethylene yield was 5% at 350 °C. Zr(OH)$_2$ shows an ethanol conversion of 43% at 350 °C and a 1,3-butadiene yield of 19.2%. Acetaldehyde and ethylene yields were 14% and 5%, respectively, at 350°C. CaO-ZrO$_2$ has a combination effect, generating an ethanol conversion of 80% at 350 °C, but ethanol conversion decreases to 67% at 400 °C. The 1,3-butadiene yield increases steadily from 20% at 250 °C up to 30% at 350 °C and, eventually, it decreases to 18% at 400 °C. The 10% SiO$_2$-ZrO$_2$ catalyst shows a greater ethanol conversion of 90% and 1,3-butadiene yield of 80% at 350 °C. Ethylene yield was 9.5% at 300 °C.

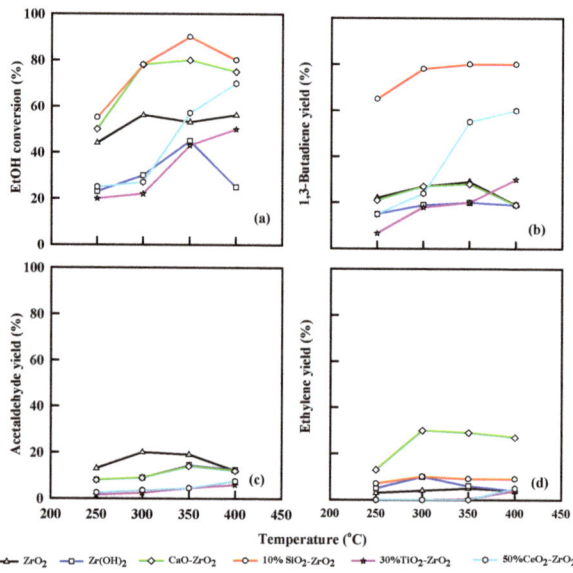

Figure 7. Plot of ethanol conversion (**a**), yields of 1,3-butadiene (**b**), acetaldehyde (**c**), and ethylene (**d**) vs. reaction temperature for different modified commercial catalysts. WHSV: 6.0 h^{-1}, catalyst: 0.4 g, N$_2$: 30 mL/min.

TiO$_2$-ZrO$_2$ has an ethanol conversion of 43% and 25% 1,3-butadiene yield. The ethanol conversion was stable as the temperature was increased from 250 to 300 °C, but it increased from 20% to 43% as the temperature was increased from 350 to 400 °C. The increase in temperature also leads to an increase in 1,3-butadiene yield. Acetaldehyde yield was 5.5%, while ethylene yield was 0.3% at 400 °C. CeO$_2$-ZrO$_2$ converted ethanol to 70% at WHSV of 6.0 h^{-1} and 400 °C, while it yielded 60% 1,3-butadiene. Acetaldehyde and ethylene yields were 7.7% and 0.5%, respectively.

2.5.4. Effect of WHSV of Ethanol at 0.75 h^{-1}

Figure 8 demonstrates the effect of 0.75 h^{-1} WHSV of ethanol during the catalytic dehydration using different modified commercial catalysts. ZrO$_2$ catalyst gave an ethanol conversion of 50% to 55% at 350 °C and 400 °C, respectively, and 1,3-butadiene yield of 22% at 350 °C. Alternatively, the yield of acetaldehyde was 19% at 300 °C, and ethylene yield was 5% at 400 °C. Zr(OH)$_2$ realized an ethanol conversion of 50% at 300 °C and a 1,3-butadiene yield of 20% at 400 °C. At 350 °C, WHSV of 0.75 h^{-1}, ZrO$_2$ catalyst yielded 8% and 6% of acetaldehyde and ethylene, respectively, while CaO-ZrO$_2$ generated 69% ethanol conversion with 23% 1,3-butadiene yield. 10% SiO$_2$-ZrO$_2$ catalyst shows an excellent ethanol conversion of 90% at 350 °C and 80% 1,3-butadiene yield.

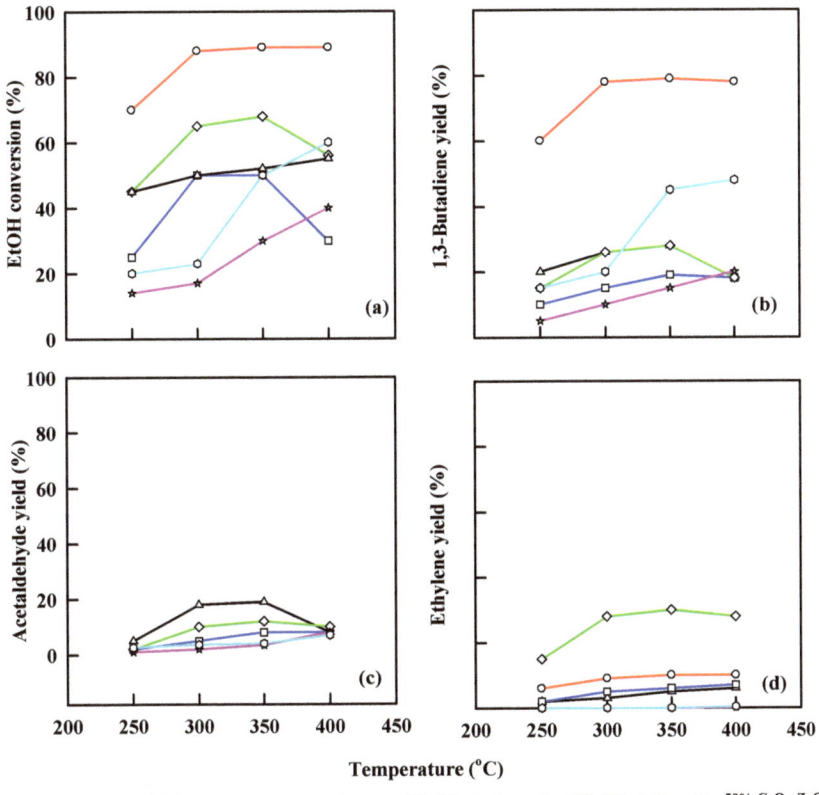

Figure 8. Plot of ethanol conversion (**a**), yields of 1,3-butadiene (**b**), acetaldehyde (**c**), and ethylene (**d**) vs. reaction temperature for different modified commercial catalysts. WHSV: 0.75 h^{-1}, catalyst: 0.4 g, N$_2$: 30 mL/min.

At 0.75 h^{-1} WHSV, 30% TiO$_2$-ZrO$_2$ yielded 40% ethanol and realized a 25% 1,3-butadiene yield, while acetaldehyde and ethylene yield were 6% and 4%, respectively, at 400 °C. CeO$_2$-ZrO$_2$ shows an ethanol conversion of 67% and 1,3-butadiene yield of 60% at 400 °C and 0.75 h^{-1} WHSV. Acetaldehyde yield increases linearly from 1.0% to 3.0% as temperature increases from 250–350 °C, while ethylene yield remains constant as temperature increases from 250–400 °C. The TiO$_2$/CeO$_2$-ZrO$_2$ catalytic system shows an increase in STY as temperature increases.

2.6. Comparing the Effect of Treated and Untreated Commercial Catalysts

The treatment of ZrO$_2$ parent catalyst with its substituent (SiO$_2$, TiO$_2$, CaO, or CeO$_2$) was conducted for ethanol dehydration to 1,3-butadiene, as shown in Table 6. The process used to treat the Zr-catalyst and its combination was conducted using 0.2 M NaOH, NH$_4$NO$_3$, and 0.5 M oxalic acid solution, as explained in the catalyst preparation methods. These reagents provide either desilication, basic and acidic sites, Lewis acidic and basic sites that are needed for the catalytic ethanol to 1,3-butadiene production [11,32,33]. By comparing the output of ethanol conversion, the treated-commercial catalyst yielded higher ethanol conversion than the un-treated commercial catalyst.

Table 6. Performance summary of treated and untreated commercial catalyst at 350 °C.

Catalyst	[a] Treated X_{EtOH} (%)	[b] Untreated X_{EtOH} (%)	[c] Treated $Y_{1,3\text{-butadiene}}$ (%)	[d] Untreated $Y_{1,3\text{-butadiene}}$ (%)
ZrO$_2$	60	34.5	30	10.6
Zr(OH)$_2$	50	30.4	20	10
2% CaO-ZrO$_2$	79.8	37.3	30	15.7
10% SiO$_2$-ZrO$_2$	88.8	40.6	75.9	20.3
30% TiO$_2$-ZrO$_2$	70.7	47	55	19.5
50% CeO$_2$-ZrO$_2$	74.5	47	65.3	16.8

[a] X_{EtOH}: ethanol conversion for treated catalyst, [b] X_{EtOH}: ethanol conversion for untreated catalyst, [c] $Y_{1,3\text{-butadiene}}$: 1,3-butadiene yield for treated catalyst, [d] $Y_{1,3\text{-butadiene}}$: 1,3-butadiene yield for untreated catalyst, WHSV = 0.75 h^{-1}.

2.7. Effect of Time on Stream on Ethanol Dehydration

Figure 9 show the stability of the following catalysts, ZrO$_2$, Zr(OH)$_2$, 2% CaO-ZrO$_2$, 30% TiO$_2$-ZrO$_2$, 50% CeO$_2$-ZrO$_2$ and 10% SiO$_2$-ZrO$_2$. The conversion of ethanol and selectivity of 1,3-butadiene decreases with time. The parent catalyst ZrO$_2$ as an acidic catalyst and Zr(OH)$_2$ as the basic catalyst show a lower ethanol conversion than 2% CaO-ZrO$_2$ and 10% SiO$_2$-ZrO$_2$. The results highlight that ZrO$_2$ has a selectivity of 57% at 10 h, and Zr(OH)$_2$ has a selectivity of 40% at 10 h time on streams. The introduction of CaO into the ZrO$_2$ catalytic system promotes the formation of extra dehydrogenation sites, making the acetaldehyde condensation site, hence the rate-limiting step. The presence of SiO$_2$ to ZrO$_2$ aids in increasing the yield of 1,3-butadiene. However, a limited amount of SiO$_2$ (10%) is needed during this reaction because it will promote the diffusion of the metals on the surface of the catalyst, and assists the formation of the active site for dehydration of ethanol and crotyl alcohol reaction to 1,3-butadiene; thus, the selectivity of 1,3-butadiene was 90%.

TiO$_2$, in combination with ZrO$_2$, as a catalytic system, gains more stability, and the selectivity of 1,3-butadiene was 87%. The presence of CeO$_2$ has a significant effect because CeO$_2$ has both Brønsted basic and Lewis basic sites, which aid in the creation of new active sites during the reaction and poisons some acidic sites at high temperatures 1,3-butadiene production to be realized. The 10% SiO$_2$-ZrO$_2$ catalytic system shows better stability than other commercial catalysts.

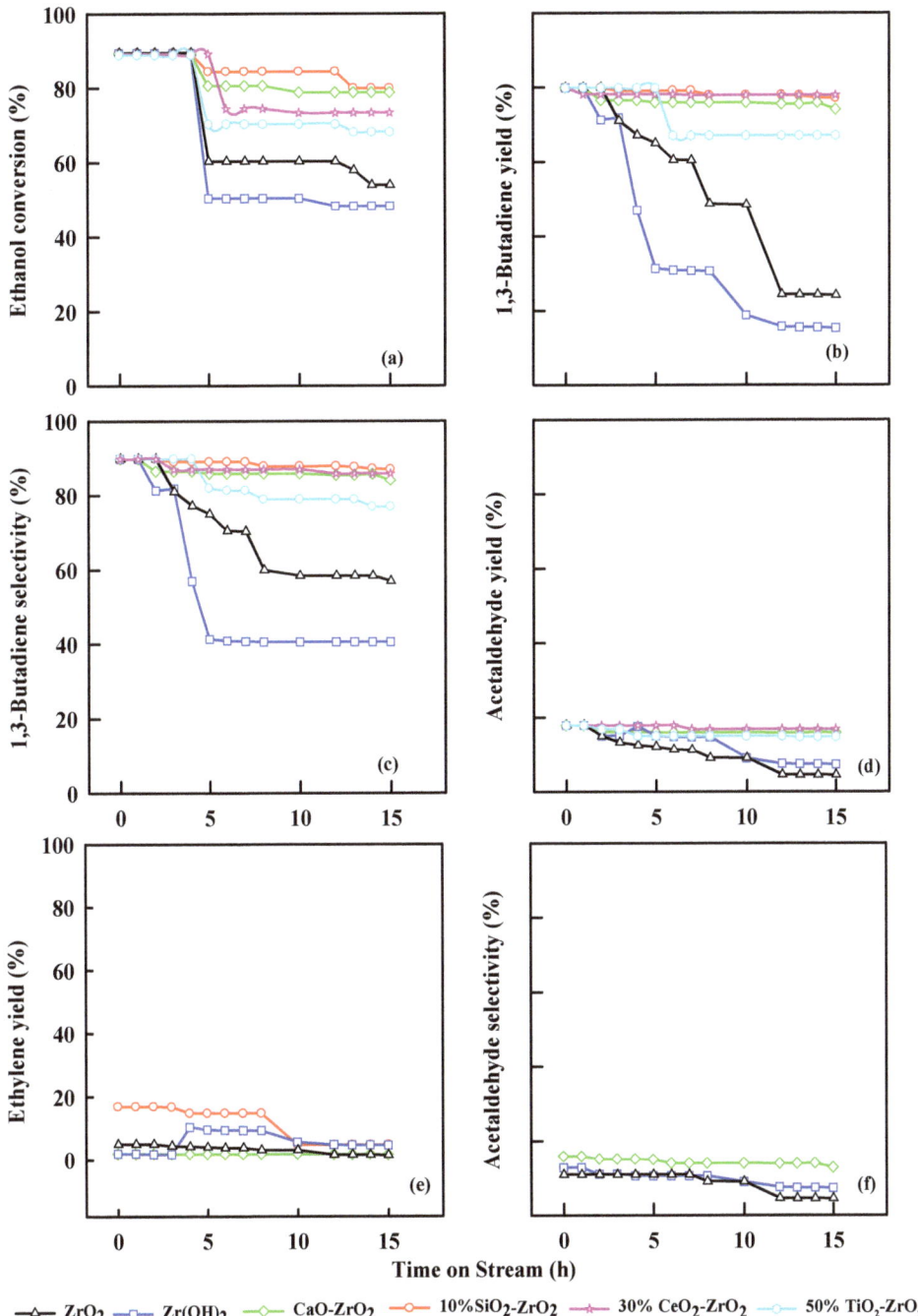

Figure 9. Conversion of ethanol (**a**), yield of 1,3-butadiene (**b**), selectivity of 1,3-butadiene (**c**), yields of acetaldehyde (**d**), yield of ethylene (**e**), and selectivity of acetaldehyde (**f**) with time on streams. WHSV: 2.5 h^{-1}, 0.4 g of catalyst, N_2 = 30 mL/min, Temp: 350 °C.

3. Materials and Methods

3.1. Materials

The catalysts of ZrO_2, $Zr(OH)_2$, $CaO-ZrO_2$, and 10% SiO_2-ZrO_2 were purchased from Daiichi Kigenso Kagaku Kogyo (Tokyo, Japan), ethanol (98%) from Echo Chemical (Miaoli, Taiwan), acetone and sodium hydroxide (NaOH) from Sigma Aldrich (St Louis, MO, USA), hydrochloric acid and ammonium phosphate from Sigma (Utah's Salt Lake, UT, USA), and 1,3-butadiene (10% in nitrogen) and ethylene (10% in nitrogen) from Ming Yang (Taoyuan, Taiwan). The commercial catalysts were treated with oxalic acid in combination with NaOH in a mole ratio of 2:1. The dried samples were then used for all the preparation of the catalysts.

3.1.1. Preparation of ZrO_2 Catalyst

Treated-ZrO_2 commercial catalyst (5 g) was mixed with 50 mL distilled water. The solution was filtered, and the residue was dried for 3 h in an oven. Later the residue was dissolved into dilute HCl (0.05 mol), filtered, and dried for 4 h. Finally, the mixture was calcined under air at 500 °C for 12 h.

3.1.2. Preparation of $Zr(OH)_2$ Catalyst

Treated-$Zr(OH)_2$ commercial catalyst (6 g) was added to 40 mL $ZrCl_4$ in a beaker and rested for 1 h, and then, 0.2 M NaOH was added and mixed vigorously for 30 min. The mixture was filtered, and the residue dried in an oven for 24 h at 60 °C. The solid white power $Zr(OH)_2$ was calcined at 550 °C for 24 h.

3.1.3. Preparation of 10% SiO_2-ZrO_2 Catalyst

Treated-10% SiO_2-ZrO_2 was pretreated using a 0.4 M of acidic oxalic solution. The commercial sample (2.0 g) was added to the NaOH aqueous solution (2.0 M, 100 mL). The solution was heated to 100 °C for 4 h. Then, the sample was dried at 150 °C for 8 h in an oven. The residue was later calcined at 500 °C for 24 h.

3.1.4. Preparation of 2% $CaO-ZrO_2$ Catalyst

Treated-$CaO-ZrO_2$ (7 g) was mixed with 0.5 M acetone in a 100 mL volumetric flask. The sample was later dried and gently added to 0.1 M of NaOH solution, and the solution rested for 3 h. Then, the solution was filtered and dried for 24 h in an oven. Finally, the dried sample was calcined at 450 °C for 24 h.

3.1.5. Preparation of 30% TiO_2-ZrO_2 Catalyst

Treated-30% TiO_2-ZrO_2 (4 g) was mixed with 100 mL distilled water and stirred for 1 h. The mixture was filtered and this was followed by the addition of 2.0 M NaOH in 100 mL aqueous solution, and the mixture was left to stand for 12 h. The solution was filtered and dried at 80 °C for 24 h. The sample was later calcined at 500 °C for 14 h.

3.1.6. Preparation of 50% CeO_2-ZrO_2 Catalyst

Treated-50% CeO_2-ZrO_2 consists of tetragonal ZrO_2, which in combination with CeO_2 produces CeO_2-ZrO_2. The commercial catalyst (5 g) was added into a $ZrO(NO_3)_2$ solution (100 mL, 0.5 M). The solution was stirred at 25 °C for 1 h and allowed to settle for 24 h. Later the precipitate was filtered and washed with deionized water and dried at 100 °C for 48 h. The sample was calcined at 500 °C for 5 h in air. Thus, the process was conducted to form CeO_2 with monoclinic ZrO_2 as support during the catalytic reaction. CeO_2 was impregnated into ZrO_2 to obtain precursors.

3.2. Dehydration of the Ethanol Using Fixed Bed Reactor

Ethanol dehydration to 1,3-butadiene is mainly an endothermic chemical reaction process involving more heat energy and higher temperatures, as shown in Figure 10. The reaction temperature is vital in this process because it gives the product selectivity. In this case, 1,3-butadiene is the target product while other products are expected, such as ethylene and acetaldehyde, mostly termed as byproducts. The main product expected for ethanol dehydration at lower temperatures was acetaldehyde or ethylene. At higher temperatures, such as 350 °C the yield of 1,3-butadiene was realized, and also the method of reactor design determines the selection of the main product. Following the specifics, dehydration of ethanol was conducted with a fixed-bed reactor.

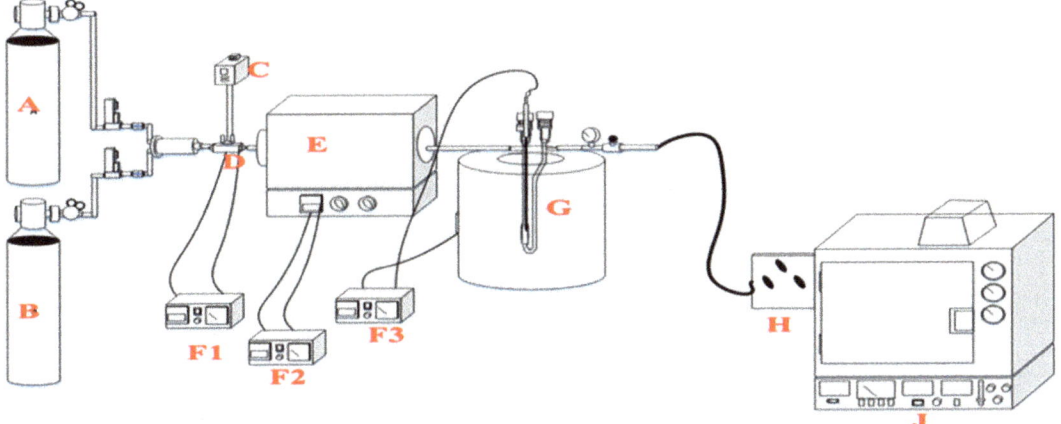

Figure 10. Schematic flow of the reactor design: A: N_2, B: H_2, C: pump, D: evaporator, E: evaporator, F1–F3: temperature controller, G: fixed-bed reactor in the furnace, H: gas sample, J: gas chromatography.

The determination of ethanol was determined with the aid of a gas chromatography–flame ionization detector (GC–FID; GC 14B, Shimadzu, Kyoto, Japan) equipped with a column of Porapak-Q-141023J (length, 3 m; diameter: 2 mm; and film, 1 µm; Quadrex, Bethany, CT, USA) under nitrogen carrier gas (30 mL/min), with N_2 as the gas carrier. The oven conditions were 60–150 °C in temperature with a 10 °C/min ramp rate. The sample was analyzed using a continuous system directly connected to the gas chromatogram. The data were collected depending on the time interval needed.

3.3. Analytical Method for Determination of Ethanol Conversion and 1,3-Butadiene Yield

The weight hourly space velocity (WHSV) is defined as the ratio of the hourly feed flow rate of the ethanol and water mixture to the catalyst weight. This work presents the catalytic performances in terms of ethanol conversion (%), product selectivity (%), and product yield (%) based on molar carbon. The ethanol conversion was calculated as

$$X(\%) = \frac{n_{0_{EtOH}} - n_{1_{EtOH}}}{n_{0_{EtOH}}} \times 100 \qquad (2)$$

where $n_{0_{EtOH}}$ is the number of moles fed into the reactor and $n_{1_{EtOH}}$ is the number of moles out of the reactor.

Selectivity is calculated by the following:

$$S_i\ (\%) = \frac{n_i \times c_i}{2(n_{0_{EtOH}} - n_{1_{EtOH}})} \times 100 \qquad (3)$$

where n_i is the number of moles of product i and c_i is the number of carbon atoms in product i (e.g., 1,3-BD is 4). The selectivity is given:

$$Y_i\ (\%) = S_i \times X \div 100 \qquad (4)$$

The WHSV is defined as the ratio of the hourly feed flow rate of ethanol and the catalyst weight:

$$\text{WHSV (h}^{-1}) = \frac{\text{Flow rate of ethanol } \left(\frac{g}{h}\right)}{\text{mass of catalyst (g)}} \tag{5}$$

The following activity measures can be used for comparative measurements such as catalyst screening, process parameter determination, optimization of catalyst production conditions, and deactivation studies. Catalysts are frequently studied in constantly operating test reactors, where conversion rates at a constant space velocity are correlated. The volume flow rate V_0 concerning the catalyst mass m_{cat} is defined as the space velocity.

$$\text{Space Velocity} = \frac{v_0}{m_{cat}} \; (\text{m}^3\text{kg}^{-1}\text{s}^{-1}) \tag{6}$$

If the catalyst mass in Equation (5) is replaced with the catalyst volume, the space velocity is proportional to the reciprocal of the residence time. A reactor's performance is often given relative to the catalyst mass or volume so that reactors of different sizes or construction can be compared. This quantity is known as the space-time yield (STY):

$$\text{STY} = \frac{\text{Desired product quantity}}{\text{Catalyst volume} \cdot \text{time}} (\text{gkg}^{-1}\text{h}^{-1}) \tag{7}$$

3.4. Catalyst Characterization

The characterization of the catalyst was aided with XPS, XRD, and TGA. X-ray diffraction (XRD) patterns were acquired on a D2 phaser X-ray diffractometer, using monochromatic Cu Kα radiation and scanning 2θ from 10° to 90°, and operated at 40 kV and 40 mA using 0.1 g of the catalyst. The relative crystallinity is calculated according to the aggregate intensities of the three peaks at 2θ of different degrees of angles. X-ray diffraction analysis is an important basis for qualitative analysis of the catalyst. This analysis could identify surface crystallinity and catalyst type. It also aids the identification of peak position and signifies the characteristic properties of each catalyst. The catalytic surface chemistry was studied using X-ray photoelectron spectroscopy (XPS). The sample analysis was undertaken with the aid of 5700C (Perkin Elmer, Akron, OH, USA) model Physical and Electronics apparatus, with MgKα radiation (1253.6 eV) using 0.1 g of the catalyst. The available data fitting of the XPS peak was achieved using Gaussian squares or the Lorentzian peak geometry. For the catalytic chemical reaction of ethanol to butadiene, the thermal property of the catalyst after the reaction was analyzed using thermogravimetric (TGA) (TA instruments Q50, USA). This suggestion is important to identify the coke deposition on the catalyst. The sample was heated from 25 °C to 800 °C at a ramp rate of 10 °C/min under nitrogen gas flow conditions.

3.5. Catalyst Testing

For the generation of 1,3-butadiene from ethanol, a fixed bed reactor system was used under standard temperature and pressure. A sample of 0.4 g of the catalyst sample was packed in the middle of the stainless tube (R 1/4 22 mm) and inserted into the furnace. Using powered SiO_2 was induced to increase the bed length, allowing the flow condition. 0.4 g of catalyst was inserted into the fixed bed. Later, ethanol was induced with nitrogen gas with a 30 cm^3/min flow rate bubbling into saturated ethanol vapor, carried into the fixed bed reactor. The temperature of the bubbler was alternated by varying the WHSV (0.75 h^{-1}, 1.25 h^{-1}, 2.5 h^{-1}, and 6.0 h^{-1}). Before the chemical reaction, the catalyst was activated by heating to 400 °C. This suggestion was made to prevent both reactant and product condensation. This heating process was conducted at 15 °C/min and kept at 400 °C for 1 h under a 30 cm^3/min flow rate under nitrogen conditions. The main products (1,3-butadiene, ethylene, and acetaldehyde) were monitored using GC-14B, FID detector.

4. Conclusions

ZrO_2 combined with SiO_2 and CaO showed high catalytic activity and stable ethanol dehydration to 1,3-butadiene in terms of yield and selectivity. At WHSV of 2.5 h^{-1}, SiO_2-ZrO_2 has a selectivity of 94% of 1,3-butadiene at 350 °C and 14 h time on stream. 2% CaO-ZrO_2 has 90% selectivity of 1,3-butadiene at 350 °C and 12 h time on stream. When the reaction conditions were at WHSV of 1.25 h^{-1} and a temperature of 300 °C, the ethanol conversion was 87.9% for 10% SiO_2-ZrO_2 with a 79% yield and 85% selectivity of 1,3-butadiene. The 10% SiO_2-ZrO_2 catalyst shows no ethylene yield, as the catalytic reaction tends to forward the yield of 1,3-butadiene. The $Zr(OH)_2$ catalyst shows a low ethanol conversion at 14 h time on stream with a 20% ethanol conversion. $Zr(OH)_2$ exposed the basic nature of the Zr-containing catalyst, and for the yield of 1,3-butadiene to be realized, a catalyst should possess a balance between the acidic sites and Lewis sites.

The 2% CaO-ZrO_2 catalyst showed a low coke formation, which means it was more thermal stable than the rest, the order of thermal stability is as follows: 2% CaO-ZrO_2 > ZrO_2 > 10% SiO_2-ZrO_2 > $Zr(OH)_2$. XPS shows the carbon particle deposition in the catalyst affects the coke formation process, as $Zr(OH)_2$ has the highest carbon deposition (28.95%), hence the lowest stability. While 10% SiO_2-ZrO_2 has the lowest carbon deposition; hence, the highest instability and highest ethanol conversion.

Supplementary Materials: The following supporting information can be downloaded at: https://www.mdpi.com/article/10.3390/catal12070766/s1, Figure S1 Low-temperature N_2 adsorption-desorption curves. Figure S2 TGA curves of 50% CeO_2-ZrO_2 and 30% TiO_2-ZrO_2.

Author Contributions: Conceptualization, H.-S.W.; investigation, A.A.B.; writing—original draft preparation, A.A.B.; writing—review and editing, H.-S.W.; supervision, H.-S.W. All authors have read and agreed to the published version of the manuscript.

Funding: The ministry of science and technology of Taiwan: MOST 109-2221-E-155-009.

Institutional Review Board Statement: Not applicable.

Informed Consent Statement: Not applicable.

Data Availability Statement: No applicable.

Conflicts of Interest: The authors declare no conflict of interest.

References

1. Ochoa, J.V.; Bandinelli, C.; Vozniuk, O.; Chieregato, A.; Malmusi, A.; Recchi, C.; Cavani, F. An analysis of the chemical, physical and reactivity features of MgO–SiO_2 catalysts for butadiene synthesis with the Lebedev process. *Green Chem.* **2016**, *18*, 1653–1663. [CrossRef]
2. Talalay, A.; Talalay, L.S.K. The Russian synthetic rubber from alcohol. A survey of the chemistry and technology of the lebedev process for producing sodiumbutadiene polymers. *Rubber Chem. Technol.* **1942**, *15*, 403–429. [CrossRef]
3. Bhattacharyya, S.K.; Ganguly, N.D. One-step catalytic conversion of ethanol to butadiene in the fixed bed. I. Single-oxide catalysts. *J. Appl. Chem.* **1962**, *12*, 97–104. [CrossRef]
4. Boronat, M.; Concepción, P.; Corma, A.; Navarro, M.T.; Renz, M.; Valencia, S. Reactivity in the confined spaces of zeolites: The interplay between spectroscopy and theory to develop structure–activity relationships for catalysis. *Phys. Chem. Chem. Phys.* **2009**, *11*, 2876–2884. [CrossRef]
5. Boronat, M.; Concepción, P.; Corma, A.; Renz, M.; Valencia, S. Determination of the catalytically active oxida-tion Lewis acid sites in Sn-beta zeolites, and their optimisation by the combination of theoretical and experimental studies. *J. Catal.* **2005**, *234*, 111–118. [CrossRef]
6. Boukha, Z.; Fitian, L.; López-Haro, M.; Mora, M.; Ruiz, J.R.; Jiménez-Sanchidrián, C.; Blanco, G.; Calvino, J.J.; Cifredo, G.A.; Trasobares, S.; et al. Influence of the calcination temperature on the nano-structural properties, surface basicity, and catalytic behavior of alumina-supported lanthana samples. *J. Catal.* **2010**, *272*, 121–130. [CrossRef]
7. Cabello González, G.M.; Villanueva Perales, A.L.; Martínez, A.; Campoy, M.; Vidal-Barrero, F. Conversion of aqueous ethanol/acetaldehyde mixtures into 1,3-butadiene over a mesostructured Ta-SBA-15 catalyst: Effect of reaction conditions and kinetic modelling. *Fuel Process. Technol.* **2022**, *226*, 107092. [CrossRef]
8. Aimon, D.; Panier, E. La mise en pratique de l'économie circulaire chez michelin. *Ann. Mines-Responsab. Environ.* **2014**, *76*, 38–44. [CrossRef]

9. Budagumpi, S.; Kim, K.-H.; Kim, I. Catalytic and coordination facets of single-site non-metallocene organometallic catalysts with N-heterocyclic scaffolds employed in olefin polymerization. *Coord. Chem. Rev.* **2011**, *255*, 2785–2809. [CrossRef]
10. Prillaman, J.T.; Miyake, N.; Davis, R.J. Calcium Phosphate Catalysts for Ethanol Coupling to Butanol and Butadiene. *Catal. Lett.* **2020**, *151*, 648–657. [CrossRef]
11. Pomalaza, G.; Arango, P.; Capron, M.; Dumeignil, F.Y. Ethanol-to-Butadiene: A Review. *Catal. Sci. Technol.* **2020**, *10*, 4860–4911. [CrossRef]
12. Gao, M.; Jiang, H.; Zhang, M. Influences of Interactive Effect Between ZrO_2 and nano-SiO_2 on the formation of 1,3-butadiene from ethanol and acetaldehyde. *Catal. Surv. Asia* **2020**, *24*, 115–122. [CrossRef]
13. Zhang, M.; Guan, X.; Zhuang, J.; Yu, Y. Insights into the mechanism of ethanol conversion into 1,3-butadiene on Zr-β zeolite. *Appl. Surf. Sci.* **2021**, *579*, 152212. [CrossRef]
14. Kyriienko, P.I.; Larina, O.V.; Balakin, D.Y.; Vorokhta, M.; Khalakhan, I.; Sergiienko, S.A.; Soloviev, S.O.; Orlyk, S.M. The effect of lanthanum in Cu/La(-Zr)-Si oxide catalysts for aqueous ethanol conversion into 1,3-butadiene. *Mol. Catal.* **2022**, *518*, 112096. [CrossRef]
15. Sushkevich, V.L.; Ivanova, I.I. Ag-Promoted ZrBEA Zeolites Obtained by Post-Synthetic Modification for Conversion of Ethanol to Butadiene. *ChemSusChem* **2016**, *9*, 2216–2225. [CrossRef]
16. Sushkevich, V.L.; Ivanova, I.I.; Ordomsky, V.V.; Taarning, E. Design of a metal-promoted oxide catalyst for the selective synthesis of butadiene from ethanol. *ChemSusChem* **2014**, *7*, 2527–2536. [CrossRef]
17. Kyriienko, P.I.; Larina, O.V.; Balakin, D.Y.; Stetsuk, A.O.; Nychiporuk, Y.M.; Soloviev, S.O.; Orlyk, S.M. 1,3-Butadiene production from aqueous ethanol over ZnO/MgO-SiO_2 catalysts: Insight into H_2O effect on catalytic performance. *Appl. Catal. A Gen.* **2021**, *616*, 118081. [CrossRef]
18. Cheong, J.L.; Shao, Y.; Tan, S.J.R.; Li, X.; Zhang, Y.; Lee, S.S. Highly Active and Selective Zr/MCF Catalyst for Production of 1,3-Butadiene from Ethanol in a Dual Fixed Bed Reactor System. *ACS Sustain. Chem. Eng.* **2016**, *4*, 4887–4894. [CrossRef]
19. Pomalaza, G.; Capron, M.; Ordomsky, V.; Dumeignil, F. Recent breakthroughs in the conversion of ethanol to butadiene. *Catalysts* **2016**, *6*, 203. [CrossRef]
20. Guan, Y.; Hensen, E.J.M. Ethanol dehydrogenation by gold catalysts: The effect of the gold particle size and the presence of oxygen. *Appl. Catal. A Gen.* **2009**, *361*, 49–56. [CrossRef]
21. Han, Z.; Li, X.; Zhang, M.; Liu, Z.; Gao, M. Sol–gel synthesis of ZrO_2–SiO_2 catalysts for the transformation of bioethanol and acetaldehyde into 1,3-butadiene. *RSC Adv.* **2015**, *5*, 103982–103988. [CrossRef]
22. Baylon, R.A.L.; Sun, J.; Wang, Y. Conversion of ethanol to 1,3-butadiene over Na doped ZnxZryOz mixed metal oxides. *Catal. Today* **2016**, *259*, 446–452. [CrossRef]
23. Kuznetsov, P.; Belskaya, N.A.; Tomilova, T.A.; Beregovtsova, N.G.; Sharypov, V. Peculiarities in methanol action during liquefaction of different coal type. *Fuel* **1989**, *68*, 1580–1583. [CrossRef]
24. Kongwudthiti, S.; Praserthdam, P.; Tanakulrungsank, W.; Inoue, M. The influence of Si–O–Zr bonds on the crys-tal-growth inhibition of zirconia prepared by the glycothermal method. *J. Mater. Process. Technol.* **2003**, *136*, 186–189. [CrossRef]
25. Feng, X.-J.; Lu, X.-B.; He, R. Tertiary amino group covalently bonded to MCM-41 silica as heterogeneous cata-lyst for the continuous synthesis of dimethyl carbonate from methanol and ethylene carbonate. *Appl. Catal. A Gen.* **2004**, *272*, 347–352. [CrossRef]
26. Zou, H.; Lin, Y. Structural and surface chemical properties of sol–gel derived TiO_2–ZrO_2 oxides. *Appl. Catal. A Gen.* **2004**, *265*, 35–42. [CrossRef]
27. Suda, A.; Yamamura, K.; Morikawa, A.; Nagai, Y.; Sobukawa, H.; Ukyo, Y.; Shinjo, H. Atmospheric pressure sol-vothermal synthesis of ceria–zirconia solid solutions and their large oxygen storage capacity. *J. Mater. Sci.* **2008**, *43*, 2258–2262. [CrossRef]
28. Chen, Z.; Zhang, X.; Yang, F.; Peng, H.; Zhang, X.; Zhu, S.; Che, L. Deactivation of a Y-zeolite based catalyst with coke evolution during the catalytic pyrolysis of polyethylene for fuel oil. *Appl. Catal. A Gen.* **2021**, *609*, 117873. [CrossRef]
29. Zhang, M.; Qin, Y.; Jiang, H.; Wang, L. Protective desilication of β zeolite: A mechanism study and its applica-tion in ethanol-acetaldehyde to 1, 3-butadiene. *Microporous Mesoporous Mater.* **2021**, *326*, 111359. [CrossRef]
30. Wu, C.-Y.; Wu, H.-S. Ethylene formation from ethanol dehydration using ZSM-5 catalyst. *ACS Omega* **2017**, *2*, 4287–4296. [CrossRef]
31. Sekiguchi, Y.; Akiyama, S.; Urakawa, W.; Koyama, T.r.; Miyaji, A.; Motokura, K.; Baba, T. One-step catalytic conversion of ethanol into 1,3-butadiene using zinc-containing talc. *Catal. Commun.* **2015**, *68*, 20–24. [CrossRef]
32. Jones, M.D.; Keir, C.G.; Di Iulio, C.; Robertson, R.A.; Williams, C.V.; Apperley, D.C. Investigations into the con-version of ethanol into 1, 3-butadiene. *Catal. Sci. Technol.* **2011**, *1*, 267–272. [CrossRef]
33. Lesmana, D.; Wu, H.-S. Modified oxalic acid co-precipitation method for preparing Cu/ZnO/Al_2O_3/Cr_2O_3/CeO_2 catalysts for the OR (oxidative reforming) of M (methanol) to produce H2 (hydrogen) gas. *Energy* **2014**, *69*, 769–777. [CrossRef]

Review

Current Developments in the Effective Removal of Environmental Pollutants through Photocatalytic Degradation Using Nanomaterials

Chandhinipriya Sivaraman [1], Shankar Vijayalakshmi [2], Estelle Leonard [3], Suresh Sagadevan [4,*] and Ranjitha Jambulingam [2,*]

1. School of Advanced Sciences, Vellore Institute of Technology, Vellore 632014, India; schandhinipriya@gmail.com
2. CO2 Research and Green Technologies Centre, Vellore Institute of Technology, Vellore 632014, India; vijimicro21@gmail.com
3. Laboratoire TIMR UTC-ESCOM, Centre de Recherche de Royallieu, Rue du Docteur Schweitzer, CS 60319, CEDEX, F-60203 Compiègne, France; e.leonard@escom.fr
4. Nanotechnology & Catalysis Research Centre, University of Malaya, Kuala Lumpur 50603, Malaysia
* Correspondence: drsureshsagadevan@um.edu.my (S.S.); jranji16@gmail.com (R.J.)

Citation: Sivaraman, C.; Vijayalakshmi, S.; Leonard, E.; Sagadevan, S.; Jambulingam, R. Current Developments in the Effective Removal of Environmental Pollutants through Photocatalytic Degradation Using Nanomaterials. *Catalysts* **2022**, *12*, 544. https://doi.org/10.3390/catal12050544

Academic Editor: Hugo de Lasa

Received: 25 February 2022
Accepted: 11 May 2022
Published: 17 May 2022

Publisher's Note: MDPI stays neutral with regard to jurisdictional claims in published maps and institutional affiliations.

Copyright: © 2022 by the authors. Licensee MDPI, Basel, Switzerland. This article is an open access article distributed under the terms and conditions of the Creative Commons Attribution (CC BY) license (https://creativecommons.org/licenses/by/4.0/).

Abstract: Photocatalysis plays a prominent role in the protection of the environment from recalcitrant pollutants by reducing hazardous wastes. Among the different methods of choice, photocatalysis mediated through nanomaterials is the most widely used and economical method for removing pollutants from wastewater. Recently, worldwide researchers focused their research on eco-friendly and sustainable environmental aspects. Wastewater contamination is one of the major threats coming from industrial processes, compared to other environmental issues. Much research is concerned with the advanced development of technology for treating wastewater discharged from various industries. Water treatment using photocatalysis is prominent because of its degradation capacity to convert pollutants into non-toxic biodegradable products. Photocatalysts are cheap, and are now emerging slowly in the research field. This review paper elaborates in detail on the metal oxides used as a nano photocatalysts in the various type of pollutant degradation. The progress of research into metal oxide nanoparticles, and their application as photocatalysts in organic pollutant degradation, were highlighted. As a final consideration, the challenges and future perspectives of photocatalysts were analyzed. The application of nano-based materials can be a new horizon in the use of photocatalysts in the near future for organic pollutant degradation.

Keywords: organic contaminants; nanomaterials; wastewater; photocatalyst; degradation; optimizing parameters

1. Introduction

Massive advancements in nanoscience and technology mean they emerged as promising solutions for environmental clean-up and the production of energy in recent decades. Nanomaterials (NMs) have opened up many new possibilities for a variety of manufacturing/industrial applications over the years, including wastewater treatment and the removal of hazardous contaminants from the atmosphere. The advancement in industrialization leads to the release of toxins, with the emission of hazardous chemicals into the atmosphere. In this regard, methods such as immobilization, biological and chemical oxidation, and incineration were widely used to treat a variety of organic and toxic industrial contaminants. Nanomaterials have the peculiarity of changing the characteristics of materials through their optical, magnetic, and electrical properties, and are helpful in many processes and applications [1]. Nanomaterials are used in many fields, from electricity to medicine, because of their unique physicochemical and biological properties [2]. Recently, visible light-induced heterogeneous photocatalysis developed rapidly, due to its advantage in the implementation of environmental remediation, particularly in wastewater treatment [3]. Nowadays, anthropogenic chemicals used in agriculture, medicine, the military, and industry directly enter the water stream easily, which causes an adverse effect on the environment, and risks the contamination of both surface and groundwater [4]. As a result of this water contamination, endocrine disruptors interfere with the normal hormonal system, which causes adverse health effects such as birth defects and developmental disorders in children, infertility, and cancerous tumors, and also causes several water-borne diseases. According to the World Health Organization, half of the world's population will suffer due to the water crisis by 2025. Environmental pollution is one of the most consequential issues currently faced all over the world and could be resolved by creating the conditions to achieve a clean and healthy environment for a better life in the world. In recent decades, population and global production growth resulted in much higher production of chemicals, due to their daily entry into the environment and resistance to biodegradation, resulting in the generation of hazards for various species. To prevent water and environmental pollution caused by the arrival of polluted industrial effluents, appropriate strategies for their treatment and reuse must be developed. Today, safe and hygienic drinking water is a unique requirement of the global health community. The clustering of densely populated and industrial areas close to water resources magnified global issues. The new approach, e.g., via an oxidative pathway, makes a distinguishing change in the removal of environmental pollutants. Photocatalysis is applicable at room temperature and pressure consumes less energy and profits from process simplicity. Several technologies are available in the wastewater treatment process, such as electrodialysis [5], membrane filtration [6], precipitation adsorption [7], electrochemical reduction [8], and electrodeionization [9]. They are very expensive and complicated, and by transferring pollutants between fluids, various wastes and by-products are generated that make it difficult to treat wastewater. Recently, photocatalysis became a viable technology for the treatment of various pollutants present in wastewater [10].

Photocatalytic reactors may play an increasingly important role in new technologies for the filtration of organic-polluted water [11–13]. The degradation of different organic contaminants with better competence utilizes heterogeneous photocatalyst-based nanoparticles. Photocatalysis gained a lot of interest in recent years, because of green energy and environmental cleanup. As a result, there are numerous reviews on the subject, focused on various types of photocatalysts and photocatalyst applications. Furthermore, there were few basic developments in the concept, and no notable breakthroughs were observed in photocatalyst plans in the past five years. There is still much work to do, both in terms of making these materials practical (which is debatable for some applications), and in terms of improving our understanding of the complex processes, particularly in some of the more complicated ternary or quaternary photocatalysts proposed. Since key works by Honda and Fujishima in 1972, and Reiche and Bard in 1979 [14,15], there was a surge in interest in photocatalysis. Many other photocatalyst materials and uses were studied, but commercial photocatalysis applications were uncommon. Low photocatalytic activity, particularly under visible or solar illumination, is usually blamed for the lack of commercial uses. As a result, significant resources were invested in the development of improved photocatalyst materials. Material development techniques for different types of photocatalysts were focused on maximizing efficiency by targeting one or more phases in the photocatalytic reaction. Several studies reported on the treatment of wastewater using different photocatalysts. The photocatalytic degradation of pollutants is mainly focused on the formation of highly reactive hydroxyl radical ions. These photocatalytic reactions are triggered by the free radical mechanism initiated by the interaction of photons using the catalysts. Therefore, in the present review paper, we mainly focused on the recent advances in photocatalytic pollutant removal from wastewater, and also elaborated in detail on the factors affecting the performance of the photocatalytic degradation of pollutants to remove them from wastewater.

2. Photocatalytic Degradation Mechanism of Dyes

A photocatalytic reaction is primarily determined by the wavelength of light (photon) energy and the catalyst. Nanomaterials used as catalysts, such as NiO, TiO_2, ZnO, ZnS, and others, are referred to as nanocatalysts. The light can be irradiated directly or indirectly, as a result of the catalyst reacting with the dye. The photocatalytic mechanism associates dye degradation with the redox capabilities, or potential, of dyes and the energy level of the conduction band of the semiconductor, or nanomaterial, used. Photocatalysis degrades dyes via photosensitization, i.e., direct, or self, dye degradation and photo-oxidation by a reactive species (catalyst), or indirect dye degradation. Both photocatalytic mechanisms rely on electronic structures, specifically the band structure of the catalyst and the dye. Due to their electronic structure, which is described by a filled valence band and an empty conduction band, nanocatalysts act as chemical activators for the illumination of light-animated redox processes. The photosensitization mechanism, also known as the direct mechanism of dye degradation, absorbs visible light. The dye is excited from the ground state to the triplet excited state, using visible light photons with wavelengths greater than 400 nm. An electron addition into the conduction band of nanocatalysts, transferred from the valence band, converts this excited state of the original dye species into a semi-oxidized radical cation (Dye^+). The reaction between these trapped e-/h+ pairs and the dissolved oxygen in the system results in the formation of superoxide radical anions (O_2^-), and the formation of hydroxyl radicals (OH). In nature, this hydroxyl radical is non-reactive, and is primarily responsible for the oxidation of the organic compounds represented by Equations (1) and (2) below:

$$\text{Dye (ground state)} + h\nu \text{ (visible light)} \rightarrow \text{dye*} \text{ (triplet state)} \quad (1)$$

$$\text{Dye*} + \text{nanocatalyst} \rightarrow \text{dye}^+\text{(cation)} + \text{nanocatalyst} - \text{(anion)} \quad (2)$$

According to several researchers, visible light acts as a driving source in photosensitization, which occurs at a very slow rate. In contrast, an indirect mechanism, known as photo-oxidation/photocatalysis, in which a catalyst sensitizes the chemical reaction for dye degradation, is found to be more prevalent than a direct mechanism. The mechanism of dye degradation is based primarily on oxidation and the reduction of the photocatalyst, as shown in Figure 1. When photons of light strike a material, they excite electrons from the valence band to the conduction band, which results in the development of electron-hole pairs. The electrons in the conduction band react with the oxygen molecule to form superoxide radical anions; however, the holes in the valence band react with effluent water to form hydroxyl radicals.

Basic Principles of Z-Scheme Photocatalysis

In photocatalysis, the Z-scheme represents/mimics the natural photosynthesis system, which has advantages such as charge separation and delayed recombination, which increases the light harvest and improves redox ability [16], as shown in Figure 2. The light absorption and production of photogenerated electron-hole pairs are the first steps. The carriers of photogenerated electrons then move to the surface, where they recombine or participate in surface redox processes. By ensuring effective charge separation, and increasing surface redox reactions, photocatalytic performance improves to maximize light absorption, maximize charge transfer at the surface, and minimize recombination. Common techniques investigated for performance improvement include morphology optimization (which can affect the surface active sites as well as charge separation), doping (which can reduce the bandgap, and sometimes has negative effects on recombination losses), using sensitizers and/or co-catalysts (to increase visible absorption, as well as provide more active sites and affect carrier dynamics), and using different materials. Photocatalytic reactors can play an efficient role in novel technologies for the purification of water polluted with organic chemicals [13]. The degradation of various organic pollutants, using nanoparticle-based heterogeneous photocatalysts with higher efficiency, is reported, and shown in Table 1.

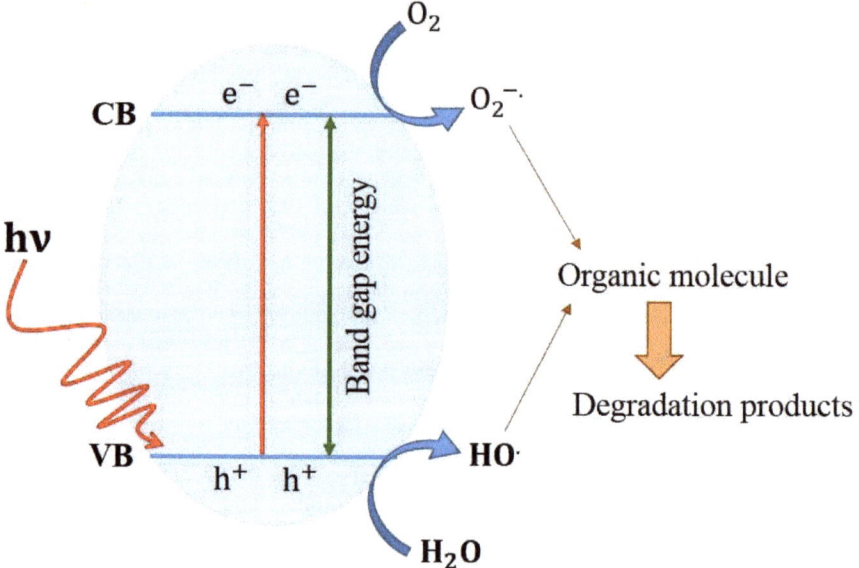

Figure 1. Schematic diagram of photocatalytic process.

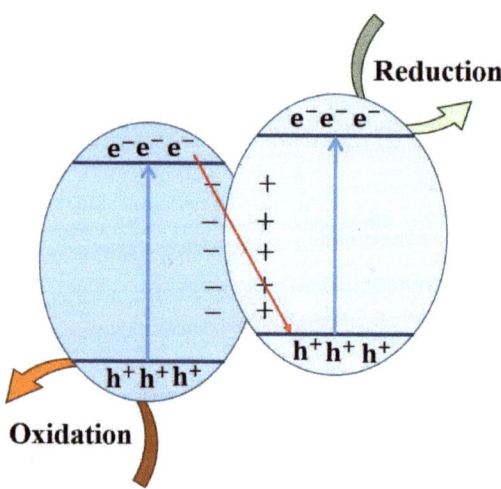

Figure 2. Schematic diagram of Z-scheme in photocatalysis.

Table 1. Degradation of organic pollutants using photocatalyst-based nanomaterial.

Materials	Pollutant	Degradation Efficiency	References
CuO nanosheets	Allura Red AC	~96.99% in 6 min	[17]
MFe_2O_4: (M = Co, Ni, Cu, Zn)	Methyl red Methyl orange Methylene blue Bromo green	78% in 50 min 92% in 50 min 89% in 50 min 93% in 50 min	[18]
α-$Bi_4V_2O_{11}$	Rhodamine B	100% in 45 min	[19]
m3D–TiO_2–HP Graphene	Hexavalent Chromium	96% in 70 min	[20]
TiO_2–graphene	Acid Black 1 dye	96% in 40 min	[21]
TiO_2–graphite composite	Paracetamol	100% in 120 min	[22]
Mg–ZnO–Al_2O_3	Caffeine	98.9% after 70 min	[23]
Zr/Ag–ZnO	Acid Black 1 dye	99.3% after 40 min	[24]
CeO_2	Yellow 6G dye	100% within 30 min	[25]
Al_2O_3–NP/SnO_2	Methyl orange	93.95% in 50 min	[26]
TiO_2 Degussa P25	Rhodamine B	33% in 180 min	[27]
CuO–GO/TiO_2	2-Chlorophenol	86% in 210 min	[28]
Copper nanoparticles	Methylene blue	91.53% in 30 min	[29]
Copper nanoparticles	Congo red	84.89% in 30 min	[29]
CuO nanorods	Reactive Black Dye	98% in 300 min	[30]
Cu/Cu (OH)$_2$	Rhodamine B	99.99% in 120 min	[31]
CuO–Cu_2O/GO	Tetracycline	90% after 120 min	[32]
Copper nanoparticles	Phenyl red	99.62% in 15 min	[33]
Cds/CuS	Methyl orange	93% in 150 min	[34]
Bismuth-doped copper aluminate	Methylene blue	99.9% in 60 min	[35]

3. Removal of Polybrominated Diphenyl Ethers (PBDEs)

Polybrominated diphenyl ethers (PBDEs) are the second most commonly used BFRs, and their molecular structure is similar to polychlorinated biphenyls (PCBs). In general, these BDEs are available commercially as mixtures in three different forms, namely, penta-, octa-, and deca-mixes. Penta-BDE is used in polyurethane foams and textiles; octa-BDE is used in styrenes, polycarbonates, and thermosets; and deca-BDE is used in synthetic textiles and electronics. As a successful replacement for PCBs, these PBDEs are found in all levels of ecosystems, and are able to redistribute globally among these ecosystems. They pose a threat to the human population, indigenous peoples, and fish consumers, as they bio-accumulate in the food chain and are highly lipophilic, similar to dioxins and PCBs [36]. Unfortunately, this accumulation of PBDEs affects motor skills and disturbs the metabolism of the thyroid hormone; hence, it is classified as a high-risk pollutant that causes serious environmental pollution. Dietary intake and dust ingestion are the dominant human exposure pathways. PBDEs were widely detected in human samples, especially in human serum and human milk. Data shows that PBDEs are generally declining in human samples worldwide, as a result of their phasing out. Due to the common use of PBDEs, their levels in humans from the USA are generally higher than that in other countries. High concentrations of PBDEs were detected in humans from PBDE production regions and e-waste recycling sites. BDE-47, -153, and -99 were proven to be the primary congeners in humans. Human toxicity data demonstrates that PBDEs have extensive endocrine disruption effects, developmental effects, and carcinogenic effects among different populations, as shown in Figure 3. Besides bio-accumulation, exposure to this toxic chemical during its production, processing, and recycling causes adverse effects in human beings. In fact, air and dust are proven to show measurable PBDE concentrations, and inhaling it could account for up to one-quarter of total exposure. Generally, several remediation techniques are followed for remediating this harmful chemical; and these include hydrothermal, adsorption, photolysis, advanced oxidative processes, and photocatalytic degradation, etc. Specifically, photocatalysis and photocatalytic degradation are regarded as the most common and reported methods for remediating these PBDEs. The most commonly studied PBDEs include their congeners BDE-47 and BDE-209, owing to both their toxicity and their intermediate products. To begin with, Azri et al. (2016) [37] use a tri-metallic catalyst, $Cu/Ni/TiO_2/PVC$, prepared using sol-gel and a hydrothermal method, and report the rate of degradation of PBDE as 65.82% [37]. Likewise, Wang et al. (2019) [38] use a metal-doped TiO_2 photocatalyst for degrading dibrominated diphenyl ethers under photocatalytic degradation; while, Li et al. (2014) [39] carry out the photocatalytic debromination of PBDEs using a Pd/TiO_2 catalyst, and conclude that TiO_2 enhances the rate of debromination upon increasing the loading of palladium. Similarly, Lei et al. (2016) [40] prepare debrominated PBDEs using $Ag-TiO_2$ under the influence of UV light, and note that the debromination is rapid. This study concludes that the effectiveness of debromination PBDEs using metal-doped TiO_2 is enhanced based on the metal additive; but reduces drastically upon using metal-doped TiO_2 catalysts exposed to air [39,40]. In addition, replacing Pd with Cu enhances the rate of electron transfer from the conduction band of TiO_2 to PBDEs [41]. It is worth mentioning that the degradation of PBDE is carried out in two different processes, with electrons sourced from striking photons favoring reduction debromination; holes or •OH generated as a result of photocatalytic reaction favor oxidation debromination. Moreover, the redox photoreduction of PBDEs using nanomaterial-based catalysts is adversely affected by the recombination of holes and electrons. However, adding water and irradiation using UV light simply enhances the rate of oxidative degradation of PBDEs, especially BDE-209 [42]. The degradation of various polybrominated diphenyl ethers, using nanoparticle-based heterogeneous photocatalysts with higher efficiency, is reported and shown in Table 2.

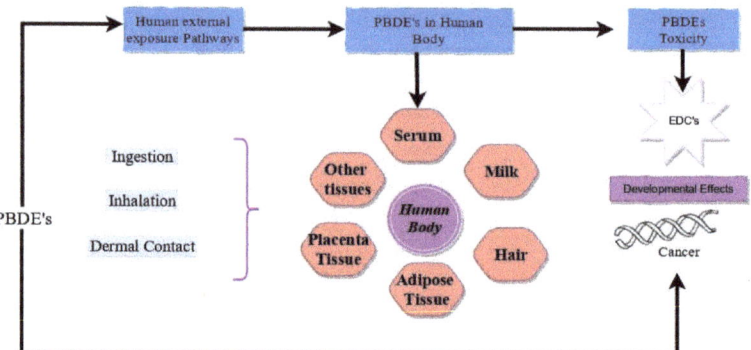

Figure 3. Human exposure to PBDEs and their health outcomes.

Table 2. Degradation of various polybrominated diphenyl ethers using nanoparticle-based heterogeneous photocatalysts.

Pollutant	Photocatalyst	Light	Results	References
BDE-209	TiO_2	UV lamp	Debromination efficiency of BDE-209 achieved, up to 95.6%	[43]
BDE-209	RGO/TiO_2	UV lamp	Debromination efficiency of BDE-209 achieved, up to 59.4%	[44]
BDE-47	RGO/TiO_2	Xe lamp	Debromination efficiency of BDE-47 achieved, up to 25%	[45]
BDE-209	CuO/TiO_2	Xe lamp	Debromination of ten PBDEs was achieved under anaerobic conditions	[46]
BDE-209	FeOCN-x	Visible Xe lamp	Higher photocatalytic activity for debromination of PBDEs was achieved	[47]
BDE-209	$AgI-TiO_2$	Xe lamp	The addition of silver iodide to the surface of TiO_2 increased the debromination efficiency of BDE-209	[48]
BDE-47	$Ag@Ag_3PO_4/g-C_3N_4/rGO$	UV lamp	Debromination efficiency of BDE-47 was achieved, up to 93.4%	[49]
BDE-47	Nickel nanoparticles	Visible lamp	Debromination of BDE-47 was achieved completely under visible irradiation	[50]
BDE-47	Ag/TiO_2	UV lamp	Ag/TiO_2 addition accelerates BDE-47 photodegradation efficiency	[51]

4. Removal of Phthalates and Their Derivative

Phthalates, or phthalate esters (PAEs), are di-esters of phthalic acid (1,2-benzene dicarboxylic acid) and are used as plasticizers for polymers to reduce their glass transition temperature, in order to induce softness and workability. In terms of classification, low-molecular-weight phthalates (dimethyl phthalate (DMP), diethyl phthalate (DEP), and dibutyl phthalate (DBP)) are used in small to medium scale commercial applications (plastic containers, materials packaging, personal care products, solvents, adhesives, lubricants, coatings, and varnishes); high-molecular-weight phthalates (such as di-n-octyl phthalate (DOP) and di-(2-ethylhexyl) phthalate (DEHP)) are used in construction and furniture industries, as shown in Figure 4. Interestingly, these phthalates are chemically bonded during polymer manufacturing, and remain inert in leaching out into the environment; they create physical bonds upon being used as plasticizers, thereby causing them to leak into the environment. As a result, these phthalates are found all around the globe and are treated as harmful environmental pollutants, as they disrupt the endocrine glands, causing severe disturbances in the functioning of hormones inside the human body, in addition to causing genetic and reproductive abnormalities in different living organisms [52–54]. The aforementioned phthalate compounds are deemed as high-risk pollutants and are cited as the predominant source of phthalate exposure, via inhalation, dermal contacts, and consuming contaminated foods.

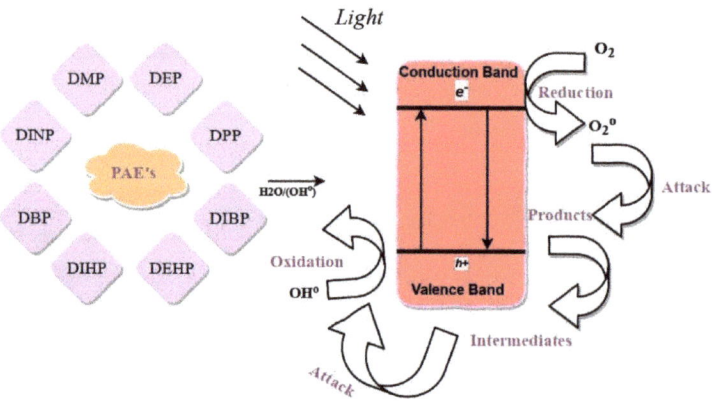

Figure 4. Photocatalytic degradation of phthalate-based compounds.

Numerous methods were suggested by different researchers, with the aim of remediating this toxic organic chemical; however, photocatalytic degradation is considered the optimum, as the catalysts provide alternative charge transfer pathways, instead of electron-hole pair recombination, and increase the surface area for adsorption. Supporting this, Kaur et al. (2019) [55] study the photocatalytic degradation of DEP using transition-metal-doped (Ni, Mn, and Co) TiO_2 nanoparticles based on their degradation rate; they recommend Mn-doped TiO_2 as an ideal catalyst, as it requires minimal excitation energy (visible light activation), owing to the lowest optical band gap being 2.47 eV. Moreover, this study concludes that doped TiO_2 catalysts perform well when compared to their undoped counterparts [55]. Likewise, Akbari-Adergani et al. (2018) [56] suggest another set of transition metals (Fe, Ag, and Co), doped in ZnO, as an effective photocatalyst for remediating DBP; they perform well (90% degradation) under visible LEDs as a light source. Similarly, Motlagh et al. (2020) [57] fabricate ZnFe-layered double hydroxides, using sulfate-intercalated anion (ZnFe-SO4−LDH) modified with graphene oxide (GO) as a photocatalyst for degrading phenazopyridine hydrochloride (PhP) under visible light irradiation; they report a maximum rate of degradation as 60.01% [57]. In addition to immobilizing photocatalysts, nanocomposites with magnetic nanoparticles, such as zero-valent iron (ZVI) were developed, which simplified the post-degradation separation simply by using magnetic properties [58,59]. Another method suggests the removal of TiO_2 from the reaction solution by an electrocoagulation technique using iron electrodes and reports 95% of TiO_2 removal under a neutral pH and 100 mA current supply. Here, the electrochemical sludge is taken as a catalyst for activating peroxymonosulfate (PMS) in order to degrade emerging contaminants because of the presence of iron species (i.e., Fe_3O_4) [60]. The degradation of various phthalates and their derivatives, using nanoparticle-based heterogeneous photocatalysts with higher efficiency, is reported and shown in Table 3.

Table 3. Degradation of various phthalates and their derivatives using nanoparticle-based heterogeneous photocatalysts.

Phthalates	Photocatalyst	Light Source	Key Results	References
DMP	Bifunctional TiO_2 {001}	UV light	Nearly 76% of DMP is degraded within 120 min	[61]

Table 3. Cont.

Phthalates	Photocatalyst	Light Source	Key Results	References
DEP	Bifunctional TiO_2 {001}	UV light	Nearly 85% of DEP is degraded within 120 min	[61]
DEP	TiO_2 (anatase)	Xenon lamp	Photocatalytic degradation of DEP was achieved up to 90% within 50 min	[62]
DEP	Zinc oxide	Hg lamp	Photocatalytic degradation of DEP was achieved up to 80% within 30 min	[63]
DMP	Hydrothermal (h-t) and TiO_2	UV lamp	DMP removal under h-t TiO_2 (62.1%) s–g TiO_2 (33.6%)	[64]
DBP	TiO_2 (P25)	Xe lamp	DBP removal from wastewater was achieved up to 90% within 30 min	[65]
DBP	α-Fe_2O_3 nanoparticles	Mercury lamp	Photocatalytic degradation of DBP was achieved up to 90% within 120 min	[66]
BBP	P25 TiO_2	UV lamp	Photocatalytic degradation of BBP was achieved up to 80% within 60 min	[67]
BBP	Cl-doped TiO_2	Xe lamp	Up to 92% of BBP was degraded within 240 min	[68]
BBP	P-doped TiO_2 thin-films	Xe lamp	98% of BBP degradation efficiency was achieved within 180 min	[69]
DEHP	N_x–TiO_{2-x}	Xenon	When compared to TiO_2 Degussa P25, N-doped TiO_2 shows a faster rate of DEHP degradation up to 90%	[70]
DEHP	Fe-Ag/ZnO	Visible lamp	About 90% of DEHP was removed within 150 min	[71]
DMP	N-doped TiO_2 (UN/TiO_2)	Visible light	DMP is removed at a degradation rate of 41% and 58% using N/TiO_2 and UN/TiO_2 within 5 h	[72]
DEP	Ni/TiO_2; Mn/TiO_2; Co/TiO_2	Hg lamp	DEP was degraded up to 92% within 1hr 30 min	[73]

Table 3. Cont.

Phthalates	Photocatalyst	Light Source	Key Results	References
DEP	WO_3/TiO_2	UV light	The photodegradation of DEP under visible light is achieved, up to 90% within 60 min	[74]
DBP	Bi, Cu co-doped $SrTiO_3$	Metal halide lamp	Nanosized Bi, Cu co-doped $SrTiO_3$ showed significant degradation efficiency than bi-doped $SrTiO_3$	[75]
DBP	m-TiO_2-NTs (mesoporous TiO_2 nanotubes)	Mercury lamp	DBP removal degradation rate constant for m-TiO_2-NTs; 7.7 times greater than that of TiO_2	[76]
DBP	Carboxymethyl β-cyclodextrin Fe_3O_4–TiO_2	Mercury lamp	In comparison to Fe_3O_4–TiO_2, CMCD–Fe_3O_4–TiO_2 shows accelerated DBP degradation within 1 h	[77]
DBP	$gC_3N_4/Bi_2O_2CO_3$; g-$C_3N_4/BiOCl$	Halogen tungsten lamp	DBP is removed up to 60% within 3 h	[78]
DEP	Nanorod ZnO/SiC nanocomposite	UV and visible lamp	DEP degradation was achieved up to 90% within 1h	[79]

5. Removal of Phenol and Phenolic Compounds

Phenol and phenolic compounds are primary toxic water pollutants; thereby requiring effective remediation techniques to reduce their harmful effects on both humans and the environment. Numerous studies were carried out in recent times to understand and optimize the degradation of phenol and phenolic compounds. Accordingly, Hassan et al. (2020) [80] report an efficiency of 90% upon degrading phenol using acetylacetonate, rather than graphene nanocomposites, as the photocatalyst, assisted by visible light irradiation. Likewise, Tang et al. (2021) [81] report 100% removal of phenol from different water sources, including sewage wastewaters, upon using a bismuth-doped TiO_2-based photocatalyst. For better effectiveness in environmental degradation, the use of the Z-scheme photocatalytic system is widely encouraged. Supporting this, Xu et al. (2021) [82] note enhanced photocatalytic degradation of phenolic compounds carried out using a Z-scheme charge transfer, with $LaFeO_3/WO_3$ as the photocatalyst. In recent times, several researchers focused on advanced oxidation processes (AOPs) for the complete mineralization of phenols, citing its rapid rate of degradation with the active participation of the hydroxyl radical, and phenol degrading into CO_2 and water, instead of any harmful by-products. The process associated with heterogeneous photocatalysis is a widely recognized AOP and requires (i) a semiconductor photocatalyst, (ii) a light energy (UV or visible or solar) source, and (iii) an electron donor or hole acceptor. In this process, hydroxyl radicals are generated upon producing sufficient charge carriers (i.e., electron-hole pair) by supplying energy greater than the bandgap of the semiconductor photocatalyst using the light energy source [83,84].

Specifically, titanium dioxide (TiO_2) is the commonly preferred and highly performing photocatalyst amongst ZnO, CuO, and β-Ga_2O_3, for degrading phenols under the influence of UV light irradiation, due to its non-toxicity, photo-stability, cost-effectiveness, inertness towards chemical and biological systems, and insolubility [85]. Moreover, UV irradiation on a lab-scale can be efficient; however, it is not recommended for commercial and large-scale degradation of phenol, due to a lack of feasibility and cost-effectiveness [85].

In addition, identifying an efficient, yet sustainable, source for UV light is very challenging, as sunlight itself contains only a fraction of UV light (4% of solar spectrum) compared to visible light (46% of solar spectrum) [86]. Hence, photocatalysts responding to sunlight and any visible light must be developed for degradation, and this is achieved by modifying existing photocatalysts using simple, known techniques such as doping, composite semiconductors, dye sensitization, and synthesizing novel, undoped, single-phase mixed oxide photocatalysts [87].

Furthermore, adding dopants such as iodine, nitrogen, sulfur, praseodymium, and iron with TiO_2 photocatalysts improves the photoresponses of the latter into the visible spectrum, thus, making the degradation of phenol highly viable using visible light [88]. For composite semiconductors, a large bandgap semiconductor is coupled with a small bandgap semiconductor, with a more negative conduction band level, thereby allowing the injection of conduction band electrons from the small bandgap semiconductor for better charge carrier separation, as shown in Figure 5. A few examples of composite photocatalysts proven effective in degrading p-nitro phenols under visible light include Co_3O_4, $Bi_4O_5I_2$, and Bi_5O_7I [89]. Chowdhury et al. [90,91] showcase effective phenol degradation using an eosin Y-sensitized Pt-loaded TiO_2 photocatalyst; Qin et al. [92] report the degradation of 4-chlorophenol using an N719 dye-sensitized TiO_2 photocatalyst. Likewise, 4-nitrophenol is degraded upon using two different photocatalysts, namely, (i) Cu(II)-porphyrin and (ii) Cu(II)-phthalocyanine-sensitized TiO_2, under visible light irradiation. It is worth mentioning that dyes are active in visible light by nature, but become excited upon illumination by any other light source. The degradation of various phenols and their derivatives, using nanoparticle-based heterogeneous catalysts, is shown in Table 4.

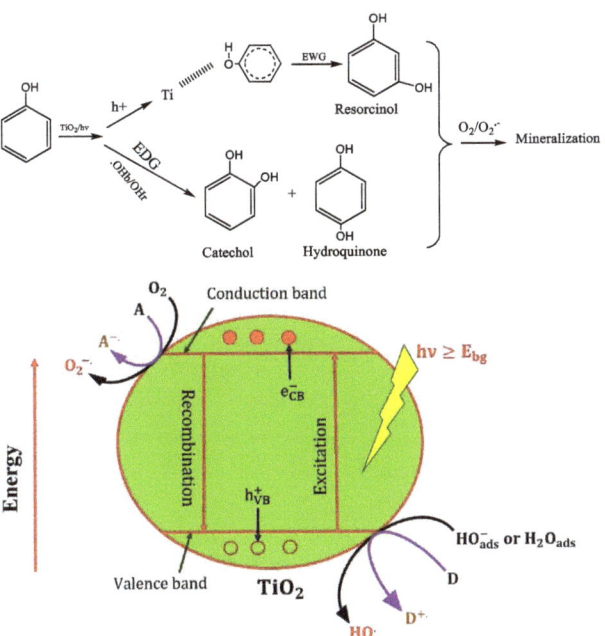

Figure 5. Photocatalytic degradation of phenol and its derivatives.

Table 4. Degradation of various phenols and their derivatives using nanoparticle-based heterogeneous catalysts.

Phenol and Phenolic Compounds	Photocatalyst and Light Source	Degradation Efficiency	References
Phenol	TiO_2/CMK-3, UV lamps, 150 min	74%	[93]
Phenol	PAN-CNT/TiO_2–NH_2, UV lamp, 7 min	99%	[94]
Phenol	UV/TiO_2 and Vis/N, C–TiO_2, 60 min	76%	[95]
Phenol	NIR irradiation, Nd-Er co-doped tetragonal $BiVO_4$, 150 min	96%	[96]
Phenol	Visible light, N-BiOBr/$NiFe_2O_4$-15 nanocomposite, 60 min	87.5%	[97]
Bisphenol A	UV-light, $BiVO_4$/CHCOO(BiO), 5h	99%	[98]
Phenol	$BiVO_4$/carbon, Xenon lamp, 5 h	80%	[99]
Phenol	$Bi_4V_2O_{10}$/$BiVO_4$, Xenon lamp, 60 min	95%	[100]
Phenol	CdS/TiO_2 Xenon lamp, 3 h	78%	[101]
4-chlorophenol	TiO_2/Zr, Xenon lamp, 4 h	95%	[102]
4-fluorophenol	Ag_3PO_4/$H_3PW_{12}O_4$ Xenon lamp, 10 min	100%	[103]
Phenol	Au/BiOBr/Graphene, Xenon lamp, 180 min	64%	[104]
Phenol	α-Fe_2O_3 nanorod/rGO, visible light, 120 min	67%	[105]
Phenol	$Bi_7O_9I_3$/rGO, visible light, 150 min	78.3%	[106]
Phenol	AgBr/BiOBr/graphene, visible light, 120 min	98%	[107]
Phenol	ZnO/TiO_2, visible light, 160 min	100%	[108]
4-chlorophenol	RGO/$CaFe_2O_4$/Ag_3PO_4, visible light, 160 min	90%	[109]
2,4 dichlorophenol	Graphene/ZnO/Co_3O_4, visible light, 150 min	91%	[110]

6. Removal of Drugs and Antibiotics and Their Derivatives

In general, drugs are discharged as pollutants into the atmosphere in form of excreta from individuals and animals, in addition to effluents discharged from pharmaceutical industries. Specifically, these drug molecules have adverse effects on the ecosystem, affecting aquatic life in terms of their lifecycle, growth retardation, and a reduction in friendly microbes. Apart from these impacts, excessive consumption of these drugs causes kidney problems in humans, in addition to increasing the immunity of pathogens towards these drugs. Moreover, these antibiotics form unknown complex compounds with heavy metals or other organic pollutants. To understand their severity, numerous researchers focus on degrading these antibiotics and drugs using photocatalysts; recently, the use of TiO$_2$ and Au-infused TiO$_2$ nanoparticles can be used to degrade a total of eight antibiotics with higher conversion efficiency [111]. Though many studies report 100% removal efficiency, Bekkali et al. (2017) [112] report 80% removal efficiency for sulfadiazine, amoxicillin, and anthramycin, upon using photosensitive TiO$_2$ irradiated using UV light. In another study, the effect of degradation on natural and synthetic antibiotics is studied, and Mohammad et al. (2020) [94] report the degradation efficiency, upon using immobilized TiO$_2$ under UV light, as 92.81% for synthetic, and 86.57% for natural, ciprofloxacin. Another catalyst, ZnO, is also used in UV-irradiated photocatalytic degradation, where better efficiencies are reported for prolonged reaction durations [113]. They are encouraged for large and commercial-scale applications and these catalysts were used for degrading sulfamethazine in different shapes, forming flower-shaped, tetra-needle-shaped, and regular ZnO nanoparticles, with T-ZnO reporting 100% conversion [113]. Likewise, 20 ppm of levofloxacin was degraded under visible light irradiation using BiVO$_4$, and reported 85% efficiency within 90 min of the conversion [114]. Another set of drug pollutants includes antineoplastic drugs used in anti-cancer treatments, which enter water bodies through excretion and effluent discharge from pharma industries, thus, increasing the levels of toxicity [115]. The schematic representation of photocatalytic degradation of pharmaceutical pollutants is shown in Figure 6.

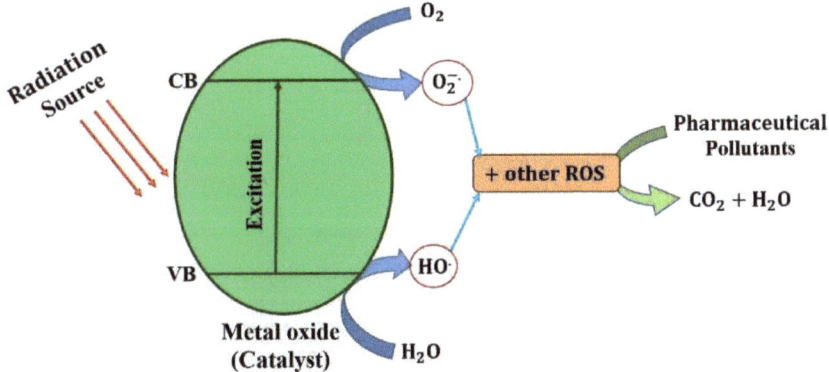

Figure 6. Photocatalytic degradation of pharmaceutical pollutants.

Interestingly, Hank Hui-Hsiang Lin and Angela Yu-Chen Lin (2013) [116] comment that high conversion efficiency is reported for photocatalysis upon degrading 5-fluorouracil and cyclophosphamide via a UV/TiO$_2$-irradiated light source. UV–visible light-irradiated photocatalytic degradation was carried out on a common antibiotic named tetracycline, using iron oxide nanoparticles as the potential photocatalyst [117]. Similarly, double-shelled ZnSnO$_3$ hollow cube nanoparticles were then used in degrading ciprofloxacin and sulfamonomethoxine [118]. Some of the commonly used photocatalysts for degrading these antibiotics include Ag$_3$PO$_4$/Ag/BiVO$_4$ Z-scheme photocatalysts [119], g-C$_3$N$_4$-doped porous carbon nitride [120], Ag$_I$/Bi$_{12}$O$_{17}$C$_{l2}$, Ag$_2$Mo$_2$O$_7$/MoS$_2$ Z-scheme 1D/2D photocatalysts [121], WO$_3$/gC$_3$N$_4$ Z-scheme photocatalysts [122], and BiOCl/g-C$_3$N$_4$/Cu$_2$O/Fe$_3$O$_4$ [123]. It is worth mentioning that, in some cases, the simultaneous degradation of antibiotic pollutants and generation of hydrogen is noted. In certain cases, non-toxic and less harmful antibiotics were also remediated; supporting this, naproxen, a common non-steroidal anti-inflammatory drug, was degraded using photocatalysts such as g-C$_3$N$_4$, carbon quantum dots, and single-atom dispersed silver [120]. Similarly, a carbon nitride-based heterojunction photocatalyst was used to degrade ibuprofen, while TiO$_2$ remains highly effective in degrading carbamazepine, diclofenac, and ibuprofen upon irradiation using a visible light source [124]. The degradation of various drugs and antibiotics, using nanoparticle-based heterogeneous catalysts, is shown in Table 5.

Table 5. Degradation of various drugs and antibiotics using nanoparticle-based heterogeneous catalysts.

Pollutant	Photocatalyst	Light Source	Drugs and Antibiotics Degradation Efficiency	References
Acetaminophen Levofloxacin	CdS sub-microspheres	Visible light, 4 h	85% 70%	[125]
Paracetamol	TiO_2–graphite composites	UV lamp, 60 min	100%	[126]
Doxycycline	$BiOBr/FeWO_4$	Xenon light, 60 min	90.4%	[127]
Tetracycline hydrochloride	$CdTe/TiO_2$	Halogen lamp, 60 min	78%	[128]
Tetracycline Fe-based	Metal-organic frameworks	Xenon lamp, 180 min	96.6%	[129]
Tetracycline hydrochloride	$ZnFe_2O_4$ porous hollow cube	Xenon lamp, 60 min	84.08%	[130]
Tetracycline	$ZnWO_4-x$ nanorods	Xenon lamp, 80 min	91%	[131]
Ofloxacin	TiO_2	Visible light, 60 min	98%	[132]
Norfloxacin	TiO_2	Visible light, 60 min	99%	[132]
Ciprofloxacin	TiO_2	Visible light, 120 min	91%	[132]
Nitrofurantoin	$Nd_2Mo_3O_9$	Tungsten incandescent, 45 min	99%	[133]

7. Removal of Dyes and Their Derivative

Dyes are a group of chemicals used as coloring agents in the textile industry, but when left untreated, these chemicals contaminate the aquatic ecosystem, causing mutation and sterility in aquatic organisms. Presently, numerous conventional techniques are practiced in wastewater treatment, but are found to be ineffective; however, the use of nanoparticle-based heterogeneous photocatalysts reports better degradation efficiency [134]. Confirming this, the rate of degradation of methylene blue using SnO_2 as a photocatalyst, assisted with UV light irradiation, has a maximum efficiency of 80–97%, along with high catalytic stability and reusability [135]. Another study uses TiO_2/WO_3-coated magnetic nanoparticles for degrading sixteen organic dyes, and notes a rapid rate of degradation, with complete decolorization noted in the case of ten dyes upon irradiation in direct sunlight [136]. Another type of photocatalyst includes copper-based nanoparticles, which exhibit effective degradability of organic dyes, as proven by Rao et al. (2019), who report 98% efficiency upon degrading Reactive Black dye using copper oxide nanorods [30]. Furthermore, these nanoparticles report similar degradation efficiencies upon remediating methyl orange, methylene blue, and Congo red, simultaneously [137]. Other photocatalytic degradation of dyes includes the degradation of rhodamine B using $Cu/Cu(OH)_2$ nanoparticles [137], the degradation of methyl orange using CdS/CuS nanoparticles [138], the degradation of methylene blue via Hummers' method using reduced graphene oxide doped with copper nanoparticles [139], the degradation of methyl red, methyl orange, and phenyl red using biologically prepared copper nanoparticles [140], and the degradation of methyl orange using graphene oxide-doped $CuO-Cu_2O$ and Cu_3N/MoS_2 [141]. The degradation of various dyes, using nanoparticle-based heterogeneous catalysts, is shown in Table 6.

Table 6. Degradation of various dyes using nanoparticle-based heterogeneous catalysts.

Dye	Photocatalyst	Light Source	Results	References
Methylene blue	ZnO/CuO	Visible lamp, 25 min	96.57%	[142]
Methylene blue	Carbon/BiVO$_4$	Xenon light, 180 min	95%	[99]
Rhodamine B			80%	
Rhodamine B	CdS-reduced graphene oxide	Xenon light, 60 min	97.2%	[143]
acid Chrome blue			65.7%	
Rhodamine B	WO$_3$	Metal halide lamp, 180 min	95%	[144]
Rhodamine B	rGO/RP-MoS$_2$	Xenon light, 30 min	99.3%	[145]
Acid Black 1 dye	TiO$_2$–graphene	Visible light, 60 min	96%	[146]

8. Factors Affecting the Degradation of Photocatalysis

The main factors that influence the photocatalytic degradation of organic pollutants are (i) the load of the catalyst, (ii) doping (iii) pH, (iv) light intensity, and (v) the lifetime and regeneration of the photocatalyst. The efficiency of the photocatalytic degradation of pollutants is highly dependent on the number of operational parameters. Several studies show the enhancement of the efficiency of the photocatalytic degradation of organic pollutants. In this review paper, the major factors influencing the efficiency of the photocatalytic degradation of pollutants are discussed below.

8.1. Catalyst Loading

The load of the catalysts is one of the key factors that influence the photocatalytic degradation of pollutants. Due to the increase in the active sites, the rate of the photocatalytic degradation of organic pollutants increases with photocatalysts dosage [147–151]. This is because when the photocatalysts are irradiated in the presence of light, it results in the formation of hydroxyl radical ions. When the concentration of the photocatalyst is low, it affects the efficiency of the photocatalytic degradation of pollutants, since more light is transmitted into the photocatalytic reactor, and less transmitted radiation is utilized in the degradation of pollutants [152,153]. The optimization process of the catalyst load is one of the major parameters that impact the whole catalytic process and its efficiency. Most of the researchers concentrated on the process optimization parameters of the photocatalytic degradation of pollutants. Based on the extensive literature survey, it is found that increasing the catalyst load increases the degradation of pollutants, which proportionally produces more hydroxyl ions and positive holes, and absorbs more photons, due to the availability of a large number of catalyst surfaces. As a result, this increases the degradation rate at a higher concentration, causing interference of the light to penetrate the solution, which restricts the light in passing through the solution [154,155]. This reduces the degradation percentage, and the phenomenon is known as the scattering of light [156]. In some cases, a certain amount of catalyst loading results in solution turbidity and, thus, blocks the UV radiation for the reaction to proceed, and finally decreases the degradation rate of the pollutants [157–160]. Beyond the optimum amount of catalysts, loading may affect the pollutant degradation rate, which is due to the increase in the opacity of the photocatalyst suspension. Increasing the light scattering and infiltration depth of the photons results in diminishing, meaning fewer photocatalysts may be activated, and also results in the agglomeration of nanoparticles at a higher concentration of photocatalysts. The agglomeration of nanoparticles occurs due to the activation of a lower number of surface-active sites during the photocatalytic degradation process, and also results in the deactivation of the activated molecules, leading to the collision of the activated molecules in the ground state [161]. Table 7 describes the dependency of the photocatalytic activity on the catalysts loading and their conversion efficiency.

Table 7. Dependency of the photocatalytic activity on the catalysts loading and their conversion efficiency.

Pollutant	Photocatalyst	Catalyst Load	Efficiency	Reference
Sulfamethoxazole	$BiVO_4/CHCOO(BiO)$	1 g/L	85%	[98]
Methylene blue	CuO/ZnO	1 g/L	96.57%	[142]
Methylene blue	$BiVO_4$/carbon	1.0 g/L	95%	[99]
Rhodamine B	CdS-reduced graphene oxide	0.4 g/L	97.2%	[143]
Rhodamine B	$Bi_4V_2O_{10}/BiVO_4$	1 g/L	100%	[100]
4-fluorophenol	$Ag_3PO_4/H_3PW_{12}O_{40}$	3 g/L	100%	[103]
Methyl orange		3 g/L	100%	
Rhodamine B	WO_3	1 g/L	95%	[144]
Rhodamine B	RP-MoS_2/rGO	0.4 g/L	99.3%	[145]
4-chloro Phenol	$ZrTiO_2$	0.1 g/L	95%	[102]

8.2. Doping

The efficiency of the photocatalyst can be increased by doping in the following ways: bad gap narrowing; oxygen vacancies; formation of impurity energy levels; unique surface area; electron trapping, etc. [162]. Generally, a catalyst with smaller bandgap energy is an effective photocatalyst to produce more electron-hole pairs. The doping process prevents the recombination of electrons and holes, and enhances photocatalytic activity by trapping the photoinduced electrons [163]. The incorporation of dopant ions in the catalysts reduces the radius of lattice ions and lattice space. Similarly, dopant ions are incorporated into the catalyst crystal lattice to enhance the electronic property of the photocatalysts, and also to improve the light absorption ability in the visible light region [164]. Increasing the optimum level by adding a dopant to the catalyst reduces the photocatalytic activity. Narrowing the charge space area increases the recombination of higher dopants than the optimum level, reducing the surface area. The dopant then turns to the recombination center, which ultimately decreases the activity of the photocatalyst [165]. In the case of TiO_2, adding too much dopant reduces the thermal stability of TiO_2, which causes phase transformation of TiO_2 anatase to turn to rutile. A higher level of doping forms clusters in the surface of the photocatalyst, which decreases the photocatalytic activity by reducing the light penetration into the actual photocatalyst surface, and shields the surface area, which causes agglomeration [162]. However, some reports suggest that adding a dopant to the surface increases the mesoporous structure of the catalyst. In the case of using a noble metal as a dopant, which separates the electron and hole pair by the phenomenon of surface plasmon resonance under visible light, and also increases the adsorption of the pollutant onto the surface of the catalyst, the traps are formed by the noble metal when it acts as a dopant, which reduces the recombination by trapping. In metals, it increases the lifetime of the catalyst by preventing corrosion, due to the organic metal reaction with the surface of the photocatalyst [166]. Doping controls the specific surface area, morphology, crystallinity of the photocatalyst, and particle size. There are different types of dopants, such as anionic and cationic dopants; adding anionic dopants makes the process simple by working under visible light with better stability, and gives a better yield under the visible region compared to UV radiation [166]. Nitrogen, sulfur, phosphorous, carbon, and fluorine are the anionic species that form impurity energy levels near the valence band, and give greater efficiency. In both substitution and interstitial mode, nitrogen is incorporated into the lattice of the photocatalyst. By adding activated carbon, the surface area of the catalyst increases and its efficiency increases. The crystalline property of the semiconductor photocatalyst is reduced when adding a dopant to the surface; adding large amounts of cerium and nitrogen decreases the crystallinity. Consequently, Chen et al. [167] study the composition of rutile, increasing the doping concentration by reducing the thermal stability of anatase. This results in the phase transformation of anatase into rutile. In the case of mesoporous nanoparticles, doping materials decrease the photocatalytic activities, due to the surface site being blocked by doping material. Under visible light irradiation, higher photocatalyst efficiency is achieved by introducing impurity energy levels, which narrow the bandgap and form oxygen-deficient sites and more electron-hole pairs [168]. The photocatalytic activity of the prepared catalyst depends on how well the recombination of the photoinduced hole–electron pairs is prevented. Doping prevents the recombination of electrons and holes and improves photocatalytic activity by trapping the photoinduced electrons. By adding the dopant substitutionally and interstitially to the photocatalyst, degradation of organic pollutants is enhanced [169]. Table 8 describes the removal of various pollutants using different types of doped photocatalysts.

Table 8. Removal of various pollutants using different types of doped photocatalysts.

Pollutant	Light Source	Dopant Material	Removal Efficiency	Reference
Methylene blue	Visible light	Ag–TiO_2	96%	[170]
Rhodamine B	Solar light	CeO_2-doped TiO_2	99.8%	[171]
POME	Visible light	Ag–TiO_2	26.77%	[172]
Phenol	UV–Visible light	Ag–TiO_2	Up to 50%	[173]
Malachite green	UV–Visible light	Fe–TiO_2	75.81%	[174]
Nitrobenzene	UV light	Fe–TiO_2	99.7%	[175]
Reactive red-198	Visible light	Cu–TiO_2	13%	[176]
Bisphenol A	Visible light	Cu–TiO_2	77%	[177]
Methylene blue	Visible light	Ni–TiO_2	63%	[178]
Phenol	Visible light	Co–TiO_2	81.72%	[179]
Rhodamine B	Solar light	Bi–TiO_2	97%	[180]
4-Chlorophenoxy acetic acid	Visible light	N–TiO_2	73%	[181]
Phenol	Visible light	N–TiO_2	Up to 25%	[182]
Phenol	Visible light	B–TiO_2	12.33%	[183]
Phenol	UV light	F–TiO_2	78%	[184]

8.3. pH of the Solution

In the photocatalytic degradation of pollutants in wastewater, the pH of the wastewater significantly influences the photocatalytic efficiency process. Several studies report on the effect of pH on the photocatalytic reaction. In the photocatalytic degradation process, the pH of the reaction mixture mainly depends on the catalyst surface charge and the chemical-charged particle present in the samples. In the case of wastewater treatment, the pH is mainly dependent on the charge of the photocatalyst, the size of the aggregates, and the position of the conductance and valence bands [185]. If the surface charge and adsorbate have similar charges, resulting in a decrease in the rate of the photodegradation process, the pH of the solution should be maintained to stabilize the photocatalytic degradation of the pollutant [186,187]. It is reported that ZnO with SnO_2 nanoparticles shows better catalytic properties at a neutral pH than at acidic (pH = 4) or alkaline (pH = 10) pH levels [188]. Similarly, if the material's surface charge opposes the adsorption process because the adsorbate contains the same charge, the pH conditions reveal the optimal adsorption [189]. The photocatalyst Mg–ZnO–Al$_2$O$_3$ was observed to degrade 20 mg/L caffeine solution at pH 9.5 [23]. According to the researchers, changes in the surface charge, and the ionization of caffeine molecules, increase the generation of hydroxyl radicals and enhance the photocatalytic degradation of pollutants under varied pH conditions of the reaction substrate [190]. Generally, pH parameters include many factors, such as decomposition, and the non-favorable adsorption dissolution of the photocatalyst [191]. However, the pH of the solution is 9 in the optimal condition for the photodegradation of the Acid Black 1 dye solution with a photodegradation efficiency of up to 90.1% using a ZnO photocatalyst [24]. The removal effectiveness is lower at acidic pH levels, due to photocatalyst dissolution. As the photocatalyst surface charge is inversely proportional to the solution charge, the pH solution must be evaluated [191]. It was recently reported that at a lower pH level, the maximum oxidizing capacity of titanium-based photocatalysts lowers the rate of the reaction, due to the presence of excess H$^+$ in the reactant solution. To enhance the photocatalytic degradation of pollutants, pH optimization is very important, in order to determine the rate of the reaction [192]. Kiomars Zargoosh et al. (2020) [193] report the use of a nanocomposite of CaAl$_2$O$_4$:Eu^{2+}:Nd^{3+} photocatalyst for the removal of methylene blue dye, by varying the pH of the solution from 7 to 10. At a higher concentration of hydroxyl ions under alkaline conditions, the rate of degradation is faster than in acidic conditions. At pH values higher than 10, the reduction in the removal efficiency may be due to the inactivation of the photocatalyst [193]. A. F. Alkaim et al. (2014) [194] report that photocatalytic dye degradation efficiency is enhanced by varying the pH from 4 to 11. The photodegradation of pollutants is enhanced at pH 6 when using TiO$_2$ as a photocatalyst, as shown in Figure 7a. Generally, dye pollutants are negatively charged in the base medium, and their adsorption may also be affected by an increase in the density of the Ti–O group on the surface. This type of mechanism always occurs, due to coulombic repulsion of the dye pollutant. Similarly, changing the pH from 6 to 7 demonstrates a decrease in the photocatalytic degradation of dye in both acidic and alkaline pH, whereas at high pH values the hydroxyl radicals are rapidly scavenged, and they do not have the opportunity to react with dye pollutants [195]. Using zinc oxide as a photocatalyst significantly enhances the photodegradation of dye at the high pH of 11, whereas at a low pH, the photodecomposition of ZnO into Zn^{2+} in acidic, neutral, and alkaline conditions results in the formation of hydroxyl radical ions, as shown in Figure 7b. In the case of Co$_3$O$_4$, the photocatalytic dye degradation efficiency is reduced at a high pH of 11. The same trend is observed in a CdS photocatalyst, as shown in Figure 7c,d. Based on the above results, it is concluded that the pH of the reactant solution plays an important role in the degradation of pollutants [196].

8.4. Light Intensity

The light intensity affects the efficiency of photocatalysts. When the intensity of the light source is low, between 0–20 mW/cm^2, the rate of the reaction increases in light intensity. At the middle range of light intensity (25 mW/cm^2), the rate of the reaction depends on the square root of the light intensity, and at high intensities the rate of the reaction is independent of light intensity. At low light intensity in the photocatalytic reaction, an electron-hole pair separates and then recombines, which reduces the formation of free radicals and results in the degradation of organic pollutants. Neppolian et al. (2003) [150] report the photocatalytic degradation of Reactive Yellow 17, Reactive Red 2, and Reactive Blue 4 dyes using titanium dioxide photocatalyst, under solar (or) UV radiation as a light source. Compared with solar radiation, UV radiation is more effective in the degradation of the selected dyes [150]. The energy of the UV radiation is large compared to the bandgap energy of the catalysts. The reason behind this is a recombination of the electron-hole pair, which is completely avoided in the presence of a UV source. Using sunlight as a source of light energy in the photocatalytic degradation of pollutants means that only 5% of the total radiation energy is used for the bandgap excitation of electrons. Hence, the percentage degradation is found to be less in solar radiation compared with the UV source of photocatalytic pollutant degradation [196]. Hung and Yuan (2000) [197] study the effect of light intensity on the photocatalytic degradation of pollutants. The light intensity ranges from 215 to 586 µW/cm^2, increasing the degradation efficiency of the pollutants with the increasing light intensity. In the study of Chanathaworn et al. (2012) [198], the intensity of the black light lamp is varied between the ranges of 0–114 W/m^2, and the impact of light intensity radiation efficiency on the degradation of the rhodamine B pollutant is analyzed. Based on the experimental results, when the intensity of the light source increases, it enhances the efficiency of the pollutant degradation. Under three different light intensities (1.24 mW/cm^2, 2.04 mW/m^2, and 3.15 mW/m^2) the decolorization of acid yellow 17 degradation is studied, using a photocatalyst to enhance the pollutant efficiency [199]. For the enhancement of the degradation rate of the pollutant using different types of photocatalyst, the light intensity may be increased from a lower frequency to a higher frequency [200,201]. Similarly, Rao et al. (2004) demonstrate that the rate of the photocatalytic degradation of acid orange 7 dye pollutant is increased 1.5 times in a peak sunlight source compared to the artificial UV light sources [202]. An overall observation of the stated research results is that the wavelength of the irradiation affects the efficiency of the photocatalytic degradation process. The scientific evidence clearly states that a shorter wavelength of irradiation stimulates the electron–hole generation, and subsequently enhances the efficiency of the catalyst [203].

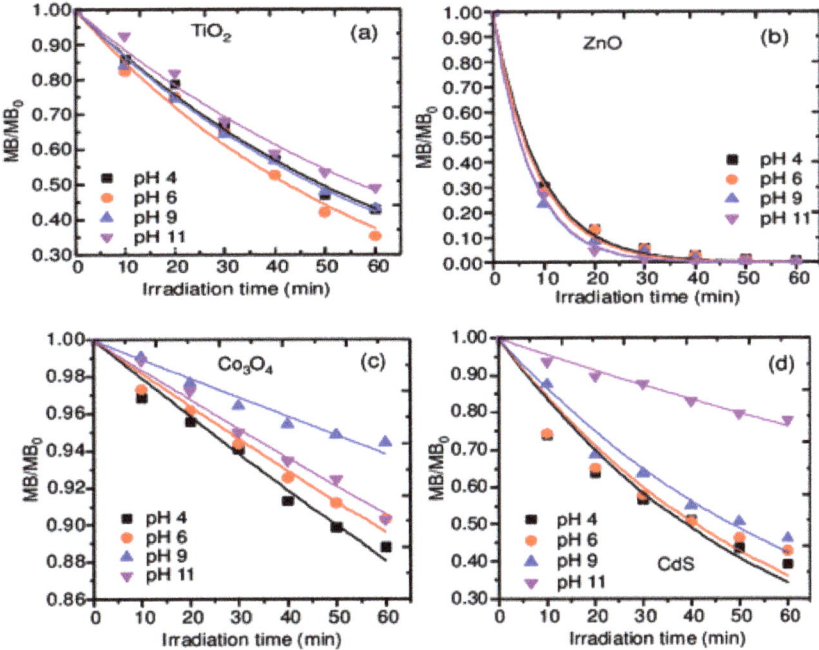

Figure 7. Photocatalytic degradation of methylene blue in the presence of different pH levels of the solution; (**a**) TiO$_2$, (**b**) ZnO, (**c**) Co$_3$O$_4$, and (**d**) CdS [194].

8.5. Lifetime and Regeneration of Photocatalyst

The lifetime and regeneration of the photocatalyst are important to ensure the efficiency and the quality of the synthesized catalyst. The photocatalytic degradation efficiency affects the lifetime and regeneration of the photocatalyst. The quality of the photocatalyst is important for the photocatalytic degradation of the contaminants, as it mainly depends on the process of the catalyst synthesis, dopants, and selection of the catalyst. Several catalyst performances reduce after a certain cycle, due to their corrosion properties. Photocatalyst must maintain good stability and reusability, which is an essential factor for the pollutant degradation process. Most photocatalysts are deactivated after several cycles of the reaction process. Many factors are responsible for the deactivation of the photocatalysts, such as the loss of photocatalyst mass, especially during the washing/purification process. It has been reported that some amounts of rod-shaped ZnO stacking occur on Cu_2O/Ag_3PO_4 during filtration [204]. Leaching of dopants usually occurs during the reaction due to photoetching [205]. A synthesized ZnO catalyst undergoes photodecomposition after three cycles in the degradation of rhodamine B, and the nanosized photocatalyst has a high porosity surface, because of the deactivation of catalysts. The highly stable, novel photocatalyst $Ag_2Nb_4O_{11}$ has 40 times the reusable capacity for the degradation of rhodamine B, methylene blue, and methyl orange [206]. The researchers discovered that the methylene blue of organic intermediates is adsorbed on the photocatalyst surface during every cycle [207]. In summary, the strength, durability, and stability of the catalyst are based on the reaction parameters involved in the photocatalytic degradation process. This is to ensure that the prepared catalyst has high reusability, and recycling the ability minimizes the cost of the process.

9. Conclusions

The review mainly focused on photocatalysis technology used to solve environmental problems. In the present paper, we explained the degradation of pollutants present in wastewater. The major pollutants present in wastewater are polybrominated diphenyl ether and its derivatives, phthalates and its derivatives, phenolic compounds, drugs and antibiotics, and dye compounds, etc. These compounds were effectively removed using nanomaterials in the presence of light. The study shows that metal oxides of titanium, barium, copper, zinc, lanthanum, nickel, silver, cerium, iron, and others are effectively used as photocatalysts in the removal of various pollutants. The removal of pollutants from wastewater is based on the reaction conditions in the presence of various types of photocatalysts. A potential photocatalyst is capable of degrading all types of pollutants with higher efficiency, but using photocatalysts has some disadvantages. The recombination rate is high and it works efficiently under UV light irradiation, but a catalyst that works in the visible light to utilize the solar energy without any restraint is needed. This can be rectified by doping and the Z-scheme effect, as both of these effectively rectify this in the various types of heterogeneous nanoparticle-based photocatalysts with common photocatalytic errors such as higher recombination rate, large bandgap, and the inability to harvest visible light. Generally, titanium, copper, zinc, iron, and all other compounds are also effective in degradation, with metal-doping enhancing the degradation efficiency. The efficiency of the photocatalyst is mainly based on fundamental factors such as pH, doping, catalysts loading, light intensity, and stability. These factors have a significant part in the degradation of pollutants. The major functions of these parameters, and their impact on the rate of photocatalytic degradation efficiency, were elaborated in detail in the present paper. Finally, a photocatalyst is cheap and more stable compared to other conventional catalysts, as well as being economical and environmentally friendly. With the growth and extension of the research, photocatalysis technology was extended to many fields, such as energy, health, environment, pollution control, and the synthesis of value-added chemicals. As a result, the relevance of photocatalysis to human life is increasing steadily. The grand challenge of photocatalysis today is to further expand the practical application of photocatalytic technology in the industrial field.

Author Contributions: C.S.: Conceptualization, Writing—original draft, S.V.: Data curation, Visualization, and Validation. E.L.: Funding acquisition. S.S.: Conceptualization, Writing—review & editing. R.J.: Conceptualization, Writing—original draft. All authors have read and agreed to the published version of the manuscript.

Funding: This research received no external funding.

Data Availability Statement: All relevant data are within the paper.

Conflicts of Interest: The authors have no conflict of interest to declare regarding the publication of this work.

References

1. Khan, I.; Saeed, K.; Khan, I. Nanoparticles: Properties, applications and toxicities. *Arab. J. Chem.* **2019**, *12*, 908–931. [CrossRef]
2. Tharani, K.; Jegatha Christy, A.; Sagadevan, S.; Nehru, L.C. Photocatalytic and antibacterial performance of iron oxide nanoparticles formed by the combustion method. *Chem. Phys. Lett.* **2021**, *771*, 138524. [CrossRef]
3. Sibhatu, A.K.; Weldegebrieal, G.K.; Sagadevan, S.; Tran, N.N.; Hessel, V. Photocatalytic activity of CuO nanoparticles for organic and inorganic pollutants removal in wastewater remediation. *Chemosphere* **2022**, *300*, 134623. [CrossRef] [PubMed]
4. Sibhatu, A.K.; Weldegebrieal, G.K.; Sagadevan, S.; Tran, N.N.; Hessel, V. Synthesis and Process Parametric Effects on the Photocatalyst Efficiency of CuO Nanostructures for Decontamination of Toxic Heavy Metal Ions. *Chem. Eng. Processing-Process Intensif.* **2022**, *173*, 108814. [CrossRef]
5. Zularisam, A.W.; Ismail, A.F.; Salim, M.R. Behaviours of natural organic matter in membrane filtration for surface water treatment—A review. *Desalination* **2006**, *194*, 211–231. [CrossRef]
6. Zazouli, M.A.; Kalankesh, L.R. Removal of precursors and disinfection by-products (DBPs) by membrane filtration from water: A review. *J. Environ. Health Sci. Eng.* **2017**, *15*, 25. [CrossRef]
7. Azimi, A.; Azari, A.; Rezakazemi, M.; Ansarpour, M. Removal of Heavy Metals from Industrial Wastewaters: A Review. *ChemBioEng Rev.* **2017**, *4*, 37–59. [CrossRef]
8. Yagub, M.T.; Sen, T.K.; Afroze, S.; Ang, H.M. Dye and its removal from aqueous solution by adsorption: A review. *Adv. Colloid Interface Sci.* **2014**, *209*, 172–184. [CrossRef]
9. Mousset, E.; Doudrick, K. A review of electrochemical reduction processes to treat oxidized contaminants in water. *Curr. Opin. Electrochem.* **2020**, *22*, 221–227. [CrossRef]
10. Arar, Ö.; Yüksel, Ü.; Kabay, N.; Yüksel, M. Various applications of electrodeionization (EDI) method for water treatment—A short review. *Desalination* **2014**, *342*, 16–22. [CrossRef]
11. Sagadevan, S.; Lett, J.A.; Weldegebrieal, G.K.; Imteyaz, S.; Johan, M.R. Synthesis, characterization, and photocatalytic activity of PPy/SnO$_2$ nanocomposite. *Chem. Phys. Lett.* **2021**, *783*, 139051. [CrossRef]
12. Nida, Q.; Singh, P.; Sabir, S.; Umar, K.; Sagadevan, S.; Oh, W. Synthesis of Polyaniline Supported CdS/CdS-ZnS/CdS-TiO$_2$ Nanocomposite for Efficient Photocatalytic Applications. *Nanomaterials* **2022**, *12*, 1355.
13. Saravan, R.S.; Muthukumaran, M.; Mubashera, S.M.; Abinaya, M.; Prasath, P.V.; Parthiban, R.; Mohammad, F.; Oh, W.C.; Sagadevan, S. Evaluation of the photocatalytic efficiency of cobalt oxide nanoparticles towards the degradation of crystal violet and methylene violet dyes. *Optik* **2020**, *207*, 164428. [CrossRef]
14. Fujishima, A.; Honda, K. Electrochemical Photolysis of Water at a Semiconductor Electrode. *Nature* **1972**, *238*, 37–38. [CrossRef]
15. Reiche, H.; Dunn, W.W.; Bard, A.J. Heterogeneous photocatalytic and photosynthetic deposition of copper on Titanium dioxide and tungsten (VI) oxide powders. *J. Phys. Chem.* **1979**, *83*, 2248–2251. [CrossRef]
16. Xu, Q.; Zhang, L.; Yu, J.; Wageh, S.; Al-Ghamdi, A.A.; Jaroniec, M. Direct Z-scheme photocatalysts: Principles, synthesis, and applications. *Mater. Today* **2018**, *21*, 1042–1063. [CrossRef]
17. Nazim, M.; Khan, A.A.P.; Asiri, A.M.; Kim, J.H. Exploring Rapid Photocatalytic Degradation of Organic Pollutants with Porous CuO Nanosheets: Synthesis, Dye Removal, and Kinetic Studies at Room Temperature. *ACS Omega* **2021**, *6*, 2601–2612. [CrossRef]
18. Gupta, N.K.; Ghaffari, Y.; Kim, S.; Bae, J.; Kim, K.S.; Saifuddin, M. Photocatalytic degradation of organic pollutants over MFe$_2$O$_4$ (M = Co, Ni, Cu, Zn) nanoparticles at neutral pH. *Sci. Rep.* **2020**, *10*, 4942. [CrossRef]
19. Kumar, S.; Sahare, P.D. Photocatalytic activity of Bismuth Vanadate for the degradation of organic compounds. *Nano* **2013**, *8*, 1350007. [CrossRef]
20. Santhosh, C.; Malathi, A.; Daneshvar, E.; Kollu, P.; Bhatnagar, A. Photocatalytic degradation of toxic aquatic pollutants by novel magnetic 3D-TiO$_2$@ HPGA nanocomposite. *Sci. Rep.* **2018**, *8*, 15531. [CrossRef]
21. Saygi, B.; Tekin, D. Photocatalytic degradation kinetics of reactive black 5 (RB 5) dyestuff on TiO$_2$ modified by pretreatment with untrasound energy. *React. Kinet. Mech. Catal.* **2013**, *110*, 251–258. [CrossRef]
22. Vaiano, V.; Sacco, O.; Matarangolo, M. Photocatalytic degradation of paracetamol under UV irradiation using TiO$_2$-graphite composites. *Catal. Today* **2018**, *315*, 230–236. [CrossRef]
23. Elhalil, A.; Elmoubarki, R.; Farnane, M.; Machrouhi, A.; Sadiq, M.; Mahjoubi, F.Z.Z.; Qourzal, S.; Barka, N. Photocatalytic degradation of caffeine as a model pharmaceutical pollutant on Mg doped ZnO-Al$_2$O$_3$ heterostructure. *Environ. Nanotechnol. Monit. Manag.* **2018**, *10*, 63–72.
24. Subash, B.; Krishnakumar, B.; Swaminathan, M.; Shanthi, M. Highly active Zr co-doped Ag-ZnO photocatalyst for the mineralization of Acid Black 1 under UV-A light illumination. *Mater. Chem. Phys.* **2013**, *141*, 114–120. [CrossRef]
25. Tambat, S.; Umale, S.; Sontakke, S. Photocatalytic degradation of metamitron using CeO$_2$ and Fe/CeO$_2$. *Integr. Ferroelectr.* **2018**, *186*, 54–61. [CrossRef]
26. Ateş, S.; Baran, E.; Yazıcı, B. Fabrication of Al$_2$O$_3$ nanopores/SnO$_2$ and its application in photocatalytic degradation under UV irradiation. *Mater. Chem. Phys.* **2018**, *214*, 17–27. [CrossRef]
27. Soares, E.T.; Lansarin, M.A.; Moro, C.C. A study of process variables for the photocatalytic degradation of rhodamine B. *Braz. J. Chem. Eng.* **2007**, *24*, 29–36. [CrossRef]

28. Alafif, Z.O.; Anjum, M.; Ansari, M.O.; Kumar, R.; Rashid, J.; Madkour, M.; Barakat, M.A. Synthesis and characterization of S-doped-rGO/ZnS nanocomposite for the photocatalytic degradation of 2-chlorophenol and disinfection of real dairy wastewater. *J. Photochem. Photobiol. A Chem.* **2019**, *377*, 190–197. [CrossRef]
29. Fathima, J.B.; Pugazhendhi, A.; Oves, M.; Venis, R. Synthesis of eco-friendly copper nanoparticles for augmentation of catalytic degradation of organic dyes. *J. Mol. Liq.* **2018**, *260*, 1–8. [CrossRef]
30. Rao, M.P.C.; Kulandaivelu, K.; Ponnusamy, V.K.; Wu, J.J.; Sambandam, A. Surfactant-assisted synthesis of copper oxide nanorods for the enhanced photocatalytic degradation of Reactive Black 5 dye in wastewater. *Environ. Sci. Pollut. Res.* **2020**, *27*, 17438–17445. [CrossRef]
31. Akram, N.; Guo, J.; Ma, W.; Guo, Y.; Hassan, A.; Wang, J. Synergistic catalysis of $Co(OH)_2$/CuO for the degradation of organic pollutant under visible light irradiation. *Sci. Rep.* **2020**, *10*, 1939. [CrossRef] [PubMed]
32. Zhang, Y.; Zhou, J.; Chen, J.; Feng, X.; Cai, W. Rapid degradation of tetracycline hydrochloride by heterogeneous photocatalysis coupling with persulfate oxidation with MIL-53(Fe) under visible light irradiation. *J. Hazard. Mater.* **2020**, *392*, 122315. [CrossRef] [PubMed]
33. Deng, X.; Wang, C.; Yang, H.; Shao, M.; Zhang, S.; Wang, X.; Ding, M.; Huang, J.; Xu, X. One-pot hydrothermal synthesis of CdS decorated CuS microflower-like structures for enhanced photocatalytic properties. *Sci. Rep.* **2017**, *7*, 3877. [CrossRef] [PubMed]
34. Kirankumar, V.S.; Sumathi, S. Photocatalytic and antibacterial activity of bismuth and copper co-doped cobalt ferrite nanoparticles. *J. Mater. Sci. Mater. Electron.* **2018**, *29*, 8738–8746. [CrossRef]
35. Kirankumar, V.S.; Sumathi, S. Structural, optical, magnetic and photocatalytic properties of bismuth doped copper aluminate nanoparticles. *Mater. Chem. Phys.* **2017**, *197*, 17–26. [CrossRef]
36. De Wit, C.A. An overview of brominated flame retardants in the environment. *Chemosphere* **2002**, *46*, 583–624. [CrossRef]
37. Azri, N.; Bakar, W.A.W.A.; Ali, R. Optimization of photocatalytic degradation of polybrominated diphenyl ether on trimetallic oxide Cu/Ni/TiO_2/PVC catalyst using response surface methodology method. *J. Taiwan Inst. Chem. Eng.* **2016**, *62*, 283–296. [CrossRef]
38. Wang, R.; Tang, T.; Wei, Y.; Dang, D.; Huang, K.; Chen, X.; Yin, H.; Tao, X.; Lin, Z.; Dang, Z.; et al. Photocatalytic debromination of polybrominated diphenyl ethers (PBDEs) on metal doped TiO_2 nanocomposites: Mechanisms and pathways. *Environ. Int.* **2019**, *127*, 5–12. [CrossRef]
39. Li, Y.; Li, J.; Deng, C. Occurrence, characteristics and leakage of polybrominated diphenyl ethers in leachate from municipal solid waste landfills in China. *Environ. Pollut.* **2014**, *184*, 94–100. [CrossRef]
40. Lei, M.; Wang, N.; Zhu, L.; Tang, H. Peculiar and rapid photocatalytic degradation of tetrabromodiphenyl ethers over Ag/TiO_2 induced by interaction between silver nanoparticles and bromine atoms in the target. *Chemosphere* **2016**, *150*, 536–544. [CrossRef]
41. Lv, Y.; Cao, X.; Jiang, H.; Song, W.; Chen, C.; Zhao, J. Rapid photocatalytic debromination on TiO_2 with in-situ formed copper co-catalyst: Enhanced adsorption and visible light activity. *Appl. Catal. B Environ.* **2016**, *194*, 150–156. [CrossRef]
42. Yao, B.; Luo, Z.; Zhi, D.; Hou, D.; Luo, L.; Du, S.; Zhou, Y. Current progress in degradation and removal methods of polybrominated diphenyl ethers from water and soil: A review. *J. Hazard. Mater.* **2020**, *403*, 123674. [CrossRef] [PubMed]
43. Huang, A.; Wang, N.; Lei, M.; Zhu, L.; Zhang, Y.; Lin, Z.; Yin, D.; Tang, H. Efficient Oxidative Debromination of Decabromodiphenyl Ether by TiO_2-Mediated Photocatalysis in Aqueous Environment. *Environ. Sci. Technol.* **2013**, *47*, 518–525. [CrossRef]
44. Lei, M.; Wang, N.; Zhu, L.; Xie, C.; Tang, H. A peculiar mechanism for the photocatalytic reduction of decabromodiphenyl ether over reduced graphene oxide–TiO_2 photocatalyst. *Chem. Eng. J.* **2014**, *241*, 207–215. [CrossRef]
45. Lei, M.; Wang, N.; Guo, S.; Zhu, L.; Ding, Y.; Tang, H. A one-pot consecutive photocatalytic reduction and oxidation system for complete debromination of tetrabromodiphenyl ether. *Chem. Eng. J.* **2018**, *345*, 586–593. [CrossRef]
46. Guo, S.; Zhu, L.; Majima, T.; Lei, M.; Tang, H. Reductive Debromination of Polybrominated Diphenyl Ethers: Dependence on Br Number of the Br-Rich Phenyl Ring. *Environ. Sci. Technol.* **2019**, *53*, 4433–4439. [CrossRef]
47. Shao, Y.-Y.; Ye, W.-D.; Sun, C.-Y.; Liu, C.-L.; Wang, Q.; Chen, C.-C.; Gu, J.-Y.; Chen, X.-Q. Enhanced photoreduction degradation of polybromodiphenyl ethers with Fe_3O_4-g-C_3N_4 under visible light irradiation. *RSC Adv.* **2018**, *8*, 10914–10921. [CrossRef]
48. Shao, Y.Y.; Ye, W.D.; Sun, C.Y.; Liu, C.L.; Wang, Q. Visible-light-induced degradation of polybrominated diphenyl ethers with AgI-TiO_2. *RSC Adv.* **2017**, *7*, 39089–39095. [CrossRef]
49. Liang, C.; Zhang, L.; Guo, H.; Niu, C.-G.; Wen, X.-J.; Tang, N.; Liu, H.-Y.; Yang, Y.-Y.; Shao, B.-B.; Zeng, G.-M. Photo-removal of 2,2′4,4′-tetrabromodiphenyl ether in liquid medium by reduced graphene oxide bridged artificial Z-scheme system of Ag@Ag_3PO_4/g-C_3N_4. *Chem. Eng. J.* **2019**, *361*, 373–386. [CrossRef]
50. Wei, Y.; Gong, Y.; Zhao, X.; Wang, Y.; Duan, R.; Chen, C.; Song, W.; Zhao, J. Ligand directed debromination of tetrabromodiphenyl ether mediated by nickel under visible irradiation. *Environ. Sci. Nano* **2019**, *6*, 1585–1593. [CrossRef]
51. Huang, K.; Liu, H.; He, J.; He, Y.; Tao, X.; Yin, H.; Dang, Z.; Lu, G. Application of Ag/TiO_2 in photocatalytic degradation of 2,2′,4,4′-tetrabromodiphenyl ether in simulated washing waste containing Triton X-100. *J. Environ. Chem. Eng.* **2021**, *9*, 105077. [CrossRef]
52. Careghini, A.; Mastorgio, A.F.; Saponaro, S.; Sezenna, E. Bisphenol A, nonylphenols, benzophenones, and benzotriazoles in soils, groundwater, surface water, sediments, and food: A review. *Environ. Sci. Pollut. Res.* **2015**, *22*, 5711–5741. [CrossRef]
53. Benjamin, S.; Masai, E.; Kamimura, N.; Takahashi, K.; Anderson, R.C.; Faisal, P.A. Phthalates impact human health: Epidemiological evidences and plausible mechanism of action. *J. Hazard. Mater.* **2017**, *340*, 360–383. [CrossRef] [PubMed]

54. Swan, S.H.; Main, K.M.; Liu, F.; Stewart, S.L.; Kruse, R.L.; Calafat, A.M.; Mao, C.S.; Redmon, J.B.; Ternand, C.L.; Sullivan, S.; et al. Decrease in anogenital dis- tance among male infants with prenatal phthalate exposure. *Environ. Health Perspect.* **2005**, *113*, 1056–1061. [CrossRef]
55. Kaur, M.; Verma, A.; Setia, H.; Toor, A.P. Comparative Study on the Photocatalytic Degradation of Paraquat Using Tungsten-Doped TiO_2 Under UV and Sunlight. In *Sustainable Engineering*; Agnihotri, A., Reddy, K., Bansal, A., Eds.; Springer: Singapore, 2019; Volume 30, pp. 145–155. [CrossRef]
56. Akbari-Adergani, B.; Saghi, M.H.; Eslami, A.; Mohseni-Bandpei, A.; Rabbani, M. Removal of dibutyl phthalate from aqueous environments using a nanophotocatalytic Fe, Ag-ZnO/VIS-LED system: Modeling and optimization. *Environ. Technol.* **2018**, *39*, 1566–1576. [CrossRef]
57. Motlagh, P.Y.; Khataee, A.; Hassani, A.; Rad, T.S. ZnFe-LDH/GO nanocomposite coated on the glass support as a highly efficient catalyst for visible light photodegra- dation of an emerging pollutant. *J. Mol. Liq.* **2020**, *302*, 112532. [CrossRef]
58. Dong, W.; Zhu, Y.; Huang, H.; Jiang, L.; Zhu, H.; Li, C.; Chen, B.; Shi, Z.; Wang, G. A performance study of enhanced visible-light-driven photocatalysis and magnetical protein separation of multifunctional yolk–shell nanostructures. *J. Mater. Chem. A* **2013**, *1*, 10030–10036. [CrossRef]
59. Chang, C.F.; Man, C.Y. Titania-coated magnetic composites as photocatalysts for phthalate photodegradation. *Ind. Eng. Chem. Res.* **2011**, *50*, 11620–11627. [CrossRef]
60. Ghanbari, F.; Zirrahi, F.; Olfati, D.; Gohari, F.; Hassani, A. TiO_2 nanoparticles removal by electrocoagulation using iron electrodes: Catalytic activity of electrochemical sludge for the degradation of emerging pollutant. *J. Mol. Liq.* **2020**, *310*, 113217. [CrossRef]
61. Gu, X.; Qin, N.; Wei, G.; Hu, Y.; Zhang, Y.-N.; Zhao, G. Efficient photocatalytic removal of phthalates easily implemented over a bi-functional {001}TiO_2 surface. *Chemosphere* **2020**, *263*, 128257. [CrossRef]
62. Huang, W.-B.; Chen, C.-Y. Photocatalytic Degradation of Diethyl Phthalate (DEP) in Water Using TiO_2. *Water Air Soil Pollut.* **2010**, *207*, 349–355. [CrossRef]
63. Liao, W.; Zheng, T.; Wang, P.; Tu, S.; Pan, W. Efficient microwave-assisted photocatalytic degradation of endocrine disruptor dimethyl phthalate over composite catalyst ZrOx/ZnO. *J. Environ. Sci.* **2010**, *22*, 1800–1806. [CrossRef]
64. Jing, Y.; Li, L.; Zhang, Q.; Lu, P.; Liu, P.; Lü, X. Photocatalytic ozonation of dimethyl phthalate with TiO_2 prepared by a hydrothermal method. *J. Hazard. Mater.* **2011**, *189*, 40–47. [CrossRef] [PubMed]
65. Kaneco, S.; Katsumata, H.; Suzuki, T.; Ohta, K. Titanium dioxide mediated photocatalytic degradation of dibutyl phthalate in aqueous solution—kinetics, mineralization and reaction mechanism. *Chem. Eng. J.* **2006**, *125*, 59–66. [CrossRef]
66. Liu, Y.; Sun, N.; Hu, J.; Li, S.; Qin, G. Photocatalytic degradation properties of α-Fe_2O_3 nanoparticles for dibutyl phthalate in aqueous solution system. *R. Soc. Open Sci.* **2018**, *5*, 172196. [CrossRef]
67. Xu, X.-R.; Li, S.-X.; Li, X.-Y.; Gu, J.-D.; Chen, F.; Li, X.-Z.; Li, H.-B. Degradation of n-butyl benzyl phthalate using TiO_2/UV. *J. Hazard. Mater.* **2009**, *164*, 527–532. [CrossRef]
68. Wang, X.-K.; Wang, C.; Jiang, W.-Q.; Guo, W.-L.; Wang, J.-G. Sonochemical synthesis and characterization of Cl-doped TiO_2 and its application in the photodegradation of phthalate ester under visible light irradiation. *Chem. Eng. J.* **2012**, *189–190*, 288–294. [CrossRef]
69. Mohamed, R.M.; Aazam, E. Synthesis and characterization of P-doped TiO_2 thin-films for photocatalytic degradation of butyl benzyl phthalate under visible-light irradiation. *Chin. J. Catal.* **2013**, *34*, 1267–1273. [CrossRef]
70. Anandan, S.; Pugazhenthiran, N.; Lana-Villarreal, T.; Lee, G.-J.; Wu, J.J. Catalytic degradation of a plasticizer, di-ethylhexyl phthalate, using Nx–TiO_2–x nanoparticles synthesized via co-precipitation. *Chem. Eng. J.* **2013**, *231*, 182–189. [CrossRef]
71. Eslami, A.; Akbari-Adergani, B.; Mohseni-Bandpei, A.; Rabbani, M.; Saghi, M.H. Synthesis and characterization of a coated Fe-Ag@ZnO nanorod for the purification of a polluted environmental solution under simulated sunlight irradiation. *Mater. Lett.* **2017**, *197*, 205–208. [CrossRef]
72. Zhou, W.; Yu, C.; Fan, Q.; Wei, L.; Chen, J.; Yu, J.C. Ultrasonic fabrication of N-doped TiO_2 nanocrystals with mesoporous structure and enhanced visible light photocatalytic activity. *Chin. J. Catal.* **2013**, *34*, 1250–1255. [CrossRef]
73. Singla, P.; Pandey, O.P.; Singh, K. Study of photocatalytic degradation of environmentally harmful phthalate esters using Ni-doped TiO_2 nanoparticles. *Int. J. Environ. Sci. Technol.* **2016**, *13*, 849–856. [CrossRef]
74. Ki, S.J.; Park, Y.-K.; Kim, J.-S.; Lee, W.-J.; Lee, H.; Jung, S.-C. Facile preparation of tungsten oxide doped TiO_2 photocatalysts using liquid phase plasma process for enhanced degradation of diethyl phthalate. *Chem. Eng. J.* **2018**, *377*, 120087. [CrossRef]
75. Jamil, T.S.; Abbas, H.A.; Youssief, A.M.; Mansor, E.S.; Hammad, F.F. The synthesis of nano-sized undoped, Bi doped and Bi, Cu co-doped $SrTiO_3$ using two sol–gel methods to enhance the photocatalytic performance for the degradation of dibutyl phthalate under visible light. *Comptes Rendus. Chim.* **2017**, *20*, 97–106. [CrossRef]
76. He, G.; Zhang, J.; Hu, Y.; Bai, Z.; Wei, C. Dual-template synthesis of mesoporous TiO_2 nanotubes with structure-enhanced functional photocatalytic performance. *Appl. Catal. B Environ.* **2019**, *250*, 301–312. [CrossRef]
77. Chalasani, R.; Vasudevan, S. Cyclodextrin-Functionalized Fe_3O_4@TiO_2: Reusable, Magnetic Nanoparticles for Photocatalytic Degradation of Endocrine-Disrupting Chemicals in Water Supplies. *ACS Nano* **2013**, *7*, 4093–4104. [CrossRef]
78. Shan, W.; Hu, Y.; Bai, Z.; Zheng, M.; Wei, C. In situ preparation of g-C_3N_4/bismuth-based oxide nanocomposites with enhanced photocatalytic activity. *Appl. Catal. B Environ.* **2016**, *188*, 1–12. [CrossRef]
79. Meenakshi, G.; Sivasamy, A. Nanorod ZnO/SiC nanocomposite: An efficient catalyst for the degradation of an endocrine disruptor under UV and visible light irradiations. *J. Environ. Chem. Eng.* **2018**, *6*, 3757–3769. [CrossRef]

80. Hassan, H.M.; Betiha, M.A.; El-Sharkawy, E.A.; Elshaarawy, R.F.; El-Assy, N.B.; Essawy, A.A.; Tolba, A.M.; Rabie, A.M. Highly selective epoxidation of olefins using vanadium (IV) schiff base-amine-tagged graphene oxide composite. *Colloids Surf. A Physicochem. Eng. Asp.* **2020**, *591*, 124520. [CrossRef]
81. Tang, W.; Chen, J.; Yin, Z.; Sheng, W.; Lin, F.; Xu, H.; Cao, S. Complete removal of phenolic contaminants from bismuth-modified TiO_2 single-crystal photocatalysts. *Chin. J. Catal.* **2021**, *42*, 347–355. [CrossRef]
82. Xu, C.; Jin, Z.; Yang, J.; Guo, F.; Wang, P.; Meng, H.; Bao, G.; Li, Z.; Chen, C.; Liu, F.; et al. A direct Z-scheme $LaFeO_3/WO_3$ photocatalyst for enhanced degradation of phenol under visible light irradiation. *J. Environ. Chem. Eng.* **2021**, *9*, 106337. [CrossRef]
83. Prasad, C.; Liu, Q.; Tang, H.; Yuvaraja, G.; Long, J.; Rammohan, A.; Zyryanov, G.V. An overview of graphene oxide supported semiconductors based photocatalysts: Properties, synthesis and photocatalytic applications. *J. Mol. Liq.* **2020**, *297*, 111826. [CrossRef]
84. Hurtado, L.; Amado-Piña, D.; Roa-Morales, G.; Peralta, E.; Del Campo, E.M.; Natividad, R. Comparison of AOPs Efficiencies on Phenolic Compounds Degradation. *J. Chem.* **2016**, *2016*, 1–8. [CrossRef]
85. Paschoalino, F.C.S.; Paschoalino, M.P.; Jordão, E.; Jardim, W.D.F. Evaluation of TiO_2, ZnO, CuO and Ga_2O_3 on the Photocatalytic Degradation of Phenol Using an Annular-Flow Photocatalytic Reactor. *Open J. Phys. Chem.* **2012**, *2*, 135–140. [CrossRef]
86. Shet, A.; Vidya, S.K. Solar light mediated photocatalytic degradation of phenol using Ag core—TiO_2 shell ($Ag@TiO_2$) nanoparticles in batch and fluidized bed reactor. *Sol. Energy* **2016**, *127*, 67–78. [CrossRef]
87. Rueda-Marquez, J.J.; Levchuk, I.; FernándezIbañez, P.; Sillanpää, M. A critical review on application of photocatalysis for toxicity reduction of real wastewaters. *J. Clean. Prod.* **2020**, *258*, 120694. [CrossRef]
88. Chowdhury, P.; Nag, S.; Ray, A.K. Degradation of Phenolic Compounds Through UV and Visible- Light-Driven Photocatalysis: Technical and Economic Aspects. In *Phenolic Compounds—Natural Sources, Importance and Applications*; IntechOpen: London, UK, 2017. [CrossRef]
89. Malefane, M.E. $Co_3O_4/Bi_4O5I_2/Bi_5O_7I$ C-Scheme Heterojunction for Degradation of Organic Pollutants by Light-Emitting Diode Irradiation. *ACS Omega* **2020**, *5*, 26829–26844. [CrossRef]
90. Chowdhury, P.; Gomaa, H.; Ray, A.K. Sacrificial hydrogen generation from aqueous triethanolamine with Eosin Y-sensitized Pt/TiO_2 photocatalyst in UV, visible and solar light irradiation. *Chemosphere* **2015**, *121*, 54–61. [CrossRef]
91. Chowdhury, P.; Moreira, J.; Gomaa, H.; Ray, A.K. Visible-Solar-Light-Driven Photocatalytic Degradation of Phenol with Dye-Sensitized TiO_2: Parametric and Kinetic Study. *Ind. Eng. Chem. Res.* **2012**, *51*, 4523–4532. [CrossRef]
92. Qin, G.; Wu, Q.; Sun, Z.; Wang, Y.; Luo, J.; Xue, S. Enhanced photoelectrocatalytic degradation of phenols with biofunctionalizedd dye-sensitized TiO_2 film. *J. Hazard. Mater.* **2012**, *199*, 226–232. [CrossRef]
93. Rahmani, A.; Rahimzadeh, H.; Beirami, S. Photo-Degradation of Phenol Using $TiO_2/CMK-3$ Photo-Catalyst Under Medium Pressure UV Lamp. *Avicenna J. Environ. Health Eng.* **2018**, *5*, 35–41. [CrossRef]
94. Mohamed, A.; Yousef, S.; Nasser, W.S.; Osman, T.A.; Knebel, A.; Sánchez, E.P.V.; Hashem, T. Rapid photocatalytic degradation of phenol from water using composite nanofbers under UV. *Environ. Sci. Eur.* **2020**, *32*, 160. [CrossRef]
95. Górska, P.; Zaleska-Medynska, A.; Jan, H. Photodegradation of phenol by UV/TiO_2 and $Vis/N,C-TiO_2$ processes: Comparative mechanistic and kinetic studies. *Sep. Purif. Technol.* **2009**, *68*, 90–96. [CrossRef]
96. Liu, T.; Tan, G.; Zhao, C.; Xu, C.; Su, Y.; Wang, Y.; Ren, H.; Xia, A.; Shao, D.; Yan, S. Enhanced photocatalytic mechanism of the Nd-Er co-doped tetragonal $BiVO_4$ photocatalysts. *Appl. Catal. B Environ.* **2017**, *213*, 87–96. [CrossRef]
97. Sin, J.-C.; Lam, S.-M.; Zeng, H.; Lin, H.; Li, H.; Tham, K.-O.; Mohamed, A.R.; Lim, J.-W.; Qing, Z. Magnetic $NiFe_2O_4$ nanoparticles decorated on N-doped BiOBr nanosheets for expeditious visible light photocatalytic phenol degradation and hexavalent chromium reduction via a Z-scheme heterojunction mechanism. *Appl. Surf. Sci.* **2021**, *559*, 149966. [CrossRef]
98. Zhang, Y.; Li, G.; Yang, X.; Yang, H.; Lu, Z.; Chen, R. Monoclinic $BiVO_4$ micro-/nanostructures: Microwave and ultrasonic wave combined synthesis and their visible-light photocatalytic activities. *J. Alloy. Compd.* **2013**, *551*, 544–550. [CrossRef]
99. Wang, X.; Zhou, J.; Zhao, S.; Chen, X.; Yu, Y. Synergistic effect of adsorption and visible-light photocatalysis for organic pollutant removal over $BiVO_4$/carbon sphere nanocomposites. *Appl. Surf. Sci.* **2018**, *453*, 394–404. [CrossRef]
100. Li, H.; Chen, Y.; Zhou, W.; Gao, H.; Tian, G. Tuning in $BiVO_4/Bi_4V_2O_{10}$ porous heterophase nanospheres for synergistic photocatalytic degradation of organic pollutants. *Appl. Surf. Sci.* **2019**, *470*, 631–638. [CrossRef]
101. Deng, Y.; Xiao, Y.; Zhou, Y.; Zeng, T.; Xing, M.; Zhang, J. A structural engineering-inspired CdS based composite for photocatalytic remediation of organic pollutant and hexavalent chromium. *Catal. Today* **2019**, *335*, 101–109. [CrossRef]
102. Mbiri, A.; Taffa, D.H.; Gatebe, E.; Wark, M. Zirconium doped mesoporous TiO_2 multilayer thin films: Influence of the zirconium content on the photodegradation of organic pollutants. *Catal. Today* **2019**, *328*, 71–78. [CrossRef]
103. Li, K.; Zhong, Y.; Luo, S.; Deng, W. Fabrication of powder and modular $H_3PW_{12}O_{40}/Ag_3PO_4$ composites: Novel visible-light photocatalysts for ultra-fast degradation of organic pollutants in water. *Appl. Catal. B Environ.* **2020**, *278*, 119313. [CrossRef]
104. Yu, X.; Wang, L.; Feng, L.-j.; Li, C.-h. Preparation of Au/BiOBr/Graphene composite and its photocatalytic performancein phenol degradation under visible light. *J. Fuel Chem. Technol.* **2016**, *44*, 937–942. [CrossRef]
105. Pradhan, G.K.; Padhi, D.K.; Parida, K.M. Fabrication of α-Fe_2O_3 nanorod/RGO composite: A novel hybrid photocatalyst for phenol degradation. *ACS Appl. Mater. Interfaces* **2013**, *5*, 9101–9110. [CrossRef] [PubMed]
106. Liu, H.; Su, Y.; Chen, Z.; Jin, Z.; Wang, Y. $Bi_7O_9I_3$/reduced grapheme oxide composite as an efficient visible-light-driven photocatalyst for degradation of organic contaminants. *J. Mol. Catal. A Chem.* **2014**, *391*, 175–182. [CrossRef]

107. Singh, P.; Raizada, P.; Sudhaik, A.; Shandilya, P.; Thakur, P.; Agarwal, S.; Gupta, V.K. Enhanced photocatalytic activity and stability of AgBr/BiOBr/graphene heterojunction for phenol degradation under visible light. *J. Saudi Chem. Soc.* **2019**, *23*, 586–599. [CrossRef]
108. Abdullah, N.S.A.; So'aib, S.; Krishnan, J. Effect of calcination temperature on ZnO/TiO$_2$ composite in photocatalytic treatment of phenol under visible light. *Malays. J. Anal. Sci.* **2017**, *21*, 173–181.
109. Peng, W.-C.; Wang, X.; Li, X.-Y. The synergetic effect of MoS$_2$ and graphene on Ag$_3$PO$_4$ for its ultra-enhanced photocatalytic activity in phenol degradation under visible light. *Nanoscale* **2014**, *6*, 8311–8317. [CrossRef]
110. Hayati, F.; Isari, A.A.; Fattahi, M.; Anvaripour, B.; Jorfibc, S. Photocatalytic decontamination of phenol and petrochemical wastewater through ZnO/TiO$_2$ decorated on reduced graphene oxide nanocomposite: Influential operating factors, mechanism, and electrical energy consumption. *RSC Adv.* **2018**, *8*, 40035–40053. [CrossRef]
111. Teixeira, I.; Quiroz, J.; Homsi, M.; Camargo, P. An Overview of the Photocatalytic H$_2$ Evolution by Semiconductor-Based Materials for Nonspecialists. *J. Braz. Chem. Soc.* **2020**, *31*, 211–229. [CrossRef]
112. Bobirică, C.; Bobirică, L.; Râpă, M.; Matei, E.; Predescu, A.M.; Orbeci, C. Photocatalytic Degradation of Ampicillin Using PLA/TiO$_2$ Hybrid Nanofibers Coated on Different Types of Fiberglass. *Water* **2020**, *12*, 176. [CrossRef]
113. Li, M.; Li, G.; Jiang, J.; Zhang, Z.; Dai, X.; Mai, K. Ultraviolet resistance and antimicrobial properties of ZnO in the polypropylene materials: A review. *J. Mater. Sci. Technol.* **2015**, *31*, 331–339. [CrossRef]
114. Saidu, U. Synthesis and Characterization of BiVO$_4$ nanoparticles and its Photocatalytic Activity on Levofloxacin Antibiotics. *ChemSearch J.* **2019**, *10*, 104–111.
115. Yang, X.; Chen, Z.; Zhao, W.; Liu, C.; Qian, X.; Zhang, M.; Wei, G.; Khan, E.; Ng, Y.H.; Ok, Y.S. Recent advances in photodegradation of antibiotic residues in water. *Chem. Eng. J.* **2021**, *405*, 126806. [CrossRef] [PubMed]
116. Lin, H.H.-H.; Lin, A.Y.-C. Photocatalytic oxidation of 5-fluorouracil and cyclophosphamide via UV/TiO$_2$ in an aqueous environment. *Water Res.* **2014**, *48*, 559–568. [CrossRef]
117. Olusegun, S.J.; Larrea, G.; Osial, M.; Jackowska, K.; Krysinski, P. Photocatalytic Degradation of Antibiotics by Superparamagnetic Iron Oxide Nanoparticles. Tetracycline Case. *Catalysts* **2021**, *11*, 1243. [CrossRef]
118. Dong, S.; Cui, L.; Zhang, W.; Xia, L.; Zhou, S.; Russell, C.; Fan, M.; Feng, J.; Sun, J. Double-shelled ZnSnO$_3$ hollow cubes for efficient photocatalytic degradation of antibiotic wastewater. *Chem. Eng. J.* **2020**, *384*, 123279. [CrossRef]
119. Wang, H.; Ye, Z.; Liu, C.; Li, J.; Zhou, M.; Guan, Q.; Lv, P.; Huo, P.; Yan, Y. Visible light driven Ag/Ag$_3$PO$_4$/AC photocatalyst with highly enhanced photodegradation of tetracycline antibiotics. *Appl. Surf. Sci.* **2015**, *353*, 391–399. [CrossRef]
120. Guo, F.; Li, M.; Ren, H.; Huang, X.; Shu, K.; Shi, W.; Lu, C. Facile bottom-up preparation of Cl-doped porous g-C$_3$N$_4$ nanosheets for enhanced photocatalytic degradation of tetracycline under visible light. *Sep. Purif. Technol.* **2019**, *228*, 115770. [CrossRef]
121. Zhou, C.; Lai, C.; Xu, P.; Zeng, G.; Huang, D.; Zhang, C.; Cheng, M.; Hu, L.; Wan, J.; Liu, Y.; et al. In Situ Grown AgI/Bi$_{12}$O$_{17}$Cl$_2$ Heterojunction Photocatalysts for Visible Light Degradation of Sulfamethazine: Efficiency, Pathway, and Mechanism. *ACS Sustain. Chem. Eng.* **2018**, *6*, 4174–4184. [CrossRef]
122. Xiao, T.; Tanga, Z.; Yang, Y.; Tang, L.; Zhou, Y.; Zou, Z. In situ construction of hierarchical WO$_3$/g-C$_3$N$_4$ composite hollow microspheres as a Z-scheme photocatalyst for the degradation of antibiotics. *ACS Sustain. Chem. Eng.* **2018**, *220*, 417–428. [CrossRef]
123. Kumar, A.; Kumar, A.; Sharma, G.; Al-Muhtaseb, A.H.; Naushad, M.; Ghfar, A.A.; Stadler, F.J. Quaternary magnetic BiOCl/g-C$_3$N$_4$/Cu$_2$O/Fe$_3$O$_4$ nano-junction for visible light and solar powered degradation of sulfamethoxazole from aqueous environment. *Chem. Eng. J.* **2018**, *334*, 462–478. [CrossRef]
124. Kumar, A.; Khan, M.; He, J.; Lo, I.M. Visible–light–driven magnetically recyclable terephthalic acid functionalized g–C$_3$N$_4$/TiO$_2$ heterojunction nanophotocatalyst for enhanced degradation of PPCPs. *Appl. Catal. B Environ.* **2020**, *270*, 118898. [CrossRef]
125. Al Balushi, B.S.; Al Marzouqi, F.; Al Wahaibi, B.; Kuvarega, A.T.; Al Kindy, S.M.Z.; Kim, Y.; Selvaraj, R. Hydrothermal synthesis of CdS sub-microspheres for photocatalytic degradation of pharmaceuticals. *Appl. Surf. Sci.* **2018**, *457*, 559–565. [CrossRef]
126. Alberti, S.; Locardi, F.; Sturini, M.; Speltini, A.; Maraschi, F.; Costa, G.A.; Ferretti, M.; Caratto, V. Photocatalysis in Darkness: Optimization of Sol-Gel Synthesis of NP-TiO$_2$ Supported on a Persistent Luminescence Material and its Application for the Removal of Ofloxacin from Water. *J. Nanomed. Nanotechnol.* **2018**, *9*, 1–6. [CrossRef]
127. Gao, J.; Gao, Y.; Sui, Z.; Dong, Z.; Wang, S.; Zou, D. Hydrothermal synthesis of BiOBr/FeWO$_4$ composite photocatalysts and their photocatalytic degradation of doxycycline. *J. Alloy. Compd.* **2018**, *732*, 43–51. [CrossRef]
128. Gong, Y.; Wu, Y.; Xu, Y.; Li, L.; Li, C.; Liu, X.; Niu, L. All-solid-state Z-scheme CdTe/TiO$_2$ heterostructure photocatalysts with enhanced visible-light photocatalytic degradation of antibiotic wastewater. *Chem. Eng. J.* **2018**, *350*, 257–267. [CrossRef]
129. Wang, D.; Jia, F.; Wang, H.; Chen, F.; Fang, Y.; Dong, W.; Zeng, G.; Li, X.; Yang, Q.; Yuan, X. Simultaneously efficient adsorption and photocatalytic degradation of tetracycline by Fe-based MOFs. *J. Colloid Interface Sci.* **2018**, *519*, 273–284. [CrossRef]
130. Cao, Y.; Lei, X.; Chen, Q.; Kang, C.; Li, W.; Liu, B. Enhanced photocatalytic degradation of tetracycline hydrochloride by novel porous hollow cube ZnFe$_2$O$_4$. *J. Photochem. Photobiol. A Chem.* **2018**, *364*, 794–800. [CrossRef]
131. Osotsi, M.I.; Macharia, D.K.; Zhu, B.; Wang, Z.; Shen, X.; Liu, Z.; Zhang, L.; Chen, Z. Synthesis of ZnWO$_{4-x}$ nanorods with oxygen vacancy for efficient photocatalytic degradation of tetracycline. *Prog. Nat. Sci.* **2018**, *28*, 408–415. [CrossRef]
132. Suwannaruang, T.; Hildebrand, J.P.; Taffa, D.H.; Wark, M.; Kamonsuangkasem, K.; Chirawatkul, P.; Wantala, K. Visible light-induced degradation of antibiotic ciprofloxacin over Fe–N–TiO$_2$ mesoporous photocatalyst with anatase/rutile/brookite nanocrystal mixture. *J. Photochem. Photobiol. A Chem.* **2020**, *391*, 112371. [CrossRef]

133. Kumar, J.V.; Karthik, R.; Chen, S.-M.; Chen, K.-H.; Sakthinathan, S.; Muthuraj, V.; Chiu, T.-W. Design of novel 3D flower-like neodymium molybdate: An efficient and challenging catalyst for sensing and destroying pulmonary toxicity antibiotic drug nitrofurantoin. *Chem. Eng. J.* **2018**, *346*, 11–23. [CrossRef]
134. Thongam, D.D.; Chaturvedi, H. Advances in nanomaterials for heterogeneous photocatalysis. *Nano Express* **2021**, *2*, 012005. [CrossRef]
135. Tammina, S.K.; Mandal, B.K.; Kadiyala, N.K. Photocatalytic degradation of methylene blue dye by nonconventional synthesized SnO_2 nanoparticles. *Environ. Nanotechnol. Monit. Manag.* **2018**, *10*, 339–350. [CrossRef]
136. Liua, H.; Guoa, W.; Lia, Y.; Heb, S.; Hea, C. Photocatalytic degradation of sixteen organic dyes by TiO_2/WO_3-coated magnetic nanoparticles under simulated visible light and solar light. *J. Environ. Chem. Eng.* **2018**, *6*, 59–67. [CrossRef]
137. Huang, H.; Zhang, J.; Jiang, L.; Zang, Z. Preparation of cubic Cu_2O nanoparticles wrapped by reduced graphene oxide for the efficient removal of rhodamine B. *J. Alloy. Compd.* **2017**, *718*, 112–115. [CrossRef]
138. Aggarwal, S. Photo Catalytic Degradation of Methyl Orange by Using CdS Semiconductor Nanoparticles Photo catalyst. *Int. Res. J. Eng. Technol.* **2016**, *3*, 451–455.
139. Aragaw, B.A.; Dagnaw, A. Copper/reduced graphene oxide nanocomposite for high performance photocatalytic methylene blue dye degradation. *Ethiop. J. Sci. Technol.* **2019**, *12*, 125–137. [CrossRef]
140. Raina, S.; Roy, A.; Bharadvaja, N. Degradation of dyes using biologically synthesized silver and copper nanoparticles. *Environ. Nanotechnol. Monit. Manag.* **2020**, *13*, 100278. [CrossRef]
141. Zhang, Z.; Sun, L.; Wu, Z.; Liu, Y.; Li, S. Facile hydrothermal synthesis of $CuO–Cu_2O/GO$ nanocomposites for the photocatalytic degradation of organic dye and tetracycline pollutants. *New J. Chem.* **2020**, *44*, 6420–6427. [CrossRef]
142. Bharathi, P.; Harish, S.; Archana, J.; Navaneethan, M.; Ponnusamy, S.; Muthamizhchelvan, C.; Shimomura, M.; Hayakawa, Y. Enhanced charge transfer and separation of hierarchical CuO/ZnO composites: The synergistic effect of photocatalysis for the mineralization of organic pollutant in water. *Appl. Surf. Sci.* **2019**, *484*, 884–891. [CrossRef]
143. Wei, X.-N.; Ou, C.-L.; Guan, X.-X.; Peng, Z.-K.; Zheng, X.-C. Facile assembly of CdS-reduced graphene oxide heterojunction with enhanced elimination performance for organic pollutants in wastewater. *Appl. Surf. Sci.* **2019**, *469*, 666–673. [CrossRef]
144. Adhikari, S.; Chandra, K.S.; Kim, D.-H.; Madras, G.; Sarkar, D. Understanding the morphological effects of WO_3 photocatalysts for the degradation of organic pollutants. *Adv. Powder Technol.* **2018**, *29*, 1591–1600. [CrossRef]
145. Bai, X.; Du, Y.; Hu, X.; He, Y.; He, C.; Liu, E.; Fan, J. Synergy removal of Cr(VI) and organic pollutants over $RP-MoS2/rGO$ photocatalyst. *Appl. Catal. B Environ.* **2018**, *239*, 204–213. [CrossRef]
146. Grzechulska, J.; Morawski, A.W. Photocatalytic decomposition of azo-dye acid black 1 in water over modified titanium dioxide. *Appl. Catal. B Environ.* **2001**, *36*, 45–51. [CrossRef]
147. Gnanaprakasam, A.; Sivakumar, V.M.; Thirumarimurugan, M. Influencing Parameters in the Photocatalytic Degradation of Organic Effluent via Nanometal Oxide Catalyst: A Review. *Indian J. Mater. Sci.* **2015**, *2015*, 601827. [CrossRef]
148. Senthilvelan, S.; Chandraboss, V.L.; Karthikeyan, B.; Natanapatham, L.; Murugavelu, M. TiO_2, ZnO and nanobimetallic silica catalyzedphotodegradation of methyl green. *Mater. Sci. Semicond. Process.* **2013**, *16*, 185–192. [CrossRef]
149. Karimi, L.; Zohoori, S.; Yazdanshenas, M.E. Photocatalytic degradation of azo dyes in aqueous solutions under UV irradiation using nano-strontium titanate as the nanophotocatalyst. *J. Saudi Chem. Soc.* **2014**, *18*, 581–588. [CrossRef]
150. Neppolian, B.; Kanel, S.R.; Choi, H.C.; Shankar, M.V.; Arabindoo, B.; Murugesan, V. Photocatalytic degradation of reactive yellow 17 dye in aqueous solution in the presence of TiO_2 with cement binder. *Int. J. Photoenergy* **2003**, *5*, 45–49. [CrossRef]
151. Mai, F.D.; Lu, C.S.; Wu, C.W.; Huang, C.H.; Chen, J.Y.; Chen, C.C. Mechanisms of photocatalytic degradation of Victoria Blue R using nano-TiO_2. *Sep. Purif. Technol.* **2008**, *62*, 423–436. [CrossRef]
152. Neppolian, B.; Choi, H.C.; Sakthivel, S.; Arabindoo, B.; Murugesan, V. Solar/UV-induced photocatalytic degradation of three commercial textile dyes. *J. Hazard. Mater.* **2002**, *89*, 303–317. [CrossRef]
153. Pouretedal, H.R.; Norozi, A.; Keshavarz, M.H.; Semnani, A. Nanoparticles of zinc sulfide doped with manganese, nickel and copper as nanophotocatalyst in the degradation of organic dyes. *J. Hazard. Mater.* **2009**, *162*, 674–681. [CrossRef] [PubMed]
154. Mathialagan, A.; Manavalan, M.; Venkatachalam, K.; Mohammad, F.; Oh, W.C.; Sagadevan, S.; Sagadevan, S. Fabrication and physicochemical characterization of g-C3N4/ZnO composite with enhanced photocatalytic activity under visible light. *Opt. Mater.* **2020**, *100*, 109643. [CrossRef]
155. Qutub, N.; Singh, P.; Sabir, S.; Sagadevan, S.; Oh, W. Enhanced photocatalytic degradation of Acid Blue dye using CdS/TiO_2 nanocomposite. *Sci. Rep.* **2022**, *12*, 5759. [CrossRef] [PubMed]
156. Muthukumaran, M.; Prasath, P.V.; Kulandaivelu, R.; Mohammad, F.; Oh, W.C. Fabrication of nitrogen-rich graphitic carbon nitride/Cu_2O ($gC3N4@Cu_2O$) composite and its enhanced photocatalytic activity for organic pollutants degradation. *J. Mater. Sci. Mater. Electron.* **2020**, *31*, 2257–2268. [CrossRef]
157. Priya, R.; Stanly, S.; Dhanalekshmi, S.B.; Mohammad, F.; Al-Lohedan, H.A.; Oh, W.C.; Sagadevan, S. Comparative studies of crystal violet dye removal between semiconductor nanoparticles and natural adsorbents. *Optik* **2020**, *206*, 164281. [CrossRef]
158. Sagadevan, S.; Lett, J.A.; Weldegebrieal, G.K.; Garg, S.; Oh, W.-C.; Hamizi, N.A.; Johan, M.R. Enhanced Photocatalytic Activity of rGO-CuO Nanocomposites for the Degradation of Organic Pollutants. *Catalysts* **2021**, *11*, 1008. [CrossRef]
159. Priya, R.; Stanly, S.; Kavitharani, T.; Mohammad, F.; Sagadevan, S. Highly effective photocatalytic degradation of methylene blue using PrO_2-MgO nanocomposites under UV light. *Optik* **2020**, *206*, 164318.

160. Muthukumaran, M.; Gnanamoorthy, G.; Prasath, P.V.; Abinaya, M.; Dhinagaran, G.; Sagadevan, S.; Mohammad, F.; Oh, W.C.; Venkatachalam, K. Enhanced photocatalytic activity of Cuprous Oxide nanoparticles for malachite green degradation under the visible light radiation. *Mater. Res. Express* **2020**, *7*, 015038. [CrossRef]
161. Li, G.; Lv, L.; Fan, H.; Ma, J.; Li, Y.; Wan, Y.; Zhao, X. Effect of the agglomeration of TiO_2 nanoparticles on their photocatalytic performance in the aqueous phase. *J. Colloid Interface Sci.* **2010**, *348*, 342–347. [CrossRef]
162. Yousefi, A.; Allahverdi, A.; Hejazi, P. Effective dispersion of nano-TiO_2 powder for enhancement of photocatalytic properties in cement mixes. *Constr. Build. Mater.* **2013**, *41*, 224–230. [CrossRef]
163. Pradeev Raj, K.; Sadaiyandi, K.; Kennedy, A.; Sagadevan, S. Photocatalytic and antibacterial studies of indium-doped ZnO nanoparticles synthesized by co-precipitation technique. *J. Mater. Sci. Mater. Electron.* **2017**, *28*, 19025–19037. [CrossRef]
164. Shie, J.-L.; Lee, C.-H.; Chiou, C.-S.; Chang, C.-T.; Chang, C.-C.; Chang, C.-Y. Photodegradation kinetics of formaldehyde using light sources of UVA, UVC and UVLED in the presence of composed silver titanium oxide photocatalyst. *J. Hazard. Mater.* **2008**, *155*, 164–172. [CrossRef] [PubMed]
165. Sobana, N.; Selvam, K.; Swaminathan, M. Optimization of photocatalytic degradation conditions of Direct Red 23 using nano-Ag doped TiO_2. *Sep. Purif. Technol.* **2008**, *62*, 648–653. [CrossRef]
166. Huang, S.; Chen, C.; Tsai, H.; Shaya, J.; Lu, C. Photocatalytic degradation of thiobencarb by a visible light-driven MoS_2 photocatalyst. *Sep. Purif. Technol.* **2018**, *197*, 147–155. [CrossRef]
167. Chen, C.; Liu, J.; Liu, P.; Yu, B. Investigation of Photocatalytic Degradation of Methyl Orange by Using Nano-Sized ZnO Catalysts. *Adv. Chem. Eng. Sci.* **2011**, *1*, 9–14. [CrossRef]
168. Schneider, J.; Matsuoka, M.; Takeuchi, M.; Zhang, J.; Horiuchi, Y.; Anpo, M.; Bahnemann, D.W. Understanding TiO_2 Photocatalysis: Mechanisms and Materials. *Chem. Rev.* **2014**, *114*, 9919–9986. [CrossRef]
169. Yu, T.; Tan, X.; Zhao, L.; Yin, Y.; Chen, P.; Wei, J. Characterization, activity and kinetics of a visible light driven photocatalyst: Cerium and nitrogen co-doped $TiO2$ nanoparticles. *Chem. Eng. J.* **2010**, *157*, 86–92. [CrossRef]
170. Margarita Skiba, V. Vorobyova Synthesis of Ag/TiO_2 nanocomposite via plasma liquid interactions and degradation methylene blue. *Appl. Nanosci.* **2020**, *10*, 4717–4723. [CrossRef]
171. Kasinathan, K.; Kennedy, J.; Elayaperumal, M.; Henini, M.; Malik, M. Photodegradation of organic pollutants RhB dye using UV simulated sunlight on ceria based TiO_2 nanomaterials for antibacterial applications. *Sci. Rep.* **2016**, *6*, 38064. [CrossRef] [PubMed]
172. Ng, K.H.; Lee, C.H.; Khan, M.R.; Cheng, C.K. Photocatalytic degradation of recalcitrant POME waste by using silver doped titania: Photokinetics and scavenging studies. *Chem. Eng. J.* **2016**, *286*, 282–290. [CrossRef]
173. Zielińska-Jurek, A.; Kowalska, E.; Sobczak, J.; Łącka, I.; Gazda, M.; Ohtani, B.; Hupka, J.; Zaleska, A. Silver-doped TiO_2 prepared by microemulsion method: Surface properties, bio- and photoactivity. *Sep. Purif. Technol.* **2010**, *72*, 309–318. [CrossRef]
174. Asiltürk, M.; Sayılkan, F.; Arpaç, E. Effect of Fe^{3+} ion doping to TiO_2 on the photocatalytic degradation of Malachite Green dye under UV and vis-irradiation. *J. Photochem. Photobiol. A* **2009**, *203*, 64–71. [CrossRef]
175. Crişan, M.; Mardare, D.; Ianculescu, A.; Drăgan, N.; Niţoi, I.; Crişan, D.; Voicescu, M.; Todan, L.; Oancea, P.; Adomniţei, C.; et al. Iron doped TiO_2 films and their photoactivity in nitrobenzene removal from water. *Appl. Surf. Sci.* **2018**, *455*, 201–215. [CrossRef]
176. Krishnakumar, V.; Boobas, S.; Jayaprakash, J.; Rajaboopathi, M.; Han, B.; Louhi-Kultanen, M. Effect of Cu doping on TiO_2 nanoparticles and its photocatalytic activity under visible light. *J. Mater. Sci. Mater. Electron.* **2016**, *27*, 7438–7447. [CrossRef]
177. Chiang, L.F.; Doong, R. Cu-TiO_2 nanorods with enhanced ultraviolet- and visible-light photoactivity for bisphenol A degradation. *J. Hazard. Mater.* **2014**, *277*, 84–92. [CrossRef] [PubMed]
178. Nakhate, G.G.; Nikam, V.S.; Kanade, K.G.; Arbuj, S.; Kale, B.; Baeg, J.O. Hydrothermally derived nanosized Ni-doped TiO_2: A visible light driven photocatalyst for methylene blue degradation. *Mater. Chem. Phys.* **2010**, *124*, 976–981. [CrossRef]
179. Jiang, P.; Xiang, W.; Kuang, J.; Liu, W.; Cao, W. Effect of cobalt doping on the electronic, optical and photocatalytic properties of TiO_2. *Solid State Sci.* **2015**, *46*, 27–32. [CrossRef]
180. Natarajan, T.S.; Natarajan, K.; Bajaj, H.C.; Tayade, R.J. Enhanced photocatalytic activity of bismuth-doped TiO_2 nanotubes under direct sunlight irradiation for degradation of Rhodamine B dye. *J. Nanoparticle Res.* **2013**, *15*, 1–18. [CrossRef]
181. Abdelhaleem, A.; Chu, W. Photodegradation of 4-chlorophenoxyacetic acid under visible LED activated N-doped TiO_2 and the mechanism of stepwise rate increment of the reused catalyst. *J. Hazard. Mater.* **2017**, *338*, 491–501. [CrossRef]
182. Boningari, T.; Inturi, S.N.R.; Suidan, M.; Smirniotis, P.G. Novel continuous single-step synthesis of nitrogen-modified TiO_2 by flame spray pyrolysis for photocatalytic degradation of phenol in visible light. *J. Mater. Sci. Technol.* **2018**, *34*, 1494–1502. [CrossRef]
183. Grabowska-Musiał, E.; Zaleska-Medynska, A.; Sobczak, J.; Gazda, M.; Hupka, J. Boron-doped TiO_2: Characteristics and photoactivity under visible light. *Procedia Chem.* **2009**, *1*, 1553–1559. [CrossRef]
184. Yu, C.; Fan, Q.; Xie, Y.; Chen, J.; Shu, Q.; Yu, J. Sonochemical fabrication of novel square-shaped F doped TiO_2 nanocrystals with enhanced performance in photocatalytic degradation of phenol. *J. Hazard. Mater.* **2012**, *237–238*, 38–45. [CrossRef]
185. Mrowetz, M.; Selli, E. Photocatalytic degradation of formic and benzoic acids and hydrogen peroxide evolution in TiO_2 and ZnO water suspensions. *J. Photochem. Photobiol. A Chem.* **2006**, *180*, 15–22. [CrossRef]
186. Wang, C.; Li, J.; Mele, G.; Yang, G.-M.; Zhang, F.-X.; Palmisano, L.; Vasapollo, G. Efficient degradation of 4-nitrophenol by using functionalized porphyrin-TiO_2 photocatalysts under visible irradiation. *Appl. Catal. B Environ.* **2007**, *76*, 218–226. [CrossRef]
187. Xiao, Q.; Zhang, J.; Xiao, C.; Si, Z.; Tan, X. Solar photocatalytic degradation of methylene blue in carbon-doped TiO_2 nanoparticles suspension. *Sol. Energy* **2008**, *82*, 706–713. [CrossRef]

188. Abbasi, S.; Hasanpour, M. The effect of pH on the photocatalytic degradation of methyl orange using decorated ZnO nanoparticles with SnO$_2$ nanoparticles. *J. Mater. Sci. Mater. Electron.* **2016**, *28*, 1307–1314. [CrossRef]
189. Gusain, R.; Gupta, K.; Joshi, P.; Khatri, O.P. Adsorptive removal and photocatalytic degradation of organic pollutants using metal oxides and their composites: A comprehensive review. *Adv. Colloid Interface Sci.* **2019**, *272*, 102009. [CrossRef]
190. Nosaka, Y.; Nosaka, A. Understanding Hydroxyl Radical (•OH) Generation Processes in Photocatalysis. *ACS Energy Lett.* **2016**, *1*, 356–359. [CrossRef]
191. Wang, W.-Y.; Ku, Y. Effect of solution pH on the adsorption and photocatalytic reaction behaviors of dyes using TiO$_2$ and Nafion-coated TiO$_2$. *Colloids Surf. A Physicochem. Eng. Asp.* **2007**, *302*, 261–268. [CrossRef]
192. Etacheri, V.; Di Valentin, C.; Schneider, J.; Bahnemann, D.; Pillai, S.C. Visible-light activation of TiO$_2$ photocatalysts: Advances in theory and experiments. *J. Photochem. Photobiol. C Photochem. Rev.* **2015**, *25*, 1–29. [CrossRef]
193. Zargoosh, K.; Rostami, M.; Aliabadi, H.M. Eu^{2+}- and Nd^{3+}-Doped CaAl$_2$O$_4$/WO$_3$/polyester nanocomposite as a sunlight-activated photocatalyst for fast removal of dyes from industrial wastes. *J. Mater. Sci. Mater. Electron.* **2020**, *31*, 11482–11495. [CrossRef]
194. Alkaim, A.; Aljeboree, A.; Alrazaq, N.; Baqir, S.; Hussein, F.; Lilo, A. Effect of pH on Adsorption and Photocatalytic Degradation Efficiency of Different Catalysts on Removal of Methylene Blue. *Asian J. Chem.* **2014**, *26*, 8445–8448. [CrossRef]
195. Fatin, S.O.; Lim, H.N.; Tan, W.T.; Huang, N.M. Comparison of photocatalytic activity and cyclic voltammetry of zinc oxide and titanium dioxide nanoparticles toward degradation of methylene blue. *Int. J. Electrochem. Sci.* **2012**, *7*, 9074–9084.
196. Bahnemann, D. Photocatalytic detoxification of polluted waters. In *The Handbook of environmental Chemistry 2. Part L: Environmental Photochemistry*; Boule, P., Ed.; Springer: Berlin, Germany, 1999; pp. 285–351.
197. Hung, C.H.; Yuan, C. Reduction of Azo-dye via TiO$_2$–photocatalysis. *J. Chin. Inst. Environ. Eng.* **2000**, *10*, 209–216.
198. Chanathaworn, J.; Bunyakan, C.; Wiyaratn, W.; Chungsiriporn, J. Photocatalytic decolorization of basic dye by TiO$_2$ nanoparticle in photoreactor. *Songklanakarin J. Sci. Technol.* **2012**, *34*, 203–210.
199. Liu, C.C.; Hsieh, Y.H.; Lai, P.F.; Li, C.H.; Kao, C.L. Photodegradation treatment of azo dye wastewater by UV/TiO$_2$ process. *Dye. Pigment.* **2006**, *68*, 191–195. [CrossRef]
200. Sakthivel, S.; Neppolian, B.; Shankar, M.V.; Arabindoo, B.; Palanichamy, M.; Murugesan, V. Solar photocatalytic degradation of azo dye: Comparison of photocatalytic efficiency of ZnO and TiO$_2$. *Sol. Energy Mater. Sol. Cells* **2003**, *77*, 65–82. [CrossRef]
201. So, C.M.; Cheng, M.Y.; Yu, J.C.; Wong, P.K. Degradation of azo dye procion red MX-5B by photocatalytic oxidation. *Chemosphere* **2002**, *46*, 905–912. [CrossRef]
202. Rao, K.V.S.; Subrahmanyam, M.; Boule, P. Immobilized TiO$_2$ photocatalyst during long-term use: Decrease of its activity. *Appl Catal B* **2004**, *49*, 239–249. [CrossRef]
203. Nguyen, V.H.; Shawn, D.L.; Wu, J.C.S.; Bai, H. Artificial sunlight and ultraviolet light induced photo-epoxidation of propylene over V-Ti/MCM-41 photocatalyst. *J. Nanotechnol.* **2014**, *5*, 566–576. [CrossRef]
204. Taddesse, A.M.; Alemu, M.; Kebede, T. Enhanced photocatalytic activity of p-n-n heterojunctions ternary composite Cu$_2$O/ZnO/Ag$_3$PO$_4$ under visible light irradiation. *J. Environ. Chem. Eng.* **2020**, *8*, 104356. [CrossRef]
205. Zhang, J.Y.; Mei, J.Y.; Yi, S.S.; Guan, X.X. Constructing of Z-scheme 3D g-C$_3$N$_4$-ZnO@graphene aerogel heterojunctions for high-efficient adsorption and photodegradation of organic pollutants. *Appl. Surf. Sci.* **2019**, *492*, 808–817. [CrossRef]
206. Tayebee, R.; Esmaeili, E.; Maleki, B.; Khoshniat, A.; Chahkandi, M.; Mollania, N. Photodegradation of methylene blue and some emerging pharmaceutical micropollutants with an aqueous suspension of WZnO-NH$_2$@H$_3$PW$_{12}$O$_{40}$ nanocomposite. *J. Mol. Liq.* **2020**, *317*, 113928. [CrossRef]
207. Priya, R.; Stanly, S.; Anuradha, R.; Sagadevan, S. Evaluation of photocatalytic activity of copper ferrite nanoparticles. *Mater. Res. Express* **2019**, *6*, 095014. [CrossRef]

Article

One-Pot Synthesis of Benzopyrano-Pyrimidine Derivatives Catalyzed by P-Toluene Sulphonic Acid and Their Nematicidal and Molecular Docking Study

Mehtab Parveen [1,*], Mohammad Azeem [1], Azmat Ali Khan [2], Afroz Aslam [1], Saba Fatima [3], Mansoor A. Siddiqui [3], Yasser Azim [4], Kim Min [5] and Mahboob Alam [5,*]

1. Division of Organic Synthesis, Department of Chemistry, Aligarh Muslim University, Aligarh 202002, India; azam23961@gmail.com (M.A.); afrozaslam10@gmail.com (A.A.)
2. Pharmaceutical Biotechnology Laboratory, Department of Pharmaceutical Chemistry, College of Pharmacy, King Saud University, Riyadh 11451, Saudi Arabia; azkhan@ksu.edu.sa
3. Section of Plant Pathology and Nematology, Botany Department, Aligarh Muslim University, Aligarh 202002, India; fatimasaba8272@gmail.com (S.F.); mansoor_bot@yahoo.co.in (M.A.S.)
4. Department of Applied Chemistry, Z.H. College of Engineering & Technology, Aligarh Muslim University, Aligarh 202002, India; yasser.azim@gmail.com
5. Department of Safety Engineering, Dongguk University, 123 Dongdae-ro, Gyeongju 780714, Korea; kimmin@dongguk.ac.kr
* Correspondence: mehtab.organic2009@gmail.com (M.P.); mahboobchem@gmail.com (M.A.)

Citation: Parveen, M.; Azeem, M.; Khan, A.A.; Aslam, A.; Fatima, S.; Siddiqui, M.A.; Azim, Y.; Min, K.; Alam, M. One-Pot Synthesis of Benzopyrano-Pyrimidine Derivatives Catalyzed by P-Toluene Sulphonic Acid and Their Nematicidal and Molecular Docking Study. *Catalysts* **2022**, *12*, 531. https://doi.org/10.3390/catal12050531

Academic Editors: Sagadevan Suresh and Is Fatimah

Received: 28 March 2022
Accepted: 2 May 2022
Published: 9 May 2022

Publisher's Note: MDPI stays neutral with regard to jurisdictional claims in published maps and institutional affiliations.

Copyright: © 2022 by the authors. Licensee MDPI, Basel, Switzerland. This article is an open access article distributed under the terms and conditions of the Creative Commons Attribution (CC BY) license (https://creativecommons.org/licenses/by/4.0/).

Abstract: A cost-effective and environmentally benign benzopyrano-pyrimidine derivative synthesis has been established with the condensation of different salicylaldehyde derivatives, piperidine/morpholine with malononitrile, in the presence of a catalyst containing p-toluene sulphonic acid (PTSA) at 80 °C temperature. This procedure offers a new and enriched approach for synthesizing benzopyrano-pyrimidine derivatives with high yields, a straightforward experimental method, and short reaction times. The synthesized compounds were investigated for their nematocidal activity, and the result shows that among the four compounds, compounds **4** and **5** showed strong nematocidal activity against egg hatching and J2s mortality. The nematocidal efficacy of the compounds might be due to the toxicity of chemicals which are soluble in ethanol. The nematocidal effectiveness was directly related to the concentration of ethanolic dilutions of the compounds, i.e., the maximum treatment concentration, the higher the nematocidal action, or the higher the mortality and egg hatching inhibition. In the present study, with support from docking analysis, the relation between chemical reactivity and nematocidal activity of compound **4** was inferred.

Keywords: benzopyrano-pyrimidine; malononitrile; piperidine; PTSA; molecular docking

1. Introduction

Heterocyclic compounds have been prepared from many methods which contain significant biological activities in multicomponent reactions [1–5]. Selectivity, atom economy, rapid reaction times, and ability are critical characteristics of multicomponent reactions. Multicomponent reactions have recently been shown to be a significant development for synthesizing structurally varied chemical collections of the drug since the products are formed in a single step, and the variety can be achieved by simply moving each component [6]. Nitrogen-containing heterocyclic pyrimidines and their fused derivatives serve an essential function in medicinal chemistry and have been employed as drug development scaffolds [7–15]. Benzopyrano-pyrimidine is an important pharmacore that exhibits anti-thrombotic, anti-inflammatory, anti-aggregating, anti-platelet, and analgesic properties [16–20]. Many benzopyrano-pyrimidines contain anti-tumor activity and cytotoxic activity against cancer cell lines [18]. Some quinazolines and pyrimidine derivatives [21] are shown in various activities in Figure 1.

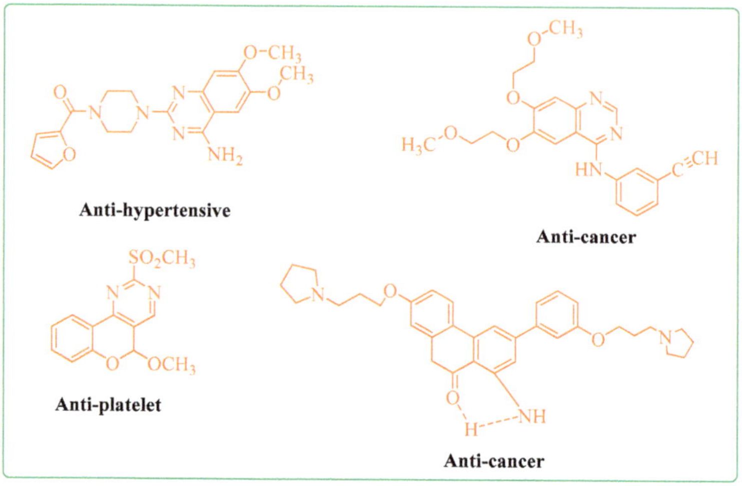

Figure 1. Biologically important derivatives of pyrimidine.

In the last decades, benzopyrano-pyrimidine one has been prepared via a three-component reaction of malononitrile, salicylaldehyde, and piperidine/morpholine by use of a catalytic amount of LiClO$_4$ [22], Na$_2$MoO$_4$·2H$_2$O [23], [Bim]BF$_4$ [24], Fe$_3$O$_4$ and SBA-15 [25], Fe(II)-benzoyl thiourea complex bound silica nanoparticles [Fe(II)-BTU-SNPs] [26], TiO$_2$–SiO$_2$ [27], p-toluenesulfonic acid supported by polystyrene in a solvent-free sonochemical multicomponent synthesis [28], and Brønsted acidic ionic liquid [29]. Solid acid catalysts have long been used in the oil refining industry, for example, in cracking processes and chemical manufacturing. In contrast, a substantial variety of acid-catalyzed reactions, including Friedel–Crafts reactions, esterification, hydration, and hydrolysis, are still catalyzed by conventional acids such as AlCl$_3$, H$_2$SO$_4$, and so on [30]. Chemical reactions employing traditional acids, on the other hand, are frequently linked with issues such as catalyst waste, corrosion, high toxicity, the use of huge volumes of catalyst, and separation and recovery challenges. Similarly, prolonged reaction times, elevated temperatures, high solvent costs, and the difficulty of separating conventional acids from the product are all disadvantages of using them as homogeneous catalysts [31] in laboratory trials for benzopyrano-pyrimidines synthesis. Additionally, organocatalysts such as proline and its derivatives and chiral phosphoric acids can be used to selectively synthesize heterocyclic compounds due to their achiral or chiral nature. Besides that, they have a number of advantages, not only due to their synthetic range, but also due to lower price. The absence of metals in organocatalysts is undeniably advantageous from both a green chemistry and economic standpoint. However, the high catalyst loading, the time and cost associated with removing and recycling excess catalyst from the reaction mixture, as well as the relatively young field, all work against widespread use of organocatalysts [32,33]. Consequently, a more efficient, ecologically friendly, and practical method of production of benzopyrano-pyrimidines was considered.

The use of p-toluene sulphonic acid (PTSA) as a solid catalyst for benzopyrano-pyrimidines synthesis via a three-component reaction in ethanol at 80 °C has been recommended as a non-explosive, non-toxic, and easily accessible option. After forming benzopyrano-pyrimidine derivatives, various spectroscopic techniques, including X-ray crystallographic, ^{1}H NMR, ^{13}C NMR, FT-IR, elemental analysis, and mass spectrophotometry, were employed to confirm the structure of the synthesized compounds. The 2-(4-(piperidine-1-yl)-5*H*-chromeno[2,3-d]pyrimidin-2-yl) phenol (**2**) structure was further confirmed by single-crystal X-ray diffraction with good conformity with earlier reports [32]. The synthesized compounds were also investigated for their nematocidal activity. The results show that compounds **4** and **5** showed strong nematocidal activity against egg hatching and J2s mortality among the four compounds. The nematocidal efficacy of the compounds might be due to the toxicity of chemicals which are soluble in ethanol. The nematocidal activity of the compounds in ethanolic dilutions was directly proportional to their concentration, i.e., the maximum treatment concentration, the higher the nematocidal action, or the more significant the mortality and egg hatching inhibition. In the current work, molecular docking of the active compound obtained from the experimental investigation was used to understand the mechanistic approach of non-bonding interactions with receptors and determine active amino acids' participation in receptors. The computer-generated 3D structure of ligands is docked into a receptor structure in various orientations, conformations, and sites via molecular docking. A molecular recognition strategy can help with medicine innovation and medicinal chemistry [34].

2. Results and Discussion

A small series of benzopyrano-pyrimidine derivatives were synthesized in this paper using p-toluene sulphonic acid in ethanol under reflux conditions. This approach outperforms other available synthetic methods in terms of yield, reaction timings, product purity, and catalyst stability. These synthetic benzopyrano-pyrimidine derivatives possess different applicability and are well-matched with several other functional groups.

2.1. Chemistry

Based on FT-IR, NMR (^{1}H & ^{13}C), and mass spectra analyses, the structure of all synthesized benzopyrano-pyrimidine derivatives was determined and found to be in good agreement with the anticipated structure. Furthermore, the spectroscopic data of compounds **2** and **4** matches those described in the literature quite well [27,28]. The reaction occurs at the carbonyl and hydroxy moieties of one mole of salicylaldehyde, as evidenced by the FT-IR spectra, which reveal that the produced molecule has no aldehyde group frequency. Furthermore, the entire compound showed a characteristic peak for the group, appearing at approximately 3418, 2857.4, 1619, and 1600.83 cm^{-1}, indicating the formation of benzopyrano-pyrimidine derivatives. The saturated proton in each synthesized molecule resonance had a sharp singlet at approximately δ 4.35 ppm, at around δ 9.0–12.5 ppm with a broad peak accounted to the-OH proton of a benzene ring, and the benzene ring proton displayed a multiplet at around 6.98–8.75 ppm in the ^{1}H NMR spectra of the synthesized compounds. ^{13}C NMR spectra display signals at about δ 119–165, which have been shown aromatic carbon, with a peak display at around δ 155–159 to -C=N and 161–165 to –C-O of pyrano moiety. Similarly, the signal resonated at δ 22–50 has been ascribed to a saturated carbon. The mass analysis of the prepared series was very suitable in conformity with the design structure.

2.2. Crystal Structure

Compound **2** crystallizes in the asymmetric unit (ASU) in the monoclinic P21/n space group (Figure 2). All of the bonds in the ASU have a considerable range of bond lengths, viz. N2–C1 (1.476(3) Å), N2–C5 (1.452(2) Å), N2–C6 (1.373(3) Å), N1–C6 (1.336(2) Å), N1–C16 (1.341(3) Å), N3–C15 (1.329(2) Å), N3–C16 (1.316(3) Å), O2–C14 (1.398(3) Å), O2–C15 (1.359(3) Å) and O1–C18 (1.313(4) Å). The molecule is non-planar with a dihedral angle of 33.17° between the mean planes of C1–N2–C5 and C7–C6–N1. Meanwhile, the dihedral angle between the mean planes of N3–C16–N1 and C9–C10–C11 is 22.80°.

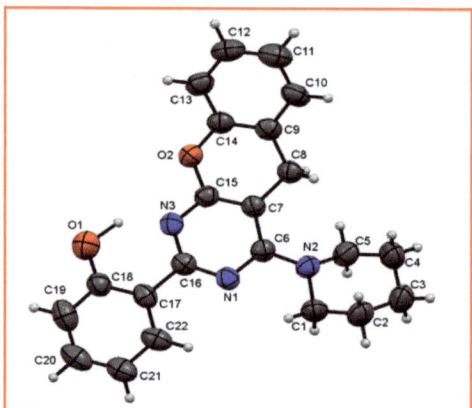

Figure 2. The asymmetric unit of compound **2** with atom labeling and displacement ellipsoids are drawn at the 50% probability level.

2.3. Mechanism

A tentative mechanism pathway for the p-toluene sulphonic acid (PTSA)-catalyzed synthesis of the benzopyrano-pyrimidine derivative has been described based on the literature [28] as shown in Scheme 1. First, the reaction is started by protonating the carbonyl group from the p-toluene sulphonic acid catalyst, which produces an active electrophilic intermediate **I** and makes the carbonyl carbon more electrophilic, lowering its pKa value. Further, the conjugate base of the catalyst generated in situ in the reaction mixture acts as a nucleophile which abstracts a proton from the active methylene carbon of malononitrile. This step facilitates the formation of a tetrahedral intermediate **II**. In the next step, elimination of the water molecule forms intermediate **III**. Then, intermolecular cyclization occurs by attaching the phenolic group of salicylaldehyde to the cyanide group, and intermediate **IV** is obtained. In the next step, piperidine attaches onto intermediate **IV**, and intermediate **V** is formed. In the last step, the second molecule of salicylaldehyde attaches to intermediate **V**. Finally, the formation targets the benzopyrano-pyrimidine derivatives with the removal of the catalyst (Scheme 1).

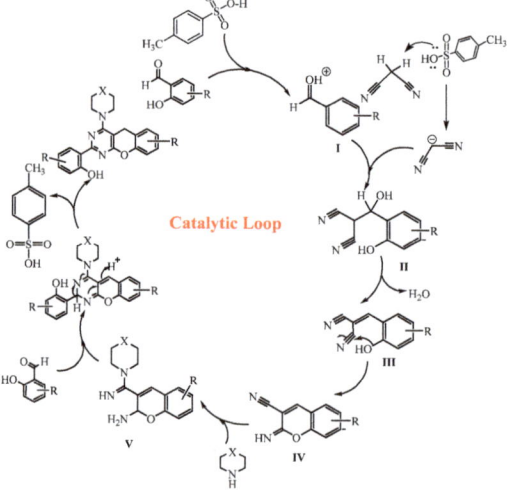

Scheme 1. A tentative mechanism for the synthesis of benzopyrano-pyrimidine derivatives.

2.4. Optimization of Reaction Conditions

Firstly, the focus was on optimizing the reaction conditions for the current protocol concerning the reaction temperature, the amount of catalyst, and the choice of solvent in our study, and selecting a suitable catalyst for a chosen model reaction using salicylaldehyde, malononitrile, and hetero/aromatic aldehyde from various catalysts to provide the best possible reaction condition for the synthesis of the benzopyrano-pyrimidine derivatives.

2.4.1. Effect of Different Solvent

In the presence of p-toluene sulphonic acid, the effect of other solvents on the reaction rate and yield of the product was investigated. Solvents such as $CHCl_3$, CH_3NO_2, CH_2Cl_2, and CH_3CN were unsatisfactory. In water, the reaction did not proceed. In methanol, DMF, and THF, the reaction completed in 8 h. In ethanol, the reaction produced the best results, the minimum time for completion, and gave a good yield. The results are shown below in Table 1.

Table 1. Effect of different solvents on the reaction [a].

Entry [a]	Solvent	Condition	Time (h) [b]	Yield [c]
1	H_2O	Reflux	24	Trace
2	CH_3CN	Reflux	10	32
3	CH_3COCH_3	Reflux	8	55
4	$CHCl_3$	Reflux	8	35
5	CH_3NO_2	Reflux	8	45
6	CH_2Cl_2	Reflux	8	25
7	THF	Reflux	8	66
8	DMF	Reflux	7	55
9	MeOH	Reflux	8	73
10	EtOH	Reflux	40 min	95

[a] Reaction of 4-chloro-salicylaldehyde (1 mmol) with malononitrile (1 mmol) and piperidine (1 mmol) in the presence of 10 mol% PTSA as a catalyst. [b] Reaction progress monitored by TLC. [c] Isolated yield.

2.4.2. Effect of Different Catalysts

There are many catalysts used in the optimization of model reactions. To emphasize the efficiency of p-toluene sulphonic acid-catalyzation compared to other catalysts, the reaction was carried out with various catalysts such as pyridine, $AlCl_3$, $FeCl_3$, $ZnCl_2$, I_2, NH_4OAc, and NaOAc. Without catalysts, the reaction occurs for a long time and has a low yield. It is observed that the reaction performed with I_2 and NH_4OAc was complicated after a long reaction time with a 72–80% yield. The model reaction was also carried out in $FeCl_3$ and $AlCl_3$ with less reaction time but low yield, and the product obtained a minimal amount. The reaction was completed with zinc chloride in a short time and with a moderate yield. The model reaction was completed in a short reaction time when p-toluene sulphonic acid was used with a high yield (Entry 9) (Table 2).

Table 2. The influence of different catalysts on the model reaction under thermal solvent-free conditions.

Entry [a]	Catalyst	Time (h) [b]	Yield [c]
1	None	23	25
2	Pyridine	3.5	50
3	$FeCl_3$	3	40
4	$AlCl_3$	2.5	35
5	I_2	12	72
6	NH_4OAc	8	75
7	NaOAc	6	80
8	$ZnCl_2$	5	60
9	PTSA	40 min	95

[a] Reaction of 4-chloro-salicylaldehyde (1 mmol) with malononitrile (1 mmol), piperidine (1 mmol) in the presence of 10 mol% PTSA as a catalyst. [b] Reaction progress monitored by TLC. [c] Isolated yield.

2.4.3. Effect of Catalyst Loading

The effect of loading catalyst was tested in the model reaction. In optimization, we analyzed the reaction by varying the loading amount of the catalyst in the model reaction from 2 to 10 mol%. Finally, the results show that 10 mol% of the catalyst was sufficient to give a better yield (entry 5) (Table 3).

Table 3. Effect of catalyst loading on the reaction.

Entry [a]	Catalyst (Mol%)	Time (Min) [b]	Yield [c]
1	2	2 h	60
2	3	1.5 h	75
3	5	1.0 h	80
4	10	40	88
5	10	40 min	95

[a] Reaction of 4-chloro-salicylaldehyde (1 mmol) with malononitrile (1 mmol), piperidine (1 mmol) in the presence of 10 mol% PTSA as a catalyst. [b] Reaction progress monitored by TLC. [c] Isolated yield.

2.4.4. Catalytic Reaction

With these encouraging results in hand, we turned to explore the scope of the reaction using different aromatic aldehydes (2a–g), malononitrile, and piperidine as substrates under the optimized reaction conditions (Table 4). It was observed that the aromatic aldehydes with electron-donating and electron-withdrawing groups reacted successfully to furnish the final products **1–6** in good yields (Table 4).

Table 4. Synthesis of benzopyrano-pyrimidine derivatives using PTSA at 80 °C temperature.

Entry	Reactant (1)	Reactant (2)	Product	Time (min)	Yield (%)	M.P.
1.	2-hydroxy-5-nitrobenzaldehyde	Piperidine	(product structure)	40	92	260–262
2.	salicylaldehyde	Piperidine	(product structure)	40	90	170–172
3.	5-chloro-2-hydroxybenzaldehyde	Piperidine	(product structure)		95%	255–256
4.	salicylaldehyde	Morpholine	(product structure)	40	90	220–222
5.	salicylaldehyde	Piperidine	(product structure)	40	90	230–232
6.	3,5-dichlorosalicylaldehyde	Piperidine	(product structure)	40	93	170

2.4.5. Catalyst Recycling

Effective catalyst recovery from the reaction mixture is the most important aspect for determining its usability for practical applications, from an environmental and economic standpoint. As a result, catalyst recycling studies were performed to establish the degree of recyclability of our catalytic system (Figure 3). As a model reaction, the reaction of 4-chloro-salicylaldehyde, malononitrile, and piperidine in the presence of 10 mol% PTSA was used. After the reaction was completed, the catalyst was recovered by extracting the mixture with ethyl acetate and then filtering it. After that, the catalyst was washed with ethyl acetate and reused in consecutive cycles. In ethanol, the catalyst maintained its activity for at least five reaction cycles, demonstrating excellent catalytic performance with a product yield of over 95%.

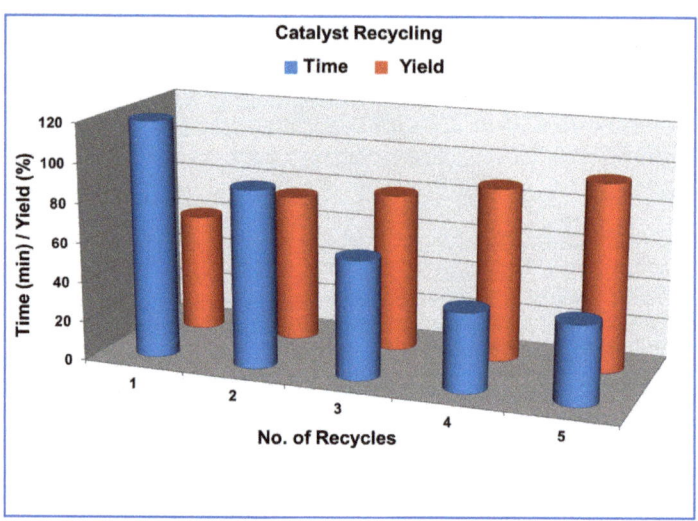

Figure 3. Recycling data of PTSA catalyst.

2.5. Nematicidal Activity

The data analysis in Table 4 indicates that mortality of juveniles of *Meloidogyne javanica* was recorded (8) in absolute alcohol (control). However, all the concentrations of each diluted compound, i.e., **2**, **3**, **4**, and **5**, significantly impacted mortality. Second stage (J2) juvenile mortality was directly correlated to the concentrations and exposure period. The highest mortality (98.9%) was observed in 100% concentration of compound **4** after 72 h of exposure time. In contrast, the lowest mortality was found in compound 2 (7.6%) at 12.5% concentration after the exposure period of 24 h (Table 5).

Table 5. Effect of different dilutions of organic chemicals on the mortality of juvenile root-knot nematode *Meloidogyne javanica* in vitro.

Compounds	Exposure Period (Hours)	Percent Mortality in Different Concentrations					Regression Equation
		100%	50%	25%	12.2%	Control	
2	24	72 (69.5)	60 (56.5)	51 (46.7)	15 (7.6)	8.00	Y = 17.19x − 13.91
	48	79 (77.1)	65 (61.9)	58 (54.3)	18 (10.8)	8.00	Y = 18.93x − 14.37
	72	83 (81.5)	79 (77.1)	70 (67.3)	23 (16.3)	8.00	Y = 20.78x − 12.30
3	24	80 (78.2)	68 (65.2)	53 (48.9)	17 (9.7)	8.00	Y = 19.59x − 16.77
	48	86 (84.7)	74 (71.7)	63 (59.7)	20 (13.0)	8.00	Y = 21.21x − 16.21
	72	91 (90.2)	85 (83.6)	76 (73.9)	29 (22.8)	8.00	Y = 22.52x − 11.86
4	24	90 (89.1)	68 (65.2)	55 (51.0)	30 (23.9)	8.00	Y = 20.35x − 13.61
	48	93 (92.3)	86 (84.7)	75 (72.8)	35 (29.3)	8.00	Y = 22.4x − 9.78
	72	99 (98.9)	90 (89.1)	81 (79.3)	55 (51.0)	8.00	Y = 21.99x − 0.71
5	24	86 (84.7)	65 (61.9)	50 (45.6)	26 (19.5)	8.00	Y = 19.58x − 14.80
	48	91 (90.2)	77 (72.8)	63 (59.7)	34 (28.2)	8.00	Y = 20.9x − 10.92
	72	95 (94.4)	82 (80.4)	67 (64.1)	39 (33.6)	8.00	Y = 21.96x − 9.78

Each value is an average of three replicates.

In vitro nematicidal activity of compounds was displayed as an LC50 value with 95% confidence limits. The effect of organic chemicals on probit output and LC 50 was calculated. As the concentrations of the compounds increased from 12.5% to 100%, juvenile mortality also increased. The toxins of compound **4** convert nematode natality into mortality with LC50 values 27.21, 17.91, and 11.89 percent after 24, 48, and 72 h of exposure time, respectively. The findings indicated that compound **4** was highly toxic to mortality of *M. javanica* at 100% concentration of 72 h time duration. Compounds **5, 3**, and **2** followed. After 24, 48, and 72 h of exposure, compound **2** showed the least toxicity in terms of nematode mortality, with LC50 values of 42.97, 34.36, and 24.44, respectively (Table 6).

Table 6. Nematicidal activity of different concentrations of compounds against juveniles of *Meloidogyne javanica*.

Compounds	Exposure Time (Hours)	LC_{50} Value in Percent (95% CL)
2	24	42.97
	48	34.36
	72	24.44
3	24	35.10
	48	27.97
	72	19.62
4	24	27.21
	48	17.91
	72	11.89
5	24	31.79
	48	21.93
	72	18.70

Similarly, dilutions of each compound were also considered effective against *M. javanica* egg hatching. After 7 days of exposure, compound **4** was the most reactive of the four compounds, whereas compound **2** was the least effective at a 12.5 percent dilution. Compound **2** had the lowest amount of hatching (95.8, 89.7, 78.8, and 54.8%). On the other hand, maximum egg hatching of *M. javanica* second-stage juveniles (J2) was shown by compound **4** (100, 96.0, 87.6, and 62.7%) at different dilutions such as 100%, 50%, 25%, and 12.5% concerning their control (ethanol). As per data analysis, the maximum percent inhibition in *Meloidogyne javanica* egg hatching was indicated by compound **4** (100%) at 100% concentration (Table 7). Alternatively, the minimum hatchability of J2s was revealed by compound **2** (54.8%) at 12.5% diluted form after seven days of time duration (Table 7).

Table 7. Effect of different ethanolic dilutions of various compounds on the egg hatching of *Meloidogyne javanica* in vitro after 7 days.

Compounds	Number of Larvae Hatched in Different Dilutions				
	100%	50%	25%	12.5%	Control
2	22 (95.8%)	55 (89.7%)	113 (78.8%)	241 (54.8%)	534 (0.00%)
3	16 (97.0%)	41 (92.3%)	87 (83.7%)	227 (57.4%)	534 (0.00%)
4	0 (100%)	21 (96.0%)	66 (87.6%)	199 (62.7%)	534 (0.00%)
5	9 (98.3%)	27 (94.9%)	78 (85.3%)	214 (59.9%)	534 (0.00%)

Each value is an average of three replicates, DW = Distilled Water (control). The value of percent inhibition in egg hatching over control is given in parentheses.

Conclusions reached that compounds **4** and **5** showed strong nematocidal activity against egg hatching and J2s mortality among the four synthesized compounds. The nematocidal efficacy of the synthesized compounds might be due to the toxicity of compounds that are soluble in ethanol. The previous findings showed that salicylaldehyde derivatives and other chemicals possess nematicidal potency against most damaging soil-borne pathogens, i.e., Phyto parasitic nematodes [35–37]. In the current in vitro testing, the ethanolic dilutions of compounds **2**, **3**, **4**, and **5** showed significant nematotoxicity or nematocidal potentiality against juvenile mortality and egg hatching of *M. javanica*. The four ethanolic doses of the synthesized compounds were the most efficient in lowering egg hatching and increasing mortality. Results analysis revealed that the nematocidal efficacy was proportionate to the concentration of compounds in ethanolic dilutions, i.e., the maximum treatment concentration, the higher the nematocidal action, or greater the mortality and egg hatching inhibition [38].

In vitro mortality investigation showed that compound **4** exhibited the highest nematocidal potency against the survival of J2s of *M. javanica* after 72 h of exposure time. Compound **4** had the least LC50 values compared to other compounds (**5**, **3**, and **2**) at 24, 48, and 72 hrs of exposure. The mortality of the second stage (J2) juveniles increases with the increase of all compound concentrations, along with exposure time initiated from 24 to 72 h. Related results were described by [39], who report that aromatic aldehydes such as salicylaldehyde, Phthaldehyde, and cinnamic aldehydes actively demonstrated nematocidal activity against the root-knot nematode, *M. incognita* in in vitro study. So, it can be concluded that the toxicity of synthesized compounds toward nematodes depends on the concentrations of treatment and the exposure period (Figure 4).

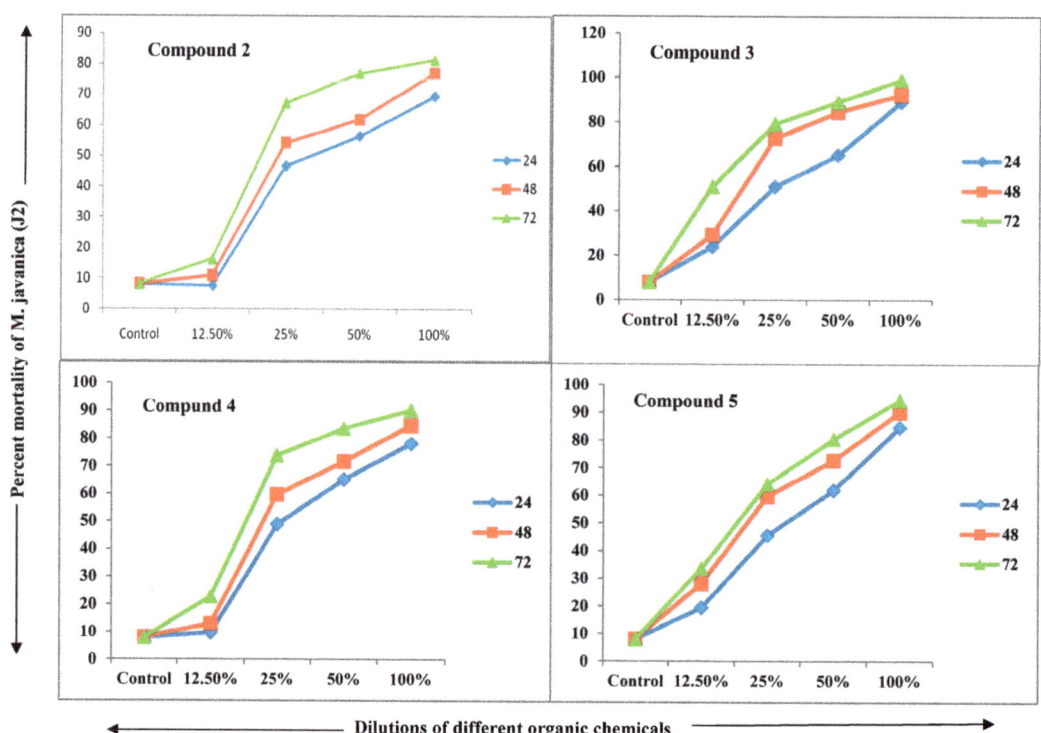

Figure 4. Regression lines show a linear relationship between different dilutions of organic chemicals against juvenile mortality *M. javanica* at another exposure period.

2.6. Docking Analysis

The crystal structure of yeast V1-ATPase in the autoinhibited form of Saccharomyces cerevisiae was chosen for the docking investigation, and docking was conducted to identify the non-bonding contacts between the compound that showed good nematocidal activity in this study and the receptor. Compound **4** was docked between β-strand β-21 and α-helix α4, containing 284–287 and 357–360 residues, in a docking experiment (Figure 5). The best docked posed established a hydrophobic pocket at the receptor site with residues PHE538, ILE541, LEU235, PRO233, PRO540, and TRP542 adjacent to compound **4** with binding energy (8.3420 kcal/mol).

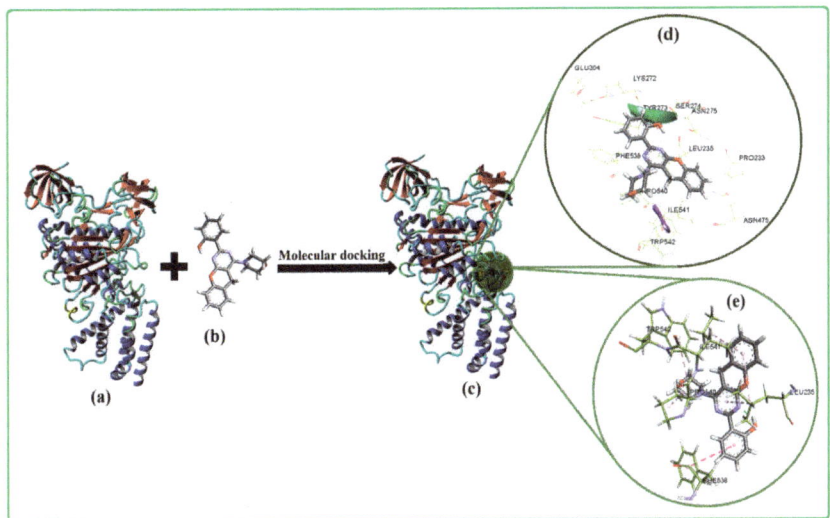

Figure 5. Molecular docking of the (**a**) receptor (PDB: 5D80) with (**b**) the compound **4**, (**c**) the docked compound into active site shown in the circle, (**d**) involvement of various amino acids interacting with compound **4**, and (**e**) compound **4** interacts with the receptor, forming non-bonding contacts with the aromatic and non-aromatic skeletons.

The envelope of these active amino acids that interact with compound **4** is positioned in the amino acids pocket in the proper orientation, to establish close contact with the receptor and stop plant-parasitic nematodes from spreading. Furthermore, apart from hydrophobic amino acids, hydrophilic amino acids are present around the ligands such as LYS 272, TYR 273, SER 274, ASN 275, ASN 475, and GLU 304 are also involved in various interactions such as van der walls, pi-pi-T-shaped, pi-alkyl, and so on, as shown in Figure 5. These interactions boosted the compound's stability as well as its biological activity.

3. Materials and Methods

All chemicals were purchased from Merck and Sigma-Aldrich (Mumbai, India) as "synthesis grade" and used without further purification. Kofler apparatus (Nageman, Germany) was used to determine melting points and are uncorrected. A Carlo Erba analyzer model 1108 (Milan, Italy) was used to analyze elemental analysis (C, H, N). The Shimadzu IR-408 instrument (Shimadzu Corporation, Kyoto, Japan) recorded the IR spectra, and the values were set in cm^{-1}. For the ^{1}H NMR and ^{13}C NMR spectra, a Bruker Avance-II 400 MHz instrument (Bruker Instruments Inc., Billerica, MA, USA) was used, and the spectra run in DMSO-d_6 with TMS as an internal standard, and the J values were measured in Hertz (Hz). Chemical shifts were reported in ppm (δ) relative to TMS. A JEOL D-300 mass spectrometer was used to record the Mass spectra. Thin-layer chromatography (TLC) glass plates (20 × 5 cm) were coated with silica gel G (Merck) (Darmstadt, Germany) and exposed to iodine vapors to check the homogeneity as well as the progress of the reaction.

*3.1. General Procedure for the Synthesis of Benzopyrano-Pyrimidines Derivatives (**1–6**)*

A mixture of salicylaldehyde **1** (1 mmol), malononitrile **2** (1 mmol), and piperidine/morpholine (1 mmol) was added to p-toluene sulphonic (10 mol%). The reaction mixture was refluxed and stirred at 80°C for the required time. The completion of the reaction was monitored by thin-layer chromatography (TLC) using n-hexane and ethyl acetate (8:2). After completing the reaction as indicated by TLC, the reaction mixture was treated with ice-cold water. The mixture was filtered to collect the crystal. Separation of the catalyst was completed by filtration and the resulting solution was extracted with ethyl acetate. The aqueous organic layer was washed with brine, poured onto anhydrous Na_2SO_4, and filtered under reduced pressure. The pure product obtained was crystallized with ethanol chloroform to afford pure crystal (Scheme 2).

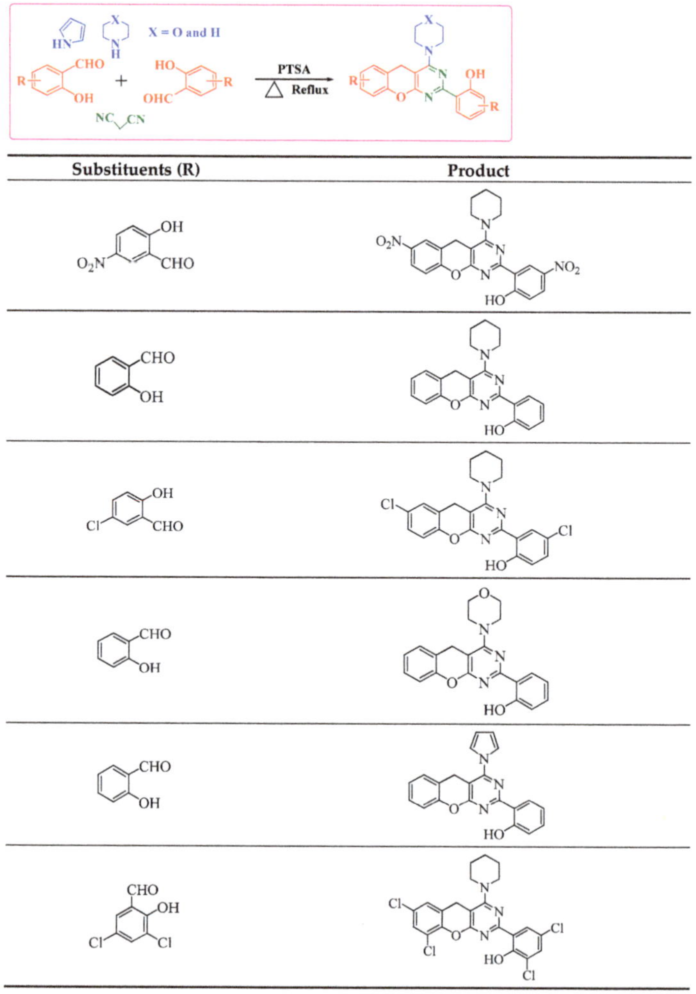

Scheme 2. Synthetic pathway for the synthesis of benzopyrano-pyrimidines derivatives **1–6**.

3.2. Spectral Data of Synthesized Compounds

Spectroscopic and elemental analysis data for the heterocyclic compounds (**1–6**) synthesized and reported in the literature are given as Supplemental Materials.

3.3. X-ray Crystallographic Studies

Additional information to the structure of compound **2** is specified in Table 8.

Table 8. Crystallographic data and experimental details compound **2**.

Parameters	Compound 2
Empirical Formula	$C_{22}H_{21}N_3O_2$
Formula weight	359.42
Crystal size/mm	$0.30 \times 0.36 \times 0.40$
Crystal system	Monoclinic
Unit cell dimensions	a = 9.9377 (2) Å
	b = 15.8917 (3) Å
	c = 12.1806 (3) Å
	$\alpha = \gamma = 900$
	$\beta = 108.7810$
Space group	$P21/n$
Z	4
Temperature (K)	293
Wavelength (Å)	0.71073 Å
Volume (Å^3)	1821.22 (7)
Density (g cm^{-3})	1.311
μ/mm^{-1}	0.086
F(000)	760
Measured reflections	44244
Independent reflections	4015 (Rint = 0.112)
Observed reflections [$I \geq 2\sigma(I)$]	2667
Goodness-of-fit on F2	1.149
Radiation type	MoKα
h, k, l max	12, 20, 15
Final R indices	$R1 = 0.078$
[$I \geq 2\sigma(I)$]	$wR2 = 0.280$

3.4. Inoculum Maintenance

The pure culture of root-knot nematode, *Meloidogyne javanica*, was multiplied on the brinjal plant under the greenhouse of the Department of Botany, Aligarh Muslim University, Aligarh (India). Infected roots were separated from the adhering soil, washed gently in tap water, and kept in a distilled water tray (DW). The root-knot nematode species *M. javanica* identification was carried out based on the technique of the perineal patterns [40]. The roots were cut into small segments, and egg masses were handpicked from the root for hatching purposes. These egg masses were transferred to Petri dishes (40 mm) containing DW at 27 ± 2 °C in a BOD incubator. The suspension containing the juveniles was collected after the fifth day of hatching, and fresh DW was added. The concentration of freshly hatched J2 juveniles of *M. javanica* was standardized as per the requirement for in vitro testing.

3.5. In Vitro Nematicidal Activity Bioassays

3.5.1. Hatching Test

Five new and uniform-size egg masses of *M. javanica* were handpicked from the infected roots of brinjal. The collected egg masses were transferred to Petri dishes (40 mm) containing 10 mL of each synthesized compound in different concentrations (100%, 50%, 25%, and 12.5%). Distilled water containing egg masses served as a control. Each treatment, including the control, had three replicates. All the Petri dishes were incubated in a BOD (biological oxygen demand) incubator at 28 °C for seven days. After seven days, hatched juveniles in the treated and control samples were recorded using a counting dish with the help of a stereomicroscope.

3.5.2. Mortality Test

For mortality, 0.4 mL of DW containing 100 juveniles of *M. javanica* was transferred into the Petri dishes containing 9.6 mL of synthesized compounds of four different concentrations (100%, 50%, 25%, and 12.5%). Petri dishes containing juveniles in DW were considered as the control. Each treatment, including the control, was replicated three times. All the Petri dishes were kept at 28 °C in BOD. The number of dead juveniles was counted with the help of a counting dish under a stereomicroscope at different time intervals (24, 48, and 72 h). The mortality of the juveniles was confirmed by transferring immobilized juveniles into fresh DW water for 1 h and observing if any movement was shown by nematode. If there was no mobility, then they were considered dead. Probit analysis was used to compute the LC50 values for all treatments based on percent mortality data and concentration [41].

The following formula was used to compute the percent inhibition in egg hatching or juvenile mortality:

$$\% \text{ inhibition or mortality} = \left(\frac{C_0 - T\alpha}{C_0}\right) \times 100\%$$

where, in terms of egg hatching, $T\alpha$ = number of juveniles hatched in each concentration of the compound dilutions, C_0 = number of juveniles hatched in the control. In the case of mortality, $T\alpha$ = number of live nematodes after 24, 48, and 72 h of exposure, C_0 = number of juveniles living in the control.

3.6. Docking Study

Molecular docking between compound **4** and the V1-ATPase crystal structure receptor downloaded from the *RCSB* Protein Data Bank was performed with YASARA software [42] using the dock_run.MCR module that was available in YASARA-Structure. Initially, the receptor's PDB file was imported into YASARA, and water molecules and ions from the receptor were removed from the structure. The missing hydrogen atoms and residues were added to the receptor. One of the multiple receptor structures was saved for docking studies. Chem Draw was used to sketch the 2D structure of compound **4**, which was then transformed to a 3D structure by Chem3D before being optimized using molecular mechanics using the MM+ force field and saved in sd format for the docking study. YASARA View was used to illustrate the best dock pose from the docking experiment, which was then converted to PDB format for Molecular graphics by a BIOVIA Discovery Studio Visualizer (Discovery Studio Visualizer, version 16.1.0; Dassault Systemes, BIOVIA Corp., San Diego, CA, USA).

4. Conclusions

The protocol provides not only a high yield of products and a shorter reaction time but also high purity, mild reaction conditions, operational simplicity, a cleaner reaction profile, increased reaction rates, and a simple workup approach. We hope that this synthetic protocol will provide a more feasible alternative to the other available methods for synthesizing benzopyrano-pyrimidine. Compounds **4** and **5** had a significant nematocidal effect on egg hatching and J2s mortality. The nematocidal efficacy of the compounds might be due to the toxicity of chemicals which are soluble in ethanol. The analysis revealed that the nematocidal efficacy was proportionate to the concentration of the compounds in ethanolic dilutions, i.e., the maximum treatment concentration, the greater the mortality or, the higher the nematocidal action, as well as egg hatching inhibition. For the investigation of close contacts with active amino acids, the molecular docking of compound **4** against the 3D structure of V1-ATPase obtained from Saccharomyces cerevisiae was employed.

Supplementary Materials: The following supporting information can be downloaded at: https://www.mdpi.com/article/10.3390/catal12050531/s1, Characterization data for compounds (1–6) and copies of NMR spectra (^{1}H and ^{13}C) for two new compounds (3–4). Figure S1: ^{1}H NMR Spectra of Compound 3; Figure S2: ^{13}C NMR Spectra of Compound 3; Figure S3: ^{1}H NMR Spectra of Compound 4; Figure S4: ^{13}C NMR Spectra of Compound 4.

Author Contributions: Supervision, formal analysis, writing—review and editing, M.P.; conceptualization, methodology, writing—original draft preparation, M.A. (Mohammad Azeem), A.A.K., K.M., A.A.; software, validation, visualization, M.A. (Mahboob Alam), Y.A.; investigation, resources, M.A. (Mahboob Alam), S.F., M.A.S. All authors have read and agreed to the published version of the manuscript.

Funding: This study was funded by the Researcher Supporting Project Number (RSP-2021/339), King Saud University, Riyadh, Saudi Arabia.

Data Availability Statement: Data obtained during this study are included in the main text and Supplementary Materials.

Acknowledgments: M. Azeem thanks the Chairman, Department of Chemistry, A.M.U, Aligarh, for providing the necessary research facilities; University Sophisticated Instrumentation Facility (USIF), AMU, Aligarh is credited for spectral analysis and single-crystal X-ray analysis. UGC is also gratefully acknowledged for research fellowship to M. Azeem.

Conflicts of Interest: The authors declare no conflict of interest.

References

1. Heiner, E. Diversity Oriented Syntheses of Conventional Heterocycles by Smart Multi-Component Reactions (MCRs) of the Last Decade. *Molecules* **2012**, *17*, 1074–1102.
2. Chitalu, C.M.; Susan, L.; Vanessa, Y.; Kelly, C. Application of multicomponent reactions to antimalarial drug discovery. Part 3: Discovery of aminoxazole 4-aminoquinolines with potent antiplasmodial activity in vitro. *Bioorg. Med. Chem. Lett.* **2007**, *17*, 4733–4736.
3. Mabkhot, Y.N.; Barakat, A.; Al-Majid, A.M.; Alshahrani, S.; Yousuf, S.; Choudhary, M.I. Synthesis, reactions, and biological activity of some new bis-heterocyclic ring compounds containing sulfur atom. *Chem. Cent. J.* **2013**, *7*, 112. [CrossRef] [PubMed]
4. Aslam, A.; Parveen, M.; Alam, M.; Silva, M.R.; Silva, P.S.P. Silica bonded N-(propylcarbamoyl)sulfamic acid (SBPCSA) as a highly efficient and recyclable solid catalyst for the synthesis of Benzylidene Acrylate derivatives: Docking and reverse integrated docking approach of network pharmacology. *Biophys. Chem.* **2020**, *266*, 106443. [CrossRef]
5. Parveen, M.; Aslam, A.; Nami, S.A.A.; Ahmad, M. Z-Acrylonitrile Derivatives: Improved Synthesis, X-ray Structure, and Interaction with Human Serum Albumin. *Curr. Org. Synth.* **2019**, *16*, 1157–1168. [CrossRef]
6. Domling, A.; Ugi, I.A. Multicomponent reaction of isocyanide. *Angew. Chem. Int. Ed. Engl.* **2000**, *39*, 3168–3210. [CrossRef]
7. Jain, K.S.; Arya, N.; Inamdar, N.N.; Auto, P.B.; Unawane, S.A.; Puranik, H.H.; Sanap, M.S.; Inamke, A.D.; Mahale, V.J.; Prajapati, C.S.; et al. The Chemistry and Bio-Medicinal Significance of Pyrimidines & Condensed Pyrimidines. *Curr. Top. Med. Chem.* **2016**, *16*, 3133–3174.
8. Parveen, M.; Aslam, A.; Alam, M.; Siddiqui, M.F.; Bano, B.; Azaz, S.; Silva, M.R.; Silva, P.S.P. Synthesis and Characterization of Benzothiophene-3-carbonitrile Derivative and Its Interactions with Human Serum Albumin (HSA). *ChemistrySelect* **2019**, *4*, 11979–11986. [CrossRef]
9. Parveen, M.; Aslam, A.; Ahmad, A.; Alam, M.; Silva, M.R.; Silva, P.S.P. A facile & convenient route for the stereoselective synthesis of Z-isoxazole-5(4H)-ones derivatives catalyzed by sodium acetate: Synthesis, multispectroscopic properties, crystal structure with DFT calculations, DNA-binding studies, and molecular docking studies. *J. Mole. Struct.* **2020**, *1200*, 127067.
10. Aslam, A.; Parveen, M.; Singh, K.; Azeem, M. Green Synthesis of Fused Chromeno-pyrazolo-phthalazine Derivatives with Silica-supported Bismuth Nitrate under Solvent-free Conditions. *Curr. Org. Synth.* **2021**, *18*, 1–8. [CrossRef]
11. Selvam, T.P.; James, C.R.; Dniandev, P.V.; Valzita, S.K. A mini-review of pyrimidine and fused pyrimidine marketed drugs. *J. Res. Pharm.* **2012**, *2*, 1–9.
12. Jain, K.S.; Chitre, T.S.; Miniyar, P.B.; Kathiravan, M.K.; Bendre, V.S.; Veer, V.S.; Shahane, S.R.; Shishoo, C.J. Biological and medicinal significance of pyrimidines. *Curr. Sci.* **2006**, *90*, 6.
13. Schenone, S.; Radi, M.; Musumeci, F.; Brullo, C.; Botta, M. Biologically Driven Synthesis of Pyrazolo[3,4-d]pyrimidines as a Protein Kinase inhibitors: An scaffold As a New Tool for Medicinal chemistry and Chemical Biological Studies. *Chem. Rev.* **2014**, *114*, 7189–7238. [CrossRef] [PubMed]
14. Rane, J.S.; Pandey, P.; Chatterjee, A.; Khan, R.; Kumar, A.; Prakash, A.; Ray, S. Targeting virus-host interaction by novel pyrimidine derivative: An in silico approach towards discovery of potential drug against COVID-19. *J. Biomol. Struct. Dyn.* **2021**, *39*, 5768–5778. [CrossRef]

15. Rani, J.; Kumar, S.; Saini, M.; Mundlia, J.; Verma, P.K. Biological potential of pyrimidine derivatives in a new era. *Res Chem. Intermed.* **2016**, *42*, 6777–6804. [CrossRef]
16. Mohana, K.M.; Prasanna Kumar, B.N.; Mallesha, L. Synthesis and biological activity of some pyrimidine derivatives. *Drug Invent. Today* **2013**, *5*, 216–232. [CrossRef]
17. Bruno, O.; Schenone, S.; Ranise, A.; Barocelli, E.; Chiavarini, M.; Ballabeni, V.; Bertoni, S. Synthesis and Pharmacological Screening of Non–acidic Gastroprotective Antipyretic Anti-inflammatory Agents with Anti-platelet properties. *Arzneim-Forsch./Drug Res.* **2000**, *50*, 140–147. [CrossRef]
18. Bruno, O.; Schenone, S.; Ranise, A.; Bondavalli, F.; Barocelli, E.; Ballabeni, V.; Chiavarini, M.; Bertoni, S.; Tognolini, M.; Impicciatore, M. New polycyclic pyrimidine derivatives with antiplatelet in vitro activity: Synthesis and pharmacological screening. *Bioorg. Med. Chem.* **2001**, *9*, 629–636. [CrossRef]
19. Bruno, O.; Brullo, C.; Schenone, S.; Bondavalli, F.; Ranise, A.; Tognolini, M.; Ballabeni, V.; Barocelli, E. Synthesis and pharmacological evaluation of 5H-[1]benzopyrano[4,3-d]pyrimidines were effective as anti-platelet/analgesic agents. *Bioorg. Med. Chem.* **2004**, *12*, 553–561. [CrossRef]
20. Bruno, O.; Brullo, C.; Ranise, A.; Schenone, S.; Bondavalli, F.; Barocelli, E.; Ballabeni, V.; Chiavarinib, M.; Tognolini, M.; Impicciatore, M. Synthesis and pharmacological evaluation of 2,5-cycloamino-5H-[1]benzopyrano[4,3-d]pyrimidines endowed with in vitro antiplatelet activity. *Bioorg. Med. Chem. Lett.* **2001**, *11*, 1397–1400. [CrossRef]
21. Ghorbani-Vaghei, R.; Shirzadi-Ahodashti, M.; Eslami, F.; Malaekehpoor, S.M.; Salimi, Z.; Toghraei-Semiromi, Z.; Noori, S. Efficient One-Pot Synthesis of Quinazoline and Benzopyrano[2,3-d]pyrimidine Derivatives Catalyzed by N-Bromosulfonamides. *J. Heterocycl. Chem.* **2017**, *54*, 215–225. [CrossRef]
22. Moafi, L.; Ahadi, S.; Bazgir, A. New HA 14-1 analog: Synthesis of 2-amino-4-cyano-4H-chromenes. *Tetrahedron Lett.* **2010**, *51*, 6270–6274. [CrossRef]
23. Heravi, M.M.; Moradi, R.; Malmir, M. Recent Advances in the Application of the Heck Reaction in the Synthesis of Heterocyclic Compounds: An Update. *Curr. Org. Chem.* **2018**, *22*, 165–198. [CrossRef]
24. Roozifar, M.; Roozifar, M.; Niya, H.F. Application of salicylic acid as an eco-friendly and efficient catalyst for synthesizing 2, 4, 6-triaryl pyridine, 2-amino-3-cyanopyridine, and polyhydroquinoline derivatives. *J. Heterocycl. Chem.* **2021**, *58*, 1117–1129. [CrossRef]
25. Shaterian, H.R.; Aghakhanizadeh, M. Aminopropyl coated on magnetic Fe_3O_4, and SBA-15 nanoparticles catalyzed mild preparation of chromeno[2,3-d] pyrimidines under ambient and solvent-free conditions. *Catal. Sci. Technol.* **2013**, *3*, 425–428. [CrossRef]
26. Amirnejat, S.; Movahedi, F.; Masrouri, H.; Mohadesi, M.; Kassaee, M.Z. Silica nanoparticles immobilized benzoylthiourea ferrous complex as an efficient and reusable catalyst for one-pot synthesis of benzopyranopyrimidines. *J. Mol. Catal. A Chem.* **2013**, *378*, 135–141. [CrossRef]
27. Kabeer, S.A.; Reddy, G.R.; Sreelakshmi, P.; Manidhar, D.M.; Reddy, C.S. TiO_2–SiO_2 Catalyzed Eco-friendly Synthesis and Antioxidant Activity of Benzopyrano[2,3-d]pyrimidine Derivatives. *J. Heterocycl. Chem.* **2017**, *54*, 2598–2604. [CrossRef]
28. Thirupathaiah, B.; Reddy, M.V.; Jeong, Y.T. Solvent-free sonochemical multicomponent synthesis of benzopyranopyrimidines catalyzed by polystyrene supported p-toluenesulfonic acid. *Tetrahedron* **2015**, *71*, 2168–2176. [CrossRef]
29. Karami, S.; Momeni, A.R.; Albadi, J. Preparation and application of triphenyl (propyl-3-hydrogen sulfate) phosphonium bromide as a new efficient ionic liquid catalyst for synthesizing 5-arylidene barbituric acids and pyrano[2,3-d]pyrimidine derivatives. *Res. Chem. Intermed.* **2019**, *45*, 3395–3408. [CrossRef]
30. Enferadi-Kerenkan, A.; Trong-On, D.; Kaliaguine, S. Heterogeneous catalysis by tungsten-based heteropoly compounds. *Catal. Sci. Technol.* **2018**, *8*, 2257–2284. [CrossRef]
31. Baghernejad, B. Application of p-toluenesulfonic Acid (PTSA) in Organic Synthesis. *Curr. Org. Chem.* **2011**, *15*, 3091–3097. [CrossRef]
32. Shaikh, I.R. Organocatalysis: Key Trends in Green Synthetic Chemistry, Challenges, Scope towards Heterogenization, and Importance from Research and Industrial Point of View. *J. Catal.* **2014**, *2014*, 402860. [CrossRef]
33. Krishnan, G.R.; Sreekumar, K. Supported and reusable organocatalysts. In *New and Future Developments in Catalysis-Hybrid Materials, Composites, and Organocatalysts*; Suib, S.L., Ed.; Elsevier: Amsterdam, The Netherland, 2013; Chapter 14; pp. 343–364.
34. Sharma, N.; Brahmachari, G.; Das, S.; Kant, R.; Gupta, V.K. 2-[4-(Piperidin-1-yl)-5H-chromeno[2,3-d]pyrimidin-2-yl]phenol. *Acta Cryst.* **2014**, *70*, 447–448. [CrossRef] [PubMed]
35. Tocco, G.; Eloh, K.; Laus, A.; Sasanelli, N.; Caboni, P. Electron-Deficient Alkynes as Powerful Tools against Root-Knot Nematode Melodogyne incognita: Nematicidal Activity and Investigation on the Mode of Action. *J. Agric. Food Chem.* **2020**, *48*, 11088–11095. [CrossRef] [PubMed]
36. Chavarria-Carvajal, J.A.; Rodriguez-Kabana, R.; Kloepper, J.W.; Morgan-Jones, G. Changes in populations of microorganisms associated with organic amendments and benzaldehyde to control plant-parasitic nematodes. *Nematropica* **2001**, *31*, 165–180.
37. Jabbar, A.; Javed, N.; Munir, A.; Khan, S.A.; Ahmed, S. In vitro and Field Evaluation of Nematicidal Potential of Synthetic Chemicals against Root-Knot Nematode Meloidogyne graminicola in rice. *Int. J. Agric. Biol.* **2019**, *22*, 381–387.
38. Oka, Y. Screening of chemical attractants for second-stage juveniles of Meloidogyne species on agar plates. *Plant Pathol.* **2021**, *70*, 912–921. [CrossRef]

39. Caboni, P.; Aissani, N.; Cabras, T.; Falqui, A.; Marotta, R.; Liori, B.; Tocco, G. Potent Nematicidal Activity of Phthalaldehyde, Salicylaldehyde, and Cinnamic Aldehyde against Meloidogyne incognita. *J. Agric. Food Chem.* **2013**, *61*, 1794–1803. [CrossRef]
40. Eisenback, J.D. Detailed morphology, and anatomy of second-stage juveniles, males, and females of the genus Meloidogyne (root-knot nematodes), An Advanced Treatise on Meloidogyne. In *An Advance Treatise on Meloidogyne*; North Carolina Department of Plant Pathology and US Agency for International Development: Raleigh, NC, USA, 1985; Volume 1, pp. 47–77.
41. Liu, G.; Lai, D.; Liu, Q.Z.; Zhou, L.; Liu, Z.L. Identification of Nematicidal Constituents of Notopterygium incisum Rhizomes against Bursaphelenchus xylophilus and Meloidogyne incognita. *Molecules* **2016**, *21*, 1276. [CrossRef]
42. Krieger, E.; Vriend, G. YASARA View-molecular graphics for all devices from smartphones to workstations. *Bioinformatics* **2014**, *20*, 2981–2982. [CrossRef]

Article

Sugarcane Bagasse Ash as a Catalyst Support for Facile and Highly Scalable Preparation of Magnetic Fenton Catalysts for Ultra-Highly Efficient Removal of Tetracycline

Natthanan Rattanachueskul [1], Oraya Dokkathin [1], Decha Dechtrirat [2,3,4], Joongjai Panpranot [5], Waralee Watcharin [6], Sulawan Kaowphong [7] and Laemthong Chuenchom [1,*]

1. Division of Physical Science (Chemistry Program) and Center of Excellence for Innovation in Chemistry, Faculty of Science, Prince of Songkla University, Hat-Yai District, Songkhla 90110, Thailand; natthanan.rc@gmail.com (N.R.); orayamod.d@gmail.com (O.D.)
2. Department of Materials Science, Faculty of Science, Kasetsart University, Bangkok 10900, Thailand; fscidcd@ku.ac.th
3. Specialized Center of Rubber and Polymer Materials for Agriculture and Industry (RPM), Faculty of Science, Kasetsart University, Bangkok 10900, Thailand
4. Laboratory of Organic Synthesis, Chulabhorn Research Institute, Bangkok 10210, Thailand
5. Department of Chemical Engineering, Faculty of Engineering, Chulalongkorn University, Bangkok 10330, Thailand; joongjai.p@chula.ac.th
6. Faculty of Biotechnology (Agro-Industry), Assumption University, Hua Mak Campus, Bangkok 10240, Thailand; waraleewtc0@gmail.com
7. Department of Chemistry and Center of Excellence in Materials Science and Technology, Faculty of Science, Chiang Mai University, Chiang Mai 50200, Thailand; sulawank@gmail.com
* Correspondence: laemthong.c@psu.ac.th; Tel.: +66-74-288416; Fax: +66-74-558841

Citation: Rattanachueskul, N.; Dokkathin, O.; Dechtrirat, D.; Panpranot, J.; Watcharin, W.; Kaowphong, S.; Chuenchom, L. Sugarcane Bagasse Ash as a Catalyst Support for Facile and Highly Scalable Preparation of Magnetic Fenton Catalysts for Ultra-Highly Efficient Removal of Tetracycline. Catalysts 2022, 12, 446. https://doi.org/10.3390/catal12040446

Academic Editors: Sagadevan Suresh and Is Fatimah

Received: 14 March 2022
Accepted: 14 April 2022
Published: 18 April 2022

Publisher's Note: MDPI stays neutral with regard to jurisdictional claims in published maps and institutional affiliations.

Copyright: © 2022 by the authors. Licensee MDPI, Basel, Switzerland. This article is an open access article distributed under the terms and conditions of the Creative Commons Attribution (CC BY) license (https://creativecommons.org/licenses/by/4.0/).

Abstract: Sugarcane bagasse ash, which is waste from the combustion process of bagasse for electricity generation, was utilized as received as a catalyst support to prepare the magnetic sugarcane bagasse ash (MBGA) with different iron-to-ash ratios using a simple co-precipitation method, and the effects of NaOH and iron loadings on the physicochemical properties of the catalyst were investigated using various intensive characterization techniques. In addition, the catalyst was used with a low amount of H_2O_2 for the catalytic degradation of a high concentration of tetracycline (800 mg/L) via a Fenton system. The catalyst exhibited excellent degradation activity of 90.43% removal with good magnetic properties and high stabilities and retained good efficiency after four cycles with NaOH as the eluent. Moreover, the hydroxyl radical on the surface of catalyst played a major role in the degradation of TC, and carbon-silica surface of bagasse ash significantly improved the efficiencies. The results indicated that the MBGA catalyst shows the potential to be highly scalable for a practical application, with high performance in the heterogeneous Fenton system.

Keywords: heterogeneous Fenton; tetracycline removal; magnetite; sugarcane bagasse ash

1. Introduction

Tetracycline (TC), the pharmaceutical antibiotic, has been widely used in many fields such as human and livestock feeding. The TC contaminative wastewater from hospitals and pharmaceutical industries is considered a threat to the aquatic system and human health due to its exposure to microorganisms causing drug resistance [1–3]. Therefore, it is imperative to efficiently remove TC from wastewater before discharge.

Various methods have been used to remove TC from aqueous systems. Adsorption is among the most popular techniques [4–6]; however, although it shows high TC removal efficiency, TC residue is still trapped inside the adsorbents, and so it only changes the location of TC. Coagulation is another method to remove TC from wastewater [7,8]. Nevertheless, it requires large amounts of expensive chemicals, resulting in overall costly operation.

Furthermore, after adsorption and coagulation processes, both the TC-loaded adsorbents and TC-contaminated coagulants are defined as hazardous wastes, which leads to difficulty in post-processing.

For the above reasons, alternative methods to efficiently remove TC from the water system are still highly required. Advance oxidation processes (AOPs) are a kind of chemical technology which degrades organic pollutants using highly reactive species such as hydroxyl radicals (·OH). The OH radicals generated from AOPs can convert toxic organic compounds in a non-selective way into the non-toxic inorganic products. Therefore, the AOPs have become more attractive [1,9,10]. One of the most attractive AOPs which can be operated at ambient temperatures and pressures is the Fenton process because of its high degradation efficiency, simple and low-cost operation, and mild and green reaction conditions [1,9–11].

The classical Fenton reagent consists of iron salts and hydrogen peroxide (H_2O_2) as an oxidant in a homogeneous solution. However, the iron sludge produced after the treatment is the main drawback of this homogenous catalyst, which is difficult to separate and can be secondary contaminant to the treated water [9,10,12].

To solve this problem, heterogeneous Fenton processes using magnetic solid iron-based catalysts have received broad attention for the elimination of various types of organic pollutants [13–16].

The inverse spinel Fe_3O_4 (magnetite) is one of the most studied magnetic particles, which can be used for a wide range of applications such as adsorption [17–21] and catalysis [15,22–25]. Many previous reports revealed its highly efficient catalytic performance in a heterogeneous Fenton system, where Fe^{2+} in magnetite structure can play a main role in initiating the Fenton reaction in the presence of H_2O_2 according to the classical Haber–Weiss mechanism [11,26–28]. With the presence of both Fe^{2+} and Fe^{3+}, this allows the Fe ions to be oxidized and reduced in a reversible loop and be kept reactive in the Fenton system. Additionally, the magnetite particles possess strong magnetic properties; therefore, they are considered suitable as a heterogeneous Fenton catalyst with rapid and effective separation from aqueous solution using an external magnet.

However, the surface area is considered an important factor in a heterogeneous reaction, while the surface area of magnetite particles is generally low. To enhance the catalytic activity of magnetite particles, the magnetic solid iron-based catalysts on supports such as silica [1,9,29,30], and carbonaceous materials [1,13,31–33] have widely been investigated due to their high stability, low toxicity, wide availability, and low-cost materials. Nevertheless, the preparation techniques of those catalysts are complex and rely on multi-step processes, including use of harmful chemicals and high temperatures [34–38]. Moreover, the concentration of TC studied by many works is still comparatively low (<50 ppm), while the catalyst dosages are comparatively high (0.5–1.0 g/L). This indicates that the actual amount of TC in the solution for those studies was considered very low (0.2–10 mg).

The fact that no one reports on the removal efficiency for catalytic degradation in terms of real absolute TC amount in mg leads to difficulty when comparing the performance of catalyst with other works. Herein, the amount of TC removal per catalyst dosage was calculated in this work to be compared clearly among literature reviews.

Sugarcane bagasse ash (BGA) is the remaining waste from the combustion process of sugarcane bagasse for generation of electricity, which is considered as the simplest and most cost-effective utilization of bagasse [39,40]. Because Thailand is one of the largest sugar producers in the world, there is still a huge amount of sugarcane bagasse ash left from the combustion process (around 365,000 tons/year) [41–43]. In fact, a common utilization of this solid waste is mainly landfill disposal, which is unsustainable for the circular economy concept [44]. Moreover, without the right care, the obtained fly ash—which is considered particulate matter—can cause air pollution, leading to respiratory diseases [45,46].

Many researchers have used it as a substitute component for building material [47–49] or as fertilizer to improve agricultural productivity [50,51], or directly used it as an adsorbent for wastewater treatment [52–54].

However, the obtained sugarcane bagasse ash with a high surface area and chemical stabilities is a promising material for catalyst support, and to the best of our knowledge, only a few studies have been found on using ash as catalyst supports [24,30,55]. Moreover, Fe_3O_4@ash from sugarcane bagasse has not been previously studied as a Fenton catalyst for the removal of organic pollutants.

Generally, catalyst support can be obtained from multiple-step synthesis, not from as-received precursors. This escalates more steps into the whole synthesis pathway. These supports should be environmentally friendly, cost effective, sustainable, and have high stability. BGA is typically obtained as waste through incomplete combustion of sugarcane bagasse during the process of generation of electricity in biomass power plants. Unlike other types of ashes, the main characteristic of BGA is the presence of high carbon and silica content and suitable textural morphology. The presence of those elements, as well as the high surface area of non-toxic BGA, can provide properties similar to those of good catalyst supports without any pretreatment process. With the simple one-step co-precipitation method, magnetic sugarcane bagasse ash can be prepared, which is easy and scalable. Moreover, high temperatures and highly toxic chemicals are no longer required, with only water-based reaction involved. This incorporation of BGA in the production of the catalyst can mitigate the environmental impact of waste disposal while decreasing the production cost and time of the magnetic Fenton catalysts.

In this work, magnetic sugarcane bagasse ash (MBGA) can be prepared via a simple co-precipitation route using iron ions and BGA as precursors. The obtained MBGA was intensively characterized and its catalytic Fenton efficiency was analyzed for the degradation of tetracycline (TC), one of the toxic antibiotics generating problems nowadays. For the first time, we report on the use of our catalysts in degradation of TC with extremely high concentration (800 mg/L). The results indicate that nearly 100% of that TC concentration, (or degradation of 40 mg of TC) could be achieved. This value is far higher than those reported in many publications. Furthermore, our catalyst exhibited high degradation of TC even upon four cycles with excellent magnetic properties being retained. The catalysts purposed here are highly scalable.

2. Results and Discussion

2.1. Optimization of the Preparation Conditions

The raw sugarcane bagasse ash (BGA) was black, unlike the fly ash usually used as catalyst supports. The black color of BGA indicated the presence of unburned carbon particles, as confirmed by the elemental analysis. Morphological investigation using SEM on the raw BGA revealed the existence of macropores even after combustion from the sugarcane-based biomass power plant (Supplementary Materials, Figure S1). Such microporosity in BGA could allow for the facile infiltration and deposition of iron ions (both Fe^{2+} and Fe^{3+}) into the inner pores and on the outside surfaces. Moreover, the polar-oxygenated functional groups on the BGA surface can enhance immobilization of the iron ions onto the BGA surfaces [43]. For this reason, BGA was selected as the catalyst support for this work. Upon the deposition of Fe^{2+} and Fe^{3+} ions, the addition of NaOH with heating at 80 °C can co-precipitate and transform them into magnetite (Fe_3O_4) particles deposited on the BGA surfaces. After washing with DI water, magnetic sugarcane bagasse ash composites (MBGA) are obtained. The effects of NaOH concentration and expected Fe_3O_4 to BGA ratio on the appearance and the physicochemical properties of MBGA were investigated. For the effect of NaOH concentration, all the samples (MBGA1-1, MBGA2-1, and MBGA5-1) were similarly brown-black in color (Figure S2). However, only MBGA2-1 was easily attracted by an external magnet when dispersed in DI water (Figure S3), while MBGA1-1 and MBGA5-1 showed only slight magnetic properties. For control, synthetic Fe_3O_4 was glossy black and strongly attracted by a magnet. This agrees with the saturation magnetization (M_s) from VSM analysis (Figure 1). The magnetization values for all samples were summarized in Table 1. All the investigated samples show hysteresis loops in the ±10 kOe range at 300 K, exhibiting superparamagnetic characteristics. The M_s of MBGA1-1 was only 1.02 emu/g, which was considered almost non-magnetic and is unsuitable for practical applications. The reason for this involves the use of a low concentration of NaOH to precipitate all iron ions into the magnetic particles, according to the stoichiometric ratio in the preparation of Fe_3O_4. In contrast, MBGA2-1 possessed M_s as high as 28.71 emu/g. Table 1 shows that the yields of the prepared catalysts were quite low for MBGA1-1 and much lower for MBGA5-1. The reason for the very low yield of only 6.15% of MBGA5-1 resulted from the excessive use of NaOH, leading to the digestion of carbon and silica (the main components in the BGA) under alkali conditions [56,57]. Although the magnetic properties of MBGA5-1 were acceptable (M_s = 10.94 emu/g), their lowest yield must be a problem for the real production and cost efficiency. According to the above reasons, in this study, the optimum NaOH concentration was found to be 2 M.

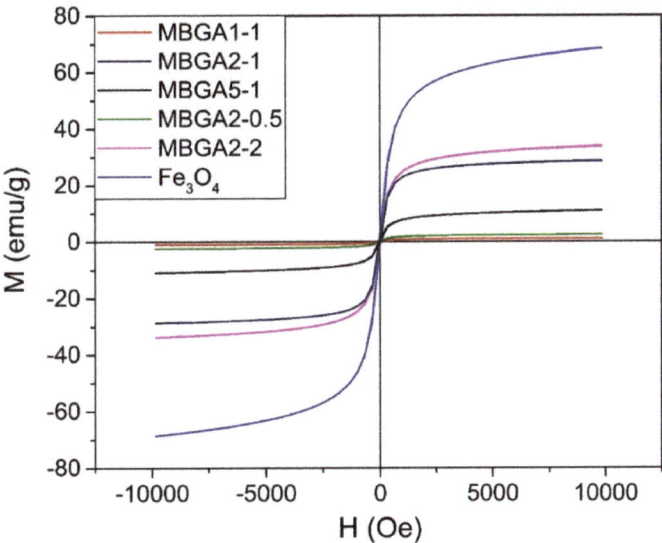

Figure 1. VSM hysteresis loops for all samples.

Table 1. Optimization of the preparation conditions and their physical and chemical properties.

Sample	NaOH (M)	Ratio of Fe_3O_4/BGA	Overall Yield (%)	M_s (emu/g)	%C (wt%)	pH_{PZC}	S_{BET} (m²/g)
BGA	-	-	-	-	11.68	2.62	55.2817
Fe_3O_4	5	-	53.93 ± 3.41	68.603	0.14	3.98	62.0286
MBGA2-0.5	2	1:2	26.37 ± 4.69	2.4508	7.45	3.62	85.4644
MBGA2-1	2	1:1	68.67 ± 5.49	28.705	5.53	4.14	96.0017
MBGA2-2	2	2:1	79.94 ± 3.91	33.878	3.44	4.32	60.0747
MBGA1-1	1	1:1	19.42 ± 2.54	1.0193	6.12	-	-
MBGA5-1	5	1:1	6.15 ± 2.82	10.943	5.81	-	-

The effect of expected Fe_3O_4 to BGA ratio on the physical appearances indicates that all samples (MBGA2-0.5, MBGA2-1, MBGA2-2) show the same almost-black color as others and can be attracted by a magnet (Figure S3). M_s values of this sample series were found to be correlated with the amount of the irons used. M_s values of MBGA2-1 and MBGA2-2 were similar (~30 emu/g), but MBGA2-2 required two times the amount of iron (Figure 1). MBGA2-0.5 showed M_s of only 2.45 emu/g due to its lowest iron sources content. Moreover, the carbon content for all samples (Table 1) confirms there was carbon left from the raw BGA, even after co-precipitation of the iron ions. This agrees with black color as a precursor for BGA and all MBGA samples.

2.2. Characterization of MBGA2-0.5, MBGA2-1, and MBGA2-2

To study in detail the morphology, surface chemistry, and the mechanisms for the formation of iron oxides on BGA, various intensive characterization techniques were employed.

The FESEM-EDX results for all samples (Figure S1) show macropores retained from the natural xylem and phloem of raw bagasse. This fact suggests that the combustion process in the power plant and subsequent magnetization with a co-precipitation method did not change the macropore morphology of the bagasse, implying the existence of a stable inorganic skeleton (mostly oxide compounds). These remaining macropores in BGA could be the advantage which enhances the diffusion of gigantic molecules such as TC. Besides, it is clearly seen that there were some nearly spherical silica particles in all samples. These nearly spherical particles of about ~5–30 μm diameter originated from the combustion process of bagasse and the inorganic components were changed to oxide compounds such as SiO_2, CaO, and Al_2O_3 [55,58]. This silica–carbon composite could enhance the stability of catalyst support [59,60]. Moreover, the BGA surface was smoother than that of MBGA catalyst samples because the iron oxide particles precipitated on the BGA surface, resulting in a rough surface. Furthermore, the macropores may increase the diffusion rates of pollutants into the inner pores, while the iron oxide particles act as active sites for a heterogeneous Fenton reaction. The iron particles distributions for all samples were homogeneous, as shown in the EDX mapping (Figure S1), and the average iron content was 4.14, 9.84, and 13.70 wt% for MBGA2-0.5, MBGA2-1, and MBGA2-2, respectively. This agrees well with FESEM images, which showed that the BGA surface tended to be rougher when the iron-to-BGA ratio increased as a result of more iron oxide particles being deposited on BGA surface. Furthermore, the carbon content was decreased as this iron ratio increased, according to the trend determined by elemental analysis (Table S1). This is also confirmed by XRD patterns in Figure 2A. The XRD patterns of magnetite (Fe_3O_4, JCPDS No. 01-084-2782) and silica (SiO_2, JCPDS No. 01-070-3755) can clearly be seen. These patterns agree with the XRD of BGA and Fe_3O_4 prepared with the same method but without BGA. Moreover, it is noted that MBGA2-0.5, MBGA1-1, and MBGA5-1 showed low relative intensity of the characteristic peak of magnetite at 35.6° (311) which agreed well with their low M_s values from VSM results.

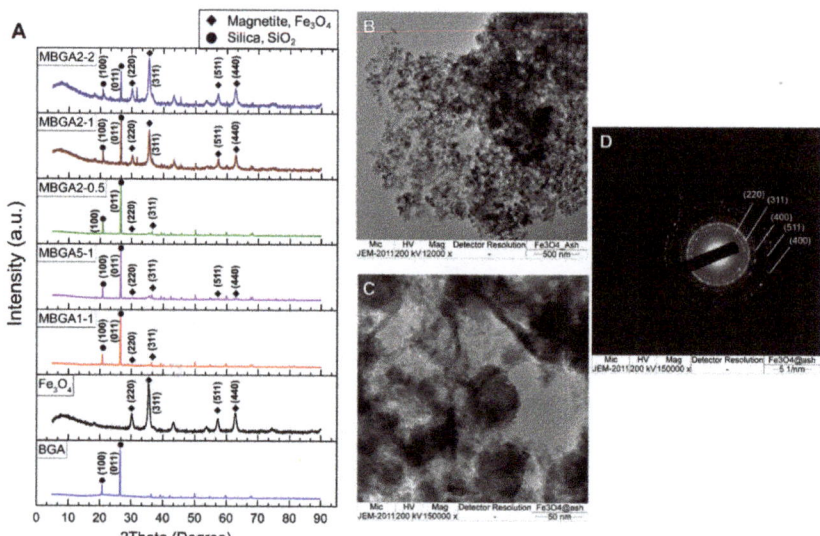

Figure 2. (**A**) XRD patterns for all samples (**B,C**) TEM image of MBGA2-1 at 12,000× and 150,000×, respectively (**D**) SAED pattern showing diffraction rings corresponding to Fe_3O_4.

The surface chemistry of samples was studied by FTIR spectroscopy as shown in Figure S4, and the band assignments are detailed in Table S2. The characteristic band of cellulose around ~1055 cm^{-1}.[43] and considerable amounts of oxygenated functional groups of carbonaceous surface from the unburned materials were retained on the surfaces of BGA after the combustion process, which displays the absorption bands at 3435, 2924, 1631, and ~1055 cm^{-1}, assigned to stretching vibrations of O-H, C-H, aromatic C=C or C=O, and C-O and Si-O-Si bonds, respectively. Moreover, pH$_{PZC}$ measured with the Zeta potential technique in Table 1 suggests that the surface of BGA and all MBGA samples was acidic. This is in agreement with the oxygenated oxidic functional groups on the BGA surface.

The similarity of the FTIR spectra and pH$_{PZC}$ for all samples suggests that the oxygenated functional groups on the surface show no difference regarding the heterogeneous Fenton reaction. It was found that the main oxygenated functional groups of MBGA samples from C 1s and O 1s high-resolution spectra were all similar, including carboxylic, hydroxyl, and ether groups. This provides hydrophilic character, and the peak at 530 eV binding energy that corresponds to magnetite (Fe$_3$O$_4$) [61]. The wide-scan XPS and high-resolution C 1s and O 1s spectra of MBGA2-1 were shown in Figure 3. This is consistent with the band for Fe-O bonds around ~604 cm^{-1} in the FTIR spectra and the magnetite pattern in the XRD results. These confirm the presence of magnetite particles on the surface of BGA. The provided detail of peak assignments for deconvoluted XPS peaks and the elemental compositions from survey scan XPS spectra of MBGA2-1 were shown in Tables S3 and S5, along with the binding energies.

To further investigate the presence of Fe$_3$O$_4$ particles in MBGA2-1, TEM-EDS analysis was performed. The magnetite particles were homogeneously distributed on the BGA support as observed in the TEM images as dark spherical particles with sizes of 15–40 nm (Figure 2B,C). Furthermore, the SAED pattern for MBGA2-1 is shown in Figure 2D. The five rings obtained from the SAED image are indexed to magnetite, which agrees with the XRD results. Moreover, the EDS spectra in Figure S5 shows C, O, Fe, and Si elements distributed homogeneously all over the samples, which agree well with EDX mapping.

For the chemical equation of the preparation of MBGA using a co-precipitation method, it can be concluded that hydroxide ions react with ferrous and ferric ions as shown in reaction (R1) [17,30,62]

$$Fe^{2+} + 2Fe^{3+} + 8OH^- \rightarrow Fe_3O_4 + 4H_2O \tag{R1}$$

After that, the precipitated magnetite particles were deposited on the BGA surface, resulting in the magnetic properties of a whole sample piece. This carbon–silica support raised more chemical and mechanical stabilities of the catalyst [59,63]. Moreover, this method provides a facile single-step and scalable technique to utilize the waste precursor (BGA) as received from power plants without additional complex steps to manufacture functional catalysts.

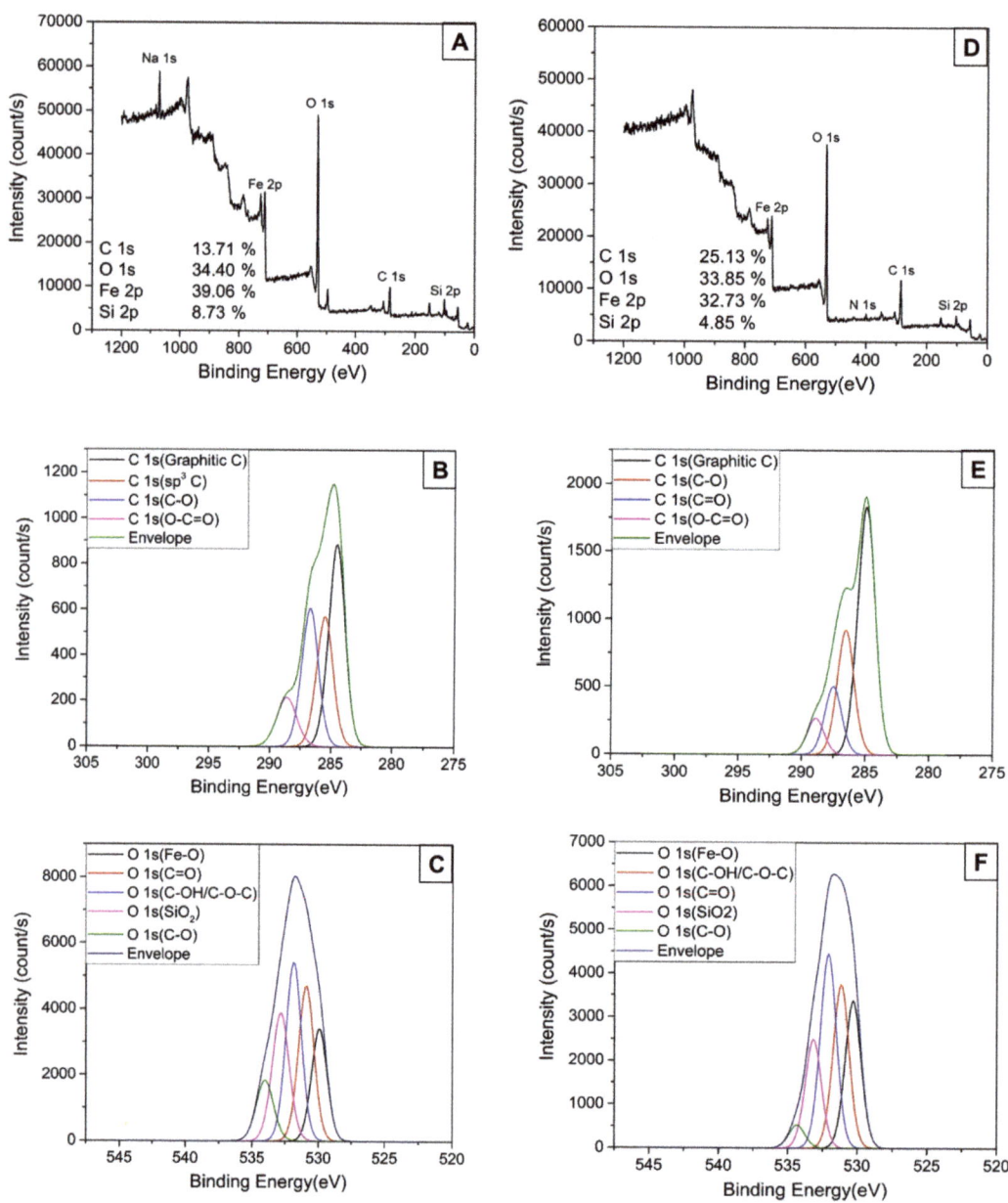

Figure 3. (**A**) The wide-scan XPS, (**B**,**C**) high-resolution C 1s and O 1s spectra of fresh MBGA2-1 (**D**) The wide-scan XPS, (**E**,**F**) high-resolution C 1s and O 1s spectra of MBGA2-1 after 4 cycles using 0.1 M NaOH as eluent.

2.3. Preliminary Catalytic Heterogeneous Fenton Reaction Test of Magnetic Sugarcane Bagasse Ash

To study the possibilities of the MBGA samples as the heterogeneous Fenton catalyst, the catalytic degradation test of tetracycline (TC) was investigated. The reaction was started by adding 5 mM of H_2O_2 into the 50 mL of 800 mg/L TC solution at natural pH (pH 3.2) with a catalyst concentration of 1 g/L. The result, as shown in Figure S6A, suggests that MBGA2-1 and MBGA2-2 have similar removal efficiency for TC after 12 h reaction time; around ~90%. Even the iron content in MBGA2-2 was higher, while the removal efficiency for MBGA2-0.5 was slightly lower due to a lower iron amount on the surface of the catalyst. For control, only TC degraded with 5 mM of H_2O_2 without a catalyst was also studied, and only 12.56% of TC was degraded. Furthermore, pure Fe_3O_4 (ground into power with the particle size of ~50 μm) revealed only 47.66% removal. This confirms that use of BGA as a catalyst support can significantly increase the removal efficiency. Furthermore, the Fe leaching tests for all samples showed that MBGA2-2 had the highest Fe leaching of 4.58 mg/L, while it was only 0.61 and 0.28 mg/L for MBGA2-1 and MBGA2-0.5, respectively. The high leaching of iron from MBGA2-2 was due to the excess amount of iron precursors which could not strongly attach to the surface of catalyst support and could easily be dissolved in natural acidic TC solution. When taking BET surface area into consideration, as shown in Table 1, as is known to all, the high BET specific surface area is beneficial for more guest molecules to access active sites. The MBGA2-1 also showed the highest BET surface area of 96 m^2/g. According to all information and intensive characterization techniques, MBGA2-1 was selected as a promising Fenton catalyst for the catalytic degradation of TC because of the high removal efficiency with low Fe leaching, and high stabilities with strong magnetic properties. Moreover, it also has the highest BET surface area and suitable surface characters with low-cost preparation. For the effect of catalyst concentration, as shown in Figure S6B, the degradation efficiency increased with increasing catalyst loading. Therefore, it was fixed at 1 g/L for all other experiments because it was the minimum amount of the catalyst which can be a representative for the whole sample in the catalytic degradation experiments. This can be considered by the standard deviation, in which 1 g/L shows the lowest value with ~90% efficiency, while 2 g/L shows only 5% higher efficiency but uses double the amount of catalyst.

2.4. Catalytic Degradation of TC by MBGA2-1: Effect of pH

Generally, solution pH can significantly influence the catalytic degradation in the Fenton system [10,64]. The effect of pH on the catalytic degradation of TC by MBGA2-1 was investigated. The experiment was conducted as shown in Figure 4A. The studied pH solution was 3.2 (natural pH), 5, 7, 9. The natural pH at 3.2 showed the highest removal efficiency of 90.43% for MBGA2-1. It is worth noting that when pH is higher, the efficiency decreased due to the reduced reactivity of radicals and the precipitation of iron ions [10]. Furthermore, MBGA2-1 shows over two times greater removal efficiency than pure Fe_3O_4, again confirming the advantage of BGA as catalyst support. In absence of the catalyst, H_2O_2 alone in natural pH can degrade TC with only ~14%. In addition, 6 h is the equilibrium time for all samples in this catalytic degradation of 800 mg/L TC solution. Therefore, the unadjusted natural pH (3.2) was chosen to be the optimum pH for further experiments.

To evaluate the kinetics of TC catalytic degradation, the data were fitted with a linear pseudo-first-order model shown as Equation (1):

$$\ln C_t = -k_1 t + \ln C_0 \qquad (1)$$

where k_1 is a rate constant for catalytic degradation in the pseudo-first-order model, and C_t is the TC concentration at time t. The kinetics of the adsorption-co-catalytic degradation of the MBGA samples were fitted well with pseudo-first-order reaction with good R^2 and the calculated pseudo-first-order rate constants, as shown in Figure S7. On the other hand, under basic conditions, the degradation rate decreased, and the removal efficiencies at pH 7 and 9 were almost the same. Moreover, it is noted that the pseudo rate constant of the MBGA2-1 at natural pH was 0.1047 h^{-1}, which was 2.3 and nearly 4 times higher than pH 5 and pH 7, respectively. This is due to Fenton reactions being favored under acidic conditions.

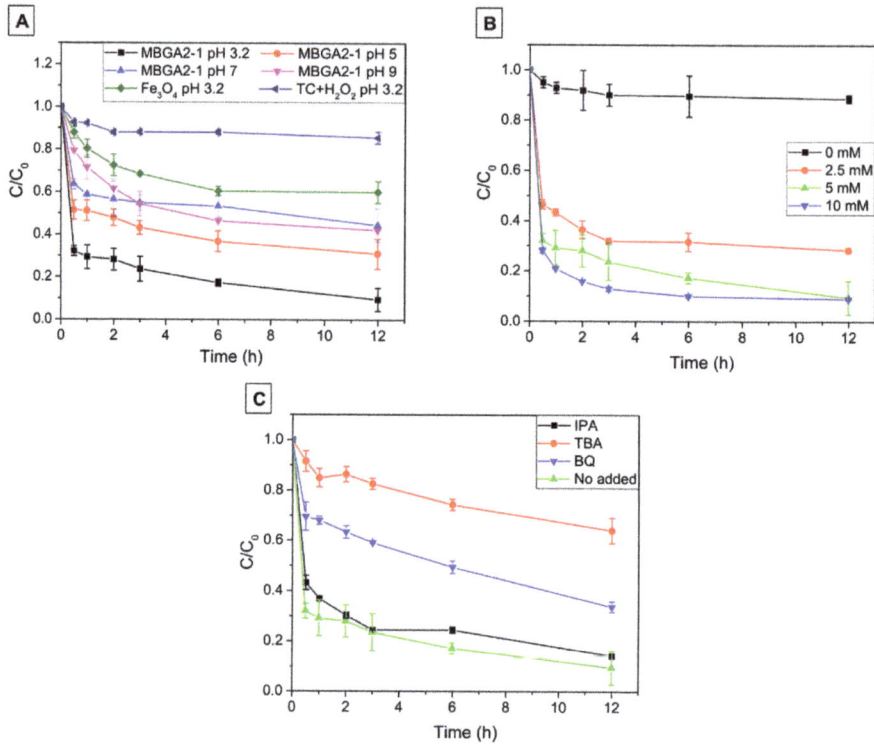

Figure 4. The catalytic degradation of TC (**A**) Effect of pH (**B**) Effect of H_2O_2 concentration (**C**) Effect of scavenger (conditions: C_0 = 800 mg/L, pH 3.2 (natural), H_2O_2 concentration = 5 mM, catalyst concentration = 1 g/L, T = 28 ± 2 °C).

2.5. Catalytic Degradation of TC by MBGA2-1: Effect of H_2O_2 Concentration

To investigate the effect of H_2O_2 concentration on the catalytic degradation of TC, the experiments with different H_2O_2 concentrations (0, 2.5, 5, 10 mM) were conducted as shown in Figure 4B. In absence of the H_2O_2, the adsorption of MBGA2-1 alone can remove only 11.4% of TC (actual adsorption capacity of 90.32 mg/g) due to the limited adsorption sites. Note that even without addition of H_2O_2, the removal of TC could reach up to 90 mg/g due to the pure adsorption mechanism. This strongly suggests the importance of using BGA as a catalyst support because of the ability of the carbon left from combustion process to attract the TC molecules. The removal efficiency for MBGA2-1 increased with the increasing H_2O_2 concentration. The removal efficiency can reach up to 80% at 6 h for 5 mM H_2O_2 concentrations. Nevertheless, while the H_2O_2 dosage rose from 5 to 10 mM, the removal efficiency barely increased, and the same efficiencies were reached at 12 h. This is because the excess H_2O_2 can compete with TC for ·OH as shown in reaction (R2–R3) [64].

$$H_2O_2 + ·OH \rightarrow HO_2· + H_2O \quad \text{(R2)}$$

$$HO_2· + ·OH \rightarrow O_2 + H_2O \quad \text{(R3)}$$

Even though 10 mM shows the highest rate in the first few hours, the same efficiencies were reached at 12 h. Therefore, to minimize the cost of operation, 5 mM of H_2O_2 was selected as the optimum concentration.

2.6. Catalytic Degradation of TC by MBGA2-1: Effect of Scavenger and Its Mechanism

To identify free radicals in Fenton oxidation experiments, different types of free-radical-trapping agents were added into the reaction systems to identify the difference of contributions to the degradation efficiency between different free radicals. In this experiment, 0.3 M tert-butyl alcohol (TBA) was used as a hydroxyl radical ($\cdot$OH) scavenger, while 20 mM p-benzoquinone (BQ) was used as a superoxide radical ($\cdot O_2^-$) scavenger [65,66]. As shown in Figure 4C, the introduction of TBA remarkably inhibited TC degradation and the removal efficiency was only 35.87% at 12 h, while adding 20 mM BQ caused a slight drop in the degradation efficiency from 90.43% without scavenger to 66.13% at 12 h. This demonstrated that $\cdot$OH was the major reactive oxygen species in this heterogeneous Fenton system.

To explain the mechanisms of the AOP by $\cdot$OH radicals in detail, two types of hydroxyl radicals were used to distinguish them. First, 0.3 M isopropanol (IPA) was utilized to individually trap free hydroxyl radicals ($\cdot OH_{free}$), while it was well known that TBA could quench both free $\cdot$OH in solution ($\cdot OH_{free}$) and $\cdot$OH adsorbed onto the surface of a catalyst ($\cdot OH_{ads}$) [65,66]. Notably, the addition of IPA slightly suppressed the degradation of TC and the removal efficiency showed only a 4.65% drop at 12 h, indicating that only a low amount of $\cdot OH_{free}$ was present in the system. Moreover, it can be concluded that $\cdot OH_{free}$, $\cdot OH_{ads}$, and $\cdot O_2^-$ contributed to the degradation of TC by a heterogeneous Fenton system. However, $\cdot OH_{ads}$ on the MBGA2-1 surface was the only major reactive species responsible for the catalytic degradation of TC.

To confirm that the major mechanisms of the removal of TC mainly involved catalytic degradation rather than adsorption, the total organic carbon (TOC) of the solution before and after the catalytic degradation of TC was determined, and the TOC removal was 56.70%, indicating that a huge amount of TC in the solution was transformed into inorganic compounds (CO_2 and H_2O). Moreover, the dramatic decrease in peak at m/z at 443.15 after the complete catalytic degradation of TC analyzed by LC-MS in Figure S8 suggests that the intermediates generated from the Fenton system were quickly oxidized into inorganic compounds. It is worth noting that the low TOC after catalytic testing also suggests that the BGA as a support released no organic compounds, unlike the release of polyaromatic hydrocarbon molecules (PAHs) in aqueous systems from various solids obtained through the combustion process. This further confirms the chemical stability of the support in this work and could eliminate the concern over the toxic chemicals released from the BGA.

The catalytic mechanism of H_2O_2 via MBGA2-1 composite is proposed in Figure 5. Firstly, the surface of the MBGA2-1 will rapidly form iron ions under acidic conditions as shown in (R4), then the surface of catalyst was adsorbed by H_2O_2, and it was quickly decomposed to $\cdot OH_{ads}$ by the Fe(II) on the surface of MBGA2-1, as shown in reaction (R5), while Fe(III) on the surface can react with H_2O_2 to generate perhydroxyl radicals ($\cdot$OOH) or superoxide radicals, and the regeneration of Fe(II) by the reduction of Fe(III) can continue the catalytic cycle (R6–R9) [11,67,68].

$$MBGA \equiv Fe-O + 2H^+ \rightarrow MBGA \equiv Fe(II)/Fe(III) + H_2O \tag{R4}$$

$$\equiv Fe^{2+} + H_2O_2 \rightarrow \equiv Fe^{3+} + OH^- + \cdot OH_{ads} \tag{R5}$$

$$\equiv Fe^{3+} + H_2O_2 \rightarrow \equiv Fe^{2+} + \cdot OOH + H^+ \tag{R6}$$

$$\equiv Fe^{3+} + \cdot OOH \rightarrow \equiv Fe^{2+} + O_2 + H^+ \tag{R7}$$

$$\equiv Fe^{3+} + H_2O_2 \rightarrow \equiv Fe^{2+} + \cdot O_2^- + 2H^+ \tag{R8}$$

$$\equiv Fe^{3+} + \cdot O_2^- \rightarrow \equiv Fe^{2+} + O_2 \tag{R9}$$

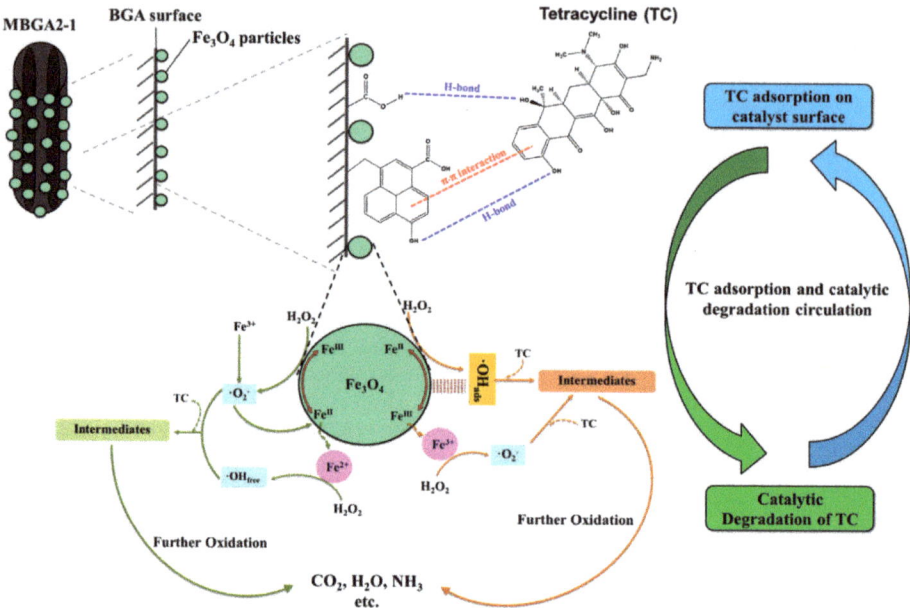

Figure 5. The TC catalytic degradation mechanism.

However, the dissolution of iron ions in the solution was inevitable due to the acidity of the solution, and the deposition of iron particles was on the surface of BGA, not in the carbon matrix. Hence, the same mechanism mentioned above also occurred to generate a little amount of ·OH$_{free}$ indicating the low Fe leaching in the solution. Moreover, the presence of ferrous ions from Fe$_3$O$_4$ on the surface of catalyst is especially favored under acidic conditions, and they played the most important role in the formation of ·OH$_{ads}$ in the reaction, which was the main reactive oxygen species.

2.7. Catalytic Degradation of TC by MBGA2-1: Effect of TC Concentration

The effect of TC concentration on TC degradation is presented in Figure 6. It was observed that the initial rate of degradation increased with the increase in the initial concentration of TC, but it barely increased when the initial TC concentrations were higher than 800 mg/L, indicating that the active sites of catalyst were occupied by TC molecules, and TC molecules cannot access the active sites at higher concentrations. This was confirmed by the removal efficiencies at 12 h, at which the high initial concentrations of TC showed lower efficiencies. Moreover, the removal capacity (Q_e) also increased with the increased initial TC concentration.

Furthermore, it is known that any heterogeneous reaction, as in the case of a Fenton reaction, is highly affected by the interactions between pollutant molecules and catalysts, including two consecutive steps. Firstly, the reactant molecules are adsorbed on the surface of the catalyst and followed by immediate degradation. To confirm this mechanism, the Langmuir–Hinshelwood equation was applied for MBGA2-1 as shown in the inset of Figure 6, and the linear expression of the Langmuir–Hinshelwood plot was as follows [69]:

$$\frac{1}{v_0} = \frac{1}{kK_{ads}C_0} - \frac{1}{k} \tag{2}$$

where v_0 is the initial rate of TC degradation, mg/L·s; k is reaction rate constant, mg/L·s, and K_{ads} is the Langmuir–Hinshelwood adsorption equilibrium constant, L/mg. The good fitness of the Langmuir–Hinshelwood equation with R^2 of 0.9974 confirmed the above assumption and also agreed with the mechanism where ·OH$_{ads}$ played the main role as the reactive oxygen species for TC degradation. The rate constant and adsorption equilibrium constant were reported in Table S5.

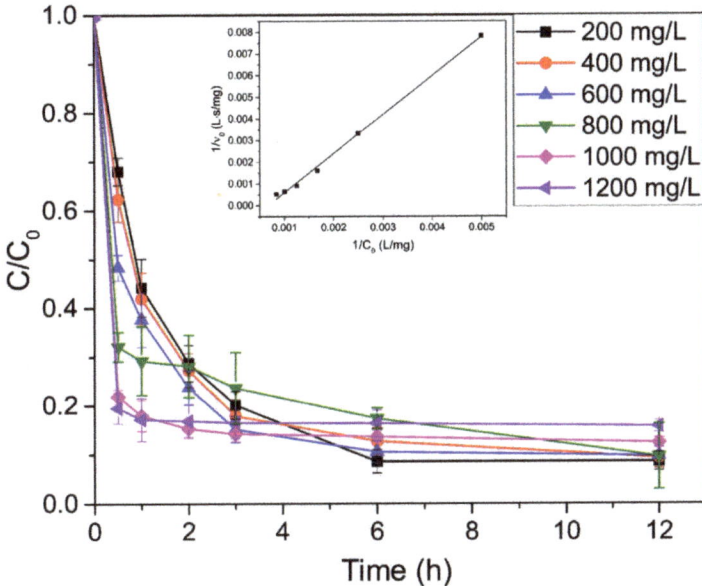

Figure 6. Effect of initial TC concentration on the catalytic degradation of TC and (inset) the Langmuir–Hinshelwood (L-H) plot (conditions: C_0 = 200–1200 mg/L, pH 3.2 (natural), H_2O_2 concentration = 5 mM, catalyst concentration = 1 g/L, T = 28 ± 2 °C).

The agreement of this model assumption with data indicates the important role of the silica–carbon support, which not only had high chemical stability but also enhanced TC adsorption by H-bond and π–π stacking, leading to the increase in the Fenton reaction's efficiency [43,60,70,71]. It could be proposed that after TC molecules were adsorbed on the surface of catalyst, the catalytic degradation of TC occurred and degraded to inorganic products, then the new incoming TC molecules adsorbed again on the surface of catalyst, creating the circulation as also shown in Figure 5. This could explain why this catalyst was able to degrade the high amount of TC almost completely in the solution (800 ppm TC).

Moreover, the TC level in real wastewater is generally low [72]. It is interesting to note that all 40 mg/L TC in 50 mL solution was removed and could not be measured by UV-Vis spectrophotometer. Therefore, 1000 mL of 40 ppm TC solution with the same amount of TC molecules compared to 50 mL of 800 ppm TC was used in the catalytic experiment with the same conditions as before. The result in Figure S9 indicates a removal efficiency of 85.47% which was only 5% lower than the catalytic degradation of 800 ppm TC with the volume of 50 mL. This was because the natural pH of lower concentration of TC (40 ppm) slightly increased (pH 4.4), leading to the decrease in the catalytic activity of the heterogeneous Fenton reaction [10], and the lower initial concentration of the reactant (TC) can decrease the initial reaction rate, according to the principles of rate law. Furthermore, the high reaction volume can also slow down the reaction in which the catalyst has limited active sites to be in contact with pollutant molecules. Moreover, the comparison studies were also investigated, as shown in Figure S10. The adsorption process in dark conditions was first carried out for 3 h to ensure that the TC molecules were fully adsorbed by the catalyst, then the Fenton reaction started with the addition of H_2O_2. The results showed that the similar removal efficiency was obtained for the first 3 h compared to the experiments without

H_2O_2 in Figure 4B. Finally, the removal efficiency after 12 h was achieved at 90.24%, which was very close to the prior experiment in the same conditions, but the H_2O_2 and catalyst were added into TC solution simultaneously. This confirms that there is no significant difference in the final efficiencies between both experiments.

2.8. The Regeneration of MBGA2-1

The recyclability of the catalyst is a crucial criterion for evaluating its practicability. The catalytic degradation experiment of 800 ppm TC (50 mL) with MBGA2-1 was first conducted using 5 mM H_2O_2 at natural pH. After that, the catalyst was regenerated by using DI water and 0.1 M NaOH as eluting agents. As shown in Figure 7A, the initial degradation efficiency of MBGA2-1 was 90%. After four regeneration cycles with NaOH and DI water as the eluents, the degradation efficiencies of MBGA2-1 were dropped to 68.97% and 37.86%, respectively. The results show that the recyclability of the catalyst was acceptable, with high efficiency in removal of TC. Additionally, 0.1 M NaOH was also found to be a better eluent than DI water. More importantly, some of TC was detected by UV-vis spectrophotometer from the eluate, confirming that simple adsorption also contributes to the TC removal. This was consistent with the elemental composition from the XPS result in Table S4. It was found that after regeneration for four cycles, MBGA2-1 had 1.73% wt nitrogen content, while it was not found on fresh MBGA2-1. Moreover, it was found that NaOH is a great eluent for TC on carbon materials, leading to better recyclability [43,73–75]. The alkaline solution can make the catalyst surface become more basic and can suppress the iron dissolution, as confirmed in Figure 7B. The iron leaching for each cycle using both DI water and 0.1 M NaOH as eluents shows the remarkable low Fe leaching in range of 0.2–0.8 mg/L when using NaOH as an eluting agent, which was lower than the EU discharge standards (<2 mg/L) [15] and also consistent with the recyclability results. Furthermore, the used MBGA2-1 was also characterized by high-resolution XPS and VSM analysis, as shown in Figure 3D–F and Figure S11. The surface chemistry based on C 1s and O 1s (Table S3) of used MBGA2-1 was found to be the similar to that of fresh MBGA2-1, while the iron content dropped insignificantly (Table S4), and the saturation magnetization of used MBGA2-1 slightly decreased. Although the M_s decreased for used MBGA2-1, the value was considered high, and the catalyst could still strongly be attracted by an external magnet. This proved that the catalyst developed in this work is stable.

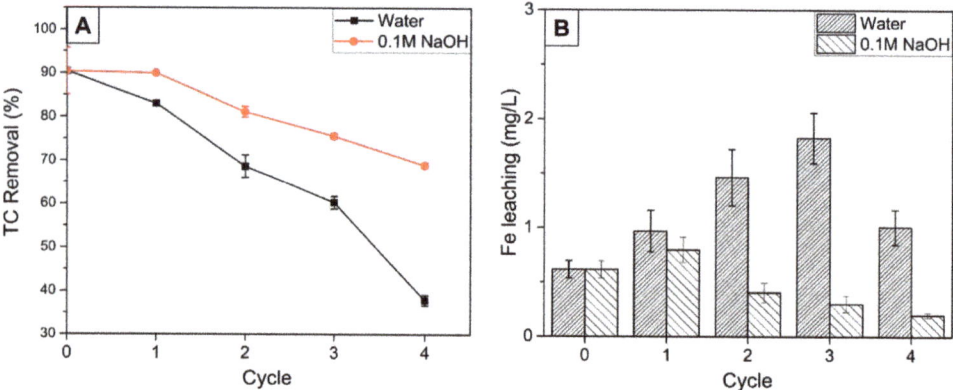

Figure 7. (**A**) The recycled experiments with different eluting agents; (**B**) the iron leaching after the degradation experiments for each cycle (conditions: C_0 = 800 mg/L, pH 3.2 (natural), H_2O_2 concentration = 5 mM, catalyst concentration = 1 g/L, T = 28 ± 2 °C).

2.9. Comparison Studies

In order to justify the MBGA2-1 as a good Fenton catalyst for the catalytic degradation of TC, a comparison with previously published studies is shown in Table S6. The magnetic Fenton catalyst in this work prepared using a simple co-precipitation method used low toxic chemicals in water-based reaction. In contrast, others previous works have employed complicated steps and large amount of chemicals, including the preparation of supports themselves. The use of such synthetic catalyst supports has also led to difficulty in scaling up, burdening the practical uses. Furthermore, in some works, toxic metals were introduced to accelerate the Fenton cycle, leading to secondary contamination with the remaining, highly toxic metals. In the present work, the utilization of Fe alone as active metal renders the easily scalable application to pilot scale in industries because the Fe precursor is relatively low cost, abundant, and non-toxic. With our catalyst, at 3 h, the removal efficiency increased rapidly to 78%, and up to 90% within 12 h. In terms of the removal efficiency, the removal amount of TC per catalyst dosage (Q_e) was calculated and normalized to compare the performance of TC degradation due to differences in initial TC concentration and reaction volume from various works. Although the pseudo-rate constant of MBGA2-1 was lower compared to those of other catalysts, our catalyst showed the highest Q_e of 721.32 mg/g among similar catalysts, both using a Fenton system and a photocatalytic system. This value is considered to be extremely high. To the best of our knowledge, its removal amount of TC per catalyst dosage (Q_e) was the highest among all of the Fenton catalysts so far. Moreover, the H_2O_2 concentration used in this work was considered low, and only a simple Fenton process is needed without the requirement of visible, UV light, or ultrasound, leading to cost-effectiveness in a real operation. In addition, the catalyst can easily be recovered from the solution by an external magnet after the process. Therefore, this MBGA2-1 is a promising Fenton catalyst to apply in practical industrial applications.

3. Experimental Section

3.1. Materials

Ferrous sulfate heptahydrate ($FeSO_4 \cdot 7H_2O$, 99.5%) was purchased from Sigma-Aldrich (Burlington, MA, USA), Ferric Chloride Hexahydrate ($FeCl_3 \cdot 6H_2O$, 99.0%) was purchased from Loba Chemie (Colaba, Mumbai). NaOH (97.0%) was purchased from RCI Labscan. HCl (36.5–38.0%) was purchased from J.T. Baker (Phillipsburg, NJ, USA). Tetracycline Hydrochloride (TC·HCl, >98.0%) was purchased from TCI America (Portland, OR, USA). Sugarcane bagasse ash (BGA) was obtained from sugar plant of Eastern Sugar and Cane Public Company Limited (Sa Kaeo, Thailand) and was thoroughly washed to remove dirt and used as catalyst support without any further purification. All experiments were performed using deionized (DI) water and all chemicals were AR grade.

3.2. Characterization of Samples

Field Emission Scanning Electron Microscope with energy dispersive X-ray spectroscopy (FESEM-EDX, Apreo, FEI, Brno-Černovice, Czech Republic) was operated to study surface and porous morphology and magnetic particles. Fourier transform infrared spectroscopy (FT-IR, Spectrum GX, Perkin Elmer, Waltham, MA, USA) with a KBr pellets technique in wavenumber ranges of 4000–400 cm^{-1} and X-ray photoelectron spectroscopy (XPS, AXIS Ultra DLD, Kratos Analytical Ltd., Manchester, UK) were performed to determine functional groups and iron species on the surface, both quantitatively and qualitatively. N_2 adsorption–desorption isotherms (ASAP2460, Micromeritics, Norcross, GA, USA) were used to determined specific surface area. The samples were degassed at 120 °C for 16 h prior to the measurements. X-ray Powder Diffraction (XRD, Philips, X'Pert MPD, Almelo, The Netherlands) was performed to study crystal structure and the form of iron in samples. The scan was run from 5° to 90° with a step size of 0.05°. A vibrating sample magnetometer (VSM, Lakeshore, Westerville, OH, USA) was used to study magnetic properties of samples at 298 K. Transmission electron microscopy with Energy Dispersive Spectroscopy (TEM-EDS, JEOL JEM-2010, Tokyo, Japan) was operated at 200 kV to study the uniformity of distribution, sizes, and other characteristics of the magnetic particles. Zeta potential (Zeta Potential Analyzer, ZetaPALS, Brookhaven, Holtsville, NY, USA) was performed to evaluate the point of zero charge (pH_{PZC}) to study the acid base or total charge on the surface of samples. The stability of the magnetic composite was studied by collecting the composite after the reaction under different conditions. After that, the total number of Fe ions left in the solution was measured by ICP-OES (AVIO 500, Perkin Elmer, Waltham, MA, USA) to confirm the iron leaching from the composite. The solution after the experiments was measured using a UV–Vis spectrophotometer and liquid chromatography-mass spectra (ESI$^-$ mode, Agilent Technologies, Santa Clara, CA, USA). The decay of the total organic carbon (TOC) in the solution was also evaluated by TOC analyzer (multi N/C 3100, Analytik jena, Jena, Germany).

3.3. Preparation of Magnetic Fe_3O_4@ash Composite

Magnetic carbon materials were synthesized by the following steps. $FeSO_4 \cdot 7H_2O$ and $FeCl_3 \cdot 6H_2O$ in 1:2 mol ratio were separately dissolved in 200 mL of DI water. Then, 5.00 g sugarcane bagasse ash (BGA) with the particle size of 50 μm was added and the mixture was stirred at 200 rpm and 80 °C in a magnetic stirring water bath. When the temperature reached 80 °C, 100 mL of 1, 2, or 5 M NaOH was added dropwise into the suspensions, then the mixture was aged for 3 h at 80 °C. After the suspensions were cooled down to RT, they were washed with DI water until the pH was ~7. Next, the magnetic solid was separated with an external magnet, then dried at 110 °C for 3 h. The magnetic sugarcane bagasse ash composite (MBGA) was obtained. The impregnation ratios of Fe_3O_4/BGA were 1:1, 2:1, and 1:2, varying the amount of iron precursor. The samples were designated with the labeling type MBGAx-y, where x represents the concentration of NaOH in M, y is the scaled ratio of Fe_3O_4/BGA; y = 1 for Fe_3O_4/BGA = 1:1; y = 0.5 for Fe_3O_4/BGA = 1:2; and y = 2 for Fe_3O_4/BGA = 2:1. For comparison, Fe_3O_4 was also prepared as a control experiment with the same ratio and under the same conditions as MBGA5-1 but without the BGA. All the prepared materials were ground to ~100 μm particle size (as determined by FESEM) before further experiments or characterization. The preparation of all samples was replicated 5 times and standard errors are reported in Table 1.

3.4. Experimental Procedure

Kinetics: The batch kinetics experiments were performed in 100 mL conical flasks with 0.05 g of the magnetic catalyst (particle size of ~100 μm) and 100 mL of 800 mg/L TC solution, on a thermostat shaker water bath (Model TOL09-FTSH-01, SCIFINETECH) at 28 ± 2 °C, at the natural (unadjusted) pH 3.2, for 0–12 h durations.

A certain amount of H_2O_2 was added to trigger the catalytic reaction and the pH was adjusted with 0.1 M HCl or 0.1 M NaOH while investigating effect of pH on the tetracycline removal.

To study the effect of TC concentration on the TC removal efficiency, the Fenton catalytic degradation experiments were performed in 100 mL conical flasks with 0.05 g of magnetic carbon material and 50 mL of 0–1200 mg/L TC solutions The mixture was shaken for 12 h in the thermostat shaker water bath at 28 ± 2 °C, at the natural pH 3.2.

All the experiments were carried out under dark conditions. After complete reaction, the solution was removed while the composite was separated from the mixture using a magnet. The concentrations of TC before and after Fenton reaction were measured by a calibrated method, using a UV-Vis spectrophotometer (UV 2600, Shimadzu, Kyoto, Japan) at λ = 357 nm. The removal amount of TC (Q_e) was calculated with the following equation (Equation (3))

$$Q_e = \frac{(C_0 - C_e)V}{m} \quad (3)$$

where C_0 and C_e are the initial and equilibrium concentrations of TC (mg/L), respectively, m is mass of the composite (g), and V is volume of the reaction solution (L).

The degradation efficiency was calculated by the following formula (Equation (4))

$$\%\text{Efficiency} = \frac{(C_0 - C_t)}{C_0} \times 100 \tag{4}$$

where C_t are the concentrations of TC at desired time intervals (mg/L)

Catalyst regeneration: The recyclability of the catalyst was performed the with MBGA2-1, 50 mL of 800 mg/L TC solution for 24 h on the thermostat shaker water bath at 28 ± 2 °C and natural pH (pH 3.2). The concentration of H_2O_2 added in the mixture was 5 mM. The composite was then collected by an external magnet. The adsorbed TC molecules were eluted from the composite by stirring the 0.5 g of the used composite in 50 mL of 0.1 M NaOH at 200 rpm for 30 min, and the eluted composite was washed with DI water until neutral pH was obtained. For comparison, the eluting agent was changed to DI water as a control. Then, the composite was dried at 110 °C for 3 h. The adsorbent was subjected to 4 consecutive experiment cycles with the same volume and concentration as the first cycle. It is noted that there was no color release observed from the composite.

4. Conclusions

Magnetic Fe_3O_4@ash composites were successfully prepared with a simple co-precipitation method using low-cost sugarcane bagasse ash received as it was from the biomass electricity-generation plants. The preparation protocol and characterization were systematically studied with various techniques. The developed catalyst was used for catalytic degradation of an ultra-high concentration of TC (800 ppm) using a Fenton system. The catalytic performance showed that the MBGA2-1 catalyst had the highest degradation activity, with good magnetic properties and high stabilities, compared to pure Fe_3O_4 and other samples. It was discovered that the carbon left on the surface of the ash helped attract TC molecules, enabling enhancement of the catalytic degradation process. The catalyst also had good recyclability with low Fe leaching using 0.1 M NaOH as the eluent. Moreover, the removal mechanism included two steps, starting with the TC adsorption on the catalyst support with oxygenated functionalities. After that, $\cdot OH_{ads}$—which was the main reactive oxygen species—attacked the molecules, leading to the degradation of TC. Additionally, we developed a new potential approach for the utilization of low-cost BGA with the circular economy concept. Considering its catalytic performance, magnetic properties, stabilities, and easily scalable preparation, the MBGA2-1 has great potential for a practical application with sustainability in wastewater remediation.

Supplementary Materials: The following supporting information can be downloaded at: https://www.mdpi.com/article/10.3390/catal12040446/s1, Figure S1: FESEM-EDX analysis; Figure S2: Physical images of all samples; Figure S3: Physical image of magnetic separation; Figure S4: FTIR spectra for all samples; Figure S5: (A) TEM image in electron mode (B-E) EDS mapping of C, O, Fe, Si, respectively; Figure S6: (A) Preliminary Catalytic Heterogeneous Fenton Reaction Test with Fe leaching for each samples, (B) The effect of catalyst concentration on the catalytic degradation of TC for MBGA2-1; Figure S7: Linear pseudo-first-order kinetics model for the catalytic degradation of TC for MBGA2-1 under different pH; Figure S8: LC-MS of the TC solution before- and after- catalytic degradation by MBGA2-1; Figure S9: Comparative studies of the catalytic degradation of low concentration of TC for MBGA2-1; Figure S10: The catalytic degradation of TC for MBGA2-1. The Fenton reaction starting by the addition of H_2O_2 after the adsorption process in dark condition for 3 h; Figure S11: VSM hysteresis loop of (black) fresh MBGA2-1, and MBGA2-1 after used 4 cycles by using (red) water and (blue) 0.1M NaOH as eluents; Table S1: The elemental compositions obtained from EDX spectra; Table S2: Assignment of the peaks in FTIR spectra for all samples; Table S3: Assignment of the peaks in XPS results (C 1s and O 1s) for MBGA2-1; Table S4: The elemental compositions (in wt%) for all samples, as obtained from XPS results; Table S5: Fitted parameters in the Langmuir-Hinshelwood model; Table S6: Literature curated data on the catalyst for TC removal and results from the current studies.

Author Contributions: N.R.: Formal analysis, writing—original draft, designed the experiments, conducted the experiments, data analysis, drafted the manuscript, prepared the manuscript, revised the manuscript. O.D.: Formal analysis, conducted experiments, data analysis; D.D.: proofread the manuscript, provided conceptual ideas; W.W.: proofread the manuscript, provided conceptual ideas; J.P.: proofread the manuscript.; S.K.: performed TEM measurements and analysis.; L.C.: Formal analysis, writing—original draft, designed the experiments, conducted the experiments, data analysis, drafted the manuscript, prepared the manuscript, revised the manuscript, funding acquisition, supervision for the whole project, corresponding author. All authors have read and agreed to the published version of the manuscript.

Funding: This work was funded by Development and Promotion of Science Technology Talents; (DPST) Research Grant (Grant No. 017/2559 and partially supported by 2021 TTSF Science & Technology Research Grant (Thailand Toray Science Foundation) for L. Chuenchom. N. Rattanachueskul thanks the Graduate School, Prince of Songkla University, for PSU-PhD. scholarship (Contract No. PSU_PHD2562-003); Partial financial support from the Research Team Promotion grant—National Research Council of Thailand (Joongjai Panpranot) is acknowledged.

Data Availability Statement: All data are available from the corresponding author on reasonable request.

Acknowledgments: L. Chuenchom. would like to acknowledge the partial support from the Center of Excellence for Innovation in Chemistry (PERCH-CIC), Ministry of Higher Education, Science, Research, and Innovation. S. Kaowphong would like to thank Chiang Mai University. We thank Titilope John Jayeoye for English proofreading.

Conflicts of Interest: The authors declare no conflict of interest.

References

1. Wang, J.; Zhuan, R. Degradation of antibiotics by advanced oxidation processes: An overview. *Sci. Total Environ.* **2020**, *701*, 135023. [CrossRef] [PubMed]
2. Zhang, Y.; Shi, J.; Xu, Z.; Chen, Y.; Song, D. Degradation of tetracycline in a schorl/H_2O_2 system: Proposed mechanism and intermediates. *Chemosphere* **2018**, *202*, 661–668. [CrossRef] [PubMed]
3. Xiang, Y.; Huang, Y.; Xiao, B.; Wu, X.; Zhang, G. Magnetic yolk-shell structure of $ZnFe_2O_4$ nanoparticles for enhanced visible light photo-Fenton degradation towards antibiotics and mechanism study. *Appl. Surf. Sci.* **2020**, *513*, 145820. [CrossRef]
4. Dutta, J.; Mala, A.A. Removal of antibiotic from the water environment by the adsorption technologies: A review. *Water Sci. Technol.* **2020**, *82*, 401–426. [CrossRef] [PubMed]
5. Priya, S.S.; Radha, K.V. A Review on the Adsorption Studies of Tetracycline onto Various Types of Adsorbents. *Chem. Eng. Commun.* **2017**, *204*, 821–839. [CrossRef]
6. Krasucka, P.; Pan, B.; Ok, Y.S.; Mohan, D.; Sarkar, B.; Oleszczuk, P. Engineered biochar—A sustainable solution for the removal of antibiotics from water. *Chem. Eng. J.* **2021**, *405*, 126926. [CrossRef]
7. Choi, K.-J.; Kim, S.-G.; Kim, S.-H. Removal of antibiotics by coagulation and granular activated carbon filtration. *J. Hazard. Mater.* **2008**, *151*, 38–43. [CrossRef]
8. Saitoh, T.; Shibata, K.; Fujimori, K.; Ohtani, Y. Rapid removal of tetracycline antibiotics from water by coagulation-flotation of sodium dodecyl sulfate and poly(allylamine hydrochloride) in the presence of Al(III) ions. *Sep. Purif. Technol.* **2017**, *187*, 76–83. [CrossRef]
9. Munoz, M.; de Pedro, Z.M.; Casas, J.A.; Rodriguez, J.J. Preparation of magnetite-based catalysts and their application in heterogeneous Fenton oxidation—A review. *Appl. Catal. B Environ.* **2015**, *176–177*, 249–265. [CrossRef]
10. Wang, N.; Zheng, T.; Zhang, G.; Wang, P. A review on Fenton-like processes for organic wastewater treatment. *J. Environ. Chem. Eng.* **2016**, *4*, 762–787. [CrossRef]
11. Ameta, R.; Chohadia, A.K.; Jain, A.; Punjabi, P.B. Fenton and Photo-Fenton Processes. In *Advanced Oxidation Processes for Waste Water Treatment*; Chapter 3; Ameta, S.C., Ameta, R., Eds.; Academic Press: Cambridge, MA, USA, 2018; pp. 49–87.
12. Martínez, F.; Molina, R.; Rodríguez, I.; Pariente, M.I.; Segura, Y.; Melero, J.A. Techno-Economical assessment of coupling Fenton/biological processes for the treatment of a pharmaceutical wastewater. *J. Environ. Chem. Eng.* **2018**, *6*, 485–494. [CrossRef]
13. Li, X.; Cui, K.; Guo, Z.; Yang, T.; Cao, Y.; Xiang, Y.; Chen, H.; Xi, M. Heterogeneous Fenton-like degradation of tetracyclines using porous magnetic chitosan microspheres as an efficient catalyst compared with two preparation methods. *Chem. Eng. J.* **2020**, *379*, 122324. [CrossRef]
14. Lian, J.; Ouyang, Q.; Tsang, P.E.; Fang, Z. Fenton-like catalytic degradation of tetracycline by magnetic palygorskite nanoparticles prepared from steel pickling waste liquor. *Appl. Clay Sci.* **2019**, *182*, 105273. [CrossRef]
15. Nie, M.; Li, Y.; He, J.; Xie, C.; Wu, Z.; Sun, B.; Zhang, K.; Kong, L.; Liu, J. Degradation of tetracycline in water using Fe_3O_4 nanospheres as Fenton-like catalysts: Kinetics, mechanisms and pathways. *New J. Chem.* **2020**, *44*, 2847–2857. [CrossRef]

16. Du, D.; Shi, W.; Wang, L.; Zhang, J. Yolk-Shell structured Fe_3O_4@void@TiO_2 as a photo-Fenton-like catalyst for the extremely efficient elimination of tetracycline. *Appl. Catal. B Environ.* **2017**, *200*, 484–492. [CrossRef]
17. Xu, J.; Liu, Z.; Zhao, D.; Gao, N.; Fu, X. Enhanced adsorption of perfluorooctanoic acid (PFOA) from water by granular activated carbon supported magnetite nanoparticles. *Sci. Total Environ.* **2020**, *723*, 137757. [CrossRef]
18. Du, C.; Song, Y.; Shi, S.; Jiang, B.; Yang, J.; Xiao, S. Preparation and characterization of a novel Fe3O4-graphene-biochar composite for crystal violet adsorption. *Sci. Total Environ.* **2020**, *711*, 134662. [CrossRef]
19. Li, Y.; Zimmerman, A.R.; He, F.; Chen, J.; Han, L.; Chen, H.; Hu, X.; Gao, B. Solvent-Free synthesis of magnetic biochar and activated carbon through ball-mill extrusion with Fe3O4 nanoparticles for enhancing adsorption of methylene blue. *Sci. Total Environ.* **2020**, *722*, 137972. [CrossRef]
20. Qu, L.; Han, T.; Luo, Z.; Liu, C.; Mei, Y.; Zhu, T. One-Step fabricated Fe_3O_4@C core–shell composites for dye removal: Kinetics, equilibrium and thermodynamics. *J. Phys. Chem. Solids* **2015**, *78*, 20–27. [CrossRef]
21. Ai, L.; Zhang, C.; Liao, F.; Wang, Y.; Li, M.; Meng, L.; Jiang, J. Removal of methylene blue from aqueous solution with magnetite loaded multi-wall carbon nanotube: Kinetic, isotherm and mechanism analysis. *J. Hazard. Mater.* **2011**, *198*, 282–290. [CrossRef]
22. Dong, Y.; Cui, X.; Lu, X.; Jian, X.; Xu, Q.; Tan, C. Enhanced degradation of sulfadiazine by novel β-alaninediacetic acid-modified Fe_3O_4 nanocomposite coupled with peroxymonosulfate. *Sci. Total Environ.* **2019**, *662*, 490–500. [CrossRef] [PubMed]
23. Ma, M.; Hou, P.; Zhang, P.; Cao, J.; Liu, H.; Yue, H.; Tian, G.; Feng, S. Magnetic Fe_3O_4 nanoparticles as easily separable catalysts for efficient catalytic transfer hydrogenation of biomass-derived furfural to furfuryl alcohol. *Appl. Catal. A Gen.* **2020**, *602*, 117709. [CrossRef]
24. Ma, C.; Jia, S.; Yuan, P.; He, Z. Catalytic ozonation of 2,2′-methylenebis (4-methyl-6-tert-butylphenol) over nano-Fe_3O_4@cow dung ash composites: Optimization, toxicity, and degradation mechanisms. *Environ. Pollut.* **2020**, *265*, 114597. [CrossRef] [PubMed]
25. Yu, X.; Lin, X.; Li, W.; Feng, W. Effective Removal of Tetracycline by Using Biochar Supported Fe_3O_4 as a UV-Fenton Catalyst. *Chem. Res. Chin. Univ.* **2019**, *35*, 79–84. [CrossRef]
26. Plakas, K.V.; Mantza, A.; Sklari, S.D.; Zaspalis, V.T.; Karabelas, A.J. Heterogeneous Fenton-like oxidation of pharmaceutical diclofenac by a catalytic iron-oxide ceramic microfiltration membrane. *Chem. Eng. J.* **2019**, *373*, 700–708. [CrossRef]
27. Poza-Nogueiras, V.; Rosales, E.; Pazos, M.; Sanromán, M.Á. Current advances and trends in electro-Fenton process using heterogeneous catalysts—A review. *Chemosphere* **2018**, *201*, 399–416. [CrossRef]
28. Xu, L.; Wang, J. Magnetic Nanoscaled Fe_3O_4/CeO_2 Composite as an Efficient Fenton-Like Heterogeneous Catalyst for Degradation of 4-Chlorophenol. *Environ. Sci. Technol.* **2012**, *46*, 10145–10153. [CrossRef]
29. Do, Q.C.; Kim, D.-G.; Ko, S.-O. Catalytic activity enhancement of a Fe_3O_4@SiO_2 yolk-shell structure for oxidative degradation of acetaminophen by decoration with copper. *J. Clean. Prod.* **2018**, *172*, 1243–1253. [CrossRef]
30. Fatimah, I.; Amaliah, S.N.; Andrian, M.F.; Handayani, T.P.; Nurillahi, R.; Prakoso, N.I.; Wicaksono, W.P.; Chuenchom, L. Iron oxide nanoparticles supported on biogenic silica derived from bamboo leaf ash for rhodamine B photodegradation. *Sustain. Chem. Pharm.* **2019**, *13*, 100149. [CrossRef]
31. Chen, W.-H.; Huang, J.-R.; Lin, C.-H.; Huang, C.-P. Catalytic degradation of chlorpheniramine over GO-Fe_3O_4 in the presence of H_2O_2 in water: The synergistic effect of adsorption. *Sci. Total Environ.* **2020**, *736*, 139468. [CrossRef]
32. Hu, X.; Liu, B.; Deng, Y.; Chen, H.; Luo, S.; Sun, C.; Yang, P.; Yang, S. Adsorption and heterogeneous Fenton degradation of 17α-methyltestosterone on nano Fe_3O_4/MWCNTs in aqueous solution. *Appl. Catal. B Environ.* **2011**, *107*, 274–283. [CrossRef]
33. Yoo, S.H.; Jang, D.; Joh, H.-I.; Lee, S. Iron oxide/porous carbon as a heterogeneous Fenton catalyst for fast decomposition of hydrogen peroxide and efficient removal of methylene blue. *J. Mater. Chem. A* **2017**, *5*, 748–755. [CrossRef]
34. Yang, Y.; Zhang, X.; Chen, Q.; Li, S.; Chai, H.; Huang, Y. Ultrasound-Assisted Removal of Tetracycline by a Fe/N–C Hybrids/H_2O_2 Fenton-like System. *ACS Omega* **2018**, *3*, 15870–15878. [CrossRef] [PubMed]
35. Wang, X.; Xie, Y.; Ma, J.; Ning, P. Facile assembly of novel g-C_3N_4@expanded graphite and surface loading of nano zero-valent iron for enhanced synergistic degradation of tetracycline. *RSC Adv.* **2019**, *9*, 34658–34670. [CrossRef]
36. Wu, Q.; Yang, H.; Kang, L.; Gao, Z.; Ren, F. Fe-Based metal-organic frameworks as Fenton-like catalysts for highly efficient degradation of tetracycline hydrochloride over a wide pH range: Acceleration of Fe(II)/Fe(III) cycle under visible light irradiation. *Appl. Catal. B Environ.* **2020**, *263*, 118282. [CrossRef]
37. Ma, S.; Jing, J.; Liu, P.; Li, Z.; Jin, W.; Xie, B.; Zhao, Y. High selectivity and effectiveness for removal of tetracycline and its related drug resistance in food wastewater through schwertmannite/graphene oxide catalyzed photo-Fenton-like oxidation. *J. Hazard. Mater.* **2020**, *392*, 122437. [CrossRef]
38. Khodadadi, M.; Panahi, A.H.; Al-Musawi, T.J.; Ehrampoush, M.H.; Mahvi, A.H. The catalytic activity of FeNi3@SiO_2 magnetic nanoparticles for the degradation of tetracycline in the heterogeneous Fenton-like treatment method. *J. Water Process Eng.* **2019**, *32*, 100943. [CrossRef]
39. To, L.S.; Seebaluck, V.; Leach, M. Future energy transitions for bagasse cogeneration: Lessons from multi-level and policy innovations in Mauritius. *Energy Res. Soc. Sci.* **2018**, *35*, 68–77. [CrossRef]
40. Stanmore, B.R. Generation of Energy from Sugarcane Bagasse by Thermal Treatment. *Waste Biomass Valorization* **2010**, *1*, 77–89. [CrossRef]
41. Wakamura, Y. Utilization of Bagasse Energy in Thailand. *Mitig. Adapt. Strateg. Glob. Chang.* **2003**, *8*, 253–260. [CrossRef]
42. Tonnayopas, D. Green Building Bricks Made with Clays and Sugar Cane Bagasse Ash. In Proceedings of the 11th International Conference on Mining, Materials and Petroleum Engineering, Sinaia, Romania, 17–19 January 2013.

43. Rattanachueskul, N.; Saning, A.; Kaowphong, S.; Chumha, N.; Chuenchom, L. Magnetic carbon composites with a hierarchical structure for adsorption of tetracycline, prepared from sugarcane bagasse via hydrothermal carbonization coupled with simple heat treatment process. *Bioresour. Technol.* **2017**, *226*, 164–172. [CrossRef] [PubMed]
44. Novais, R.M.; Ascensão, G.; Tobaldi, D.M.; Seabra, M.P.; Labrincha, J.A. Biomass fly ash geopolymer monoliths for effective methylene blue removal from wastewaters. *J. Clean. Prod.* **2018**, *171*, 783–794. [CrossRef]
45. Le Blond, J.S.; Woskie, S.; Horwell, C.J.; Williamson, B.J. Particulate matter produced during commercial sugarcane harvesting and processing: A respiratory health hazard? *Atmos. Environ.* **2017**, *149*, 34–46. [CrossRef]
46. Rodríguez-Díaz, J.; García, J.; Sánchez, L.; Silva, M.; Silva, V.; Arteaga-Pérez, L. Comprehensive Characterization of Sugarcane Bagasse Ash for Its Use as an Adsorbent. *Bioenergy Res.* **2015**, *8*, 1885–1895. [CrossRef]
47. Madurwar, M.; Mandavgane, S.; Ralegaonkar, R. Use of sugarcane bagasse ash as brick material. *Curr. Sci.* **2014**, *117*, 1044–1051.
48. Xu, Q.; Ji, T.; Gao, S.-J.; Yang, Z.; Wu, N. Characteristics and Applications of Sugar Cane Bagasse Ash Waste in Cementitious Materials. *Materials* **2018**, *12*, 39. [CrossRef]
49. Sales, A.; Lima, S.A. Use of Brazilian sugarcane bagasse ash in concrete as sand replacement. *Waste Manag.* **2010**, *30*, 1114–1122. [CrossRef]
50. Webber, C.P., Jr.; Spaunhorst, D.; Petrie, E. Impact of Sugarcane Bagasse Ash as an Amendment on the Physical Properties, Nutrient Content and Seedling Growth of a Certified Organic Greenhouse Growing Media. *J. Agric. Sci.* **2017**, *9*, 1. [CrossRef]
51. Purnomo, C.W.; Respito, A.; Sitanggang, E.P.; Mulyono, P. Slow release fertilizer preparation from sugar cane industrial waste. *Environ. Technol. Innov.* **2018**, *10*, 275–280. [CrossRef]
52. Mane, V.S.; Mall, I.D.; Srivastava, V.C. Use of bagasse fly ash as an adsorbent for the removal of brilliant green dye from aqueous solution. *Dyes Pigment.* **2007**, *73*, 269–278. [CrossRef]
53. Gaikwad, D.R. Low cost Sugarcane Bagasse Ash as an Adsorbent for Dye Removal from Dye Effluent. *Int. J. Chem. Eng. Appl.* **2010**, *1*, 309–318.
54. Mor, S.; Negi, P.; Ravindra, K. Potential of agro-waste sugarcane bagasse ash for the removal of ammoniacal nitrogen from landfill leachate. *Environ. Sci. Pollut. Res.* **2019**, *26*, 24516–24531. [CrossRef] [PubMed]
55. Abdul Mutalib, A.A.; Ibrahim, M.L.; Matmin, J.; Kassim, M.F.; Mastuli, M.S.; Taufiq-Yap, Y.H.; Shohaimi, N.A.M.; Islam, A.; Tan, Y.H.; Kaus, N.H.M. SiO_2-Rich Sugar Cane Bagasse Ash Catalyst for Transesterification of Palm Oil. *Bioenergy Res.* **2020**, *13*, 986–997. [CrossRef]
56. Meng, Q.; Xiang, S.; Zhang, K.; Wang, M.; Bu, X.; Xue, P.; Liu, L.; Sun, H.; Yang, B. A facile two-step etching method to fabricate porous hollow silica particles. *J. Colloid Interface Sci.* **2012**, *384*, 22–28. [CrossRef]
57. Park, J.; Han, Y.; Kim, H. Formation of Mesoporous Materials from Silica Dissolved in Various NaOH Concentrations: Effect of pH and Ionic Strength. *J. Nanomater.* **2012**, *2012*, 528174. [CrossRef]
58. Clark, M.W.; Despland, L.M.; Lake, N.J.; Yee, L.H.; Anstoetz, M.; Arif, E.; Parr, J.F.; Doumit, P. High-efficiency cogeneration boiler bagasse-ash geochemistry and mineralogical change effects on the potential reuse in synthetic zeolites, geopolymers, cements, mortars, and concretes. *Heliyon* **2017**, *3*, e00294. [CrossRef]
59. Subramanian, V.; Ordomsky, V.V.; Legras, B.; Cheng, K.; Cordier, C.; Chernavskii, P.A.; Khodakov, A.Y. Design of iron catalysts supported on carbon–silica composites with enhanced catalytic performance in high-temperature Fischer–Tropsch synthesis. *Catal. Sci. Technol.* **2016**, *6*, 4953–4961. [CrossRef]
60. Ma, S.; Gu, J.; Han, Y.; Gao, Y.; Zong, Y.; Ye, Z.; Xue, J. Facile Fabrication of C–TiO_2 Nanocomposites with Enhanced Photocatalytic Activity for Degradation of Tetracycline. *ACS Omega* **2019**, *4*, 21063–21071. [CrossRef]
61. Zhu, X.; Qian, F.; Liu, Y.; Matera, D.; Wu, G.; Zhang, S.; Chen, J. Controllable synthesis of magnetic carbon composites with high porosity and strong acid resistance from hydrochar for efficient removal of organic pollutants: An overlooked influence. *Carbon* **2016**, *99*, 338–347. [CrossRef]
62. Mohan, D.; Sarswat, A.; Singh, V.K.; Alexandre-Franco, M.; Pittman, C.U. Development of magnetic activated carbon from almond shells for trinitrophenol removal from water. *Chem. Eng. J.* **2011**, *172*, 1111–1125. [CrossRef]
63. Pompe, C.E.; Slagter, M.; de Jongh, P.E.; de Jong, K.P. Impact of heterogeneities in silica-supported copper catalysts on their stability for methanol synthesis. *J. Catal.* **2018**, *365*, 1–9. [CrossRef]
64. Lai, C.; Huang, F.; Zeng, G.; Huang, D.; Qin, L.; Cheng, M.; Zhang, C.; Li, B.; Yi, H.; Liu, S.; et al. Fabrication of novel magnetic $MnFe_2O_4$/bio-char composite and heterogeneous photo-Fenton degradation of tetracycline in near neutral pH. *Chemosphere* **2019**, *224*, 910–921. [CrossRef] [PubMed]
65. Ma, X.; Cheng, Y.; Ge, Y.; Wu, H.; Li, Q.; Gao, N.; Deng, J. Ultrasound-enhanced nanosized zero-valent copper activation of hydrogen peroxide for the degradation of norfloxacin. *Ultrason. Sonochem.* **2018**, *40*, 763–772. [CrossRef] [PubMed]
66. Hou, L.; Wang, L.; Royer, S.; Zhang, H. Ultrasound-assisted heterogeneous Fenton-like degradation of tetracycline over a magnetite catalyst. *J. Hazard. Mater.* **2016**, *302*, 458–467. [CrossRef]
67. Gan, Q.; Hou, H.; Liang, S.; Qiu, J.; Tao, S.; Yang, L.; Yu, W.; Xiao, K.; Liu, B.; Hu, J.; et al. Sludge-Derived biochar with multivalent iron as an efficient Fenton catalyst for degradation of 4-Chlorophenol. *Sci. Total Environ.* **2020**, *725*, 138299. [CrossRef]
68. Tang, J.; Wang, J. Fenton-like degradation of sulfamethoxazole using Fe-based magnetic nanoparticles embedded into mesoporous carbon hybrid as an efficient catalyst. *Chem. Eng. J.* **2018**, *351*, 1085–1094. [CrossRef]

69. Nitoi, I.; Oancea, P.; Constantin, L.A.; Raileanu, M.; Crisan, M.; Cristea, I.; Cosma, C. Relationship between structure of some nitroaromatic pollutants and their degradation kinetic parameters in UV-VIS./TiO$_2$ system. *J. Environ. Prot. Ecol.* **2016**, *17*, 315–322.
70. Wang, X.; Zhuang, Y.; Zhang, J.; Song, L.; Shi, B. Pollutant degradation behaviors in a heterogeneous Fenton system through Fe/S-doped aerogel. *Sci. Total Environ.* **2020**, *714*, 136436. [CrossRef]
71. Qiu, Y.; Xu, X.; Xu, Z.; Liang, J.; Yu, Y.; Cao, X. Contribution of different iron species in the iron-biochar composites to sorption and degradation of two dyes with varying properties. *Chem. Eng. J.* **2020**, *389*, 124471. [CrossRef]
72. Javid, A.; Mesdaghinia, A.; Nasseri, S.; Mahvi, A.H.; Alimohammadi, M.; Gharibi, H. Assessment of tetracycline contamination in surface and groundwater resources proximal to animal farming houses in Tehran, Iran. *J. Environ. Health Sci. Eng.* **2016**, *14*, 4. [CrossRef]
73. Sayğılı, H.; Güzel, F. Effective removal of tetracycline from aqueous solution using activated carbon prepared from tomato (*Lycopersicon esculentum* Mill.) industrial processing waste. *Ecotoxicol. Environ. Saf.* **2016**, *131*, 22–29. [CrossRef] [PubMed]
74. Chen, Z.; Mu, D.; Chen, F.; Tan, N. NiFe$_2$O$_4$@ nitrogen-doped carbon hollow spheres with highly efficient and recyclable adsorption of tetracycline. *RSC Adv.* **2019**, *9*, 10445–10453. [CrossRef]
75. Chang, P.H.; Li, Z.; Jean, J.S.; Jiang, W.T.; Wu, Q.; Kuo, C.Y.; Kraus, J. Desorption of tetracycline from montmorillonite by aluminum, calcium, and sodium: An indication of intercalation stability. *Int. J. Environ. Sci. Technol.* **2014**, *11*, 633–644. [CrossRef]

Article

Statistical Modeling and Performance Optimization of a Two-Chamber Microbial Fuel Cell by Response Surface Methodology

Muhammad Nihal Naseer [1,*], Asad A. Zaidi [2], Hamdullah Khan [1], Sagar Kumar [1], Muhammad Taha bin Owais [1], Yasmin Abdul Wahab [3,*], Kingshuk Dutta [4], Juhana Jaafar [5], Nor Aliya Hamizi [3], Mohammad Aminul Islam [6], Hanim Hussin [7,8], Irfan Anjum Badruddin [9,10] and Hussein Alrobei [11]

[1] Department of Engineering Sciences, PN Engineering College, National University of Sciences and Technology (NUST), Islamabad 44000, Pakistan; hamdullahkhan121@gmail.com (H.K.); mr.sagar_kumar@aol.com (S.K.); Taha.owais@outlook.com (M.T.b.O.)
[2] Department of Mechanical Engineering, Faculty of Engineering Science and Technology, Hamdard University, Madinat al-Hikmah, Hakim Mohammad Said Road, Karachi 74600, Pakistan; asad.nednust@gmail.com
[3] Nanotechnology & Catalysis Research Centre, University of Malaya, Kuala Lumpur 50603, Malaysia; aliyahamizi@um.edu.my
[4] Advanced Polymer Design and Development Research Laboratory (APDDRL), School for Advanced Research in Petrochemicals (SARP), Central Institute of Petrochemicals Engineering and Technology (CIPET), Bengaluru 562149, Karnataka, India; dr.kingshukdutta@gmail.com
[5] Advanced Membrane Technology Research Centre, Universiti Teknologi Malaysia, Johor Bahru 81310, Malaysia; juhana@petroleum.utm.my
[6] Department of Electrical Engineering, Faculty of Engineering, University of Malaya, Kuala Lumpur 50603, Malaysia; aminul.islam@um.edu.my
[7] School of Electrical Engineering, College of Engineering, Universiti Teknologi MARA, Shah Alam 40450, Malaysia; hanimh@uitm.edu.my
[8] Center of Printable Electronics, Institute for Advanced Studies (IAS), University of Malaya, Kuala Lumpur 50603, Malaysia
[9] Research Center for Advanced Materials Science (RCAMS), King Khalid University, Abha 61413, Saudi Arabia; irfan@kku.edu.sa
[10] Mechanical Engineering Department, College of Engineering, King Khalid University, Abha 61421, Saudi Arabia
[11] Department of Mechanical Engineering, College of Engineering, Prince Sattam bin Abdulaziz University, AlKharj 16273, Saudi Arabia; h.alrobei@psau.edu.sa
* Correspondence: nihal.me@pnec.nust.edu.pk (M.N.N.); yasminaw@um.edu.my (Y.A.W.)

Citation: Naseer, M.N.; Zaidi, A.A.; Khan, H.; Kumar, S.; Owais, M.T.b.; Abdul Wahab, Y.; Dutta, K.; Jaafar, J.; Hamizi, N.A.; Islam, M.A.; et al. Statistical Modeling and Performance Optimization of a Two-Chamber Microbial Fuel Cell by Response Surface Methodology. Catalysts 2021, 11, 1202. https://doi.org/10.3390/catal11101202

Academic Editor: Asuncion Quintanilla

Received: 13 August 2021
Accepted: 28 September 2021
Published: 1 October 2021

Publisher's Note: MDPI stays neutral with regard to jurisdictional claims in published maps and institutional affiliations.

Copyright: © 2021 by the authors. Licensee MDPI, Basel, Switzerland. This article is an open access article distributed under the terms and conditions of the Creative Commons Attribution (CC BY) license (https://creativecommons.org/licenses/by/4.0/).

Abstract: Microbial fuel cell, as a promising technology for simultaneous power production and waste treatment, has received a great deal of attention in recent years; however, generation of a relatively low power density is the main limitation towards its commercial application. This study contributes toward the optimization, in terms of maximization, of the power density of a microbial fuel cell by employing response surface methodology, coupled with central composite design. For this optimization study, the interactive effect of three independent parameters, namely (i) acetate concentration in the influent of anodic chamber; (ii) fuel feed flow rate in anodic chamber; and (iii) oxygen concentration in the influent of cathodic chamber, have been analyzed for a two-chamber microbial fuel cell, and the optimum conditions have been identified. The optimum value of power density was observed at an acetate concentration, a fuel feed flow rate, and an oxygen concentration value of 2.60 mol m^{-3}, 0.0 m^3, and 1.00 mol m^{-3}, respectively. The results show the achievement of a power density of 3.425 W m^{-2}, which is significant considering the available literature. Additionally, a statistical model has also been developed that correlates the three independent factors to the power density. For this model, R^2, adjusted R^2, and predicted R^2 were 0.839, 0.807, and 0.703, respectively. The fact that there is only a 3.8% error in the actual and adjusted R^2 demonstrates that the proposed model is statistically significant.

Keywords: microbial fuel cell; optimization; power density; response surface methodology; green energy

1. Introduction

Microbial fuel cells (MFCs) are bio-electrochemical cells that use microbes as catalysts to produce renewable energy/bioelectricity by consuming a wide range of biodegradable organic matter [1]. MFCs have received a great deal of attention in recent years due to the multitude of benefits associated with this technology [2]. MFCs provide an environmentally friendly and low-cost green energy option by utilizing waste. To get energy from waste, different thermal technologies, such as pyrolysis, gasification and incineration, are often used; however, these technologies are also often criticized due to associated health concerns, economical imbalance, operational complications and greenhouse gas generation [3]. On the other hand, MFC is a non-thermal technology that operates at low temperatures (below 20 °C), and is referred as the safest technology to utilize various wastes and generate energy with no toxic byproducts [3,4]. Another benefit of using MFC is its efficiency towards pollutant removal from waste, with better effluent quality [5]. It must be noted here that in various anaerobic fermentation processes, subsequent anaerobic treatment of sludge is required due to the high chemical oxygen demand (COD) that increases energy consumption and operating costs [6]. However, in the case of MFCs, the COD removal efficiency is much higher, even at the low effluent rate, which makes MFC a promising future technology [5,7]. Despite being such an amazing and promising technology, MFCs are still not commercially viable. The major hurdles in MFC's way to the industrial sector includes low power density output and scaling up issues [8–10]. To cope with these issues, the techniques of mathematical modeling and optimization of MFCs can be adopted.

All engineering systems or processes can be regarded as operations in which certain parameters are provided as inputs to the systems. After passing through an operation, the input parameters provide a desired response or output. Researchers are always interested in varying input parameters in such a way that they provide an optimized environment for maximizing the output. This requires following of three steps: first, to select an admissible range to vary the values of the input parameters; second, to perform a series of experiments at some points in the admissible range and collect the results; third, to predict the results for the rest of the design points in such a way that the predicted results are accurate and reliable [11]. In this regard, the response surface methodology (RSM) provides a platform to develop a statistical model that, after a set of validation studies, can be used to accurately predict results in the entire admissible range of parameters. Basically, RSM is a mathematical tool that uses mathematical and statistical methods to develop a functional relationship between the desired response (output) and a set of control variables (input) [12]. It is one of the most commonly used techniques for optimization purpose as it allows studying effect of multiple variables and their interactions on a single variable [13]. Many studies [14–18] have used RSM for optimization purposes, hence, used in this study.

The motivation of this study comes from the fact that the highly promising MFC device technology is still unable to reach the stage of industrial application. In this regard, the major drawbacks of MFCs, namely the low power density output and scaling up issues, can be solved by carrying out optimization studies. For this purpose, RSM can be applied by developing a mathematical model, which can be further used by the research community to predict and validate experimental results. Published literature contains several studies that have been conducted for the optimization of MFCs. For instance, Geetanjali et al. [17] modified anodic carbon cloth, using $NiWO_4$ and graphene oxide, to optimize the performance of a single chamber MFC. It was observed that, with the modified electrode, the internal resistance towards electron transfer, from microbes to the electrode, got decreased that ultimately increased the power density of the MFC by 8.5 folds. Islam et al. [19] used RSM for optimization of a dual-chamber MFC. The influence of four independent parameters, namely substrate concentration, pH, residence time and co-culture composition, was studied. The observations revealed that there was weak or no influence of the following combinations of parameters: (i) substrate concentration-residence time; (ii) pH–residence time; (iii) pH–substrate concentration; and (iv) residence time–substrate concentration. On the other hand, a strong correlation was observed between the power density and the following set of parameters: (i) co-culture composition and residence time; (ii) pH and co-culture composition; and (iii) substrate concentration and co-culture composition. Another study [20] used standard composite design to understand the correlation between power density of a dual-chamber MFC, COD and cathodic aeration rate. The main purpose of this study was to simultaneously improve the efficiency of the MFC towards electricity generation and phosphorous removal. Application of the RSM technique revealed that it was not possible to optimize MFC for concurrent power production and phosphorous removal under the same conditions. Nevertheless, increasing the COD and cathodic aeration rate in MFC was found to favor both electricity generation and phosphorous removal.

In the optimization studies available in the literature, various parameters of MFCs have been optimized under certain conditions. However, the understanding of the fact that some parameters have a higher impact on the performance of an MFC is crucial to paving its way to the industrial sector. A study conducted by Zeng et al. [21] revealed a list of parameters to which MFC's power density output is most sensitive. The flow rate of the fuel feed to the anodic chamber, initial concentration of acetate in the anodic chamber and initial concentration of O_2 in the cathodic chamber were observed to be among the most sensitive parameters in MFCs. Decreasing the value of these three parameters by a ratio of 0.8 resulted in the decrement of the power density output by 10.57%, 78.42%, and 0.12%, respectively. On the other hand, increasing the value of the flow rate of the fuel feed to anode by a ratio of 1.2 resulted in a decrement of the power density by 17.74%; while, increase by the same ratio caused increment of 31.09% and 0.08%, respectively, in the power density for the other two parameters. In this study, the effect of one parameter was analyzed by keeping the others constant, which served as a limitation of the application of this study [21]. This is because, in a practical scenario, studying the combined impact of all the design parameters is crucial.

In the present paper, RSM has been utilized to study the correlation between the power density output and the flow rate of the fuel feed to the anodic chamber, initial concentration of acetate in the anodic chamber and initial concentration of O_2 in the cathodic chamber. By using RSM, the combined impact of these three parameters has been studied and an analytical model has been developed that relates power density to the abovementioned three parameters, which is the novelty of this study. The developed analytical model has been further verified using different criteria and has finally been used to find the optimal design points that enhanced the power density of MFCs.

2. Results and Discussions

2.1. Analysis of Variance (ANOVA)

Based on the control variables and their respective response values (Section 3.4), a linear polynomial function was developed to predict the power density. The coefficients of the model were estimated by using the RSM in *Design Expert* software, at a confidence level of 95%. Equation (1) represents the analytical function to predict the power density, based on the three-selected control variables.

In the analytical model of power density, shown in Equation (1), the coefficients of A and B were high as compared to the coefficient of C. This indicates that the power density is highly influenced by the changes in the flow rate of fuel feed to the anode and the concentration of acetate in the anode, as compared to the changes in the concentration of oxygen in the cathode.

$$R = +2.49 - 0.43 \times A + 0.45 \times B + 0.05 \times C \tag{1}$$

After developing an analytical model, the next step was to check the validity of the model. For this purpose, the following four basic criteria were used: (i) value of predicted R^2; (ii) value of adjusted R^2; (iii) adequate precision; and (iv) difference between adjusted R^2 and predicted R^2. Here, R^2 is the coefficient of determination that depicts the closeness between the actual data points and the fitted regression line [22]. To minimize the errors, the value of R^2 was adjusted in terms of the number of independent variables, and the new obtained value of R^2 was termed as the adjusted R^2. For a model to be significant, the value of both R^2 and adjusted R^2 must be close to 1. In the present study, the value of R^2 and adjusted R^2 were found to be 0.8397 and 0.8097, respectively, that validates the model presented in Equation (1).

Adequate precision is a term that measures the signal to noise (S/N) ratio of the model. If the value of adequate precision is greater than 4, the model is considered acceptable [23]. For the model proposed in this study, the value of adequate precision was found to be 16.62, i.e., about 4 times higher than required. This implies that the model is significant. Moreover, it was observed that the difference between the adjusted R^2 and the predicted R^2 was less than 0.2, which is an insignificant difference; and therefore, it can be safely said that the two values are in good agreement with each other.

2.2. Model Validation

After preliminary validation using ANOVA, the model was further validated graphically by plotting different graphs (Figure 1). Figure 1a depicts the relation between the actual and the predicted values, which indicates that there exists a strong correlation between the two values and that the model is valid with a negligible amount of inaccuracy for practical purposes.

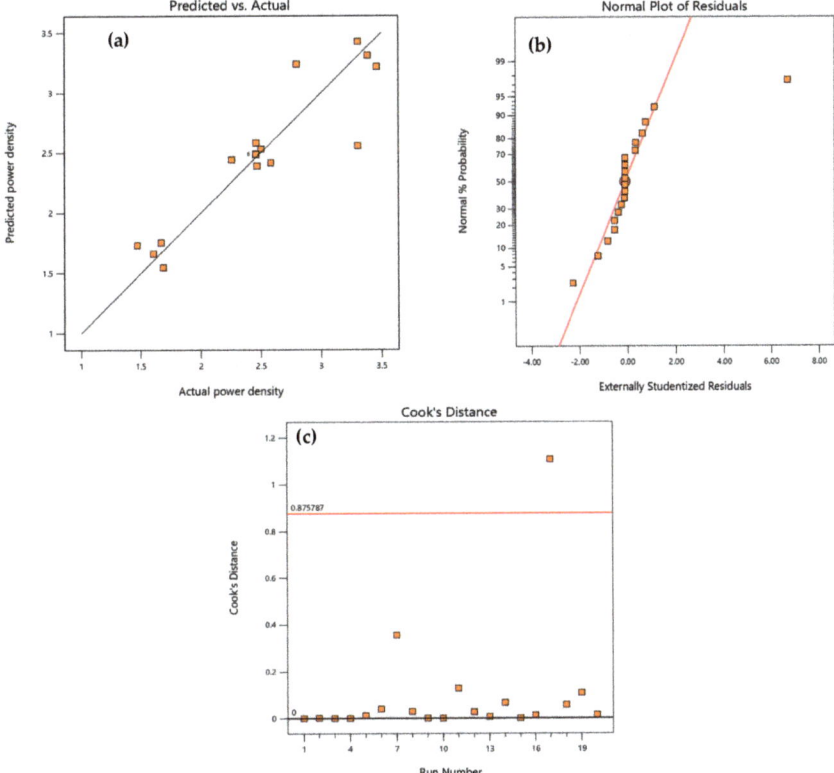

Figure 1. Analytical model validation graphs. (**a**) Predicted vs. actual. (**b**) Normal plot of residuals. (**c**) Cook's distance.

Model adequacy studentized residuals is basically the difference between the actual response value and the best fit value based on the developed model [24,25]. Figure 1b presents the model adequacy studentized residuals for the present study. From this figure, it can be concluded that the study does not include any sort of abnormality as the residual values for the model are insignificant. From the plot shown in Figure 1c, it can be observed that only one point lies outside the Cook's distance, which means that a substantial number of points lies within the acceptable region. This further implies that there are no such potentially powerful points that may affect the response abnormally owing to their location in the plot [26,27]; hence, the model can be considered significant.

2.3. Optimal Conditions

After verifying the validity of the model, the next step was to interpret the results and find the optimal design points. The developed analytical model was used to generate contour plots and response surface plots against the control variables to study the effect of independent parameters on power density. An understanding of the interactions of variables is essential for achieving the highest power density. The three-dimensional response surface helps in the visualization of the interactions between the variables. However, for simplification, contour plots can also be used. Figure 2 depicts the contour plots and the 3D response surfaces for the interaction between power density, acetate concentration, oxygen concentration, and fuel feed flow rate.

Figure 2a shows the combined effect of the interaction between oxygen concentration and fuel feed flow rate on power density. It can be observed from Figure 2 that the power density increases with the increase in oxygen concentration, while keeping the fuel feed flow rate below 1.9×10^{-5} $m^3 \cdot h^{-1}$. Accordingly, the highest power density value of 3.42 $W \cdot m^{-2}$ occurred at a fuel feed flow rate of 1.8×10^{-5} $m^3 \cdot h^{-1}$ and an oxygen concentration of 1 $mol \cdot m^{-3}$. This point, with the highest power density, was considered as the optimal design point.

Figure 2b relates the effect of fuel feed flow rate and acetate concentration on power density. The trend suggests that an increment in the acetate concentration has a significant effect on the power density. It was realized that the power density attained its maximum value at an acetate concentration of 2.6 $mol \cdot m^{-3}$ and a fuel feed flow rate of 1.8×10^{-5} $m^3 \cdot h^{-1}$. Similarly, Figure 2c depicts the response of the power density upon variation of the acetate and the oxygen concentrations. The power density was found to be dependent on the concentration of both reactants. The highest value of power density was obtained at an acetate concentration of 2.6 $mol \cdot m^{-3}$ and an oxygen concentration of 1 $mol \cdot m^{-3}$.

Based on the optimization study, the different obtained results have been depicted in Table 1. Fuel feed flow rate, initial acetate concentration and initial oxygen concentration values of 0.00 $m^3 \cdot h^{-1}$, 2.60 $mol \cdot m^{-3}$, and 1.00 $mol \cdot m^{-3}$, respectively, have been determined to be the optimum values that resulted in the generation of the maximum power density (i.e., 3.425 $W \cdot m^{-2}$). The obtained results were somewhat predictable, based on the already available literature on MFCs.

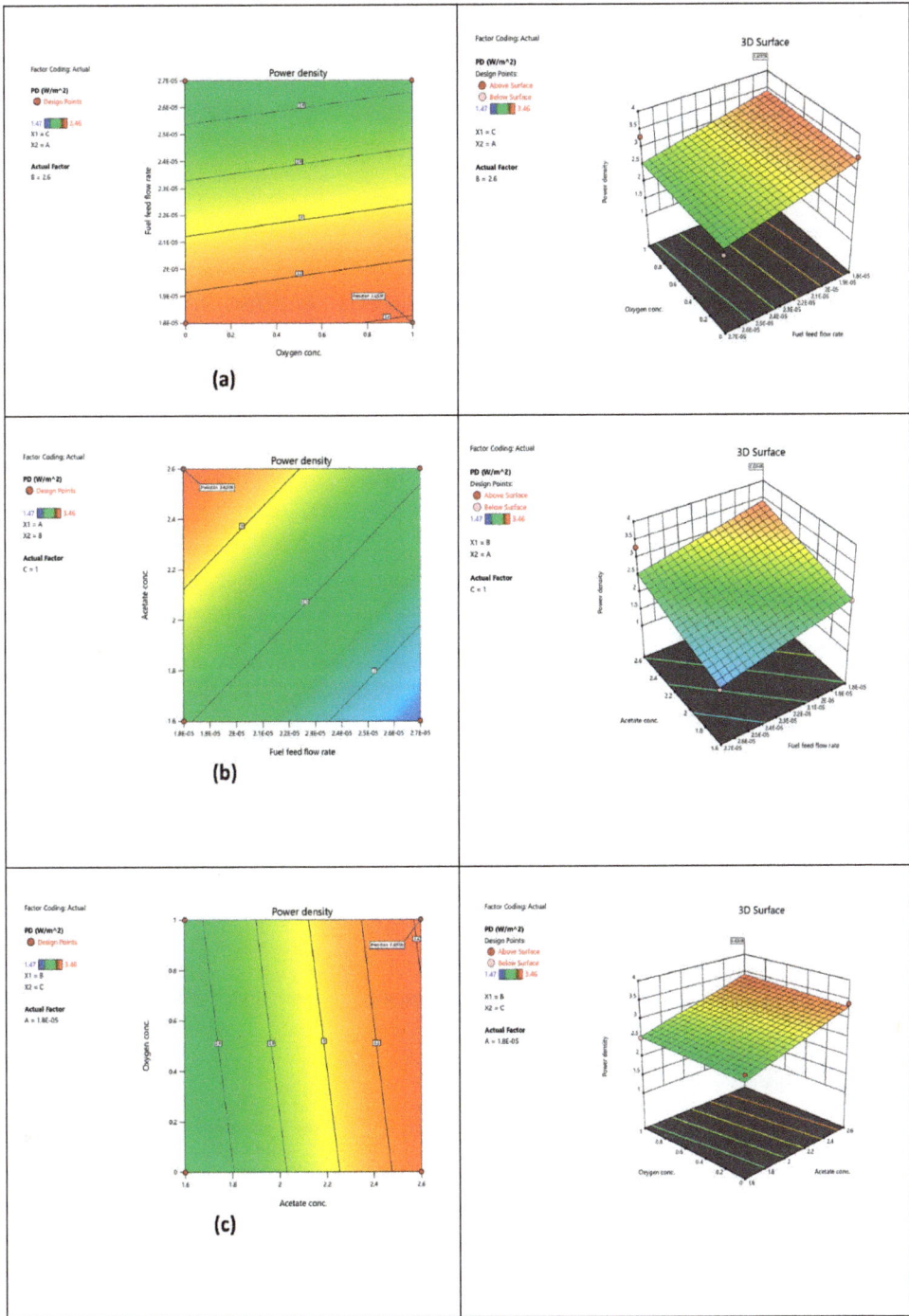

Figure 2. Contour plots of optimization to study the effect of the selected independent parameters on the MFC power density: (**a**) Effect of oxygen concentration and fuel feed flowrate, (**b**) effect of fuel feed flowrate and acetate concentration, and (**c**) effect of acetate concentration and oxygen concentration.

Table 1. Consolidated results of the optimization study.

No.	Flow Rate of Fuel Feed to Anode ($m^3 \cdot h^{-1}$)	Concentration of Acetate in Anode ($mol \cdot m^{-3}$)	Concentration of Oxygen in Cathode ($mol \cdot m^{-3}$)	Power Density ($W \cdot m^{-2}$)	Desirability	
1	0.00	2.60	1.00	3.42506	0.99411298	Selected
2	0.00	2.60	0.92	3.41641	0.99264417	-
3	0.00	2.60	0.90	3.41367	0.99217825	-
4	0.00	2.60	0.89	3.41246	0.99196836	-
5	0.00	2.60	0.88	3.41126	0.9917674	-
6	0.00	2.60	0.85	3.40772	0.9911634	-
7	0.00	2.60	0.83	3.4051	0.990717	-
8	0.00	2.60	0.79	3.40082	0.98998557	-
9	0.00	2.60	1.00	3.41726	0.9898039	-
10	0.00	2.60	0.76	3.39749	0.98941787	-

Cathodic reactions have always been of interest to the research community, as it is usually observed that the performance of an MFC is limited by cathodic reactions [28]. In this study, upon changing the cathodic concentrations, by keeping other variables constant, a substantial increment in the MFC performance was observed. With respect to the fuel feed flowrate at the anode, a recent study was performed by You et al. [29]. After a series of experiments, it was noted that a higher fuel feed flow rate does not result in an enhancement of the power density of an MFC. Therefore, a low fuel feed flow rate was recommended. This finding lies in line with the present finding, as the optimization study revealed that the maximum power density is achievable at a hypothetically low fuel feed flow rate (ideally at 0 $m^3 \cdot h^{-1}$). It must be mentioned here that in case of very low substrate concentration (say, 2.0 mM), an increase in the flow rate resulted in an initial increase in the power density; however, following this initial increase, no further change in the power density was observed with increase in the flow rate. Therefore, it can be concluded that increasing the fuel feed flowrate results in a decrease in the power density (an observation that lies in line with the published literature) [21,30]. With respect to acetate concentration, a recent study showed that a medium level (for instance, between 500 and 3000 $mg \cdot L^{-1}$ of acetate concentration, the highest power output was achieved at a concentration of 2000 $mg \cdot L^{-1}$) of acetate concentration is favorable towards the achievement of the maximum power density [31], which agrees with the findings of this study.

From Table 1, it is pertinent to observe that the value of the flow rate of the fuel feed to the anodic chamber has been recommended to be 0.00 $m^3 \cdot h^{-1}$. This implies that a steady-state operation is recommended for obtaining the highest power density output. In other words, to attain the maximum power density, a particular amount of feed must be first introduced in the MFC, followed by an allowance of time to reach the steady-state. After reaching the steady-state, the power density output will be the maximum. However, after a certain point, the power density will start to decrease. At this very point, it is recommended to empty the fuel cell and introduce a new feed. This recommendation lies in line with that suggested by Esfandyari et al. [32].

2.4. Validation of Optimization

The optimal conditions recommended by the numerical model have been highlighted in Table 1. The statistical model, Equation (1), predicts that under the mentioned optimal conditions, the power density must be 3.425 $W \cdot m^{-2}$. In order to validate these optimal results, a comparative study was conducted in which the power density of the control MFC was compared to that of the optimized MFC (see Figure 3). The control cell was operated at a fuel feed flow rate of 2.25×10^{-5} $m^3 \cdot h^{-1}$, an influent acetate concentration of 1.56 $mol \cdot m^{-3}$ and an influent oxygen concentration of 0 $mol \cdot m^{-3}$; whereas, the optimized MFC was operated at the proposed optimized conditions. All the experiments were performed at least twice and average values are stated. On average, a marginally insignificant value of deviation (2%) was observed. It was observed that the optimized MFC produced a much higher power density compared to the control MFC. Hence, the objective of the study was achieved. The obtained results, when compared to the literature, revealed that the maximum power density reported in an earlier study with the same experimental setup, but without optimization, was 2.039 W·m^{-2} [21]. Therefore, it is evident that, because of the optimization, an open circuit maximum power density of 3.425 W·m^{-2} could be achieved in the present study, with an average deviation of 2%.

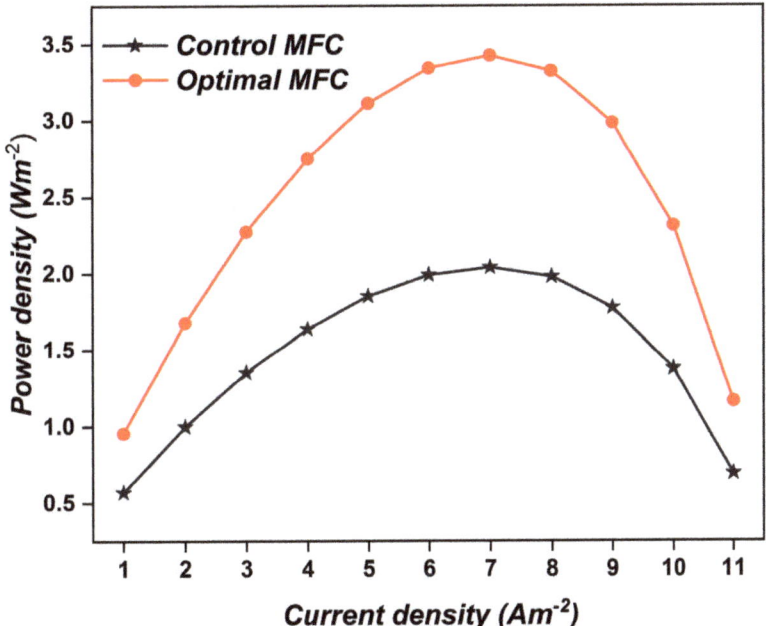

Figure 3. Comparative analysis of the control and the optimized MFCs.

3. Materials and Methods

3.1. MFC Design

To conduct this study, a dual chamber microbial fuel cell was constructed with the following specifications. A transparent sheet of polyacrylic plastic was used to construct both chambers of the MFC. The two chambers, i.e., the anodic and the cathodic chambers, were constructed with an equal volume of 5.5 cm^3 and equal dimensions of 5.5 cm × 1.0 cm × 1.0 cm. An anode, having dimensions of 4.5 cm × 1.0 cm × 0.5 cm and made of graphite felt was used in this experiment. For this purpose, analytical grade polyacrylonitrile-based graphite felt was purchased from Beijing Sanys Carbon (Beijing, China) Co., Ltd. Both the anodic and the cathodic chambers were modeled as continuously stirred tank reactors. The cathode also possessed the same dimensions and materials, apart from the fact that it was coated with 0.3 mg·cm^{-2} of platinum powder. The catalyst was first magnetically stirred with isopropanol and 5 wt% Nafion solutions, followed by sonication for a period of 30 min to obtain a fine ink for the catalyst. The obtained ink was then carefully deposited onto the surface of the electrode by employing the spray coating technique [33]. To separate both the chambers, Nafion 117 membrane, having an area of 5 cm^2, a thickness of 0.1778 mm and a conductivity of 5 S·m^{-1}, was used. The anode and the cathode chambers were connected by using a platinum wire that was externally connected to a variable resistance box of 10–10 k Ω, that was systematically varied during the experiments and results were obtained by using Ohm's law. The maximum power density value was obtained against an external resistance of 1 kΩ. Upon increasing the resistance value, rapid drop of voltage was observed. The design of this MFC was inspired by the design reported by Zeng et al. [21].

3.2. MFC Feed: Inoculation and Anolyte Preparation

For this study, we used an electrogenic-mixed firmicute consortium. The consortium was isolated from the anode electrode biofilm of an MFC operated with wastewater collected from the Karachi Water & Sewerage Board (KWSB), operated in fed-batch mode over a period of 6 months. Upon isolation consortium it was revealed that the isolated microbes belonged to the *Lysinibacillus* species, that contributes to make MFC cost effective [34]. For DNA isolation, PCR amplification and anolyte preparation, we closely followed the techniques described elaborately by Kumar et al. [35–37] and Rudra et al. [38]. These mixed strains were suspended in the buffer solution and subsequently transferred to synthetic feed wastewater to make the final volume of the microbial enriched anolyte. Inoculation of the microbe colonies was performed in sterilized solution that contained nutrient broth. Growth was allowed under laboratory conditions (temperature: 37 °C, time: 24 to 48 h), by maintaining anoxic conditions the whole process. The synthetic feed was prepared using 1.56 mM acetate buffer solution, and it was used to conduct all the experiments. The buffer was prepared using 4 mM sodium acetate in conjunction with acetic acid. The pH of the buffer was maintained at 7. To maintain anaerobic conditions in the anodic chamber, N_2 was purged through the solution, while the cathodic chamber was directly in contact with atmospheric air [16]. The external resistance of the circuit during the inoculation procedure was 10 kΩ to ensure high biomass yield.

3.3. Working Conditions

All the experiments were conducted under a steady temperature of 25–30 °C and a pressure of 1 atm. All the readings were obtained from the MFC once the steady state condition was achieved, which took about 36 min from the start of the device operation.

3.4. Experimental Design

The sensitivity analysis of all the parameters of the MFC was conducted by following Zeng et al. [21]. This study was conducted by varying one parameter at a time, while keeping all other parameters constant. In this study [21], it was observed that the fuel feed flow rate, the rate of reaction in the anodic chamber, the rate of electron transfer between two chambers, the initial acetate concentration, the initial oxygen concentration, the membrane thickness, the membrane conductivity and the distance between the chambers were the most influential parameters that determines the power density of an MFC.

In this study, for performing the RSM studies, *Design Expert* software by Stat-Ease, Inc. Minneapolis, MN-55413, USA was used. The maximum and minimum values of the employed independent parameters, along with the mean have been presented in Table 2.

Table 2. Values of independent variables used for the process optimization.

Symbol	Parameter	Unit	Minimum	Maximum	Mean
A	Flow rate of fuel feed to anode	m^3 h^{-1}	0	2.25×10^{-5}	1.12×10^{-5}
B	Initial concentration of acetate in anode	mol m^{-3}	1.26	2.94	2.1
C	Initial concentration of oxygen in cathode	mol m^{-3}	0.34	1.34	0.5

To study the effects of control variables on the power density of MFCs, a total of 20 experiments were performed. The number of experiments were decided based on the formula given in Equation (2).

$$N = 2^k + 2k + 6 \qquad (2)$$

In Equation (2), N denotes the total number of experiments required, while k is the total number of independent variables selected for a given study [39]. Furthermore, the first term on the right-hand side denotes the number of factorial points, followed by the number of axial points (the second term) and the center points (the third term). In the present study, the value of k is 3. This implies a total of 20 experiments, with 8 factorial points, 6 axial points, and 6 central points.

Table 3 lists all these experiments, with their corresponding values of power densities.

Table 3. Central composite design of experiments for optimization of the MFC.

Run	Type	Factor 1 A $m^3 \cdot h^{-1}$	Factor 2 B $mol \cdot m^{-3}$	Factor 3 C $mol \cdot m^{-3}$	Response R $W \cdot m^{-2}$
1	Axial	2.25×10^{-5}	2.940896	0.5	2.797
2	Center	2.25×10^{-5}	2.1	0.5	2.461
3	Axial	3.01×10^{-5}	2.1	0.5	1.671
4	Axial	2.25×10^{-5}	1.259104	0.5	1.47
5	Axial	2.25×10^{-5}	2.1	1.340896	2.463
6	Factorial	0.000027	1.6	1	1.607
7	Factorial	0.000027	2.6	1	3.304
8	Center	2.25×10^{-5}	2.1	0.5	2.461
9	Center	2.25×10^{-5}	2.1	0.5	2.461
10	Axial	1.49×10^{-5}	2.1	0.5	3.46
11	Factorial	0.000027	2.6	0	2.261
12	Factorial	0.000018	2.6	0	3.386
13	Factorial	0.000018	1.6	1	2.503
14	Factorial	0.000018	1.6	0	2.585
15	Factorial	0.000018	2.6	1	3.304
16	Center	2.25×10^{-5}	2.1	0.5	2.461
17	Factorial	0.000027	1.6	0	1.689
18	Center	2.25×10^{-5}	2.1	0.5	2.461
19	Axial	2.25×10^{-5}	2.1	0.3409	2.47
20	Center	2.25×10^{-5}	2.1	0.5	2.461

4. Conclusions

A microbial fuel cell is a promising source of green electricity that also simultaneously reduces organic waste. Despite being an appealing technology, it is still largely laboratory-based and has so far failed to find applications in the industrial sector. This is primarily owing to its low power density output. This study uses RSM to achieve an enhanced power density of MFCs. For this purpose, a central composite design was adopted and a series of 20 experiments were conducted. Based on the experimental results, a statistical model was developed that depicted the relation of three independent variables to power density output. The model was validated using the ANOVA test and Crook distance test. Based on all the experiments and the validated model, it has been inferred that fuel feed flow rate, initial acetate concentration and initial oxygen concentration values of 0.00 $m^3 \cdot h^{-1}$, 2.60 $mol \cdot m^{-3}$, and 1.00 $mol \cdot m^{-3}$, respectively, are the optimum values that provide the maximum power density of 3.425 $W \cdot m^{-2}$. Additionally, this study successfully resulted in an improvement of the power density of an MFC from 2.039 $W \cdot m^{-2}$ (maximum value reported in the literature for a similar study) to 3.425 $W \cdot m^{-2}$.

Author Contributions: Conceptualization, M.N.N. and A.A.Z.; Data curation, H.K., S.K. and M.T.b.O.; Formal analysis, M.N.N., A.A.Z., H.K., Y.A.W. and K.D.; Funding acquisition, Y.A.W., N.A.H., I.A.B. and H.A.; Investigation, M.N.N.; Methodology, M.N.N., K.D. and J.J.; Project administration, A.A.Z., Y.A.W., K.D., J.J. and N.A.H.; Resources, H.K., Y.A.W., J.J., N.A.H. and H.H.; Software, S.K., M.T.b.O., M.A.I. and H.H.; Supervision, A.A.Z., Y.A.W., K.D. and J.J.; Validation, M.N.N., Y.A.W., M.A.I. and H.H.; Visualization, S.K., M.T.b.O., M.A.I. and H.H.; Writing—original draft, M.N.N. and H.K.; Writing—review & editing, A.A.Z., K.D., I.A.B. and H.A. All authors have read and agreed to the published version of the manuscript.

Funding: This work was supported by the University of Malaya through grant number ST030-2019. The authors also extend their appreciation to the Deanship of Scientific Research at King Khalid University for funding this work through a research group program under grant number RGP. 2/166/42.

Data Availability Statement: The authors confirm that the data supporting the findings of this study are available within the article.

Conflicts of Interest: The authors declare no conflict of interest.

References

1. Jayapiriya, U.; Goel, S. Influence of cellulose separators in coin-sized 3D printed paper-based microbial fuel cells. *Sustain. Energy Technol. Assess.* **2021**, *47*, 101535.
2. Naseer, M.N.; Zaidi, A.A.; Khan, H.; Kumar, S.; bin Owais, M.T.; Jaafar, J.; Suhaimin, N.S.; Wahab, Y.A.; Dutta, K.; Asif, M.; et al. Mapping the field of microbial fuel cell: A quantitative literature review (1970–2020). *Energy Rep.* **2021**, *7*, 4126–4138. [CrossRef]
3. Din, M.I.; Nabi, A.G.; Hussain, Z.; Khalid, R.; Iqbal, M.; Arshad, M.; Muhjahid, A.; Hussain, T. Microbial fuel cells—A preferred technology to prevail energy crisis. *Int. J. Energy Res.* **2021**, *45*, 8370–8388. [CrossRef]
4. Rossi, R.; Fedrigucci, A.; Setti, L. Characterization of electron mediated microbial fuel cell by Saccharomyces cerevisiae. *Chem. Eng. Trans.* **2015**, *43*, 337–342.
5. Li, W.-W.; Sheng, G.-P.; Yu, H.-Q. *Chapter 14-Electricity Generation from Food Industry Wastewater Using Microbial Fuel Cell Technology*, in *Food Industry Wastes*; Kosseva, M.R., Webb, C., Eds.; Academic Press: San Diego, CA, USA, 2013; pp. 249–261.
6. Cusick, R.D.; Kiely, P.D.; Logan, B.E. A monetary comparison of energy recovered from microbial fuel cells and microbial electrolysis cells fed winery or domestic wastewaters. *Int. J. Hydrogen Energy* **2010**, *35*, 8855–8861. [CrossRef]
7. Kim, J.R.; Premier, G.C.; Hawkes, F.R.; Rodríguez, J.; Dinsdale, R.M.; Guwy, A.J. Modular tubular microbial fuel cells for energy recovery during sucrose wastewater treatment at low organic loading rate. *Bioresour. Technol.* **2010**, *101*, 1190–1198. [CrossRef]
8. Trapero, J.; Horcajada, L.; Linares, J.J.; Lobato, J. Is microbial fuel cell technology ready? An economic answer towards industrial commercialization. *Appl. Energy* **2017**, *185*, 698–707. [CrossRef]
9. Gajda, I.; Greenman, J.; Ieropoulos, I.A. Recent advancements in real-world microbial fuel cell applications. *Curr. Opin. Electrochem.* **2018**, *11*, 78–83. [CrossRef]
10. Algar, C.K.; Howard, A.; Ward, C.; Wanger, G. Sediment microbial fuel cells as a barrier to sulfide accumulation and their potential for sediment remediation beneath aquaculture pens. *Sci. Rep.* **2020**, *10*, 13087. [CrossRef]
11. Sarabia, L.A.; Ortiz, M.C. 1.12-Response Surface Methodology, in *Comprehensive Chemometrics*; Brown, S.D., Tauler, R., Walczak, B., Eds.; Elsevier: Oxford, UK, 2009; pp. 345–390.
12. Khuri, A.I.; Mukhopadhyay, S. Response surface methodology. *WIREs Comput. Stat.* **2010**, *2*, 128–149. [CrossRef]
13. Aydar, A.Y. Utilization of response surface methodology in optimization of extraction of plant materials. *Stat. Approaches Emphas. Des. Exp. Appl. Chem. Process.* **2018**, *1*, 157–169.
14. Feng, R.; Zaidi, A.A.; Zhang, K.; Shi, Y. Optimisation of Microwave Pretreatment for Biogas Enhancement through Anaerobic Digestion of Microalgal Biomass. *Period. Polytech. Chem. Eng.* **2018**, *63*, 65–72. [CrossRef]
15. Zaidi, A.A.; Khan, S.Z.; Shi, Y. Optimization of nickel nanoparticles concentration for biogas enhancement from green algae anaerobic digestion. *Mater. Today: Proc.* **2020**, *39*, 1025–1028. [CrossRef]
16. Sarafraz, M.; Safaei, M.R.; Goodarzi, M.; Arjomandi, M. Experimental investigation and performance optimisation of a catalytic reforming micro-reactor using response surface methodology. *Energy Convers. Manag.* **2019**, *199*, 111983. [CrossRef]
17. Geetanjali; Rani, R.; Sharma, D.; Kumar, S. Optimization of operating conditions of miniaturize single chambered microbial fuel cell using $NiWO_4$/graphene oxide modified anode for performance improvement and microbial communities dynamics. *Bioresour. Technol.* **2019**, *285*, 121337. [CrossRef]
18. Sedighi, M.; Aljlil, S.A.; Alsubei, M.D.; Ghasemi, M.; Mohammadi, M. Performance optimisation of microbial fuel cell for wastewater treatment and sustainable clean energy generation using response surface methodology. *Alex. Eng. J.* **2018**, *57*, 4243–4253. [CrossRef]
19. Islam, M.A.; Ong, H.R.; Ethiraj, B.; Cheng, C.K.; Khan, M.R. Optimization of co-culture inoculated microbial fuel cell performance using response surface methodology. *J. Environ. Manag.* **2018**, *225*, 242–251. [CrossRef]
20. Almatouq, A.; Babatunde, A. Identifying optimized conditions for concurrent electricity production and phosphorus recovery in a mediator-less dual chamber microbial fuel cell. *Appl. Energy* **2018**, *230*, 122–134. [CrossRef]
21. Zeng, Y.; Choo, Y.F.; Kim, B.-H.; Wu, P. Modelling and simulation of two-chamber microbial fuel cell. *J. Power Sources* **2010**, *195*, 79–89. [CrossRef]
22. Cheng, C.-L.; Shalabh; Garg, G. Coefficient of determination for multiple measurement error models. *J. Multivar. Anal.* **2014**, *126*, 137–152. [CrossRef]
23. David, I.; Adubisi, O.; Ogbaji, O.; Eghwerido, J.; Umar, Z. Resistant measures in assessing the adequacy of regression models. *Sci. Afr.* **2020**, *8*, e00437. [CrossRef]
24. Zhu, L. Checking the adequacy of a partially linear model. *Nonparametric Monte Carlo Tests Appl.* **2005**, *1*, 61–83.
25. Chapter 5-Applications. In *Inference for Heavy-Tailed Data Analysis*; Academic Press: Cambridge, MA, USA, 2017; pp. 133–158.
26. Chattoraj, S.; Mondal, N.K.; Das, B.; Roy, P.; Sadhukhan, B. Biosorption of carbaryl from aqueous solution onto Pistia stratiotes biomass. *Appl. Water Sci.* **2013**, *4*, 79–88. [CrossRef]
27. Dalma, K.E.; Haydee, K.M.; Radu, T. Dynamic modelling of pesticides uptake by triticum spp. Anatomical compartments. *Agric. Food* **2016**, *4*, 215–228.

28. Zhao, F.; Harnisch, F.; Schröder, U.; Scholz, F.; Bogdanoff, P.; Herrmann, I. Challenges and Constraints of Using Oxygen Cathodes in Microbial Fuel Cells. *Environ. Sci. Technol.* **2006**, *40*, 5193–5199. [CrossRef] [PubMed]
29. You, J.; Greenman, J.; Ieropoulos, I. Novel Analytical Microbial Fuel Cell Design for Rapid in Situ Optimisation of Dilution Rate and Substrate Supply Rate, by Flow, Volume Control and Anode Placement. *Energies* **2018**, *11*, 2377. [CrossRef]
30. Moon, H.; Chang, I.S.; Kim, B.H. Continuous electricity production from artificial wastewater using a mediator-less microbial fuel cell. *Bioresour. Technol.* **2006**, *97*, 621–627. [CrossRef]
31. Ullah, Z.; Zeshan, S. Effect of substrate type and concentration on the performance of a double chamber microbial fuel cell. *Water Sci. Technol.* **2019**, *81*, 1336–1344. [CrossRef]
32. Esfandyari, M.; Fanaei, M.A.; Gheshlaghi, R.; Mahdavi, M.A. Mathematical modeling of two-chamber batch microbial fuel cell with pure culture of Shewanella. *Chem. Eng. Res. Des.* **2017**, *117*, 34–42. [CrossRef]
33. Nandy, A.; Kumar, V.; Mondal, S.; Dutta, K.; Salah, M.; Kundu, P.P. Performance evaluation of microbial fuel cells: Effect of varying electrode configuration and presence of a membrane electrode assembly. *New Biotechnol.* **2015**, *32*, 272–281. [CrossRef] [PubMed]
34. Nandy, A.; Kumar, V.; Kundu, P.P. Utilization of proteinaceous materials for power generation in a mediatorless microbial fuel cell by a new electrogenic bacteria Lysinibacillus sphaericus VA5. *Enzym. Microb. Technol.* **2013**, *53*, 339–344. [CrossRef] [PubMed]
35. Kumar, V.; Nandy, A.; Das, S.; Salahuddin, M.; Kundu, P.P. Performance assessment of partially sulfonated PVdF-co-HFP as polymer electrolyte membranes in single chambered microbial fuel cells. *Appl. Energy* **2015**, *137*, 310–321. [CrossRef]
36. Kumar, V.; Kumar, P.; Nandy, A.; Kundu, P.P. Crosslinked inter penetrating network of sulfonated styrene and sulfonated PVdF-co-HFP as electrolytic membrane in a single chamber microbial fuel cell. *RSC Adv.* **2015**, *5*, 30758–30767. [CrossRef]
37. Kumar, V.; Kumar, P.; Nandy, A.; Kundu, P.P. A nanocomposite membrane composed of incorporated nano-alumina within sulfonated PVDF-co-HFP/Nafion blend as separating barrier in a single chambered microbial fuel cell. *RSC Adv.* **2016**, *6*, 23571–23580. [CrossRef]
38. Rudra, R.; Kumar, V.; Kundu, P. Acid catalysed cross-linking of poly vinyl alcohol (PVA) by glutaraldehyde: Effect of crosslink density on the characteristics of PVA membranes used in single chambered microbial fuel cells. *RSC Adv.* **2015**, *5*, 83436–83447. [CrossRef]
39. Hosseinpour, M.; Vossoughi, M.; Alemzadeh, I. An efficient approach to cathode operational parameters optimization for microbial fuel cell using response surface methodology. *J. Environ. Health Sci. Eng.* **2014**, *12*, 33. [CrossRef]

Article

Effect of Temperature, Syngas Space Velocity and Catalyst Stability of Co-Mn/CNT Bimetallic Catalyst on Fischer Tropsch Synthesis Performance

Omid Akbarzadeh [1,*], Solhe F. Alshahateet [2], Noor Asmawati Mohd Zabidi [3], Seyedehmaryam Moosavi [4], Amir Kordijazi [5], Arman Amani Babadi [1], Nor Aliya Hamizi [1], Yasmin Abdul Wahab [1], Zaira Zaman Chowdhury [1] and Suresh Sagadevan [1,*]

[1] Nanotechnology & Catalysis Research Centre, University of Malaya, Kuala Lumpur 50603, Malaysia; ar.amani65@gmail.com (A.A.B.); aliyahamizi@um.edu.my (N.A.H.); yasminaw@um.edu.my (Y.A.W.); dr.zaira.chowdhury@um.edu.my (Z.Z.C.)
[2] Department of Chemistry, Mutah University, P.O. BOX 7, Mutah, Karak 61710, Jordan; s_alshahateet@mutah.edu.jo
[3] Department of Fundamental and Applied Sciences, Universiti Teknologi PETRONAS, Bandar Seri Iskandar 32610, Perak, Malaysia; noorasmawati_mzabidi@utp.edu.my
[4] Department of Chemistry and Bioengineering, Vilnius Gediminas Technical University, 10223 Vilnius, Lithuania; m.moosavi1987@gmail.com
[5] Department of Industrial and Manufacturing Engineering, University of Wisconsin Milwaukee, Milwaukee, WI 53211, USA; kordija2@uwm.edu
* Correspondence: omid.akbarzadeh63@gmail.com (O.A.); drsureshnano@gmail.com (S.S.)

Citation: Akbarzadeh, O.; Alshahateet, S.F.; Mohd Zabidi, N.A.; Moosavi, S.; Kordijazi, A.; Babadi, A.A.; Hamizi, N.A.; Wahab, Y.A.; Chowdhury, Z.Z.; Sagadevan, S. Effect of Temperature, Syngas Space Velocity and Catalyst Stability of Co-Mn/CNT Bimetallic Catalyst on Fischer Tropsch Synthesis Performance. Catalysts 2021, 11, 846. https://doi.org/10.3390/catal11070846

Academic Editor: Javier Ereña Loizaga

Received: 6 May 2021
Accepted: 10 July 2021
Published: 14 July 2021

Publisher's Note: MDPI stays neutral with regard to jurisdictional claims in published maps and institutional affiliations.

Copyright: © 2021 by the authors. Licensee MDPI, Basel, Switzerland. This article is an open access article distributed under the terms and conditions of the Creative Commons Attribution (CC BY) license (https://creativecommons.org/licenses/by/4.0/).

Abstract: The effect of reaction temperature, syngas space velocity, and catalyst stability on Fischer-Tropsch reaction was investigated using a fixed-bed microreactor. Cobalt and Manganese bimetallic catalysts on carbon nanotubes (CNT) support (Co-Mn/CNT) were synthesized via the strong electrostatic adsorption (SEA) method. For testing the performance of the catalyst, Co-Mn/CNT catalysts with four different manganese percentages (0, 5, 10, 15, and 20%) were synthesized. Synthesized catalysts were then analyzed by TEM, FESEM, atomic absorption spectrometry (AAS), and zeta potential sizer. In this study, the temperature was varied from 200 to 280 °C and syngas space velocity was varied from 0.5 to 4.5 L/g.h. Results showed an increasing reaction temperature from 200 °C to 280 °C with reaction pressure of 20 atm, the Space velocity of 2.5 L/h.g and H_2/CO ratio of 2, lead to the rise of CO % conversion from 59.5% to 88.2% and an increase for C_{5+} selectivity from 83.2% to 85.8%. When compared to the other catalyst formulation, the catalyst sample with 95% cobalt and 5% manganese on CNT support (95Co5Mn/CNT) performed more stable for 48 h on stream.

Keywords: carbon nanotubes; thermal treatment; cobalt; Fischer-Tropsch; catalyst; acid treatment

1. Introduction

Fischer-Tropsch Synthesis (FTS) utilizes syngas (H_2 + CO) to generate hydrocarbons which have a significant role among eco-friendly fuels and renewable energies. Due to abundant natural gas and coal resources, gas to the liquid process is appealing as a source of feed instead of declining crude oil reservoirs. Fuels produced with FTS are eco-friendly and have very low levels of greenhouse gases. Cobalt catalyst is a popular catalyst choice for FTS [1,2]. It is of most economic interest to have liquid hydrocarbons with long-chain carbon atoms, referred to as C_{5+}. We employed the same combined acid and heat pre-treatments of CNT as Tavasoli et al. [3], but the strong electrostatic adsorption (SEA) technique was used for the preparation of the Co/CNT catalyst, with the pH of the precursor solution being regulated throughout the metal deposition. Schwarz proposed that the electrostatic interactions between a metallic ion and a charged support may be used to control the metallic ion's adsorption over surfaces with two oxide fractions [4,5]. The concept of this methodology has been efficiently employed to generate highly distributed bimetallic catalysts, with a variety of oxide and carbon substrates [6–8]. Depending upon the pH solution is acidic or basic, the hydroxyl (–OH) groups on the surface of an oxide become protonated or deprotonated naturally. These charged hydroxyl groups can then absorb metal complex ions in an oppositely charged solution. The density of charged hydroxyl groups on the oxide surface is determined by the pH at which the surface is neutrally charged, which is called the Point of Zero Charge (PZC). The hydroxyl groups deprotonate above the PZC, making the surface negatively charged and allowing cationic complexes to be adsorbed onto the surface by electrostatic adsorption technique [8]. Previous research on CNT-supported cobalt catalysts is used as an impregnation approach without pH control during catalyst synthesis [9,10]. The deposition of cobalt solution because of pre-treated CNT support has been carried out in this study using the SEA principle at a specific pH. During the synthesis process, the pH has been monitored for cobalt solution. On the characteristics and performance of Co/CNTs catalysts, the impacts of combined acid and heat pre-treatments of CNT support are discussed. The activity and stability of the Co/CNT catalyst in FTS are increased by combining an acid with the thermal pre-treatment of CNT at 900 °C.

Consequently, optimizing the distribution of the reaction product is important and this can be accomplished by varying any of the reaction factors, like temperature, pressure, H_2/CO ratio, reactor type, and catalyst, etc. [11–16]. Increasing FT operating pressure for cobalt-based catalysts was reported to have a negligible impact on enhancing the reaction rate and C_{5+} selectivity [17–19]. As a part of ongoing research, the effect of reaction pressure on Co/CNT catalyst performance with different supporting materials has been studied. The selectivity of shorter molecular hydrocarbons (C_1-C_2) has been revealed to be substantially enhanced by increasing the reaction temperature and the H_2/CO ratio. However, the selection of long molecular hydrocarbons (C_{5+}) is substantially enhanced by reducing the pressure of the reaction [20–23]. Some researchers recorded the influence of operating conditions on the product selectivity for cobalt-based catalysts [20,24,25] and revealed that the olefin selectivity of the hydrocarbon product range reduced with rising pressure, which has reported in prior studies [19,21,24]. This study is a continuation of previous research that has been examined and published [26–30]. The current investigation aims to prepare cobalt manganese bimetallic catalysts on CNT substrate employing SEA technique, also to study effects of temperature, syngas space velocity, and catalyst stability through Fischer-Tropsch synthesis by performing Co-Mn/CNT bimetallic catalyst.

2. Process Result Dissection

The effects of temperature, catalyst stability, and syngas space velocity on the catalytic efficiency of monometallic and bimetallic Co-Mn were analyzed. The findings of the reaction were contrasted by product selection in terms of carbon monoxide conversion and hydrocarbon. In the reaction study part, all the reactions were performed two times and the standard deviation value was calculated to be ±1 percent for all reactions. Carbon mass balance was calculated from the moles of carbon entering the reactor relative to the moles of carbonaceous products formed. The advantage and novelty of the current studies were performed by the SEA method for synthesizing Co-Mn catalysts on CNT support for Fischer-Tropsch (FT) reaction, which has not been reported previously [26–29]. A high percentage of CO conversion and C_{5+} selectivity was obtained in the present investigation.

2.1. Influence of Reaction Temperature on Catalyst Efficiency

Table 1 revealed that the Fischer-Tropsch synthesis rates, as well as the CO conversion, are under the strong influence of reaction temperature. The results illustrate that rising Fischer-Tropsch reaction temperature from 200 to 280 °C boosts % CO conversion from 59.5 to 88.2%. Increasing FTS temperature rises the motions of hydrogen on the catalyst surface and results in greater CO conversion [31]. Simultaneously, the WGS reaction rate rises from 0.55 to 0.80. The rate of WGS reaction or CO_2 formation can be increased and related to the rise in water semi-pressure, owing to the rise in Fischer-Tropsch synthesis reaction rate [32]. A comparison of hydrocarbon product selectivity for 95Co5Mn/CNT catalysts at 220 °C and 280 °C shows a significant change with decreases in molecular weight hydrocarbons for greater reaction temperature [33]. The results indicate that methane selectivity using 95Co5Mn/CNT catalysts at 200 and 280 °C is 10.8%, and 15.2%, respectively. Further, the selectivity of C_{5+} hydrocarbons for 95Co5Mn/CNT catalysts increases from 83.2% (200 °C) to 85.8% (240 °C). The olefin to paraffin ratio reduced from 0.54 to 0.15. This finding presented that greater reaction temperature enhances the carbon monoxide conversion, but for C_{5+} selectivity, increasing reaction temperature leads to a hydrocarbon chain moves towards a shorter chain [32,34].

Table 1. Influence of reaction temperature (°C) on CO conversion%, C_1 selectivity%, C_2–C_4 selectivity%, C_{5+} selectivity%, olefinity and WGS reaction *.

CO Conversion%	200	220	240	260	280
Co/CNT	48.3	54.3	58.2	59.6	59.5
95Co5Mn/CNT	59.5	78.2	86.6	87.5	88.2
90Co10Mn/CNT	57.1	73.1	79.8	80.3	81.5
85Co15Mn/CNT	55.2	67.1	73.2	74.1	74.5
80Co20Mn/CNT	50.2	61.8	66.3	67.6	67.5
C_1 selectivity%					
Co/CNT	15.5	16.9	16.5	18.6	19.5
95Co5Mn/CNT	10.8	11.3	11.8	13.5	15.2
90Co10Mn/CNT	12.3	12.8	13.3	14.9	16.5
85Co15Mn/CNT	13.1	13.6	14.1	15.5	17.1
80Co20Mn/CNT	14.2	14.5	15.8	16.5	18.5
C_2–C_4 selectivity%					
Co/CNT	12.6	13	13.4	17.3	19.6
95Co5Mn/CNT	5.5	6.1	6.7	9.6	11.3
90Co10Mn/CNT	7.7	8.2	8.7	11.8	13.7
85Co15Mn/CNT	8.5	9.8	9.4	13.6	15.6
80Co20Mn/CNT	9.6	10.6	10.5	14.2	16.5
C_{5+} selectivity%					
Co/CNT	72.1	71.1	69.1	56.7	50.6
95Co5Mn/CNT	83.2	82.4	85.8	68.2	61.7
90Co10Mn/CNT	82.5	81.6	78.7	67.3	60.6
85Co15Mn/CNT	80.5	79.5	76.5	64.4	58.4
80Co20Mn/CNT	78.5	77.5	74.5	63.5	55.2
Olefinity					
Co/CNT	0.92	0.71	0.63	0.72	0.84
95Co5Mn/CNT	0.54	0.19	0.15	0.22	0.38
90Co10Mn/CNT	0.63	0.34	0.27	0.37	0.55
85Co15Mn/CNT	0.78	0.42	0.34	0.5	0.65
80Co20Mn/CNT	0.87	0.57	0.45	0.58	0.75
WGS selectivity					
Co/CNT	0.24	0.33	0.36	0.45	0.53
95Co5Mn/CNT	0.55	0.58	0.65	0.76	0.80
90Co10Mn/CNT	0.48	0.51	0.58	0.68	0.74
85Co15Mn/CNT	0.43	0.45	0.51	0.61	0.69
80Co20Mn/CNT	0.38	0.43	0.45	0.56	0.64

* Reaction condition: Pressure: 20 atm, Space velocity: 2.5 L/h.g, Ratio of H_2/CO:2.

Table 1 illustrates five reaction temperatures: 200, 220, 240, 260, and 280 °C, which were used at an H_2/CO feed proportion of two and a pressure of 20 atm on different Co-Mn/CNT catalyst compositions. As seen in Table 1, increasing the reaction temperature enhanced carbon monoxide conversion. It has been shown that increasing the temperature promotes CO molecule dissociation on active sites of catalysts, hence, increasing the rate of CO hydrogenation [35]. According to Vannice and coworkers [36], CO molecules become active at higher operating temperatures due to the strong interaction of C and O atoms with active metal surfaces [37]. However, the reaction temperature has been increased to 260 °C, activity dropped again, indicating that the optimum reaction temperature was 240 °C. Diffusion effects have been suggested by many studies for an increase in the activity of FTS catalysts as temperature rises [38]. It has been claimed that when the operating temperature rises, the migration of molecules away from active sites improves, increasing the number of active sites available. These findings were comparable to those of prior studies [37].

The methane is used as a fuel, the production of C_1 and C_2–C_4 should be reduced to a minimum during FTS [39]. As indicated in Table 1, increasing the reaction temperature improved methane selectivity. The increase in methane selectivity with increasing temperature has been attributed to an increase in CO molecule dissociation on the catalyst surface, resulting in more carbons available for hydrogenation by H_2 molecules [40].

C_{5+} selectivity is considered a preferred FTS output, and reaction conditions are geared toward increasing C_{5+} products. Table 1 summarizes the effects of temperature on C_{5+} selectivity. With a drop-in temperature, selectivity for C_{5+} increased. During the FTS process, it has been found that increasing the operating temperature reduces chain propagation and improves the chain termination step [41]. The FTS is a surface polymerization reaction, increasing the temperature reduces selectivity for long-chained molecules while improving selectivity for lower hydrocarbons.

As indicated in Table 1, increasing temperature resulted in a decrease in olefinity. In the temperature range of 220 to 250 °C, the Olefin to Paraffin ratio declined faster than in the temperature range of 250 to 280 °C, when it climbed again, which was consistent with thermodynamic assumptions. Olefins are generated first, then propagated to form long-chained hydrocarbons during FTS. As most olefins were hydrogenated, an increase in temperature increased CO hydrogenation and hence decreased olefinity [42]. CO hydrogenation decreased at the temperature range of 250 to 280 °C due to the high chemisorption of CO molecules on the catalyst surface, which reduced the likelihood of hydrogen molecules hydrogenating CO molecules. Hunter and coworkers [42] noticed a similar pattern. On the other hand, some researchers discovered a contrary tendency, claiming that increasing the operating temperature increased olefinity. The rate of WGS increased with increasing temperature for monometallic and bimetallic catalysts (Table 1), and other researchers have found the same pattern [42].

2.2. Influence of Space Velocity on Catalyst Efficiency

Effect of mass space velocity (Vm) on the catalytic efficiency of CNT-supported monometallic and bimetallic catalysts of Co and Mn have selected at 240 °C with a total flow rate of reactants varied between 0.5, 1.5, 2.5, 3.5, 4.5 L/g.h with H_2/CO feed ratio of 2/1 and reaction pressure of 20 atm. The results in Table 2 show that for Co-Mn/CNT, as space velocity increased, CO conversion decreased. Liu and co-workers [43] reported a similar trend for commercial Co-Mn catalysts where it was observed that conversion of CO was lowered to 30% from 82 % by an increase in space velocity from 0.46 to 1.85 L/g.h. Similar results have been previously reported [44]. Catalyst's weight was kept constant, which shows that space velocity was influenced by the total flow rate of reactants. Consequently, the variations in the selectivity of the product would be because of the residence time.

Table 2. Effect of space velocity (L/g.h) on CO conversion%, CH_4 selectivity% and C_{5+} selectivity% *.

CO Conversion%	0.5	1.5	2.5	3.5	4.5
Co/CNT	72.2	71.3	69.7	65.8	50.7
95Co5Mn/CNT	89.7	88.9	86.6	82.0	67.4
90Co10Mn/CNT	82.4	81.6	78.4	72.7	60.4

Table 2. *Cont.*

CO Conversion%	0.5	1.5	2.5	3.5	4.5
85Co15Mn/CNT	80.9	79.7	76.6	71.6	58.6
80Co20Mn/CNT	78.3	77.5	74.8	70.3	55.8
C_1 Selectivity%					
Co/CNT	12.6	13.5	13.4	16.3	19.4
95Co5Mn/CNT	5.5	6.1	6.7	9.7	11.8
90Co10Mn/CNT	7.7	8.2	8.7	11.9	13.6
85Co15Mn/CNT	8.5	9.7	9.4	12.5	15.7
80Co20Mn/CNT	9.6	10.6	10.5	14.4	16.5
C_{5+} Selectivity%					
Co/CNT	72.1	71.1	69.1	63.5	50.6
95Co5Mn/CNT	88.5	87.7	85.8	72.3	61.7
90Co10Mn/CNT	82.9	81.6	78.5	70.5	60.6
85Co15Mn/CNT	80.5	79.5	76.5	68.4	58.4
80Co20Mn/CNT	78.5	77.5	74.5	66.7	55.7

* Reaction condition: Pressure: 20 atm, Temperature: 240 °C, Ratio of H_2/CO: 2.

3. Catalyst Stability and Used Catalyst TEM

Figure 1 shows carbon monoxide conversion with time on stream (TOS) for as-received Co/CNT, 95Co5Mn/CNT, 90Co10Mn/CNT, 85Co15Mn/CNT, and 80Co20Mn/CNT catalysts samples. Catalysts showed different stability patterns within a period of 48 h.

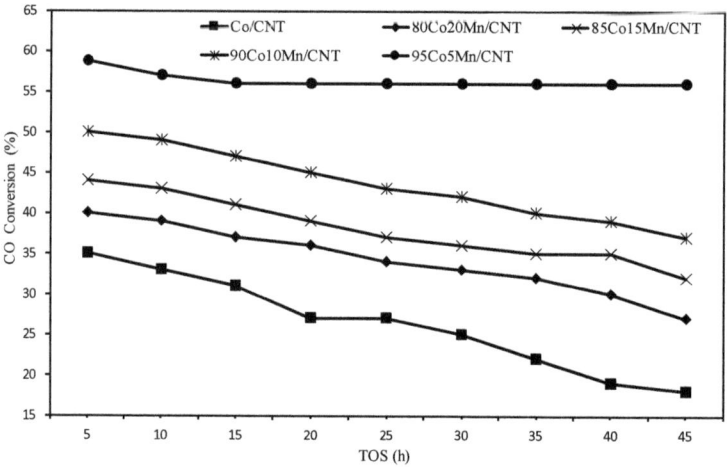

Figure 1. Time on stream (TOS) efficiency for catalysts deposited on CNT pre-treated at 240 °C, 20 atm, and H_2/CO = 2 reaction conditions.

Figure 1 shows the stability of 95Co5Mn/CNT catalyst in contrast to other formulation catalyst samples. For Co/CNT catalyst sample, CO conversion dropped drastically from 35 to 18% during 48 h. For 95Co15Mn/CNT catalyst results show a slow deactivation from 58.7% of carbon monoxide conversion to 56.9% within 48 h. The stability of catalysts may be related to Mn%, functional groups, structure, defects, and morphology of CNT substrate [33]. For catalyst prepared on 95Co5Mn/CNT pre-treated at 900 °C, CO conversion and C_{5+} selectivity were determined as 58.7%, and 59.1%, respectively. The superior efficiency of 95Co5Mn/CNT compared to other catalyst samples attributed to the higher dispersion and reducibility of cobalt-oxide nanoparticles were confined inside the CNT channels [45]. Figure 2 depicts the TEM images of the catalysts at temperatures of (a)

600 and (b) 900 °C. The particle size was found to be raised from 4.2 to 20.5 nm at 600 °C whereas 7.2 to 14.1 nm at 900 °C to indicate the treated catalyst samples [46].

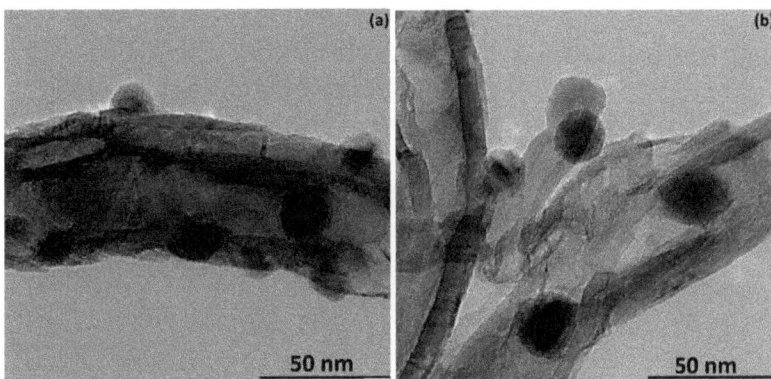

Figure 2. TEM image of the catalyst after FTS with thermal treatment (a) 600 °C (b) 900 °C.

Catalyst deactivation reveals that sintering was extremely high during FTS. The results of the TEM revealed that the active sites on the outside surface of the CNT sinter at a faster rate than the active sites inside the CNT channels. As previously stated, the majority of the cobalt active sites are enclosed within CNT.

The confinement of the reaction liquid inside the pores can improve their interaction, allowing more cobalt active sites to be exposed, thus, encouraging the growth of longer hydrocarbon chains [47]. The findings are consistent with those of other researchers [48], who found that the inside surface of CNT has an electron deficiency, which can enhance CO separation and lead to the synthesis of longer hydrocarbon chains [49]. The catalyst heated at 900 °C resulted in high nanoparticles in the channels with the decreased deactivation rate [50], according to our TEM results (Table 3), increasing the ratio of active sites enclosed inside CNT channels to active sites outside of CNT channels is thought to be a major component in improving C_{5+} selectivity and lowering CH_4 rates [51]. The difference in electron dispersion between the internal and external surfaces of the CNT, as well as the cobalt particle confinement phenomena [48]. Due to the electron deficit on the inner surface of the CNT, there is a strong interaction between cobalt oxides and the support. Since the lower sintering potential when compared to the catalyst active sites on the external surface of the CNT channel.

Table 3. Textural properties of Co/CNT catalysts at various wt% loading.

Samples	BET Surface Area (m^2/g)	Total Pore Volume (m^3/g)
Pristine CNT	138.2	1.58
CNT.A	223.2	0.88
CNT.A.T	266.4	0.54
Co/CNTs.A	198.5	0.55
95Co5Mn/CNT.A.T	217.5	0.36
90Co10Mn/CNT.A.T	220.8	0.48
85Co15Mn/CNT.A.T	223.4	0.55
80Co20Mn/CNT.A.T	225.3	0.58

4. Experimental

4.1. Functionalization of CNT Substrate

Functionalization and activation by introducing functional groups to the CNT support using nitric acid are essential before metal loading [49]. The functionalization course was aimed at improving the interaction among catalyst active sites and the CNT substrate surface. Pre-treatment with acid purifies synthesized CNT, adds oxygen-containing functional groups (–OH) on the catalyst support surface, and removes the fullerene cap from carbon nanotubes to have open CNT channels [50]. A wet chemical oxidation method is the commonly accepted process for activating and functionalizing carbon nanotubes. Around 2 g of purchased CNT (purity > 95%, CVD, length: 10–20 μm, diameter: 30–50 nm, Nanostructured and Amorphous Materials Inc.) were added to a single necked round bottom flask and add 35 vol% nitric acid (Merck) at 110 °C for 10 h [51]. After reflux, the blend was cooled down to ambient temperature, diluted with deionized water, filtered using a filter membrane of 0.2 μm pore size, and washed many times till the residue filtrate pH reached about 7 [52]. Neutralized CNT was dried overnight in an oven at 120 °C and acid-treated CNT continued with thermal treatment for 3 h at 900 °C under flowing argon gas at 20 mL min^{-1} [53].

4.2. Point of Zero Charges (PZC), Co Adsorption on CNT, and Catalyst Preparation

The common technique of impregnation was used to synthesize cobalt catalysts which produced a heterogeneous distribution of cobalt catalyst active sites on the substrate, but the Strong Electrostatic Adsorption (SEA) technique lead to greater catalyst particle dispersion and narrower distribution of catalyst size [54–57]. CNT, silica, alumina, and other metal oxides supports have hydroxyl groups on the surface. Based on the SEA technique, the point of zero charges is the pH value of the medium that the hydroxyl group on the surface remains neutral. A range of tests was carried out to find the optimum of catalyst metal active sites on CNT substrate by utilizing cationic hexamine of complexes of catalyst metal. Graph pattern shows metal adsorption increases meaningfully at pH > PZC [54,56,57]. Catalyst samples made via the SEA process [58–62] at optimal pH were found with lower particle size and higher dispersion in contrast to catalyst samples synthesized by the common impregnation technique.

Equilibrium pH at high oxide loading (EpHL) technique [58] was conducted to find the PZC of CNT substrate. The pH value was adjusted range of 2–14 by the addition of nitric acid or ammonium hydroxide to distilled water. By pouring into a conical flask, 0.5 g weighted CNT was added up with the addition of 50 mL of each solution. A rotary shaker was used to shake the mixture for 1 h before measuring the final pH value. Figure 3a performed PZC of CNT support at pH 9.5. The pH of the cobalt nitrate precursor solution was set to a range of 2–14 to study the cobalt adsorption against pH. Weighed CNT was combined into solutions and shaken for 1 h, and the final pH was then measured. The volume of 5ml of filtered cobalt solution of every sample was analyzed for the percentage of cobalt via atomic absorption spectrophotometer (AAS). Figure 3b indicates the plot of Co adsorption versus pH and showed optimal pH for Co adsorption is 14. At chosen pH = 14, the cobalt precursor was uptake by 10 wt% at Co-Mn metal loads from an excess solution on CNT substrate to avoid pH change. The sample was filtered and dried for 24 h under airflow. The dried sample was calcinated in a tubular furnace at 400 °C for 4 h under airflow to eliminate residual reactants.

Based on the SEA preparation method, the surface of functionalized CNT changed to negatively charged, the pH solution was greater than the PZC of the CNT. The PZC of the CNT support was found to be 9.5. The highest cobalt adsorbed on the CNT happened when the Co precursor solution remained at a pH of 14. Accordingly, the uptake of Co ions on the pre-treated CNT was occurred at pH 14 using a solution of $Co(NO_3)_2$. Dried catalyst samples were calcined in a tubular furnace at 400 °C for 4 h under Ar gas flow. The metal loading on CNT was performed at 10 wt% during the catalyst synthesis period.

4.3. Catalyst Characterization

Fischer-Tropsch catalyst performance is significantly affected by catalyst physicochemical properties. Therefore, it is important to characterize the catalyst's physicochemical characteristics. FTS catalyst surface physical and chemical properties, such as catalytic activity and selectivity were characterized. Figure 4 shows Transmission Electron Microscopy (TEM) of different catalyst samples conducted by a Zeiss LIBRA 200 FE TEM at 200 kV accelerating voltage. TEM results presenting catalyst samples with 5% Mn have the highest dispersion and narrow size metal particle size distribution. The rise in the Mn metal % from 5 to 20%, lead to enhance the catalyst active sites adsorption on the CNT substrate and particularly increasing from 15 and 20%, agglomeration phenomena of catalyst particles occurred and catalyst active sites agglomerate on CNT support and lead to a decline of catalyst CO conversion up to 25% and C_{5+} selectivity up to 10%.

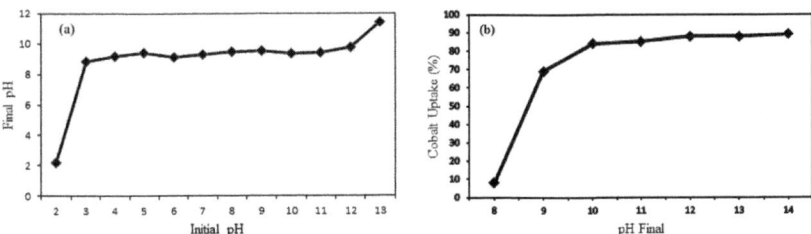

Figure 3. The finding of (a) PZC of CNT support, (b) Co-Mn adsorption versus pH survey by AAS.

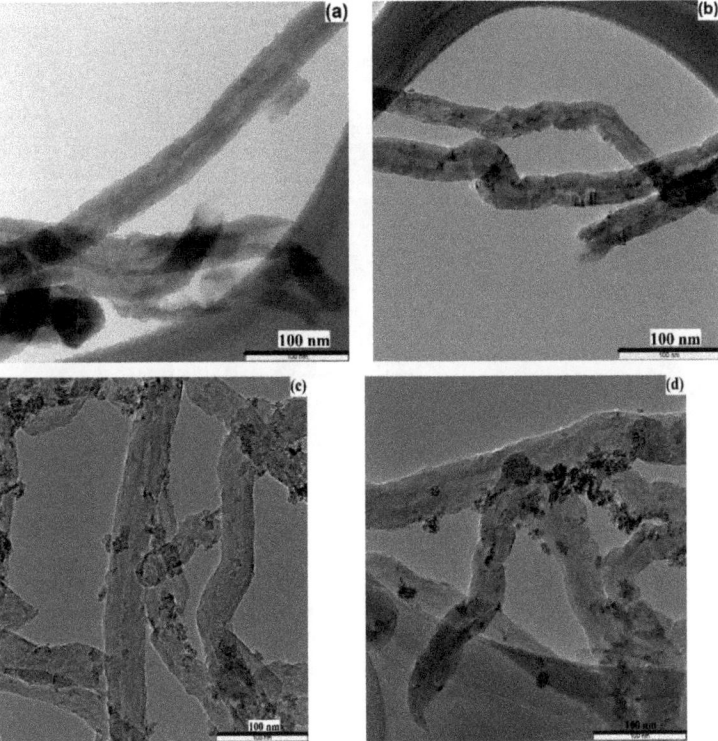

Figure 4. TEM images of (a) 95Co5Mn/CNT, (b) 90Co10Mn/CNT, (c) 85Co15Mn/CNT, (d) 80Co20Mn/CNT catalysts.

Field-Emission Scanning electron microscopy (FESEM) was used to evaluate sample morphology and elemental surface structure using a Zeiss Supra 55 VP with voltage acceleration: 5 KV, magnification: 100.00 KX, and operating distance: 4 mm. (Figure 5 FESEM images confirm and support the TEM results, demonstrating that increasing the

Mn percent from 5 to 20%, catalyst active sites agglomerate on CNT support, and lead to a decline of catalyst CO conversion and C_{5+} selectivity up to 25%, and 10%, respectively. Atomic Absorption Spectrometer (AAS) was used by Agilent Technologies GTA 120 to evaluate cobalt and manganese adsorption on the CNT substrate.

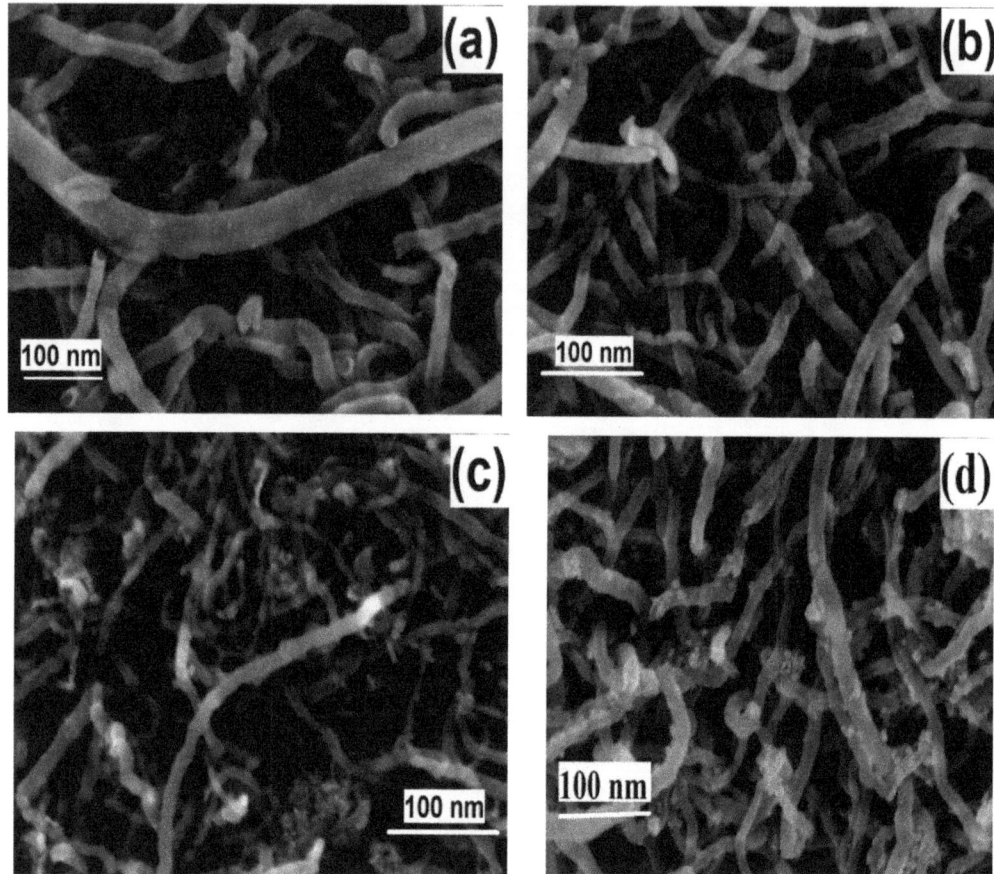

Figure 5. FESEM images of (**a**) 95Co5Mn/CNT, (**b**) 90Co10Mn/CNT, (**c**) 85Co15Mn/CNT, (**d**) 80Co20Mn/CNT catalysts.

4.4. Reactor Setup, Product Sampling, and Analysis

FTS performed in a continuous flow fixed-bed with the Micro-activity-reference reactor (Micromeritics, Norcross, GA, USA) were attached with mass flow controllers (Hi-Tec Bronkhorst, Ruurlo, The Netherlands). Carbon monoxide and H_2 were applied as reactant gases. The amount of 0.02 g catalyst was located in a stainless-steel reactor chamber (9 mm i.d. × 200 mm length) and placed in quartz tools without any dilution. Prior to the reaction, the catalysts were lowered in-situ beneath H_2 flow at 0.1 MPa and 420 °C for 10 h. The process was performed in different reaction parameters for 48 h time-on-stream (TOS). The reactor was attached to the gas chromatograph (Agilent Hewlett-Packard Series 6890, Santa Clara, CA, USA) attached with two TCD and one FID detector. Products were analyzed every 30 minutes using DB-5 column. Hydrocarbon selectivity (FID1: Methane, Ethane, Propane, Ethylene, Iso-butane, n-butane, n-pentane, n-hexane, n-heptane, TCD2: CO_2, CO, N_2, O_2, and TCD3: H_2) were calculated after reaction completion (10 h). The results were collected at a steady-state setup using a carbon balance of 99–102%. The reproducibility was checked by doing all reactions two times under the same reaction and catalyst terms. STD of experimental results were ±5.0%. The CO, methane (CH_4), and C_{5+} selectivity conversion percentages were analyzed using Equations (1)–(3) respectively [63]:

$$\text{CO conversion (\%)} = \frac{CO_{in} - CO_{out}}{CO_{in}} \times 100 \quad (1)$$

$$CH_4 \text{selectivity}(\%) = \frac{\text{Mole of } CH_4}{\text{Total moles of hydrocarbons}} \times 100 \quad (2)$$

$$C_{5+} \text{selectivity}(\%) = \frac{\text{Moles of } C_{5+}}{\text{Total moles of hydrocarbons}} \times 100 \quad (3)$$

The FTS level shown in Equation (1) and the reaction rate of the water gas change (Equation (5)) is equal to the carbon dioxide formation rate ($RFCO_2$) and can be described by [60,64,65]:

$$RFTS(\text{g HC/gcat/h}) = \text{g hydrocarbons produced/gcat} * h^{-1} \quad (4)$$

$$RWGS(gCO_2/gcat/h) = RFCO_2 = gCO_2 \text{produced/gcat} * h^{-1} \quad (5)$$

It is a significant step to ease and initiate calcined catalysts before reaction. Catalysts were reduced to 12.5 h at 420 °C under 1.8 L/g.h flow of H_2. After catalyst in-situ activation, the temperature was reduced to the required temperature of the Fischer-Tropsch reaction, and the reactor tube flushed for 10 min with helium gas. Fischer-Tropsch reaction was carried out at 2/1 H_2/CO (v/v) ratio and 20 atm pressure. Additional experiments were performed to explore the impacts of space velocity (0.5, 1.5, 2.5, 3.5, and 4.5 L/g.h), temperature (200, 220, 240, 260, 280 °C), and catalyst stability by conducting different catalysts. Figure 6 shows the schematic diagram of the micro activity-reference reactor (Micromeritics).

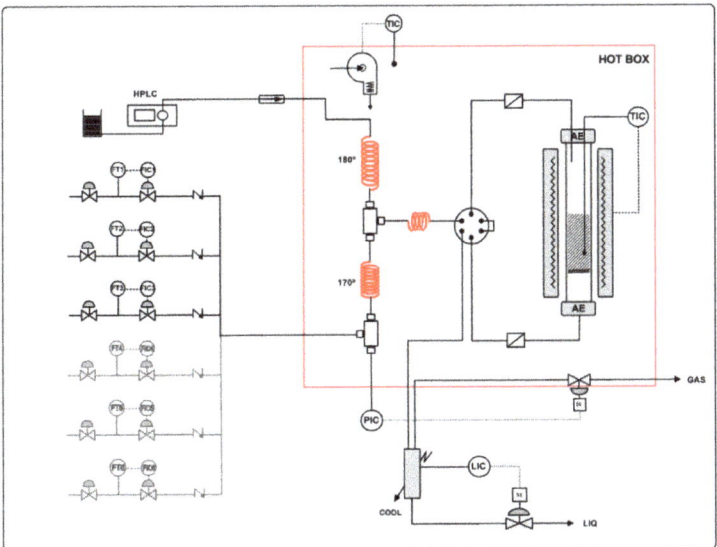

Figure 6. Schematic diagram of the micro activity-reference reactor (Micromeritics).

The textural properties of the BET surface area and total pore volume are shown in Table 3. According to the results, the total area (BET) increased from 217.5 to 225.3 m^2/g with a 5 to 20% increase in Mn load. Higher nanoparticle dispersion may be causing an increase in surface area. From the findings, the overall pore volume increased from 0.36 to

0.58 (m^3/g) as the Mn percent of catalysts raised from 5% to 20%. The addition of cobalt and manganese to CNT support increased overall pore volumes in both BET surface areas.

The XRD patterns of CNT support and catalyst samples are shown in Figure 7. The peaks at 26° and 44° correspond to carbon nanotubes [66]. Diffraction peaks of Co$_3$O$_4$ spinel appear in the monometallic Co/CNT sample in the ranges of 32° and 37.1° [66]. At two values of 32.5° and 44°, the A.T sample reveals a hematite pattern (Mn$_2$O$_3$) [67]. Co$_3$O$_4$ spinel diffraction peaks were observed at 32.5 and 37.1° in bimetallic 95Co5Mn/CNT catalyst XRD patterns. Due to the low manganese content in the catalyst, Mn$_2$O$_3$ was only linked with a weak peak at 44°.

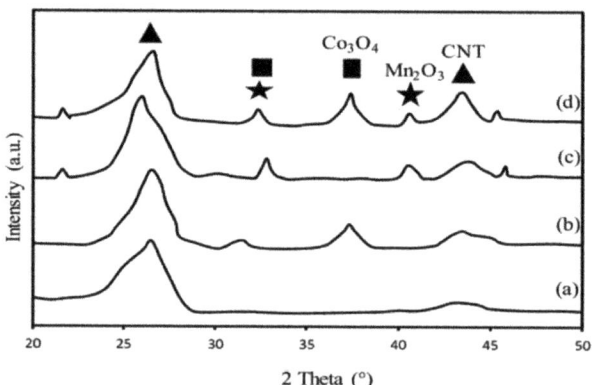

Figure 7. XRD patterns of (**a**) CNT (**b**) Co/CNT. A.T (**c**) Mn/CNT.A.T (**d**) 95Co5Mn/CNT.A.T catalysts.

Figure 8 shows XRD patterns of calcined catalysts with various Mn metal loading. The peak at 25° and 43° shows unique Co$_3$O$_4$ crystal planes [68]. For Co$_3$O$_4$, the most significant peak was seen at 36.8°. Lower intensity peaks were detected at 32.5° and 44°, showing Mn oxide diffraction peaks, due to the limited number of Mn promoters in the catalyst XRD pattern. The average particle size of the catalysts was estimated as 6-8 nm using XRD and TEM images [69–75]. Table 3 shows that as manganese load increases from 5% to 20%, the average particle size of Co$_3$O$_4$ drops from 7.5 to 6.5 nm, which is similar to the results of the TEM study (Figure 4). The agglomeration of cobalt particles raises the average particle size. The average particle size drops somewhat when Mn is added to the Co catalyst, as seen in Table 3.

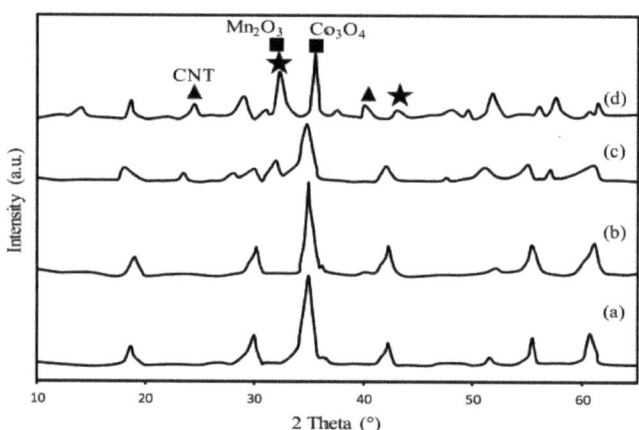

Figure 8. XRD profile of Co–Mn/CNT with Mn content (**a**) 5, (**b**) 10, (**c**) 15 and (**d**) 20%.

5. Conclusions

The Cobalt-Manganese bimetallic catalyst was synthesized by acid and thermal-treated CNT substrate using the SEA process. The efficiency of various percentage formulations of the Co-Mn catalyst supported on CNT was verified by the FTS reaction. High stability was proven by 95Co5Mn/CNT catalyst for more than 45 h. It was concluded that reaction variables created a high impact on catalytic activities and product selectivities during the FTS process. An increase in reaction temperature up to 280 °C enhanced carbon monoxide percent conversion up to 88.2% and reduced C_{5+} selectivity up to 55.2%, while increased WGS rate up to 0.8. An increase in space velocity up to 4.5 (L/g.h) decreases CO percent conversion to 55.8% and decreases C_{5+} selectivity to 55.7%. However, after optimization analysis, 95Co5Mn/CNT catalyst formulation showed a high efficiency at 240 °C with a space velocity of 2.5(L/g.h). In the mentioned condition, carbon monoxide conversion and C_{5+} selectivity were 86.6% and 85.8% respectively.

Author Contributions: Data curation, N.A.M.Z., S.M., A.A.B., N.A.H. and Z.Z.C.; Formal analysis, N.A.M.Z., S.M., A.K., Z.Z.C. and S.S.; Funding acquisition, S.F.A.; Investigation, A.K., Y.A.W. and S.S.; Validation, N.A.H. and Y.A.W.; Visualization, A.K., A.A.B., N.A.H. and Y.A.W.; Writing—original draft, O.A.; Writing—review & editing, S.S. All authors have read and agreed to the published version of the manuscript.

Funding: This research was funded by the Ministry of Education, Malaysia under the Fundamental Research Grant Scheme (FRGS/1/2012/SG01/UTP/02/01).

Data Availability Statement: All data generated or analysed during this study are included in this published article.

Acknowledgments: The authors acknowledge the Universiti Teknologi PETRONAS and University of Malaya.

Conflicts of Interest: The authors declare no conflict of interest with this work.

References

1. Iglesia, E. Design, synthesis, and use of cobalt-based Fischer-Tropsch synthesis catalysts. *Appl. Catal. A Gen.* **1997**, *161*, 59–78. [CrossRef]
2. Cho, K.M.; Park, S.; Gil Seo, J.; Youn, M.H.; Nam, I.; Baeck, S.-H.; Chung, J.S.; Jun, K.-W.; Song, I.K. Effect of calcination temperature of alumina supports on the wax hydrocracking performance of Pd-loaded mesoporous alumina xerogel catalysts for the production of middle distillate. *Chem. Eng. J.* **2009**, *146*, 307–314. [CrossRef]
3. Tavasoli, A.; Mortazavi, Y.; Khodadadi, A.A.; Mousavian, M.A.; Sadagiani, K.; Karimi, A. Effects of differen loadings of Ru and Re on physico-chemical properties and performance of 15% Co/Al2O3 FTS catalysts. *Iran. J. Chem. Eng.* **2005**, *24*, 9–17.
4. Pivehzhani, O.A.; Kordijazi, A.; Sagadevan, S.; Moosavi, S.; Babadi, A.A.; Wahab, Y.A.; Hamizi, N.A.; Chowdhury, Z.Z. Syngas to Green Fuel Conversion: Nanocatalysis Approach. In *Advanced Heterogeneous Catalysts Volume 1 Applications at the Nano-Scale*; American Chemical Society: Washington, DC, USA, 2020; pp. 545–579. [CrossRef]
5. Khodakov, A.Y.; Chu, W.; Fongarland, P. Advances in the Development of Novel Cobalt Fischer−Tropsch Catalysts for Synthesis of Long-Chain Hydrocarbons and Clean Fuels. *Chem. Rev.* **2007**, *107*, 1692–1744. [CrossRef]
6. Berge, P.; Loosdrecht, J.; Barradas, S.; Kraan, A. Oxidation of cobalt based Fischer–Tropsch catalysts as a deactivation mechanism. *Catal. Today* **2000**, *58*, 321–334. [CrossRef]
7. Jacobs, G.; Das, T.K.; Zhang, Y.; Li, J.; Racoillet, G.; Davis, B.H. Fischer–Tropsch synthesis: Support, loading, and promoter effects on the reducibility of cobalt catalysts. *Appl. Catal. A Gen.* **2002**, *233*, 263–281. [CrossRef]
8. Breejen, J.P.D.; Radstake, P.B.; Bezemer, G.L.; Bitter, J.H.; Frøseth, V.; Holmen, A.; De Jong, K.P. On the Origin of the Cobalt Particle Size Effects in Fischer−Tropsch Catalysis. *J. Am. Chem. Soc.* **2009**, *131*, 7197–7203. [CrossRef] [PubMed]
9. Sun, X.; Li, Y. Ga_2O_3 and GaN Semiconductor Hollow Spheres. *Angew. Chem. Int. Ed.* **2004**, *43*, 3827–3831. [CrossRef] [PubMed]
10. Xing, C.; Yang, G.; Wang, D.; Zeng, C.; Jin, Y.; Yang, R.; Suehiro, Y.; Tsubaki, N. Controllable encapsulation of cobalt clusters inside carbon nanotubes as effective catalysts for Fischer-Tropsch synthesis. *Catal. Today* **2013**, *215*, 24–28. [CrossRef]
11. Mirzaei, A.A.; Shirzadi, B.; Atashi, H.; Mansouri, M. Modeling and operating conditions optimization of Fischer–Tropsch synthesis in a fixed-bed reactor. *J. Ind. Eng. Chem.* **2012**, *18*, 1515–1521. [CrossRef]
12. Feyzi, M.; Irandoust, M.; Mirzaei, A.A. Effects of promoters and calcination conditions on the catalytic performance of iron–manganese catalysts for Fischer–Tropsch synthesis. *Fuel Process. Technol.* **2011**, *92*, 1136–1143. [CrossRef]
13. A Mirzaei, A.; Babaei, A.B.; Galavy, M.; Youssefi, A. A silica supported Fe–Co bimetallic catalyst prepared by the sol/gel technique: Operating conditions, catalytic properties and characterization. *Fuel Process. Technol.* **2010**, *91*, 335–347. [CrossRef]

14. Atashi, H.; Siami, F.; Mirzaei, A.; Sarkari, M. Kinetic study of Fischer–Tropsch process on titania-supported cobalt–manganese catalyst. *J. Ind. Eng. Chem.* **2010**, *16*, 952–961. [CrossRef]
15. Júnior, L.C.P.F.; Miguel, S.D.; Fierro, J.L.G.; Rangel, M.D.C. Evaluation of Pd/La2O3 catalysts for dry reforming of methane. *Stud. Surf. Sci. Catal.* **2007**, *167*, 499–504.
16. Pendyala, V.R.R.; Jacobs, G.; Mohandas, J.C.; Luo, M.; Hamdeh, H.H.; Ji, Y.; Ribeiro, M.C.; Davis, B.H. Fischer–Tropsch Synthesis: Effect of Water Over Iron-Based Catalysts. *Catal. Lett.* **2010**, *140*, 98–105. [CrossRef]
17. Gheitanchi, R.; Khodadadi, A.A.; Taghizadeh, M.; Mortazavi, A.Y. Effects of ceria addition and pre-calcination temperature on performance of cobalt catalysts for Fischer-Tropsch synthesis. *React. Kinet. Catal. Lett.* **2006**, *88*, 225–232. [CrossRef]
18. Concepción, P.; López, C.; Martínez, A.; Puntes, V. Characterization and catalytic properties of cobalt supported on delaminated ITQ-6 and ITQ-2 zeolites for the Fischer–Tropsch synthesis reaction. *J. Catal.* **2004**, *228*, 321–332. [CrossRef]
19. Arsalanfar, M.; Mirzaei, A.; Bozorgzadeh, H.; Atashi, H. Effect of process conditions on the surface reaction rates and catalytic performance of MgO supported Fe–Co–Mn catalyst for CO hydrogenation. *J. Ind. Eng. Chem.* **2012**, *18*, 2092–2102. [CrossRef]
20. Liu, Y.; Teng, B.-T.; Guo, X.-H.; Chang, J.; Tian, L.; Hao, X.; Wang, Y.; Xiang, H.-W.; Xu, Y.-Y.; Li, Y.-W. Effect of reaction conditions on the catalytic performance of Fe-Mn catalyst for Fischer-Tropsch synthesis. *J. Mol. Catal. A Chem.* **2007**, *272*, 182–190. [CrossRef]
21. Kim, S.-M.; Bae, J.W.; Lee, Y.-J.; Jun, K.-W. Effect of CO2 in the feed stream on the deactivation of Co/γ-Al2O3 Fischer–Tropsch catalyst. *Catal. Commun.* **2008**, *9*, 2269–2273. [CrossRef]
22. Li, H.; Wang, S.; Ling, F.; Li, J. Studies on MCM-48 supported cobalt catalyst for Fischer–Tropsch synthesis. *J. Mol. Catal. A Chem.* **2006**, *244*, 33–40. [CrossRef]
23. Li, T.; Yang, Y.; Tao, Z.; Zhang, C.; Xiang, H.; Li, Y. Study on an iron–manganese Fischer–Tropsch synthesis catalyst prepared from ferrous sulfate. *Fuel Process. Technol.* **2009**, *90*, 1247–1251. [CrossRef]
24. Morales, F.; De Smit, E.; De Groot, F.M.F.; Visser, T.; Weckhuysen, B.M. Effects of manganese oxide promoter on the CO and H2 adsorption properties of titania-supported cobalt Fischer–Tropsch catalysts. *J. Catal.* **2007**, *246*, 91–99. [CrossRef]
25. Zhang, C.-H.; Yang, Y.; Teng, B.-T.; Li, T.-Z.; Zheng, H.-Y.; Xiang, H.-W.; Li, Y.-W. Study of an iron-manganese Fischer–Tropsch synthesis catalyst promoted with copper. *J. Catal.* **2006**, *237*, 405–415. [CrossRef]
26. Akbarzadeh, O.; Zabidi, N.A.M.; Wahab, Y.A.; Hamizi, N.A.; Chowdhury, Z.Z.; Merican, Z.M.A.; Ab Rahman, M.; Akhter, S.; Rasouli, E.; Johan, M.R. Effect of Cobalt Catalyst Confinement in Carbon Nanotubes Support on Fischer-Tropsch Synthesis Performance. *Symmetry* **2018**, *10*, 572. [CrossRef]
27. Akbarzadeh, O.; Zabidi, N.A.M.; Hamizi, N.A.; Wahab, Y.A.; Merican, Z.M.A.; Yehya, W.A.; Akhter, S.; Shalauddin, M.; Rasouli, E.; Johan, M.R. Effect of pH, Acid and Thermal Treatment Conditions on Co/CNT Catalyst Performance in Fischer–Tropsch Reaction. *Symmetry* **2019**, *11*, 50. [CrossRef]
28. Akbarzadeh, O.; Zabidi, N.A.M.; Wahab, Y.A.; Hamizi, N.A.; Chowdhury, Z.Z.; Merican, Z.M.A.; Ab Rahman, M.; Akhter, S.; Shalauddin, M.; Johan, M.R. Effects of Cobalt Loading, Particle Size, and Calcination Condition on Co/CNT Catalyst Performance in Fischer–Tropsch Reactions. *Symmetry* **2018**, *11*, 7. [CrossRef]
29. Akbarzadeh, O.; Zabidi, N.A.M.; Merican, Z.M.A.; Sagadevan, S.; Kordijazi, A.; Das, S.; Babadi, A.A.; Ab Rahman, M.; Hamizi, N.A.; Wahab, Y.A.; et al. Effect of Manganese on Co-Mn/CNT Bimetallic Catalyst Performance in Fischer-Tropsch Reaction. *Symmetry* **2019**, *11*, 1328. [CrossRef]
30. Akbarzadeh, O.; Zabidi, N.A.M.; Wang, G.; Kordijazi, A.; SadAbadi, H.; Moosavi, S.; Babadi, A.A.; Hamizi, N.A.; Wahab, Y.A.; Ab Rahman, M.; et al. Effect of Pressure, H2/CO Ratio and Reduction Conditions on Co–Mn/CNT Bimetallic Catalyst Performance in Fischer-Tropsch Reaction. *Symmetry* **2020**, *12*, 698. [CrossRef]
31. Tavasoli, A.; Sadaghiani, K.; Khodadadi, A.A.; Mortazavi, Y. Raising distillate selectivity and catalyst life time in Fischer-Tropsch synthesis by using a novel dual-bed reactor. *Iran. J. Chem. Eng.* **2007**, *26*, 109–117.
32. Gavrilović, L.; Jørgensen, E.A.; Pandey, U.; Putta, K.R.; Rout, K.R.; Rytter, E.; Hillestad, M.; Blekkan, E.A. Fischer-Tropsch synthesis over an alumina-supported cobalt catalyst in a fixed bed reactor–Effect of process parameters. *Catal. Today* **2021**, *369*, 150–157. [CrossRef]
33. Graham, U.M.; Dozier, A.; Khatri, R.A.; Bahome, M.C.; Jewell, L.L.; Mhlanga, S.D.; Coville, N.J.; Davis, B.H. Carbon Nanotube Docking Stations: A New Concept in Catalysis. *Catal. Lett.* **2009**, *129*, 39–45. [CrossRef]
34. Abbaslou, R.M.M.; Tavassoli, A.; Soltan, J.; Dalai, A.K. Iron catalysts supported on carbon nanotubes for Fischer–Tropsch synthesis: Effect of catalytic site position. *Appl. Catal. A Gen.* **2009**, *367*, 47–52. [CrossRef]
35. Bahome, M.C.; Jewell, L.L.; Hildebrandt, D.; Glasser, D.; Coville, N.J. Fischer–Tropsch synthesis over iron catalysts sup-ported on carbon nanotubes. *Appl. Catal. A. Gen.* **2005**, *287*, 60–67. [CrossRef]
36. Vannice, M. The catalytic synthesis of hydrocarbons from H2/CO mixtures over the group VIII metals II. The kinetics of the methanation reaction over supported metals. *J. Catal.* **1975**, *37*, 462–473. [CrossRef]
37. O'Shea, V.A.D.L.P.; Alvarez-Galvan, M.C.; Campos-Martin, J.M.; Fierro, J.L.G. Fischer–Tropsch synthesis on mono- and bimetallic Co and Fe catalysts in fixed-bed and slurry reactors. *Appl. Catal. A Gen.* **2007**, *326*, 65–73. [CrossRef]
38. Komaya, T.; Bell, A. Estimates of rate coefficients for elementary processes occurring during Fischer-Tropsch synthesis over RuTiO2. *J. Catal.* **1994**, *146*, 237–248. [CrossRef]
39. Vogel, B.; Feck, T.; Grooß, J.-U. Impact of stratospheric water vapor enhancements caused by CH4 and H2O increase on polar ozone loss. *J. Geophys. Res. Space Phys.* **2011**, *116*. [CrossRef]
40. Wojciechowski, B.W. The Kinetics of the Fischer-Tropsch Synthesis. *Catal. Rev.* **1988**, *30*, 629–702. [CrossRef]

41. Tian, Z.; Wang, C.; Si, Z.; Ma, L.; Chen, L.; Liu, Q.; Zhang, Q. Huang, H. Fischer-Tropsch synthesis to light olefins over iron-based catalysts supported on KMnO$_4$ modified activated carbon by a facile method. *Appl. Catal. A Gen.* **2017**, *541*, 50–59. [CrossRef]
42. Bukur, D.B.; Sivaraj, C. Supported iron catalysts for slurry phase Fischer–Tropsch synthesis. *Appl. Catal. A Gen.* **2002**, *231*, 201–214. [CrossRef]
43. Yu, K.; Gu, Z.; Ji, R.; Lou, L.-L.; Ding, F.; Zhang, C.; Liu, S. Effect of pore size on the performance of mesoporous material supported chiral Mn(III) salen complex for the epoxidation of unfunctionalized olefins. *J. Catal.* **2007**, *252*, 312–320. [CrossRef]
44. Raje, A.P.; O'Brien, R.J.; Davis, B.H. Effect of potassium promotion on iron-based catalysts for Fischer–Tropsch synthe-sis. *J. Catal.* **1998**, *180*, 36–43. [CrossRef]
45. Tavasoli, A.; Trépanier, M.; Dalai, A.K.; Abatzoglou, N. Effects of confinement in carbon nanotubes on the activity, selectivity, and lifetime of Fischer–Tropsch Co/carbon nanotube catalysts. *J. Chem. Eng. Data* **2021**, *55*, 2757–2763. [CrossRef]
46. Bezemer, G.L.; Bitter, J.H.; Kuipers, H.P.C.E.; Oosterbeek, H.; Holewijn, J.E.; Xu, X.; Kapteijn, F.; Van Dillen, A.A.J.; De Jong, K.P. Cobalt Particle Size Effects in the Fischer–Tropsch Reaction Studied with Carbon Nanofiber Supported Catalysts. *J. Am. Chem. Soc.* **2006**, *128*, 3956–3964. [CrossRef]
47. Nguyen, T.T.; Serp, P. Confinement of Metal Nanoparticles in Carbon Nanotubes. *ChemCatChem* **2013**, *5*, 3595–3603. [CrossRef]
48. Rehman, W.U.; Merican, Z.M.A.; Bhat, A.H.; Hoe, B.G.; Sulaimon, A.A.; Akbarzadeh, O.; Khan, M.S.; Mukhtar, A.; Saqib, S.; Hameed, A.; et al. Synthesis, characterization, stability and thermal conductivity of multi-walled carbon nanotubes (MWCNTs) and eco-friendly jatropha seed oil based nanofluid: An experimental investigation and modeling approach. *J. Mol. Liq.* **2019**, *293*, 111534. [CrossRef]
49. Wang, D.; Yang, G.; Ma, Q.; Wu, M.; Tan, Y.; Yoneyama, Y.; Tsubaki, N. Confinement Effect of Carbon Nanotubes: Copper Nanoparticles Filled Carbon Nanotubes for Hydrogenation of Methyl Acetate. *ACS Catal.* **2012**, *2*, 1958–1966. [CrossRef]
50. Pan, X.; Bao, X. The Effects of Confinement inside Carbon Nanotubes on Catalysis. *Acc. Chem. Res.* **2011**, *44*, 553–562. [CrossRef]
51. Xiao, J.; Pan, X.; Guo, S.; Ren, P.; Bao, X. Toward Fundamentals of Confined Catalysis in Carbon Nanotubes. *J. Am. Chem. Soc.* **2015**, *137*, 477–482. [CrossRef]
52. Akbarzadeh, O.; Zabidi, N.A.M.; Abdullah, B.; Subbarao, D.; Bawadi, A. Synthesis and Characterization of Co/CNTs Catalysts Prepared by Strong Electrostatic Adsorption (SEA) Method. *Appl. Mech. Mater.* **2014**, *625*, 328–332. [CrossRef]
53. Akbarzadeh, O.; Zabidi, N.A.M.; Abdullah, B.; Subbarao, D. Dispersion of Co/CNTs via strong electrostatic adsorption method: Thermal treatment effect. *AIP Conf. Proc.* **2015**, *1669*, 020052.
54. Akbarzadeh, O.; Zabidi, N.A.M.; Abdullah, B.; Subbarao, D. Synthesis of Co/CNTs Catalyst via Strong Electrostatic Adsorption: Effect of Calcination Condition. *Adv. Mater. Res.* **2015**, *1109*, 1–5. [CrossRef]
55. Akbarzadeh, O.; Zabidi, N.A.M.; Abdullah, B.; Subbarao, D.; Bawadi, A. Synthesis of Co/CNTs via Strong Electrostatic Adsorption: Effect of Metal Loading. *Adv. Mater. Res.* **2014**, *1043*, 101–104. [CrossRef]
56. Akbarzadeh, O.; Zabidi, N.A.M.; Abdullah, B.; Subbarao, D.; Bawadi, A. Influence of Acid and Thermal Treatments on Properties of Carbon Nanotubes. *Adv. Mater. Res.* **2013**, *832*, 394–398. [CrossRef]
57. Elbashir, N.O.; Roberts, C.B. Enhanced Incorporation of α-Olefins in the Fischer–Tropsch Synthesis Chain-Growth Process over an Alumina-Supported Cobalt Catalyst in Near-Critical and Supercritical Hexane Media. *Ind. Eng. Chem. Res.* **2005**, *44*, 505–521. [CrossRef]
58. Maitlis, P.M.; Zanotti, V. The role of electrophilic species in the Fischer–Tropsch reaction. *Chem. Commun.* **2009**, 1619–1634. [CrossRef] [PubMed]
59. Dry, M. Chemical concepts used for engineering purposes. *Adv. Pharmacol.* **2004**, *152*, 196–257. [CrossRef]
60. Zhou, W.-G.; Liu, J.-Y.; Wu, X.; Chen, J.-F.; Zhang, Y. An effective Co/MnO$_x$ catalyst for forming light olefins via Fischer–Tropsch synthesis. *Catal. Commun.* **2015**, *60*, 76–81. [CrossRef]
61. Park, J.; Regalbuto, J.R. A Simple, Accurate Determination of Oxide PZC and the Strong Buffering Effect of Oxide Surfaces at Incipient Wetness. *J. Colloid Interface Sci.* **1995**, *175*, 239–252. [CrossRef]
62. Bartholomew, C.H.; Rahmati, M.; Reynolds, M.A. Optimizing preparations of Co Fischer-Tropsch catalysts for stability against sintering. *Appl. Catal. A Gen.* **2020**, *602*, 117609. [CrossRef]
63. Jiang, F.; Wang, S.; Zheng, J.; Liu, B.; Xu, Y.; Liu, X. Fischer-Tropsch synthesis to lower α-olefins over cobalt-based catalysts: Dependence of the promotional effect of promoter on supports. *Catal. Today* **2021**, *369*, 158–166. [CrossRef]
64. Macheli, L.; Carleschi, E.; Doyle, B.P.; Leteba, G.; Steen, E. Tuning catalytic performance in Fischer-Tropsch synthesis by metal-support interactions. *J. Catal.* **2021**, *395*, 70–79. [CrossRef]
65. Bitter, J.H.; De Jong, K.P. ChemInform Abstract: Preparation of Carbon-Supported Metal Catalysts. *ChemInform* **2009**, *40*, 157–176. [CrossRef]
66. Yahya, N. *Carbon and Oxide Nanostructures*; Springer Science and Business Media LLC: Berlin/Heidelberg, Germany, 2011.
67. Zhang, D.; Fu, H.; Shi, L.; Fang, C.; Li, Q. Carbon nanotube assisted synthesis of CeO2 nanotubes. *J. Solid State Chem.* **2007**, *180*, 654–660. [CrossRef]
68. Hazemann, P.; Decottignies, D.; Maury, S.; Humbert, S.; Meunier, F.C.; Schuurman, Y. Selectivity loss in Fischer-Tropsch synthesis: The effect of cobalt carbide formation. *J. Catal.* **2021**, *397*, 1–12. [CrossRef]
69. Zolfaghari, Z.S.; Tavasoli, A.; Tabyar, S.; Pour, A.N. Enhancement of bimetallic Fe-Mn/CNTs nano catalyst activity and product selectivity using microemulsion technique. *J. Energy Chem.* **2014**, *23*, 57–65. [CrossRef]

70. Jacobs, G.; Patterson, P.M.; Das, T.K.; Luo, M.; Davis, B.H. Fischer–Tropsch synthesis: Effect of water on Co/Al2O3 catalysts and XAFS characterization of reoxidation phenomena. *Appl. Catal. A Gen.* **2004**, *270*, 65–76. [CrossRef]
71. Li, Z.; Si, M.; Xin, L.; Liu, R.; Liu, R.; Lü, J. Cobalt catalysts for Fischer–Tropsch synthesis: The effect of support, precipitant and pH value. *Chin. J. Chem. Eng.* **2018**, *26*, 747–752. [CrossRef]
72. Tavasoli, A.; Sadagiani, K.; Khorashe, F.; Seifkordi, A.; Rohani, A.; Nakhaeipour, A. Cobalt supported on carbon nanotubes—A promising novel Fischer–Tropsch synthesis catalyst. *Fuel Process. Technol.* **2008**, *89*, 491–498. [CrossRef]
73. Jothimurugesan, K.; Goodwin, J.G.; Gangwal, S.K.; Spivey, J.J. Development of Fe Fischer–Tropsch catalysts for slurry bubble column reactors. *Catal. Today* **2000**, *58*, 335–344. [CrossRef]
74. Bezemer, G.; van Laak, A.; van Dillen, A.; de Jong, K. Cobalt supported on carbon nanofibers- a promising novel Fischer-Tropsch catalyst. *Adv. Pharmacol.* **2004**, *147*, 259–264. [CrossRef]
75. Bechara, R.; Balloy, D.; Vanhove, D. Catalytic properties of Co/Al$_2$O$_3$ system for hydrocarbon synthesis. *Appl. Catal. A Gen.* **2001**, *207*, 343–353. [CrossRef]

Review

Recent Progress in Low-Cost Catalysts for Pyrolysis of Plastic Waste to Fuels

Ganjar Fadillah [1], Is Fatimah [1,*], Imam Sahroni [1], Muhammad Miqdam Musawwa [1], Teuku Meurah Indra Mahlia [2] and Oki Muraza [3]

[1] Department of Chemistry, Faculty of Mathematics and Natural Sciences, Universitas Islam Indonesia, Jl. Kaliurang Km 14, Sleman, Yogyakarta 55584, Indonesia; ganjar.fadillah@uii.ac.id (G.F.); sahroni@uii.ac.id (I.S.); musawwa.miqdam@uii.ac.id (M.M.M.)
[2] School of Information, Systems and Modelling, Faculty of Engineering and Information Technology, University of Technology Sydney, Sydney, NSW 2007, Australia; tmindra.mahlia@uts.edu.au
[3] Research & Technology Innovation, Pertamina, Sopo Del Building, 51st Fl. Jl. Mega Kuningan Barat III, Jakarta Pusat 12950, Indonesia; omuraza@kfupm.edu.sa
* Correspondence: isfatimah@uii.ac.id

Abstract: The catalytic and thermal decomposition of plastic waste to fuels over low-cost catalysts like zeolite, clay, and bimetallic material is highlighted. In this paper, several relevant studies are examined, specifically the effects of each type of catalyst used on the characteristics and product distribution of the produced products. The type of catalyst plays an important role in the decomposition of plastic waste and the characteristics of the oil yields and quality. In addition, the quality and yield of the oil products depend on several factors such as (i) the operating temperature, (ii) the ratio of plastic waste and catalyst, and (iii) the type of reactor. The development of low-cost catalysts is revisited for designing better and effective materials for plastic solid waste (PSW) conversion to oil/bio-oil products.

Keywords: plastic; waste-to-fuels; low-cost catalysts; solid acid catalysts

1. Introduction

Pollution of plastics in the environment has become a serious issue in recent years, producing more than 300 million tons per year [1]. One of the problems is the microplastics issue which is related to the non-degradable properties of the plastic polymer. Many efforts have been reported to reduce and overcome the presence of plastics and microplastic wastes, and one of the promising alternatives is the conversion of plastic waste into renewable energy [2,3]. More than just overcoming the environmental problems, with a designed plastic waste management system, the conversion of plastic waste into renewable energy will also contribute to the vital issue of energy conservation [4,5]. As hydrocarbon is the backbone of plastics, catalytic processes of plastic structure within the pyrolysis technique can produce hydrogen and liquid fuel. The conversion mechanisms include cracking, hydrocracking, and hydrogenation can restore the energy contained in plastic, as a sustainable process for sustainable energy [6–8]. For those mechanisms, catalysts play important roles in determining the effectiveness and efficient conversion process. Although in the perspective of kinetics and thermodynamic, the catalyst's role is to accelerate the reaction, it in fact determines the dominant product of the reaction which is further called selectivity, the optimum condition for the reaction, and also the energy required for the process [9]. The use of the solid catalyst for the catalytic pyrolysis is established and favorable, as it is easily handled, efficient in mass transfer conditions to obtain a high yield. Pyrolysis can be performed by thermal or catalytic processes. However, compared to the thermal decomposition method, the catalytic method has some advantages that could be carried out at a lower temperature and reduce the solid residues such as carbonized char and volatile fraction, short time process, high product selectivity, high octane number, etc. [10].

Moreover, in pyrolysis by cracking catalytic process, the used catalyst can be easy to reuse and reproduce, which can be classified into sustainable approaches [11]. The main factor for the mechanism lays in the domination of radical propagation steps in thermal pyrolysis instead of cationic propagation occurs in the catalytic process, leading to an uncontrolled decomposition. These advantages are feasible from the economic point of view.

Moreover, the combination of plastic or plastic waste with biomass as feed for pyrolysis, also called co-pyrolysis, has also gained much attention. The combination in the catalytic pyrolysis mechanism has been reported to provide synergistic effects such as minimizing coke formation and increasing yield and selectivity towards gasoline and aromatic fraction [12,13]. Wang et al. (2021) reported that the addition of low-density polyethylene (LDPE) to biomass waste could improve the selectivity and aromatics product, including xylene, benzene, and toluene [14]. However, the selectivity conversion also depends on the different types of catalysts used. Dai et al. (2021) studied the pyrolysis process using different types and tandem catalysts [15]. Their study revealed that tandem catalysts could improve the selectivity of naphtha. Many solid catalysts are reported with specific results, and some of them are the catalysts based on zeolite, bimetallic and clay with the comparable popularity expressed in Figure 1. Among these kinds of catalysts, zeolite-based catalysts are the most popular for both pyrolysis and co-pyrolysis for the combined plastic waste/biomass. An intensive catalytic mechanism is provided by reactant migration and surface reaction over the microporous structure of zeolite. Even though a similar mechanism also occurs on clay, clay-based catalyst receives less attention for the processes [16]. By considerations of many possible modifications towards clay, clay-based catalysts are good candidates as low-cost catalysts for plastic waste pyrolysis. With many modifications for zeolite and clay framework with metals or metal oxides zeolite, the catalytic activity enhancements were attempted by increasing the effectiveness of the reaction pathways such as hydrogenation, hydro-deoxygenation, cracking, etc. Catalysts' thermal and chemical stability refers to the use of high-temperature conditions and very complex reactions involved within the mechanisms as important characters, besides the solid acidity and capability to provide efficient mass transport in the catalytic steps. Referring to their abundant sources in nature, both clay and zeolite materials are found as cheap minerals [17–19].

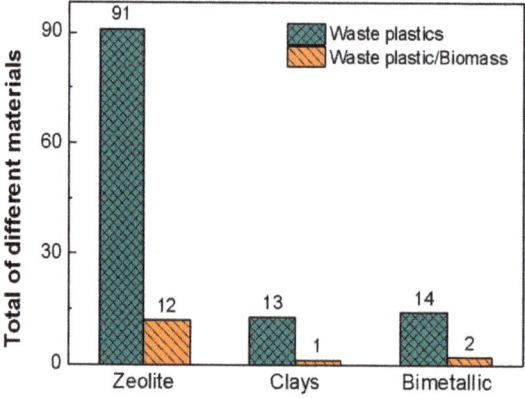

Figure 1. Popularity of different types of catalytic materials for pyrolysis of plastic waste from 2015 to 2020. (Source: Web of Knowledge, https://www.webofknowledge.com, accessed on 1 April 2020).

However, there is no clear conclusion or generalization obtained regarding reference [16]'s effect of catalyst performance. Other factors such as temperature, pressure and plastic/biomass ratio in co-pyrolysis allow for the discernment of the impact on the result within these perspectives, and bimetal catalysts are also developed, especially for improvement on gaseous products and hydrogen. In this review, we discuss some essential properties of low-cost catalysts for plastic waste pyrolysis.

2. Clay-Based Catalysts for Plastic Pyrolysis

Catalytic reactions of plastic pyrolysis depend on solid acid mechanisms on the surface which include cracking, isomerization, oligomerization, cyclisation and aromatization reactions. These various mechanisms are governed by acidity, density, porous size, and porous structure of the catalyst surface. Both Lewis and Brønsted acid sites of a clay catalyst play roles in the cracking mechanism which is initiated by the abstraction of the hydride ion from the polymer structure by Lewis acid sites of the catalyst, or the addition of a proton to the C–C bonds by Bronsted acid sites of the catalyst [20]. The higher amount of Brønsted acid sites on the surface of the catalyst provides more hydrogen ions for double bond cleavage and further propagation steps. Meanwhile, with a different role, the Lewis acid sites influence the surface interaction of catalysts with polyolefin, which is an important part of the whole surface reaction in heterogeneous catalysis.

The surface acidity and high specific surface area of the catalyst play an important role in producing liquid products instead of gas products. The availability of micropores in the clay structure has the potency to act as a heat sink and allows a greater residence time for feed molecules to absorb the heat and have interactions that result in hydrogen transfer [21]. The main role of solid catalyst in the liquid product is enhancing the ability to crack the polymer structure to form an intermediate in the mechanism. This influences by increasing the liquid product along with decreasing the wax content. Less wax from the use of bentonite clay refers to the presence of surface acidity but the surface acidity of clay is lower than zeolite as the impact of Si/Al ratio [22]. Less surface acidity led to the lower Brønsted acidity compared with zeolite which minimizes the potential of secondary reaction such as an over-cracking mechanism, so more liquid product distribution is achieved [23]. Increased conversion and selectivity in producing liquid were exhibited by modifying clay structure via pillarization using aluminum. The increasing conversion not only came from increasing the specific surface area and Lewis's surface acidity distribution, but also came from the thermal stability of the integrity of the clay structure for the sustained pores and surface area for the cracking mechanism [24,25].

Moreover, the stability of pillared clay led to the renewability properties whereas regenerated and reused catalysts showed practically identical conversion and yield values compared with the fresh catalyst. According to the identification of the composition of oil yield from pyrolysis reaction, the presence of metal oxide as a pillar minimizes the side-reaction mechanism. Two main mechanisms: (i) cyclization/aromatization of pyrolytic intermediate and (ii) cracking dominantly occur and play roles in producing diesel fraction in the liquid product together with the high yield of hydrogen gas. Another important aspect from the study on varied metal oxide impregnated onto acid-washed bentonite clay for polypropylene (PP) and high-density polyethylene (HDPE) pyrolysis is the importance of Lewis acid sites from metal ions for facilitating reactions via the formation of the hydration of a proton or the hydride ion due to its surface acidity of catalyst materials [26]. The β-scission of chain-end carbonium ion is the following reaction from acid interaction with polymer chain for the further production of gas and liquid fractions. The β-scission mechanism is presented in Figure 2.

The more effective acid–polymer interaction in the mechanism also prevents residue such as coke formation on the surface as shown by the comparison on acid-washed bentonite clay (AWBC) and metal oxide impregnated AWBC [20]. Table 1 presents the pyrolysis reaction of plastic waster over the different types of clay catalysts.

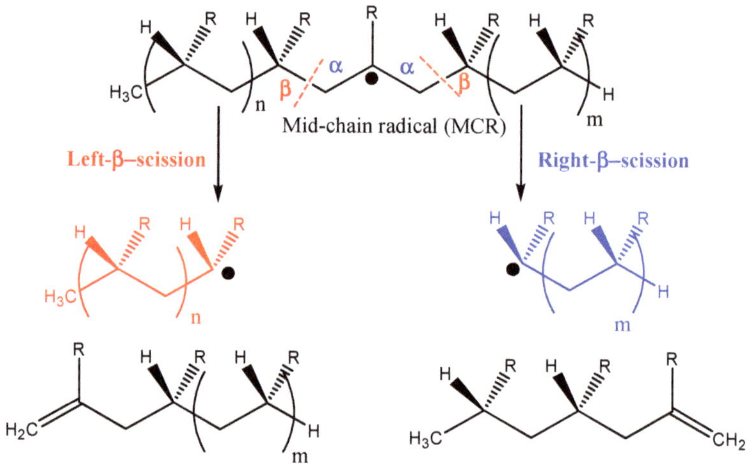

Figure 2. The mid-chain radical mechanism for obtaining low molecular product.

Table 1. Recent works on pyrolysis of plastic waste over clay catalysts.

No	Catalyst	Plastic Type	Result	Ref
1	Calcium Bentonite	Polypropylene (PP), low-density polyethylene (LDPE), and high-density polyethylene (HDPE)	The yield was influenced by temperature and ratio of effect to catalyst. The major product as condensable fraction was in the temperature range 400–550 °C and the optimum condition was at 500 °C at the catalyst to plastics ratio of 1:3	[27]
2	Kaolin	PP	Ahoko kaolin exhibited as effective as a low-cost catalyst for producing gasoline/diesel grade fuel with the PP as waste sources. The yield was influenced by catalyst to plastic ratio	[28]
3	Restructured and pillared clay	polyolefin	Restructured and pillared clay showed good selectivity towards aliphatic, produced more liquid	[29]
5	Fe, Ti, Zr- pillared clay	HDPE, polystyrene [PS], PP	Fe-pillared clay showed excellent yield of diesel fraction in liquid product and H2	[30]
6	Fe-pillared clay	Heavy gas oil(HGO)/HDPE	The presence of HGO improved the oil yield from both thermal and catalytic pyrolysis of HDPE	[31]
7	Tungstophosphoric acid (TPA)/kaolin	Low-density polyethylene (LDPE)	TPA loaded kaolin (5-TPA-K) produced higher percentages of gasoline-like hydrocarbons (C11–C14)	[32]
8	Co, Fe, Mn, Zn impregnated acid-washed bentonite clay (AWBC)	PP and HDPE	Metal oxide impregnation on acid-washed bentonite clay not only improves conversion but also yield reduce coke formation	[20]
				Manos

Referring to the product distribution of pyrolysis reaction, it can be summarized that catalyst surface acidity and pore characteristics are mainly responsible for catalytic performance. The mesoporous structure with a high surface area is closely related to the Si/Al ratio of the catalyst. The surface acidity facilitated the mechanism of the reaction by the formation of the hydride ion or the addition of a proton due to the inherent acidity, which is simultaneously incorporated with the impregnated metals. This mechanism increased the liquid yield as a substantial improvement over thermal cracking which has a tendency to produce gas as the result of the radical mechanism, also called the random scission mechanism [28,29]. The study on HDPE pyrolysis over HZSM-12 revealed that the solid acidity linearly decreased the activation energy (Ea), as proof of the important role of the acid mechanism in accelerating the reaction [33]. However, the extremely high acidity leads to increased yield by the over-cracking that leads to the formation of much smaller molecules. For this reason, aluminum-pillared clay, which has mild acidity while producing a higher liquid yield (~70%) [34], is similar to the use of ultra-stable zeolite (USY (71%) [35], which is higher compared to the yield by ZSM-5 (61%) [36]. The plot in

Figure 3 represents the relationship between the solid acidity and the liquid yield from several papers.

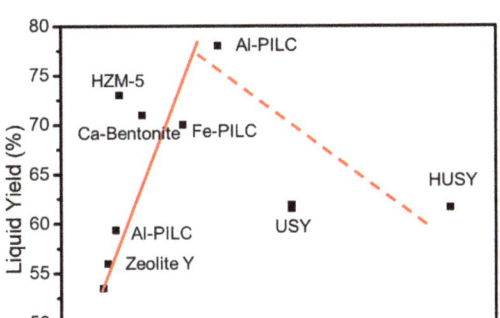

Figure 3. Relationship between solid acidity and liquid yield [27,33–37].

3. Plastic Waste-to-Fuel over Zeolite Catalysts

Zeolite (ZSM-5) has been widely reported as an effective and selective material catalyst for producing biofuel through the thermocatalytic reaction [38]. Besides its catalyst base, ZSM-5 is a low-cost catalyst for the conversion of plastic waste to biofuel. The ZSM-5 catalyst also presents excellent thermal stability, good selectivity, activity, and deactivation by coke [39,40]. In thermocatalytic reactions, the ZSM-5 effectively enhances the deoxygenation and cracking reaction to produce stable oil [41,42]. Onwudili et al. (2019) have studied the influence of temperature and type of zeolite as a catalyst for pyrolysis reaction to convert the plastic waste to biofuel liquid products [43]. Using different catalysts at temperatures of 500 and 600 °C showed no significant results of fuel-range liquid products. However, the increasing temperature resulted in the increasing gas composition of the products. Besides that, the high acidic catalyst can lead to a faster production of gases. More acidity promotes the formation of the hydrogenation steps, leading to the synthesis of other free radicals which resulted in β-scission for the gas production [16,44,45]. The related study by Kassargy et al. (2019) also reported that the yields of the liquids fractions are linearly dependent on the proportion and the type of plastic waste in the mixture [44].

Miscolczi et al. (2019) studied the effect of the loading metal to zeolite structure for catalysis plastic to fuel [45]. The authors modified the zeolite catalyst with several metal ions like $Fe^{2+/3+}$, Cu^{2+}, Ce^{2+}, H^+, Mg^{2+}, Ni^{2+}, Sn^{2+}, and Zn^{2+}. Their studies revealed that the presence of metal loading (8–10%) to zeolite structure could affect the pore diameter and the macropore surface area of the materials. The surface area is an essential factor in the pyrolysis reaction process besides the acidity of the catalyst [46,47]. Therefore, modification with several metal ions to zeolite structure can easily control the surface area of the catalyst materials which affects the decreasing temperature decomposition of the plastics during the pyrolysis process. Gorbannezhad et al. (2020) revealed that the co-pyrolysis also depended on the composition of the zeolite catalyst [48]. The authors studied the co-pyrolysis process by combining zeolite (HZSM-5) and sodium carbonate/gamma-alumina for improving the hydrocarbon products of the reaction process. Their study showed that the combination of catalysts could improve the hydrocarbon product to 8.7% at the temperature process of 700 °C. The presence of sodium salt in the composition of the catalyst can improve the deoxygenation reaction so that the breakdown process of macromolecules becomes faster and more effective to produce low molecular compounds such as hydrocarbons. Besides that, the other studies also reported that the catalytic pyrolysis process is affected by the type of zeolite catalyst [49,50]. Several types of catalyst-based zeolite have a difference in the pore size and acidity which affected the produced product [51]. For example, the larger pore size facilitates the conversion of the plastics as source materials to polyalkylaromatics while the smaller pore size only converts to aromatic compounds with small dynamic diameters. Based on their study, the use of suitable zeolite as a catalyst in the pyrolysis reaction is a crucial step to determine the produced product during the reaction process.

Susastriawan et al. (2020) studied the effect of zeolite size on pyrolysis of LDPE plastic waste at low temperatures [52]. The authors revealed that reducing zeolite size could enhance the reaction rate, pyrolysis temperature, heat transfer rate, and the oil products because the smaller size has a high surface-active area to contact with the plastics during a pyrolysis process. The zeolite size of 1 mm showed the highest value of oil yields; however, the particle size of 1–3 mm did not indicate the results of oil produced significantly. In the other studies also reported by Kim et al. (2018), the authors showed that the presence of the phenolic functional group on lignin could enhance 39% of the aromatic hydrocarbon in the totally resulted product [53]. The presence of a hydroxyl group on the chemical structure of lignin gives good selectivity on the decomposition of reactant to form an aromatic hydrocarbon as a major product [54,55]. In addition, the presence of a hydroxyl group on the surface of a zeolite-type catalyst can contribute to the condensation process during pyrolysis. Besides that, hydroxyl can accelerate the formation of aromatic hydrocarbon products through dihydroxylation, aromatization, isomerization, and oligomerization mechanism. Table 2 shows the comparison of zeolite types for pyrolysis reaction of plastic waste. All researchers confirmed that the zeolite catalyst can improve the acid-activation and thermal activation in the pyrolysis reaction of plastic waste (PE, PP, PVC, PET, and PS) [56,57].

Table 2. Pyrolysis of plastic waste over zeolite-based catalysts.

Type of Plastic	Catalyst	Reaction Condition	Conversion	Selectivity to	Remark (Catalyst/Plastic Ratio)	Ref
Polystyrene and polyolefeins (PS/PO)	Y-zeolite	600 °C for 30 min under N_2 gas	High yield valuable aromatics such as benzene and toluene	90% of the aromatic content	2 g of catalyst and 2 g of plastic	[43]
Polyethylene and polypropylene (PE/PP)	USY-zeolite	500 °C	Liquid fractions are dominated by hydrocarbon (C_5–C_7), C_3 and C_4 for gaseous products.	80% of liquid production	Catalyst/plastics ratio of 1:10	[44]
Polystyrene (PS)	Natural/Synthetic zeolite	450 °C for 75 min	60.8% conversion to ethylbenzene and 38.3% convertsion to alpha-methylstyrene for natural and synthetic zeolite respectively	54% and 50% of liquids products for natural and synthetic zeolite, respectively	Catalys/PS ratio of 0.1 kg:1 kg	[1]
High density Polyethylene (HDPE))	Co-Y–zeolite	600 °C for 30 min	40% of gas yield	68% of hydrogen production	Catalys/HDPE ratio of 2:1	[58]
Plastic mixtures (HDPE/PP/PS/PET/PVC)	Regenerated ZSM-5	440 °C for 30 min	Almost 60% of plastic waste conversion to liquids phase	97.4% of aromatics with 23% of styrene as major composition	Catalyst/plastic waste ratio of 1:10	[59]
Plastic mixtures (PE/PP/PS/PET/PVC)	ZSM-5	500 °C for 30 min	58.4% conversion to gases phase	50.7% of C_3–C_4 types and 27.9% of styrene	Catalyst/plastic waste ratio of 1:10	[60]

4. Effect of Co-Feeding with Biomass Feedstock

The biomass feedstock consists of extractive (0–14%), lignin (16%), hemicellulose (20%), and cellulose (>40%) [61–63]. Generally, the yields of the produced biofuels are dependent on several factors such as the composition of the nature of the feedstock, the moisture content of biomass feedstocks, reactor design, and operating temperature conditions [64–67]. In the pyrolysis process, the biomass feedstock can be mixed with other materials to improve the quantity and the quality of the product as shown in Table 3.

There are some pathways for the pyrolysis of biomass like the chemical, biological, and thermochemical conversion as shown in Figure 4. However, the thermochemical reaction is commonly used for the pyrolysis reaction of biomass feedstock because the process has high energy to split and stretch the rigid structure that has biomass [68]. The synergistic effect is remarkable, coming from the presence of produced free radicals from biomass decomposition that contribute to enhancing the scission of chain hydrocarbons from plastics. Decreasing activation energy and the pyrolysis index, representing the easiness of the pyrolytic reaction for producing volatile products (methane, aliphatic hydrocarbon (paraffin), carbon dioxide, aromatic hydrocarbon), are the quantitative parameters revealed to be advantageous of co-pyrolysis [69]. The characteristics of biomass, especially the ratio of hydrogen to carbon effective (H/C_{eff}), heavily influences the product distribution. Higher H/C_{eff} significantly improves olefin and aromatic yield and reduces coke formation [70–72]. In addition, Bhoi et al. (2019) reported operating conditions such as heating rate, type and particle size of biomass, temperature, carrier gas, type of catalyst, and vapor residence time [73]. The authors and the other related studies reported that the type of catalyst and temperature have major impacts on quality and biofuel yields [74–76]. However, the choice of temperature during the pyrolysis reaction depends on the main composition of the biomass sources, for example, hemicellulose and cellulose degrade at approximately 200–350 °C and 330–370 °C, respectively, while lignin occurs at 400 °C [77,78].

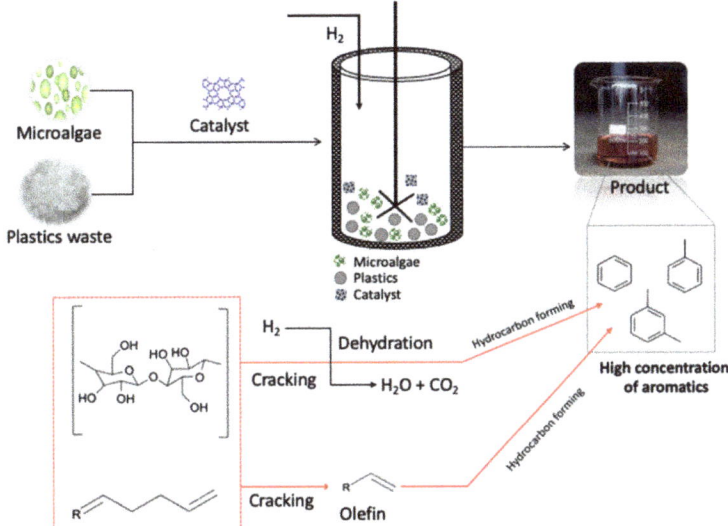

Figure 4. The illustration of pyrolysis process with the natural biomass and plastic waste as a feedstock.

Xu et al. (2020) studied the pyrolysis reaction with the mixing of microalgae (*Enteromorpha prolifera*) and HDPE plastics using the HZSM-5 catalyst [79]. The authors reported that the presence of microalgae in the plastic waste could increase the concentration of aliphatic hydrocarbons and decrease the nitrogen/oxygen-containing compounds and the acidity of the products. Algae as natural resources have received attention because they can produce a high amount of bio-oil than the other resources of biomass [80]. The related study also reported that using microalgae as a pyrolysis feedstock has many advantages like faster rate of growth and high lipid content, lower energy consumption during a process with the percent of energy recovery of 76%, increase in the carbon content to 89%, and decrease in the oxygen content to 0.3% [81–83]. Furthermore, Qari et al. (2017) reported that the characteristics of microalgae significantly affected the types and the yields of biofuel products [84]. In the produced bio-oil product, the mixing between biomass and plastic waste can reduce the moisture content because the hydroxyl group in the biomass can directly bind with the plastics which contain the high hydrogen atom to form the hydrogen bonding [85]. Figure 4 shows the process of pyrolysis reaction with the mixing feedstock between natural biomass of microalgae and plastic waste.

Table 3. Effect of co-feeding with other feeds.

Type of Biomass/Polymer	Catalyst	Reaction Condition	Conversion	Selectivity to	Remark (Catalyst/Plastic Ratio)	Ref
Laminaria japonica/polypropylene	Pt/mesoporous MFI, Mesoporous MFI, HZSM-5, Al-SBA-16	500 °C, 1 atm	60.50 59.11 58.15 52.44	20–40% of monoaromatic hydrocarbon, 5–20% of polyaromatic hydrocarbon	Biomass/plastic/catalyst ratio = 10:10:1	[86]
Corn stalk/HDPE	ZSM-5	700 °C, 1 atm	90% of hydrocarbon	20–30% of aromatic hydrocarbon	Biomass/plastic/catalyst = 1:4:1	[87]
pine sawdust/LDPE	Ni-CaO-C	-	90% of gas product	86.74% to H_2 gas	Biomass/plastic/catalyst: 1:1:2	[88]
Corn stalk/Polystyrene	ZSM-5	600 °C	90% of liquid product	78.89% of monoaromatic hydrocarbon	4:1:0.1	[89]
L. japonica/polypropylene	Al-SBA 15	500	30% of liquid product	35% of liquid product is oxygenate	Seaweed/polypropylene/catalyst = 1: 1:1	[90]
Rice husk/PE	Ni/γ-Al_2O_3	Pyrolysis at 600 °C followed by catalytic reforming at 800 °C	80% of H_2 and CO	45% of gaseous product is H_2	50~75% PE proportion	[91]

Muneer et al. (2019) studied the effect of catalyst ratio to feedstock for the pyrolysis of corn stalk (CS) and polypropylene in a bed reactor at 500 °C [89]. The increase in the catalyst could enhance the liquid oil yield to 66.5% at ZSM-5 catalyst to feedstock ratio of 1:4. ZSM-5 catalyst was found as an effective material for polymer cracking and dehydration of biomass because of its high surface area and high selectivity to produce the hydrocarbon [92]. The production of bio-oil from biomass feedstocks depends on the acidic site and the pore structure of the catalyst because the acidic site increases the rate of polymer cracking. Similar results were also reported by Balasundram et al. (2018); their study revealed that the increasing catalyst amount four times could improve the coke decomposition by 17.1% [93]. It can be summarized that the addition of biomass in the co-pyrolysis enhances the efficiency to produce liquid products and reduce activation energy. By using ZSM-5 catalyst and corn stalk in the co-pyrolysis of HPDE, it was found that increased hydrocarbon yield and H/C eff were obtained on increasing biomass/HDPE ratio, along with decreasing coke [87]. Hydrogen atoms for the co-pyrolysis process were provided by HDPE, leading to an improvement in the rate of hydrocarbon production, meanwhile, oxygenated compounds in the biomass play a role to promote the cracking of HPDE and the chain scission. However, at a certain level, the increasing catalyst dosage may affect the increased charring, which leads to reduce liquid product. For example, the co-pyrolysis of cellulose/polyethylene over montmorillonite K10. A similar trend is also identified on cellulose pyrolysis using montmorillonite on cellulose. The availability of more surface acidity in Al-SBA enhances the co-pyrolysis to produce more C1–C4 hydrocarbon compounds, CO, CO_2, and deoxygenation reactions [86].

5. Recent Reports Bimetallic Catalysts for Pyrolysis of Plastics

Bimetallic catalysts have been widely used for the pyrolysis reaction of plastic waste to biofuel. Previously, the monometallic type has been widely studied and reported for the catalytic cracking process. For example, Wen et al. (2014) prepared Ni-loaded to CNTs for polyolefin; although the material showed good performance as the catalyst, the concentration of carbon was relatively very high [94]. Therefore, bimetallic with integrating the different types of material catalyst provides some advantages in the pyrolysis reaction such as large surface area due to its smaller size, good stability, and synergy effect between combined two metals [95]. Yao et al. (2017) studied the Ni-Fe bimetallic catalyst at a ratio of 1:3 for the pyrolysis of waste plastics [96]. The authors reported that the presence of a bimetallic catalyst could enhance H_2 production to five times higher compared to the process without a catalyst. Chen et al. (2020) also reported that the Fe-Ni bimetallic modified MCM-41 could improve the produced oil to 49.9% with a percentage of single styrene hydrocarbon of 65.93% at 10%Fe-10%Ni/MCM-41 [97]. Besides that, the presence of bimetallic/MCM-41 can reduce the bromine content from 10% to 2.3% (wt.). The developed catalyst provides a large surface area; thus, the plastics can directly enter the pore structure for the cracking process. The iron metals act as the base site which converts the raw materials to styrene, while the combination with nickel–metal oxides increases the acidity of the catalyst, thus the multi-ring compounds can also be converted to a single hydrocarbon structure. The character of the external surface and pore size determines the chemisorption, and these are designable by the synthesis method. In this case, the higher pore diameter of Fe-Ni catalyst tends to give higher H_2 desorption [98], which is in line with the trend of long-chain products by the high pore size of Co/SBA-15 [51].

Li et al. (2016) reported the bi-functional Mo-Ni/SiO_2-Al_2O_3 catalyst for the thermal pyrolysis of crude oil [99]. The presence of the bimetallic of Mo-Ni on the composite catalyst could improve the catalytic reactivity and the amount of yield of fuel oil produced to 57.9%. The metallic of Mo and Ni has several advantages for catalyst in pyrolysis like low-cost adsorbent, excellent stability performance, high surface area, and also ease of regeneration [100]. However, their study reported that the reactivity of the bimetallic catalyst is dependent on the sulfurization process. The increased sulfurization process can regularly improve the percentage of conversion of crude oil to fuel oil until it reaches 86.9%. In addition, some related studies have reported that the bimetallic catalyst type has good selectivity conversion [101,102]. A more specific capability of bimetallic catalyst is shown by Fe-Ni/MCM-41 for not only the decomposition of polymer structure but also to conduct debromination mechanism for plastic waste containing brominated flame-retardant (BFR) pyrolysis. The results indicated that iron showed a satisfactory capability for debromination, in combination with Ni's ability to produce gaseous products via the hydrogenation mechanism. One of the proposed pathways of the reaction mechanism that occurs is the interaction between organobromine and some metal oxides to produce non-brominated organic compounds, elimination of β-H by Lewis acid sites and dissociative adsorption two-stage reaction [97].

Cai et al. (2020) reported the carbon-based Fe-Ni bimetallic catalyst for fast pyrolysis of plastic waste [103]. The pyrolysis reaction was carried out in a fixed-bed reactor with the ratio of catalyst to plastic waste at 1:2 (wt. ratio) at a temperature of 500 °C. The fixed-bed reactor has some advantages for pyrolysis reaction such as a simple design with the catalyst loaded into the bed column, irregularity in plastic shape, and low thermal experiments [104,105]. The authors explained that Fe-Ni as a bimetallic catalyst played a crucial role in the oxygen reduction reaction (ORR). Moreover, the developed catalyst has good methanol tolerance and stability, avoiding the aggregation and corrosion process on the surface of the catalyst due to its oxygen-containing functional groups on the surface [106]. Zhou et al. (2020) also reported that the bimetallic Ni-Fe/ZrO_2 catalyst showed excellent decomposition of polystyrene at a low-temperature process (500 °C) [107]. Their studies revealed that the presence of bimetallic catalysts could improve the catalytic activity in the decomposition of waste. The combined properties between Ni-Fe could decrease the water–gas shift reaction and the activation energy of the reforming reaction [108–110]. Table 4 presents the different types of bimetallic catalysts for the pyrolysis process.

Table 4. Recent reports bimetallic catalysts for pyrolysis of plastics.

Type of Plastic	1st Metal	2nd Metal	Condition, Pressure (atm), Temperature (°C)	Conversion	Selectivity to	Remark (Catalyst/Plastic Ratio)	Ref
Low density polyethylene (LDPE)	Mo-MgO	Fe	Atmospheric pressure, 750 and 400 °C	LDPE waste plastic to carbon nanotubes	High quality carbon nanostructures materials	0.5 g:15 g plastics	[111]
Polypropylene (PP)	La_2O_3	Ni-Cu	500 °C, 700 °C for 2.5 h	PP to Carbon nanotubes and carbon nanofibers	Carbon yields of 1458% produced	0.5 g:15 g plastics	[112]
Polypropylene (PP)	MgO	Ni/Mo	800 °C, 10 min	PP to CNT	394% of carbon product	0.15 g:5 g polymer	[113]
Polypropylene (PP)	Ni-	Al	800 °C	PP to MWCNTs	85%	Dependence on the ratio Ni/Al and the amount of Ni-Al catalyst	[114]
Polypropylene (PP)	Ni-Al	Zn, Mg, Ca, Ce, Mn	500 °C	PP to CNTs	The highest carbon deposition 62% and hydrogen 86.4% to Ni-Mn-Al	1 g:2 g waste polypropylene	[115]
Polyehtylene (PE)	Ni	Ce (Ni-Ce core by silica)	800 °C	PE to hydrogen	Hydrogen concentration 60%	Weight ratio Catalyst:plastic 1.0	[116]
LDPE	Ni	Fe	800 °C	Carbon nanotubes (CNTs)	Maximum hydrogen concentration and hydrogen yield 73.93% and 84.72 mg.g^{-1}	0.5 g:1 g waste plastic	[96]
Low-density polyethylene (LDPE) waste	MgO	Co/Mo	400 °C	High quality multi-walled Carbon Nanotubes and hydrogen	Optimum CNTs 1040% wtCoMo(6.5) MgO	0.75 g:15 g plastics	[117]
Polypropylene (PP)	Ni	Fe	500 °C	CNTs	93% filamentous carbon nanotubes	0.5 catalyst:1 g PP	[118]
HDPE	Ni	Mn-Al	800 °C	Hydrogen and carbon nanotubes	48% total carbon (with no steam), hydrogen yield 94.4% (with steam)	0.5 g catalyst:1 g waste plastic (HDPE)	[119]

6. Factor Affecting in Pyrolysis Process

In the pyrolysis process, many reaction parameters strongly affect the quantity and quality of the resulting products, such as temperature, heating rate, composition blending ratio, and the type of reactor design. However, the reactor designs play a crucial factor in obtaining a high yield quantity and the quality of the product. Previous studies have already reported several types of reactor design for the pyrolysis process, such as fixed-bed, transported bed, rotating cone, plasma pyrolysis, vortex centrifuge, circulating fluidized, entrained flow, etc. [120]. The fixed-bed reactor is commonly used for the pyrolysis process due to the simple process, a large sample quantity and a high product yield. However, the reactor experiences several drawbacks, such as the catalyst being difficult to replace during a pyrolysis process, side reaction and product, and the required high temperature and pressures. Therefore, recently, many types of reactor-based fast pyrolysis reactors have been developed. The principle of the reactors is to optimize the percentage of products in low temperatures and pressure during the pyrolysis process.

Xue et al. (2015) reported the fast pyrolysis of HDPE waste using a fluidized bed reactor [121]. Their study found that HDPE waste could increase the formation of acid and furans in the products and decrease the formation of phenol and vanillin compounds at a relatively lower temperature process. Orozco et al. (2021) studied the pyrolysis of plastic waste using a spouted bed reactor in continuous mode [122]. Their study revealed that the synergy effect between the plastic-type and the total amount of catalyst could decrease the temperature. The reaction design of fast pyrolysis reaction is based on optimizing the product yields by decreasing the size of plastic waste to less than 1 mm and flowing the carrier gas as shown in Figure 5. However, despite the lack of the reactor, there is still a high amount of oxygen content in products. A related study has been reported for removing the oxygen content in pyrolysis products by hydrotreating formate-assisted pyrolysis [123]. Reducing oxygen in the resulting product can improve the percentage of the product by up to 92%.

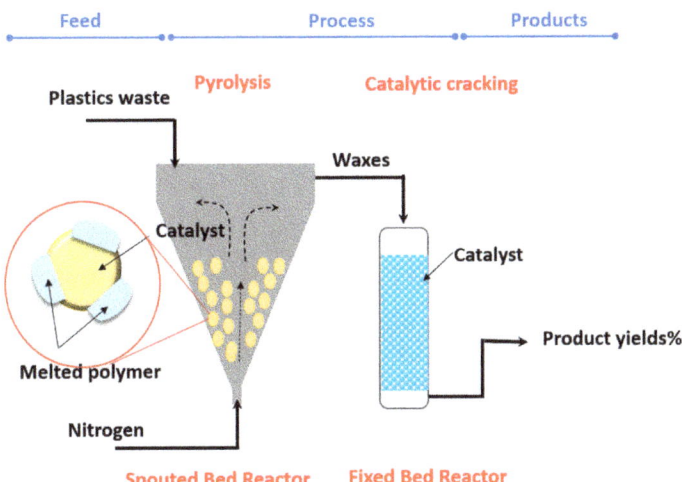

Figure 5. The illustration of combined pyrolysis reactor designs for improving the yields products.

7. Conclusions and Future Prospect

In summary, we reported and focused on the catalytic decomposition of plastic waste to produce liquid fuel using a low-cost catalyst. Several types of low-cost catalysts were summarized, such as zeolite, clay, and bimetallic. Generally, some studies reported that the low-cost catalyst could be used for catalytic cracking of plastic waste to liquid fuel. However, several factors are still required for improving the quantity and quality of the products, such as decreasing the particle size of waste in feedstocks, temperature, pressure, ratio composition between catalyst to feedstocks, type of catalyst, and type of reactor design. Unbeatably, pyrolysis and co-pyrolysis of plastic waste to produce fuel is a promising sustainable technology. Several conditions of the operation process influence the yield of the products such as (i) heating rate, (ii) the ratio of blending materials, and (iii) the pyrolysis temperature [104]. Additionally, the heterogeneous catalyst plays an important role in the conversion, and from the techno-economic point of view, exploration of the low-cost catalyst with high efficiency and lifetime is still required. Designing an appropriate catalyst is a major challenge in developing the technique, and referring to this study, the selectivity of a catalyst toward a specific product becomes important in the design of the reaction set up, as well as the lifetime of the catalyst. Catalyst solid acidity along with the specific surface area governs the optimum condition of the conversion, and these are related to the structure of solid support and their combination with active metal/bimetals on the surface. Compared to commercial solid acid catalysts, natural zeolite and clays are better options since they are cost-effective and more selective towards liquid products [1]. Referring to decoking capability, tunable performance of bimetallic catalysts and the potencies of zeolite and clay utilization, bimetallic-modified zeolite and clay catalysts are interesting for exploration. The designed Fe-Ni with zeolite or clay supports, for example, is a good candidate for a low-cost pyrolysis catalyst for bromide-containing plastic wastes which are abundantly produced from flame-retarded plastic production. In addition, some strategies to create integrative plastic waste management including co-pyrolysis of biomass/plastic wastes with these catalyst candidates need to be investigated for furthermore well-implemented.

Funding: Funding was provided by World Class Professor Program, Ministry of Research and Higher Education, Republic of Indonesia, 2020.

Data Availability Statement: The data used to support the findings of this study are available from the corresponding author upon request.

Conflicts of Interest: The authors declare no conflict of interest.

References

1. Miandad, R.; Barakat, M.A.; Rehan, M.; Aburiazaiza, A.S.; Ismail, I.M.I.; Nizami, A.S. Plastic waste to liquid oil through catalytic pyrolysis using natural and synthetic zeolite catalysts. *Waste Manag.* **2017**, *69*, 66–78. [CrossRef] [PubMed]
2. Rehan, M.; Nizami, A.-S.; Asam, Z.-U.-Z.; Ouda, O.K.M.; Gardy, J.; Raza, G.; Naqvi, M.; Mohammad Ismail, I. Waste to Energy: A Case Study of Madinah City. *Energy Procedia* **2017**, *142*, 688–693. [CrossRef]
3. Samun, I.; Saeed, R.; Abbas, M.; Rehan, M.; Nizami, A.-S.; Asam, Z.-U.-Z. Assessment of Bioenergy Production from Solid Waste. *Energy Procedia* **2017**, *142*, 655–660. [CrossRef]
4. Banu, J.R.; Sharmila, V.G.; Ushani, U.; Amudha, V.; Kumar, G. Impervious and influence in the liquid fuel production from municipal plastic waste through thermo-chemical biomass conversion technologies—A review. *Sci. Total Environ.* **2020**, *718*, 137287. [CrossRef]
5. Zakir Hossain, H.M.; Hasna Hossain, Q.; Uddin Monir, M.M.; Ahmed, M.T. Municipal solid waste (MSW) as a source of renewable energy in Bangladesh: Revisited. *Renew. Sustain. Energy Rev.* **2014**, *39*, 35–41. [CrossRef]
6. Munir, D.; Irfan, M.F.; Usman, M.R. Hydrocracking of virgin and waste plastics: A detailed review. *Renew. Sustain. Energy Rev.* **2018**, *90*, 490–515. [CrossRef]
7. Walendziewski, J. Continuous flow cracking of waste plastics. *Fuel Process. Technol.* **2005**, *86*, 1265–1278. [CrossRef]
8. Kunwar, B.; Cheng, H.N.; Chandrashekaran, S.R.; Sharma, B.K. Plastics to fuel: A review. *Renew. Sustain. Energy Rev.* **2016**, *54*, 421–428. [CrossRef]
9. Zhan, H.; Zhuang, X.; Song, Y.; Liu, J.; Li, S.; Chang, G.; Yin, X.; Wu, C.; Wang, X. A review on evolution of nitrogen-containing species during selective pyrolysis of waste wood-based panels. *Fuel* **2019**, *253*, 1214–1228. [CrossRef]
10. Al-Salem, S.M.; Antelava, A.; Constantinou, A.; Manos, G.; Dutta, A. A review on thermal and catalytic pyrolysis of plastic solid waste (PSW). *J. Environ. Manag.* **2017**, *197*, 177–198. [CrossRef] [PubMed]
11. Olaremu, A.G.; Adedoyin, W.R.; Ore, O.T.; Adeola, A.O. Sustainable development and enhancement of cracking processes using metallic composites. *Appl. Petrochem. Res.* **2021**, *11*, 1–18. [CrossRef]
12. Campuzano, F.; Brown, R.C.; Martínez, J.D. Auger reactors for pyrolysis of biomass and wastes. *Renew. Sustain. Energy Rev.* **2019**, *102*, 372–409. [CrossRef]

13. Al-Hamamre, Z.; Saidan, M.; Hararah, M.; Rawajfeh, K.; Alkhasawneh, H.E.; Al-Shannag, M. Wastes and biomass materials as sustainable-renewable energy resources for Jordan. *Renew. Sustain. Energy Rev.* **2017**, *67*, 295–314. [CrossRef]
14. Wang, Z.; Burra, K.G.; Lei, T.; Gupta, A.K. Co-pyrolysis of waste plastic and solid biomass for synergistic production of biofuels and chemicals-A review. *Prog. Energy Combust. Sci.* **2021**, *84*, 100899. [CrossRef]
15. Dai, L.; Zhou, N.; Li, H.; Wang, Y.; Liu, Y.; Cobb, K.; Cheng, Y.; Lei, H.; Chen, P.; Ruan, R. Catalytic fast pyrolysis of low density polyethylene into naphtha with high selectivity by dual-catalyst tandem catalysis. *Sci. Total Environ.* **2021**, *771*, 144995. [CrossRef]
16. Budsaereechai, S.; Hunt, A.J.; Ngernyen, Y. Catalytic pyrolysis of plastic waste for the production of liquid fuels for engines. *RSC Adv.* **2019**, *9*, 5844–5857. [CrossRef]
17. Kumar, B.S.; Dhakshinamoorthy, A.; Pitchumani, K. K10 montmorillonite clays as environmentally benign catalysts for organic reactions. *Catal. Sci. Technol.* **2014**, *4*, 2378–2396. [CrossRef]
18. Pienkoß, F.; Ochoa-Hernández, C.; Theyssen, N.; Leitner, W. Kaolin: A Natural Low-Cost Material as Catalyst for Isomerization of Glucose to Fructose. *Acs Sustain. Chem. Eng.* **2018**, *6*, 8782–8789. [CrossRef]
19. Król, M. Natural vs. Synthetic Zeolites. *Crystals* **2020**, *10*, 622. [CrossRef]
20. Ahmad, I.; Khan, M.; Khan, H.; Ishaq, M.; Tariq, R.; Gul, K.; Ahmad, W. Influence of Metal-Oxide-Supported Bentonites on the Pyrolysis Behavior of Polypropylene and High-Density Polyethylene. *J. Appl. Polym. Sci.* **2014**, *132*, 1–19. [CrossRef]
21. Patil, V.; Adhikari, S.; Cross, P. Co-pyrolysis of lignin and plastics using red clay as catalyst in a micro-pyrolyzer. *Bioresour. Technol.* **2018**, *270*, 311–319. [CrossRef]
22. Fatimah, I.; Rubiyanto, D.; Prakoso, N.I.; Yahya, A.; Sim, Y.-L. Green conversion of citral and citronellal using tris(bipyridine)ruthenium(II)-supported saponite catalyst under microwave irradiation. *Sustain. Chem. Pharm.* **2019**, *11*, 61–70. [CrossRef]
23. Tarach, K.A.; Góra-Marek, K.; Martinez-Triguero, J.; Melián-Cabrera, I. Acidity and accessibility studies of desilicated ZSM-5 zeolites in terms of their effectiveness as catalysts in acid-catalyzed cracking processes. *Catal. Sci. Technol.* **2017**, *7*, 858–873. [CrossRef]
24. Vogt, E.T.C.; Weckhuysen, B.M. Fluid catalytic cracking: Recent developments on the grand old lady of zeolite catalysis. *Chem. Soc. Rev.* **2015**, *44*, 7342–7370. [CrossRef]
25. Geng, J.; Sun, Q. Effects of high temperature treatment on physical-thermal properties of clay. *Thermochim. Acta* **2018**, *666*, 148–155. [CrossRef]
26. Ferrini, P.; Dijkmans, J.; De Clercq, R.; Van de Vyver, S.; Dusselier, M.; Jacobs, P.A.; Sels, B.F. Lewis acid catalysis on single site Sn centers incorporated into silica hosts. *Coord. Chem. Rev.* **2017**, *343*, 220–255. [CrossRef]
27. Panda, A.K. Thermo-catalytic degradation of different plastics to drop in liquid fuel using calcium bentonite catalyst. *Int. J. Ind. Chem.* **2018**, *9*, 167–176. [CrossRef]
28. Hakeem, I.G.; Aberuagba, F.; Musa, U. Catalytic pyrolysis of waste polypropylene using Ahoko kaolin from Nigeria. *Appl. Petrochem. Res.* **2018**, *8*, 203–210. [CrossRef]
29. De Stefanis, A.; Cafarelli, P.; Gallese, F.; Borsella, E.; Nana, A.; Perez, G. Catalytic pyrolysis of polyethylene: A comparison between pillared and restructured clays. *J. Anal. Appl. Pyrolysis* **2013**, *104*, 479–484. [CrossRef]
30. Li, K.; Lei, J.; Yuan, G.; Weerachanchai, P.; Wang, J.-Y.; Zhao, J.; Yang, Y. Fe-, Ti-, Zr- and Al-pillared clays for efficient catalytic pyrolysis of mixed plastics. *Chem. Eng. J.* **2017**, *317*, 800–809. [CrossRef]
31. Faillace, J.G.; de Melo, C.F.; de Souza, S.P.L.; da Costa Marques, M.R. Production of light hydrocarbons from pyrolysis of heavy gas oil and high density polyethylene using pillared clays as catalysts. *J. Anal. Appl. Pyrolysis* **2017**, *126*, 70–76. [CrossRef]
32. Attique, S.; Batool, M.; Yaqub, M.; Görke, O.; Gregory, D.; Shah, A. Highly efficient catalytic pyrolysis of polyethylene waste to derive fuel products by novel polyoxometalate/kaolin composites. *Waste Manag. Res.* **2020**, 0734242X1989971. [CrossRef]
33. Silva, A.O.S.; Souza, M.J.B.; Pedroza, A.M.G.; Coriolano, A.C.F.; Fernandes, V.J.; Araujo, A.S. Development of HZSM-12 zeolite for catalytic degradation of high-density polyethylene. *Microporous Mesoporous Mater.* **2017**, *244*, 1–6. [CrossRef]
34. Manos, G.; Yusof, I.Y.; Gangas, N.H.; Papayannakos, N. Tertiary Recycling of Polyethylene to Hydrocarbon Fuel by Catalytic Cracking over Aluminum Pillared Clays. *Energy Fuels* **2002**, *16*, 485–489. [CrossRef]
35. Kassargy, C.; Awad, S.; Burnens, G.; Kahine, K.; Tazerout, M. Experimental study of catalytic pyrolysis of polyethylene and polypropylene over USY zeolite and separation to gasoline and diesel-like fuels. *J. Anal. Appl. Pyrolysis* **2017**, *127*, 31–37. [CrossRef]
36. Ghaffar, N.; Johari, A.; Tuan Abdullah, T.A.; Ripin, A. Catalytic Cracking of High Density Polyethylene Pyrolysis Vapor over Zeolite ZSM-5 towards Production of Diesel. *IOP Conf. Ser. Mater. Sci. Eng.* **2020**, *808*, 012025. [CrossRef]
37. Seddegi, Z.S.; Budrthumal, U.; Al-Arfaj, A.A.; Al-Amer, A.M.; Barri, S.A.I. Catalytic cracking of polyethylene over all-silica MCM-41 molecular sieve. *Appl. Catal. A Gen.* **2002**, *225*, 167–176. [CrossRef]
38. Lok, C.M.; Van Doorn, J.; Aranda Almansa, G. Promoted ZSM-5 catalysts for the production of bio-aromatics, a review. *Renew. Sustain. Energy Rev.* **2019**, *113*, 109248. [CrossRef]
39. Gayubo, A.G.; Aguayo, A.T.; Atutxa, A.; Prieto, R.; Bilbao, J. Deactivation of a HZSM-5 Zeolite Catalyst in the Transformation of the Aqueous Fraction of Biomass Pyrolysis Oil into Hydrocarbons. *Energy Fuels* **2004**, *18*, 1640–1647. [CrossRef]
40. Kang, Y.-H.; Wei, X.-Y.; Liu, G.-H.; Ma, X.-R.; Gao, Y.; Li, X.; Li, Y.-J.; Ma, Y.-J.; Yan, L.; Zong, Z.-M. Catalytic Hydroconversion of Ethanol-Soluble Portion from the Ethanolysis of Hecaogou Subbituminous Coal Extraction Residue to Clean Liquid Fuel over a Zeolite Y/ZSM-5 Composite Zeolite-Supported Nickel Catalyst. *Energy Fuels* **2020**, *34*, 4799–4807. [CrossRef]

41. Wang, C.; Liu, Q.; Song, J.; Li, W.; Li, P.; Xu, R.; Ma, H.; Tian, Z. High quality diesel-range alkanes production via a single-step hydrotreatment of vegetable oil over Ni/zeolite catalyst. *Catal. Today* **2014**, *234*, 153–160. [CrossRef]
42. Hou, X.; Qiu, Y.; Zhang, X.; Liu, G. Analysis of reaction pathways for n-pentane cracking over zeolites to produce light olefins. *Chem. Eng. J.* **2017**, *307*, 372–381. [CrossRef]
43. Onwudili, J.A.; Muhammad, C.; Williams, P.T. Influence of catalyst bed temperature and properties of zeolite catalysts on pyrolysis-catalysis of a simulated mixed plastics sample for the production of upgraded fuels and chemicals. *J. Energy Ins.* **2019**, *92*, 1337–1347. [CrossRef]
44. Kassargy, C.; Awad, S.; Burnens, G.; Kahine, K.; Tazerout, M. Gasoline and diesel-like fuel production by continuous catalytic pyrolysis of waste polyethylene and polypropylene mixtures over USY zeolite. *Fuel* **2018**, *224*, 764–773. [CrossRef]
45. Miskolczi, N.; Juzsakova, T.; Sója, J. Preparation and application of metal loaded ZSM-5 and y-zeolite catalysts for thermo-catalytic pyrolysis of real end of life vehicle plastics waste. *J. Energy Ins.* **2019**, *92*, 118–127. [CrossRef]
46. Ahmed, M.H.M.; Masuda, T.; Muraza, O. The role of acidity, side pocket, and steam on maximizing propylene yield from light naphtha cracking over one-dimensional zeolites: Case studies of EU–1 and disordered ZSM–48. *Fuel* **2019**, *258*, 116034. [CrossRef]
47. Galadima, A.; Muraza, O. Zeolite catalyst design for the conversion of glucose to furans and other renewable fuels. *Fuel* **2019**, *258*, 115851. [CrossRef]
48. Ghorbannezhad, P.; Park, S.; Onwudili, J.A. Co-pyrolysis of biomass and plastic waste over zeolite- and sodium-based catalysts for enhanced yields of hydrocarbon products. *Waste Manag.* **2020**, *102*, 909–918. [CrossRef]
49. Lin, X.; Zhang, Z.; Wang, Q. Evaluation of zeolite catalysts on product distribution and synergy during wood-plastic composite catalytic pyrolysis. *Energy* **2019**, *189*, 116174. [CrossRef]
50. Kianfar, E.; Hajimirzaee, S.; Mousavian, S.; Mehr, A.S. Zeolite-based catalysts for methanol to gasoline process: A review. *Microchem. J.* **2020**, *156*, 104822. [CrossRef]
51. Han, J.; Xiong, Z.; Zhang, Z.; Zhang, H.; Zhou, P.; Yu, F. The Influence of Texture on Co/SBA–15 Catalyst Performance for Fischer–Tropsch Synthesis. *Catalysts* **2018**, *8*, 661. [CrossRef]
52. Susastriawan, A.A.P.; Purnomo; Sandria, A. Experimental study the influence of zeolite size on low-temperature pyrolysis of low-density polyethylene plastic waste. *Therm. Sci. Eng. Prog.* **2020**, *17*, 100497. [CrossRef]
53. Kim, J.-Y.; Heo, S.; Choi, J.W. Effects of phenolic hydroxyl functionality on lignin pyrolysis over zeolite catalyst. *Fuel* **2018**, *232*, 81–89. [CrossRef]
54. Li, T.; Ma, H.; Wu, S.; Yin, Y. Effect of highly selective oxypropylation of phenolic hydroxyl groups on subsequent lignin pyrolysis: Toward the lignin valorization. *Energy Convers. Manag.* **2020**, *207*, 112551. [CrossRef]
55. Ma, H.; Li, T.; Wu, S.; Zhang, X. Effect of the interaction of phenolic hydroxyl with the benzene rings on lignin pyrolysis. *Bioresour. Technol.* **2020**, *309*, 123351. [CrossRef]
56. Miandad, R.; Rehan, M.; Barakat, M.A.; Aburiazaiza, A.S.; Khan, H.; Ismail, I.M.I.; Dhavamani, J.; Gardy, J.; Hassanpour, A.; Nizami, A.-S. Catalytic Pyrolysis of Plastic Waste: Moving Toward Pyrolysis Based Biorefineries. *Front. Energy Res.* **2019**, *7*, 1–17. [CrossRef]
57. Yao, D.; Yang, H.; Chen, H.; Williams, P.T. Investigation of nickel-impregnated zeolite catalysts for hydrogen/syngas production from the catalytic reforming of waste polyethylene. *Appl. Catal. B Environ.* **2018**, *227*, 477–487. [CrossRef]
58. Akubo, K.; Nahil, M.A.; Williams, P.T. Aromatic fuel oils produced from the pyrolysis-catalysis of polyethylene plastic with metal-impregnated zeolite catalysts. *J. Energy Inst.* **2019**, *92*, 195–202. [CrossRef]
59. López, A.; de Marco, I.; Caballero, B.M.; Adrados, A.; Laresgoiti, M.F. Deactivation and regeneration of ZSM-5 zeolite in catalytic pyrolysis of plastic wastes. *Waste Manag.* **2011**, *31*, 1852–1858. [CrossRef] [PubMed]
60. López, A.; de Marco, I.; Caballero, B.M.; Laresgoiti, M.F.; Adrados, A.; Aranzabal, A. Catalytic pyrolysis of plastic wastes with two different types of catalysts: ZSM-5 zeolite and Red Mud. *Appl. Catal. B Environ.* **2011**, *104*, 211–219. [CrossRef]
61. Sharma, A.; Pareek, V.; Zhang, D. Biomass pyrolysis—A review of modelling, process parameters and catalytic studies. *Renew. Sustain. Energy Rev.* **2015**, *50*, 1081–1096. [CrossRef]
62. Kan, T.; Strezov, V.; Evans, T.J. Lignocellulosic biomass pyrolysis: A review of product properties and effects of pyrolysis parameters. *Renew. Sustain. Energy Rev.* **2016**, *57*, 1126–1140. [CrossRef]
63. Ha, J.-M.; Hwang, K.-R.; Kim, Y.-M.; Jae, J.; Kim, K.H.; Lee, H.W.; Kim, J.-Y.; Park, Y.-K. Recent progress in the thermal and catalytic conversion of lignin. *Renew. Sustain. Energy Rev.* **2019**, *111*, 422–441. [CrossRef]
64. Hossain, Z.; Johnson, E.N.; Wang, L.; Blackshaw, R.E.; Cutforth, H.; Gan, Y. Plant establishment, yield and yield components of Brassicaceae oilseeds as potential biofuel feedstock. *Ind. Crop. Prod.* **2019**, *141*, 111800. [CrossRef]
65. Umrigar, V.R.; Chakraborty, M.; Parikh, P. Catalytic activity of zeolite Hβ for the preparation of fuels' additives: Its product distribution and scale up calculation for the biofuel formation in a microwave assisted batch reactor. *J. Environ. Chem. Eng.* **2018**, *6*, 6816–6827. [CrossRef]
66. Ibn Ferjani, A.; Jeguirim, M.; Jellali, S.; Limousy, L.; Courson, C.; Akrout, H.; Thevenin, N.; Ruidavets, L.; Muller, A.; Bennici, S. The use of exhausted grape marc to produce biofuels and biofertilizers: Effect of pyrolysis temperatures on biochars properties. *Renew. Sustain. Energy Rev.* **2019**, *107*, 425–433. [CrossRef]
67. Kumar, M.; Olajire Oyedun, A.; Kumar, A. A review on the current status of various hydrothermal technologies on biomass feedstock. *Renew. Sustain. Energy Rev.* **2018**, *81*, 1742–1770. [CrossRef]

68. Brigagão, G.V.; de Queiroz Fernandes Araújo, O.; de Medeiros, J.L.; Mikulcic, H.; Duic, N. A techno-economic analysis of thermochemical pathways for corncob-to-energy: Fast pyrolysis to bio-oil, gasification to methanol and combustion to electricity. *Fuel Process. Technol.* **2019**, *193*, 102–113. [CrossRef]
69. Akyurek, Z. Sustainable Valorization of Animal Manure and Recycled Polyester: Co-pyrolysis Synergy. *Sustainability* **2019**, *11*, 2280. [CrossRef]
70. Zhang, H.; Cheng, Y.-T.; Vispute, T.; Xiao, R.; Huber, G. Catalytic Conversion of Biomass-derived Feedstocks into Olefins and Aromatics with ZSM-5: The Hydrogen to Carbon Effective Ratio. *Energy Environ. Sci.* **2011**, *4*, 2297–2307. [CrossRef]
71. Zhang, Y.; Bi, P.; Wang, J.; Jiang, P.; Wu, X.; Xue, H.; Liu, J.; Zhou, X.; Li, Q. Production of jet and diesel biofuels from renewable lignocellulosic biomass. *Appl. Energy* **2015**, *150*, 128–137. [CrossRef]
72. Zhang, L.; Bao, Z.; Xia, S.; Lu, Q.; Walters, K.B. Catalytic Pyrolysis of Biomass and Polymer Wastes. *Catalysts* **2018**, *8*, 659. [CrossRef]
73. Bhoi, P.R.; Ouedraogo, A.S.; Soloiu, V.; Quirino, R. Recent advances on catalysts for improving hydrocarbon compounds in bio-oil of biomass catalytic pyrolysis. *Renew. Sustain. Energy Rev.* **2020**, *121*, 109676. [CrossRef]
74. Guedes, R.E.; Luna, A.S.; Torres, A.R. Operating parameters for bio-oil production in biomass pyrolysis: A review. *J. Anal. Appl. Pyrolysis* **2018**, *129*, 134–149. [CrossRef]
75. Zhu, Y.; Xu, G.; Song, W.; Zhao, Y.; Miao, Z.; Yao, R.; Gao, J. Catalytic microwave pyrolysis of orange peel: Effects of acid and base catalysts mixture on products distribution. *J. Energy Inst.* **2021**. [CrossRef]
76. Sogancioglu, M.; Yel, E.; Ahmetli, G. Investigation of the Effect of Polystyrene (PS) Waste Washing Process and Pyrolysis Temperature on (PS) Pyrolysis Product Quality. *Energy Procedia* **2017**, *118*, 189–194. [CrossRef]
77. Chen, Y.; Fang, Y.; Yang, H.; Xin, S.; Zhang, X.; Wang, X.; Chen, H. Effect of volatiles interaction during pyrolysis of cellulose, hemicellulose, and lignin at different temperatures. *Fuel* **2019**, *248*, 1–7. [CrossRef]
78. Collard, F.-X.; Blin, J. A review on pyrolysis of biomass constituents: Mechanisms and composition of the products obtained from the conversion of cellulose, hemicelluloses and lignin. *Renew. Sustain. Energy Rev.* **2014**, *38*, 594–608. [CrossRef]
79. Xu, S.; Cao, B.; Uzoejinwa, B.B.; Odey, E.A.; Wang, S.; Shang, H.; Li, C.; Hu, Y.; Wang, Q.; Nwakaire, J.N. Synergistic effects of catalytic co-pyrolysis of macroalgae with waste plastics. *Process Saf. Environ. Prot.* **2020**, *137*, 34–48. [CrossRef]
80. Adnan, M.A.; Xiong, Q.; Muraza, O.; Hossain, M.M. Gasification of wet microalgae to produce H_2-rich syngas and electricity: A thermodynamic study considering exergy analysis. *Renew. Energy* **2020**, *147*, 2195–2205. [CrossRef]
81. Kositkanawuth, K.; Sattler, M.L.; Dennis, B. Pyrolysis of Macroalgae and Polysytrene: A Review. *Curr. Sustain. Renew. Energy* **2014**, *1*, 121–128. [CrossRef]
82. Milledge, J.; Benjamin, S.; Dyer, P.; Harvey, P. Macroalgae-Derived Biofuel: A Review of Methods of Energy Extraction from Seaweed Biomass. *Energies* **2014**, *7*, 7194–7222. [CrossRef]
83. Li, F.; Srivatsa, S.C.; Bhattacharya, S. A review on catalytic pyrolysis of microalgae to high-quality bio-oil with low oxygeneous and nitrogenous compounds. *Renew. Sustain. Energy Rev.* **2019**, *108*, 41–497. [CrossRef]
84. Qari, H.; Rehan, M.; Nizami, A.-S. Key Issues in Microalgae Biofuels: A Short Review. *Energy Procedia* **2017**, *142*, 898–903. [CrossRef]
85. Chen, W.; Lu, J.; Zhang, C.; Xie, Y.; Wang, Y.; Wang, J.; Zhang, R. Aromatic hydrocarbons production and synergistic effect of plastics and biomass via one-pot catalytic co-hydropyrolysis on HZSM-5. *J. Anal. Appl. Pyrolysis* **2020**, *147*, 104800. [CrossRef]
86. Kim, Y.-M.; Lee, H.W.; Choi, S.J.; Jeon, J.-K.; Park, S.H.; Jung, S.-C.; Kim, S.C.; Park, Y.-K. Catalytic co-pyrolysis of polypropylene and Laminaria japonica over zeolitic materials. *Int. J. Hydrogen Energy* **2017**, *42*, 18434–18441. [CrossRef]
87. Zhang, B.; Zhong, Z.; Ding, K.; Song, Z. Production of aromatic hydrocarbons from catalytic co-pyrolysis of biomass and high density polyethylene: Analytical Py-GC/MS study. *Fuel* **2015**, *139*, 622–628. [CrossRef]
88. Chai, Y.; Gao, N.; Wang, M.; Wu, C. H2 production from co-pyrolysis/gasification of waste plastics and biomass under novel catalyst Ni-CaO-C. *Chem. Eng. J.* **2020**, *382*, 122947. [CrossRef]
89. Muneer, B.; Zeeshan, M.; Qaisar, S.; Razzaq, M.; Iftikhar, H. Influence of in-situ and ex-situ HZSM-5 catalyst on co-pyrolysis of corn stalk and polystyrene with a focus on liquid yield and quality. *J. Clean. Prod.* **2019**, *237*, 117762. [CrossRef]
90. Lee, H.; Choi, S.; Park, S.; Jeon, J.-K.; Jung, S.-C.; Kim, S.; Park, Y.-K. Pyrolysis and co-pyrolysis of Laminaria japonica and polypropylene over mesoporous Al-SBA-15 catalyst. *Nanoscale Res. Lett.* **2014**, *9*, 376. [CrossRef] [PubMed]
91. Xu, D.; Xiong, Y.; Ye, J.; Su, Y.; Dong, Q.; Zhang, S. Performances of syngas production and deposited coke regulation during co-gasification of biomass and plastic wastes over Ni/γ-Al2O3 catalyst: Role of biomass to plastic ratio in feedstock. *Chem. Eng. J.* **2020**, *392*, 123728. [CrossRef]
92. Tan, S.; Zhang, Z.; Sun, J.; Wang, Q. Recent progress of catalytic pyrolysis of biomass by HZSM-5. *Chin. J. Catal.* **2013**, *34*, 641–650. [CrossRef]
93. Balasundram, V.; Ibrahim, N.; Kasmani, R.M.; Isha, R.; Hamid, M.K.A.; Hasbullah, H.; Ali, R.R. Catalytic upgrading of sugarcane bagasse pyrolysis vapours over rare earth metal (Ce) loaded HZSM-5: Effect of catalyst to biomass ratio on the organic compounds in pyrolysis oil. *Appl. Energy* **2018**, *220*, 787–799. [CrossRef]
94. Wen, X.; Chen, X.; Tian, N.; Gong, J.; Liu, J.; Rümmeli, M.H.; Chu, P.K.; Mijiwska, E.; Tang, T. Nanosized Carbon Black Combined with Ni_2O_3 as "Universal" Catalysts for Synergistically Catalyzing Carbonization of Polyolefin Wastes to Synthesize Carbon Nanotubes and Application for Supercapacitors. *Environ. Sci. Technol.* **2014**, *48*, 4048–4055. [CrossRef] [PubMed]

95. Kaya, B.; Irmak, S.; Hasanoğlu, A.; Erbatur, O. Developing Pt based bimetallic and trimetallic carbon supported catalysts for aqueous-phase reforming of biomass-derived compounds. *Int. J. Hydrog. Energy* **2015**, *40*, 3849–3858. [CrossRef]
96. Yao, D.; Wu, C.; Yang, H.; Zhang, Y.; Nahil, M.A.; Chen, Y.; Williams, P.T.; Chen, H. Co-production of hydrogen and carbon nanotubes from catalytic pyrolysis of waste plastics on Ni-Fe bimetallic catalyst. *Energy Convers. Manag.* **2017**, *148*, 692–700. [CrossRef]
97. Chen, T.; Yu, J.; Ma, C.; Bikane, K.; Sun, L. Catalytic performance and debromination of Fe–Ni bimetallic MCM-41 catalyst for the two-stage pyrolysis of waste computer casing plastic. *Chemosphere* **2020**, *248*, 125964. [CrossRef]
98. Yao, D.; Wang, C.-H. Pyrolysis and in-line catalytic decomposition of polypropylene to carbon nanomaterials and hydrogen over Fe- and Ni-based catalysts. *Appl. Energy* **2020**, *265*, 114819. [CrossRef]
99. Li, J.; Xia, H.; Wu, Q.; Hu, Z.; Hao, Z.; Zhu, Z. Hydrocracking of the crude oil from thermal pyrolysis of municipal wastes over bi-functional Mo–Ni catalyst. *Catal. Today* **2016**, *271*, 172–178. [CrossRef]
100. Sridhar, A.; Rahman, M.; Infantes-Molina, A.; Wylie, B.J.; Borcik, C.G.; Khatib, S.J. Bimetallic Mo-Co/ZSM-5 and Mo-Ni/ZSM-5 catalysts for methane dehydroaromatization: A study of the effect of pretreatment and metal loadings on the catalytic behavior. *Appl. Catal. A Gen.* **2020**, *589*, 117247. [CrossRef]
101. Upare, D.P.; Park, S.; Kim, M.S.; Jeon, Y.P.; Kim, J.; Lee, D.; Lee, J.; Chang, H.; Choi, S.; Choi, W.; et al. Selective hydrocracking of pyrolysis fuel oil into benzene, toluene and xylene over CoMo/beta zeolite catalyst. *J. Ind. Eng. Chem.* **2017**, *46*, 356–363. [CrossRef]
102. Hamid, S.; Niaz, Y.; Bae, S.; Lee, W. Support induced influence on the reactivity and selectivity of nitrate reduction by Sn-Pd bimetallic catalysts. *J. Environ. Chem. Eng.* **2020**, *8*, 103754. [CrossRef]
103. Cai, N.; Yang, H.; Zhang, X.; Xia, S.; Yao, D.; Bartocci, P.; Fantozzi, F.; Chen, Y.; Chen, H.; Williams, P.T. Bimetallic carbon nanotube encapsulated Fe-Ni catalysts from fast pyrolysis of waste plastics and their oxygen reduction properties. *Waste Manag.* **2020**, *109*, 119–126. [CrossRef]
104. Kasar, P.; Sharma, D.K.; Ahmaruzzaman, M. Thermal and catalytic decomposition of waste plastics and its co-processing with petroleum residue through pyrolysis process. *J. Clean. Prod.* **2020**, *265*, 121639. [CrossRef]
105. Gao, Z.; Li, N.; Wang, Y.; Niu, W.; Yi, W. Pyrolysis behavior of xylan-based hemicellulose in a fixed bed reactor. *J. Anal. Appl. Pyrolysis* **2020**, *146*, 104772. [CrossRef]
106. Lai, C.; Wang, J.; Lei, W.; Xuan, C.; Xiao, W.; Zhao, T.; Huang, T.; Chen, L.; Zhu, Y.; Wang, D. Restricting Growth of Ni_3Fe Nanoparticles on Heteroatom-Doped Carbon Nanotube/Graphene Nanosheets as Air-Electrode Electrocatalyst for Zn–Air Battery. *Acs Appl. Mater. Interfaces* **2018**, *10*, 38093–38100. [CrossRef]
107. Zhou, H.; Saad, J.M.; Li, Q.; Xu, Y. Steam reforming of polystyrene at a low temperature for high H_2/CO gas with bimetallic Ni-Fe/ZrO10.1021/acsami.8b13751 catalyst. *Waste Manag.* **2020**, *104*, 42–50. [CrossRef] [PubMed]
108. Baamran, K.S.; Tahir, M. Ni-embedded TiO_2-$ZnTiO_3$ reducible perovskite composite with synergistic effect of metal/support towards enhanced H_2 production via phenol steam reforming. *Energy Convers. Manag.* **2019**, *200*, 112064. [CrossRef]
109. Kharaji, A.G.; Shariati, A.; Takassi, M.A. A Novel γ-Alumina Supported Fe-Mo Bimetallic Catalyst for Reverse Water Gas Shift Reaction. *Chin. J. Chem. Eng.* **2013**, *21*, 1007–1014. [CrossRef]
110. Wu, P.; Sun, J.; Abbas, M.; Wang, P.; Chen, Y.; Chen, J. Hydrophobic SiO_2 supported Fe-Ni bimetallic catalyst for the production of high-calorie synthetic natural gas. *Appl. Catal. A Gen.* **2020**, *590*, 117302. [CrossRef]
111. Aboul-Enein, A.A.; Awadallah, A.E. Production of nanostructured carbon materials using Fe–Mo/MgO catalysts via mild catalytic pyrolysis of polyethylene waste. *Chem. Eng. J.* **2018**, *354*, 802–816. [CrossRef]
112. Aboul-Enein, A.A.; Awadallah, A.E. Production of nanostructure carbon materials via non-oxidative thermal degradation of real polypropylene waste plastic using La_2O_3 supported Ni and Ni-Cu catalysts. *Polym. Degrad. Stab.* **2019**, *167*, 157–169. [CrossRef]
113. Bajad, G.S.; Tiwari, S.K.; Vijayakumar, R.P. Synthesis and characterization of CNTs using polypropylene waste as precursor. *Mater. Sci. Eng. B* **2015**, *194*, 68–77. [CrossRef]
114. Shen, Y.; Gong, W.; Zheng, B.; Gao, L. Ni–Al bimetallic catalysts for preparation of multiwalled carbon nanotubes from polypropylene: Influence of the ratio of Ni/Al. *Appl. Catal. B Environ.* **2016**, *181*, 769–778. [CrossRef]
115. Nahil, M.A.; Wu, C.; Williams, P.T. Influence of metal addition to Ni-based catalysts for the co-production of carbon nanotubes and hydrogen from the thermal processing of waste polypropylene. *Fuel Process. Technol.* **2015**, *130*, 46–53. [CrossRef]
116. Wu, S.-L.; Kuo, J.-H.; Wey, M.-Y. Thermal degradation of waste plastics in a two-stage pyrolysis-catalysis reactor over core-shell type catalyst. *J. Anal. Appl. Pyrolysis* **2019**, 104641. [CrossRef]
117. Aboul-Enein, A.A.; Awadallah, A.E. Impact of Co/Mo ratio on the activity of CoMo/MgO catalyst for production of high-quality multi-walled carbon nanotubes from polyethylene waste. *Mater. Chem. Phys.* **2019**, *238*, 121879. [CrossRef]
118. Wang, J.; Shen, B.; Lan, M.; Kang, D.; Wu, C. Carbon nanotubes (CNTs) production from catalytic pyrolysis of waste plastics: The influence of catalyst and reaction pressure. *Catalysis Today* **2019**, *351*, 50–57. [CrossRef]
119. Wu, C.; Nahil, M.A.; Miskolczi, N.; Huang, J.; Williams, P.T. Processing Real-World Waste Plastics by Pyrolysis-Reforming for Hydrogen and High-Value Carbon Nanotubes. *Environ. Sci. Technol.* **2014**, *48*, 819–826. [CrossRef]
120. Saravanan, A.; Hemavathy, R.V.; Sundararaman, T.R.; Jeevanantham, S.; Kumar, P.S.; Yaashikaa, P.R. 1—Solid waste biorefineries. In *Refining Biomass Residues for Sustainable Energy and Bioproducts*; Kumar, R.P., Gnansounou, E., Raman, J.K., Baskar, G., Eds.; Academic Press: Cambridge, MA, USA, 2020; pp. 3–17.

121. Xue, Y.; Zhou, S.; Brown, R.C.; Kelkar, A.; Bai, X. Fast pyrolysis of biomass and waste plastic in a fluidized bed reactor. *Fuel* **2015**, *156*, 40–46. [CrossRef]
122. Orozco, S.; Alvarez, J.; Lopez, G.; Artetxe, M.; Bilbao, J.; Olazar, M. Pyrolysis of plastic wastes in a fountain confined conical spouted bed reactor: Determination of stable operating conditions. *Energy Convers. Manag.* **2021**, *229*, 113768. [CrossRef]
123. Khlewee, M.; Gunukula, S.; Wheeler, M.C.; DeSisto, W.J. Hydrotreating of reduced oxygen content bio-oils produced by formate-assisted pyrolysis. *Fuel* **2019**, *254*, 115570. [CrossRef]

MDPI
St. Alban-Anlage 66
4052 Basel
Switzerland
www.mdpi.com

Catalysts Editorial Office
E-mail: catalysts@mdpi.com
www.mdpi.com/journal/catalysts

Disclaimer/Publisher's Note: The statements, opinions and data contained in all publications are solely those of the individual author(s) and contributor(s) and not of MDPI and/or the editor(s). MDPI and/or the editor(s) disclaim responsibility for any injury to people or property resulting from any ideas, methods, instructions or products referred to in the content.

www.ingramcontent.com/pod-product-compliance
Lightning Source LLC
LaVergne TN
LVHW070433100526
838202LV00014B/1585